BIOLOGICAL APPLICATIONS OF MICROFLUIDICS

BIOLOGICAL APPLICATIONS OF MICROFLUIDICS

EDITED BY

FRANK A. GOMEZ
Department of Chemistry and Biochemistry
California State University, Los Angeles

WILEY-INTERSCIENCE
A JOHN WILEY & SONS, INC., PUBLICATION

Published by John Wiley & Sons, Inc., Hoboken, New Jersey
Published simultaneously in Canada

Library of Congress Cataloging-in-Publication Data:

Biological applications of microfluidics / Edited by Frank A. Gomez.
p. cm.
ISBN 978-0-470-07483-1 (cloth)
1. Fluidic devices. 2. Microfluidics. I. Title.
TJ853.G66 2008
570.28'4–dc22

2007030019

Printed in the United States of America
10 9 8 7 6 5 4 3 2 1

Dedicated to my wife Olivia,
sons Frank and Nicholas and my dog Mimis
and for all of their hours of devotion
and unconditional love.

CONTENTS

PREFACE

Microfluidics is still a field in its infancy and, like a baby, it is growing with its future ahead of it. The focus of this book is to highlight some of the more recent works in microfluidics and to demonstrate its potential in solving societal problems.

In Chapter 1, a basic overview of microfluidics is described highlighting some of the more recent seminal works in the area. Chapters 2–6 deal with the use of microfluidics in cells including cells sorting, manipulation, and analysis. Chapters 7–10 detail work on chemical (enzymatic and nonenzymatic) synthesis on a microfluidic format. Chapters 11–14 present work focusing on the use of microfluidics in chemical separations. Chapters 15–17 focus on biomedical applications of microfluidics including microengineering and point-of-care diagnostics. Chapters 18–21 deal with fabrication of microfluidics chips. Chapters 22–25 focus on hybrid microfluidic applications.

I thank the contributors for useful and interesting discussions of their works and the stimulation it has provided for my research groups own studies. I also thank my research students at California State University, Los Angeles, who have contributed to our own studies in microfluidics. I thank Maggie Pereira for her many hours of reformatting the references. I hope that this book will help readers to arrive at a better understanding of the great potential of microfluidics and the wonderful things to come.

Frank A. Gomez, Ph.D.

CONTRIBUTORS

Vinay Abhyyankar, University of Wisconsin—Madison, Department of Biomedical Engineering, Madison, WI 53706

Nancy L. Allbritton, Department of Physiology and Biophysics, University of California, Irvine, CA 92697

David J. Beebe, University of Wisconsin—Madison, Department of Biomedical Engineering, Madison, WI 53706

Paul W. Bohn, Department of Chemical and Biomolecular Engineering, University of Notre Dame, Notre Dame, IN 46556

Rattikan Chantiwas, Department of Chemistry, Louisiana State University, Baton Rouge, LA 70803

Bong Geun Chung, Department of Biomedical Engineering, University of California Irvine, Irvine, CA 92697

Erik S. Douglas, UCSF/UC Berkeley Joint Graduate Group in Bioengineering, University of California, Berkeley, CA 94720

Emad El-Giar, Department of Chemistry, Mississippi State University, MS State, MS 39759

Albert Folch, Department of Bioengineering, University of Washington, Seattle, WA 98195

Carlos D. García, Department of Chemistry, The University of Texas at San Antonio, San Antonio, TX 78249

Levi A. Gheber, Department of Biotechnology Engineering, National Institute for Biotechnology in the Negev, Ben-Gurion University of the Negev, Beer-Sheva, Israel

S. Douglass Gilman, Department of Chemistry, Louisiana State University, Baton Rouge, LA 70803

Alvaro Gomez, Department of Chemistry, California State University, Los Angeles, Los Angeles, CA 90032

Frank A. Gomez, Department of Chemistry, California State University, Los Angeles, Los Angeles, CA 90032

Joseph Harris, Department of Pathology, University of California Irvine, Irvine, CA 92697

Charles S. Henry, Department of Chemistry, Colorado State University, Fort Collins, CO 80523

Jong Wook Hong, Department of Mechanical Engineering, Auburn University, Auburn, AL 36849

Jia Sheng Hu, Department of Pathology, University of California Irvine, Irvine, CA 92697

Jamie M. Iannacone, Department of Chemistry, University of Illinois, Urbana, IL 61801

Marquita Jacobs, Department of Chemistry, Mississippi State University, MS State, MS 39759

Noo Li Jeon, Department of Biomedical Engineering and Department of Pathology, University of California Irvine, Irvine, CA 92697

Emil P. Kartalov, Department of Pathology, Keck School of Medicine, University of Southern California, CA 90033 and the Department of Applied Physics, California Institute of Technology, Pasadena, CA 91125

Kathleen Kelly, C. Eugene Bennett Department of Chemistry, West Virginia University, Morgantown, WV 26506-6045

Robert Kennedy, Department of Chemistry, University of Michigan, Ann Arbor, MI 48109-1055

Horacio Kido, Department of Mechanical and Aerospace Engineering, University of California, Irvine, Irvine, CA 92697

Bo Young Kim, Department of Chemistry, University of Illinois, Urbana, IL 61801

Travis L. King, Department of Chemical and Biomolecular Engineering, University of Notre Dame, Notre Dame, IN 46556

Jörg Lahann, Departments of Chemical Engineering, Materials Science and Engineering and Macromolecular Science and Engineering, University of Michigan, Ann Arbor, MI 48109

Nianzhen Li, Department of Bioengineering, University of Washington, Seattle, WA 98195

Wei-Yu Lin, Department of Molecular and Medical Pharmacology, University of California, Los Angeles and Crump Institute for Molecular Imaging, University of California, Los Angeles, CA 90032

Chung N. Liu, Department of Chemical Engineering, University of California, Berkeley, CA 94720

Marc Madou, Department of Mechanical and Aerospace Engineering, University of California, Irvine, Irvine, CA 92697

Gloria Thomas Magee, Department of Chemistry, Mississippi State University, MS State, MS 39759

George Manuel Maltezos, Department of Applied Physics, California Institute of Technology, Pasadena, CA 91125

Richard A. Mathies, UCSF/UC Berkeley Joint Graduate Group in Bioengineering and Department of Chemistry, University of California, Berkeley, CA 94720

Robert J. Meagher, Sandia National Laboratories, Livermore, CA 94551-0969

Geeta Mehta, Department of Biomedical Engineering, University of Michigan, Ann Arbor, MI 48108

Naresh Menon, Biotechnology Group, Physical Optics Corporation, Torrance, CA 90501

Edwin S. Monuki, Department of Pathology, University of California Irvine, Irvine, CA 92697

Maria F. Mora, Department of Chemistry, The University of Texas at San Antonio, San Antonio, TX 78249

Amy Paguirigan, University of Wisconsin—Madison, Department of Biomedical Engineering, Madison, WI 53706

Jeong Won Park, Department of Biomedical Engineering, University of California Irvine, Irvine, CA 92697

Menake E. Piyasena, Department of Chemistry, California State University, Los Angeles, Los Angeles, CA 90032

John Puccinelli, University of Wisconsin—Madison, Department of Biomedical Engineering, Madison, WI 53706

Brent Reschke, C. Eugene Bennett Department of Chemistry, West Virginia University, Morgantown, WV 26506-6045

Zechariah Sandlin, Department of Chemistry, University of Michigan, Ann Arbor, MI 48109-1055

Axel Scherer, Department of Applied Physics, California Institute of Technology, Pasadena, CA 91125

Christopher E. Sims, Department of Physiology and Biophysics, University of California, Irvine, CA 92697

Anup K. Singh, Sandia National Laboratories, Livermore, CA 94551-0969

Jonathan V. Sweedler, Department of Chemistry, University of Illinois, Urbana, IL 61801

Shuichi Takayama, Macromolecular Science and Engineering, University of Michigan, Ann Arbor, MI 48108

Clive Taylor, Department of Pathology, Keck School of Medicine, University of Southern California, CA 90033

Aaron T. Timperman, C. Eugene Bennett Department of Chemistry, West Virginia University, Morgantown, WV 26506-6045

Nicholas M. Toriello, UCSF/UC Berkeley Joint Graduate Group in Bioengineering, University of California, Berkeley, CA 94720

Yu-suke Torisawa, Department of Biomedical Engineering, University of Michigan, Ann Arbor, MI 48108

Anna Tourovskaia, Department of Bioengineering, University of Washington, Seattle, WA 98195

Hsian-Rong Tseng, Department of Molecular and Medical Pharmacology, University of California, Los Angeles and Crump Institute for Molecular Imaging, University of California, Los Angeles, CA 90032

Behrad Vahidi, Department of Biomedical Engineering, University of California Irvine, Irvine, CA 92697

Jonathon A. Vickers, Department of Chemistry, Colorado State University, Fort Collins, CO 80523

Emanuel Waddell, Department of Chemistry, University of Alabama in Huntsville, Huntsville, AL 35899

Yanju Wang, Department of Molecular and Medical Pharmacology, University of California, Los Angeles and Crump Institute for Molecular Imaging, University of California, Los Angeles, CA 90032

Paul Watts, Department of Chemistry, The University of Hull, Hull HU6 7RX, UK

Xiaoyan Yan, Department of Chemistry, Louisiana State University, Baton Rouge, LA 70803

Mutsumi Yoshida, Departments of Chemical Engineering, Materials Science and Engineering and Macromolecular Science and Engineering, University of Michigan, Ann Arbor, MI 48109

Zeta Tak For Yu, Crump Institute for Molecular Imaging, University of California, Los Angeles and Department of Mechanical and Aerospace Engineering, University of California, Los Angeles, CA 90032

Jim Zoval, Department of Mechanical and Aerospace Engineering, University of California, Irvine, Irvine, CA 92697

1

MICROFLUIDICS

Frank A. Gomez

1.1 MICROFLUIDICS

Society deems largess in the affirmative; the more money the better; the larger the house the grander; the bigger the bicep the stronger. The world of scientific discovery, invention, and development is, on the contrary, moving counter to this belief. Need for high throughput methods requiring small quantities of material have prompted the design and development of instruments with novel platforms that can accomplish the same tasks and more but in miniaturized form.

The scientific literature is abundant in statements highlighting the "coming revolution" in microfluidic technology and applications; the term microfluidics generally used to describe any technology that moves microscopic and nanoscale volumes of fluid through micro-sized channels on a microelectromechanical system (MEMS). Microfluidics can be considered a combination of fluid mechanics, surface science, chemistry, and biology (and in many cases optics, microscopy, electronics, control systems, and microfabrication as well). Analogous to the integration of processes on-chip, research in this field involves an integration of many disciplines. Applications for microfluidic devices (MDs) have proliferated at a speed reminiscent of the use of electronics shortly after the invention of the integrated circuit. The term "lab-on-a-chip" has become almost passé in the sheer numbers of papers and researchers who espouse the notion of miniaturizing everything larger than a fingernail. And yet, why not? Microfluidic lab-on-a-chip technologies represent a revolution in laboratory experimentation, bringing the benefits of miniaturization, integration, and automation to many research-based industries. The use of microfluidic chips has attracted great attention as microchips

Biological Applications of Microfluidics, Edited by Frank A. Gomez

offer several advantages over nonmicrofludic-based techniques including portability, speed to analysis, ability to multiplex, and compatibility with other techniques.

Microfluidics has been the motivation for various biochemical application advancements in point-of-care diagnostics, bioterrorism detection, and drug discovery. Potential applications include biotechnology, pharmaceuticals, life sciences, defense, public health, and agriculture, each of which has its own needs. For example, microfluidic technologies have been embraced by the pharmaceutical industry because of the benefits of miniaturization, integration, and automation to research and development. At present, nearly every company and institution doing research and development is aware of this revolution in portable instrumentation and are modifying long-term strategic plans with miniaturization in mind. The expense, inefficiency and high maintenance, and space requirements of robotic automation systems present barriers to performing experiments. By contrast, microfluidic chips are inexpensive, require little maintenance and space, and need only minute amounts of samples and chemical inputs to make experiments work, making them more efficient, and potentially cheaper to use. A query of PubMed indicates that the citation rate for the word “microfluidics” increased fivefold between 2001 and 2004 and doubled to 500 between 2003 and 2004. In 1998, the number of issued patents detailing the use of a microfluidic platform in a commercial application was less than 25. In 2004, that number had risen to over 350. The microfluidics market for life science applications reached $750 million in 2004 and is expected to reach $2 billion by 2010. Specific applications of life sciences companies include *in vitro* diagnostics, drug delivery, and drug delivery markets.

The advantages of microfluidics include, but are not limited to, small sample volumes leading to greater efficiency of chemical reagents; low production costs per device thereby allowing for disposability; high throughput synthesis and screening of biological species and drug targets; parallel processing of samples; fast sampling times; accurate and precise control of samples and reagents reducing the need for pipetting; low power consumption; and versatile format for integration of various detection schemes thereby leading to greater sensitivity.

Fabrication of devices involve techniques already cultivated in the semiconductor industry and in manufacturing and have been used in the development of very complex devices. To allow a cost-efficient and timely development of full-fledged custom solutions, these platforms comprise a set of validated microfluidic and nonmicrofluidic components, unit operations, design rules, materials, fabrication technologies, and development tools.

During the past 15 years, advances in microfluidic have been rapid, yet, much more research is required due to the drastic reduction in length scales accompanying minaturization, the increased role of electrokinetic effects, and, most importantly, the inherent difficulties in interfacing the macroworld with the microscale. A few of the main areas of research involving microfluidics are detailed in the following section.

1.1.1 Fluid Manipulation

Much of the research on microfluidics has focused on the development of the microfluidic components, and specifically micropumps, micromixers, word-to-chip microfluidic interfaces, and microvalves. One of the most important elements of a successful MD is reliable microvalves since they make possible the manipulation of liquid flow in the channels, on/off switching of fluid flow, and injection of minute volumes of solution into the separation channel. To address the issue of sample loading and manipulation, a number of valve-type techniques have been developed. A recent review by Oh and Ahn detailed several types of microvalves and classified them as active or passive employing mechanical, nonmechanical, and external systems (1).

A microfluidic valving system based on multilayer soft lithography (MSL), recently developed by Quake et al., has been fortuitously integrated into a variety of microfluidic devices (2,3). They combined soft lithography with the capability to bond multiple patterned layers of elastomer. Layered structures are constructed by binding two layers (fluid and control) of elastomer, each of which is separately cast from a micromachined mold. Still other types of valves including magnetically controlled microvalves have been detailed (4–6).

1.1.2 Single-Cell Analysis

Flow and laser scanning cytometry and automated microscopy are the most widely used techniques for single-cell analysis. In the next 10 years, microfluidic platforms have the potential to displace some of these traditional techniques in the analysis of single cells. The dimensions (10–100 μm) of microfludic channels are highly compatible with the sizes of biological cells. There is a wealth of information that can be obtained when analysis of cells is conducted one by one and not in aggregate. For example, analysis of single cells will allow for greater discrimation of different types of cells and components housed within them thereby facilitating individual examination of cellular components. The micrometer channel dimensions that are afforded by microfluidic chips makes them ideally suited for sample introduction, manipulation, reaction, separation, and detection of individual cells. Many of these devices integrate cell sampling, single-cell loading, docking, lysing, and separation configurations including capillary electrophoresis (CE). Various applications have been demonstrated including cell viability assays and ionophore-mediated flux measurements.

For example, recently a device was developed in poly(dimethylsiloxane) (PDMS) to separate a single cell from a bulk suspension and integrated valves and pumps for delivery of nanoliter volumes of reagents to the cell. In this work cell viability assays ionophore-mediated intracellular calcium flux measurements and multistep receptor-mediated calcium measurements (7). Another device was developed to examine the contraction of a single cardiac myocyte (Heart muscle cell) and integrated on-chip dye loading and quantitative analysis of calcium (8). mRNA and genomic DNA were recently isolated on a MD from small numbers of mammalian and bacterial cells, respectively (9). A follow-up to this work detailed a more robust isolation and purification scheme whereby quantitative calibrations for microfluidic mRNA isolation and cDNA synthesis were performed demonstrating sensitivity sufficient to detect medium and low copy number transcripts in single cells (10).

1.1.3 Microreactors

Microfluidic microreactors offer new opportunities due to the inherent parallel processing, integrated sensing, and automation afforded lab-on-a-chip formats. Microreactors have great potential in the screening of thousands of compounds in combinatorial chemistry and process optimization. Here, the goal is to develop robust systems that integrate synthesis, separation, and analysis all on one format. As the number and complexity of chemical reactions and separations carried out in a microfluidic format increases, the need for precise temperature monitoring and control within the microchannels increases. Microfluidics offers the advantages of enhanced heat and mass transport. The ability to incorporate sensitive and fast temperature measurement into a single chip may lead to better separations and reaction control. Miniaturization of chemical reactions has been demonstrated to generate products in higher yield and purity compared to the equivalent bulk reactions and in much shorter periods of time (11). Applications of microreactors include nucleic acid analysis, combinatorial synthesis, protein and peptide mapping, enzyme catalysis, and cell characterization (12–17).

Since physical processes can be more easily controlled when instrumental dimensions are in micrometer scale, the process by which scientists synthesize molecules will certainly change in the coming decades. A variety of microreactor designs and chip geometries have been developed that can accommodate a myriad of organic and inorganic synthetic reactions. These geometries frequently allow to hasten a reaction to product whereas standard reaction techniques yield little to no product. Recent organic reactions include the carbonylative cross-coupling reactions of arylhalides to form secondary amides (18), glycosylation reactions (19), and the use of encapsulated microsomes as immobilized nanoparticles for drug synthesis (20). In this work, yield, selectivity, and rates, as functions of composition, flow rates and temperature, were used to optimize conditions for transformation.

1.1.4 Cell Sorting

Effective methods for manipulating, isolating and sorting cells and particles are essential for the development of microfluidic-based life science research and point-of-care (POC) diagnostic platforms (21). Current techniques in high speed cell sorting are limited by the inherent coupling among three competing parameters of performance: purity, throughput, and rare cell recovery. Cell sorters have great potential in the area of high throughput screening as applied to genetics, microbiology, and cell biology. Recent advances in MEMS technology have combined fluidics, mechanics, optics, and electronics to sort a large number of cells. Integrating cell sorters onto microfluidic platforms has demonstrated the versatility of lab-on-a-chip devices (22–26). The devices can be used as stand-alone devices or part of an integrated system for point-of-care diagnostics and cytometric measurements. Besides the cost savings, an important advantage of the microfluidic cell sorters is that they will allow the implementation of sophisticated assays that are impractical to perform with traditional cell sorters or microtiter plates.

1.1.5 Point-of-Care Diagnostics in Developing and Third-World Countries

Besides life science research point-of-care diagnostics has, in recent years, driven the microfluidics market. A few years ago it was fine to demonstrate a single test on a device. Now, clinical diagnostics devices are integrating multiple sophisticated biochemical analyses (PCR amplification, cell lysis, separation, and detection) all on a single platform and in real time (27–31). These devices will lead to a revolution in the medical world. Blood tests will be done by patients themselves with handheld microfluidic-based devices that are easily able to measure the concentration of certain substances in their blood. The results of these tests will help indicate the type and amount of medication that the patient needs to take.

Developing countries is where microfluidic technologies have their greatest potential. In principle, microfluidic devices offer advantages that seem ideal to address the immense public health challenges of developing countries. It is in the third world where famine, hunger, and political strife make even subsistence living only a hope to millions of people. And nowhere on earth does the affliction of infectious disease loom and strike with such intrepid fortitude than in developing countries. Unlike the United States, Europe, and some parts of the Far East where vaccines and medicines are but a doctor's visit away, HIV/AIDS, tuberculosis, malaria, and other diseases run rampant afflicting children or, if more prevalent in adults, leave thousands of children orphan to cope in a world with little hope. It is these areas where diagnostic technologies would have its maximum benefit. Unfortunately, developing countries pose problems different from developed countries including lack of a health infrastructure, environmental problems, and not to mention financial limitations. Hence, any type(s) of microfluidic device that is (are) developed must be inexpensive, easy to use, disposable, and easily manufactured.

1.1.6 Microchip Capillary Electrophoresis

Microchip capillary electrophoresis (MCE) has great potential due to the small sample volumes and reagents needed in micron-sized channels and because the pumping mechanism is of the electroosmotic type. CE was one of the first techniques to be integrated onto a microfluidic platform and was a natural method for integration onto a chip format (32–34). There is a large body of work in the area of MCE. For example, immunoassays have been developed using MCE that exploits the small volumes required in a microfluidic format and the separations ability of CE (35). PCR has been integrated and coupled to MCE combining heaters, temperature sensors, and valves for genotyping from bacterial cells (15). A variety of detection techniques have been employed in MCE including conductivity (36), electrochemiluminescence (37), and amperometric (38).

This book is an attempt to bring together the work of scientists and researchers from across many disciplines to demonstrate the strong overlap between their efforts and how their work typifies the exciting potential of microfluidics in the biological realm. As can be seen, people with diverse educational backgrounds and training have found microfluidics to be an interesting area of study. It is my hope that this book will inspire future scientists to consider microfluidics a field to study. Only by bringing together people of varied backgrounds can the field reach its full potential.

REFERENCES

1. Oh KW, Ahn CH. J Micromech Microeng 2006;16:R13–R39.
2. Quake SR, Scherer A. Science 2000;290:1536–1540.
3. Unger MA, Chou HP, Thorsen T, Scherer A, Quake SR. Science 2000;288:113–116.
4. Bae B, Kee H, Kim S, Lee Y, Sim T, Kim Y, Park K. J Micromech Microeng 2003;13:613–619.
5. Siegel AC, Shevkoplyas SS, Weibel DB, Bruzewicz DA, Martinez AW, Whitesides GM. Angew Chem Int Ed 2006;45:6877–6882.
6. Fu C, Rummler Z, Chomburg W. J Micromech Microeng 2003;13:S96–S102.
7. Wheeler AR, Throndset WR, Whelan RJ, Leach AM, Zare RN, Liao YH, Farrell K, Manger ID, Daridon A. Anal Chem 2003;75:3581–3586.
8. Li X, Li PCH. Anal Chem 2005;77:4315–4322.
9. Hong JW, Studer V, Hang G, Anderson WF, Quake SR. Nat Biotechnol 2004;22:435–439.
10. Marcus JS, Anderson WF, Quake SR. Anal Chem 2006;78:3084–3089.
11. Watts P, Haswell SJ. Chem Soc Rev 2005;34:235–246.
12. Kerby MB, Legge RS, Tripathi A. Anal Chem 2006;78:8273–8280.
13. Watts P, Haswell SJ. Comb Chem High Throughput Screening 2004;7:397–405.
14. Palm AK, Novotny MV. Rapid Commun Mass Spectrom 2004;18:1374–1382.
15. Lagally ET, Scherer JR, Blazej RG, Toriello NM, Diep BA, Ramchandani M, Sensabaugh GR, Riley LW, Mathies RA. Anal Chem 2004;76:3162–3170.
16. Koh WG, Pishko M. Sens Actuators, B 2005;106 (1):335–342.
17. Liu A-L, Zhou T, He F-Y, Xu J-J, Lu Y, Chen H-Y, Xia X-H. Lab Chip 2006;6:811–818.
18. Miller PW, Long NJ, de Mello AJ, Vilar R, Passchier J, Gee A. Chem Commun 2006;546–548.
19. Ratner DM, Murphy ER, Jhunjhunwala M, Snyer DA, Jensen KF, Seeberger PH. Chem Commun 2005;578–580.
20. Zguris JC, Itle LJ, Hayes D, Pishko MV. Biomed Microdevices 2005;7:117–125.

21. Applegate RW, Squier J, Vestad T, Oakey J, Marr DWM, Bado P, Dugan MA, Said AA. Lab Chip 2006;6:422–426.
22. Chen CC, Zappe S, Sahin O, Zhang XJ, Fish M, Scott M, Solgaard O. Sens Actuat B 2004;102:59–66.
23. Pamme N, Wilhelm C. Lab Chip 2006;6:974–980.
24. Studer V, Jameson R, Pellereau E, Pepin A, Chen Y. Microelectronic Engineering 2004;73/74:852–857.
25. MacDonald M, Spalding G, Dholakia K. Nature 2003;426:421–424.
26. Sundararajan N, Kim D, Berlin AA. Lab Chip 2005;5:350–354.
27. Srinivasan V, Pamula VK, Fair RB. Lab Chip 2004;4:310–315.
28. Bhattacharyya A, Klapperich CM. Anal Chem 2006;78:788–792.
29. Zangmeister RA, Tarlov MJ. Anal Chem 2004,76,3655–3659.
30. Hong JW, Studer V, Hang G, Anderson WF, Quake SR. Nature Biotechnol 2004;22:435–439.
31. Bessette PH, Hu X, Soh HT, Daugherty PS. Anal Chem 2007;79:2174–2178.
32. Harrison DJ, Manz A, Fan Z, Ludi H, Widmer HM. Anal Chem 1992;64:1926–1932.
33. Harrison DJ, Fluri K, Seiler K, Fan Z, Effenhauser CS, Manz A. Science 1993;261:895–897.
34. Effenhauser CS, Manz A, Widmer HM. Anal Chem 1993;65:2637–2642.
35. Kawabata T, Watanabe M, Nakamura K, Satomura S. Anal Chem 2005;77:5579–5582.
36. Vrouwe EX, Luttge R, van den Berg A. Electrophoresis 2005;25:1660–1667.
37. Du Y, Wei H, Kang J, Yan J, Yin X, Yang X, Wang E. Anal Chem 2005;77:7993–7997.
38. Vickers JA, Henry CS. Electrophoresis 2005;26:4641–4647. Ding Y, Garcia CD. Electrophoresis 2006;27:5119–5127.

PART I

CELL ANALYSIS ON MICROFLUIDIC DEVICES

2

USING MICROFLUIDICS TO UNDERSTAND AND CONTROL THE CELLULAR MICROENVIRONMENT

Amy Paguirigan, John Puccinelli, Vinay Abhyankar and David J. Beebe

2.1 INTRODUCTION: ENGINEERING THE MICROENVIRONMENT

Engineering the cellular microenvironment is becoming an increasingly common aim for purposes as applied as tissue engineering and drug delivery, to those focusing on more basic science such as chemotaxis studies and gene expression assays. Although microfluidics holds enormous potential to provide a platform for new and more relevant cellular assays, more in-depth investigation of the engineered microenvironments will be required for this potential to be fully realized. As new devices are designed, fabricated, and applied in new ways, new controls, assays, and experiments must also be designed taking into account the limitations and potential artifacts of each device. Aspects of the engineered microenvironment that are completely new and different from the more traditional issues associated with traditional cell culture and treatment now must be appreciated and understood.

A cellular microenvironment consists of both chemical and mechanical components (Fig. 2.1). Locally derived chemical stimuli can be produced by similar or dissimilar cell types allowing a field of cells to function cooperatively in normal tissues. Cells can communicate locally not only via species secreted into the extracellular space, but also via gap junctions that connect the cytoplasms of neighboring cells (allowing direct solute transport) and other cell–cell junctions that connect to intracellular signaling pathways and the cytoskeleton. These stimuli

Biological Applications of Microfluidics, Edited by Frank A. Gomez

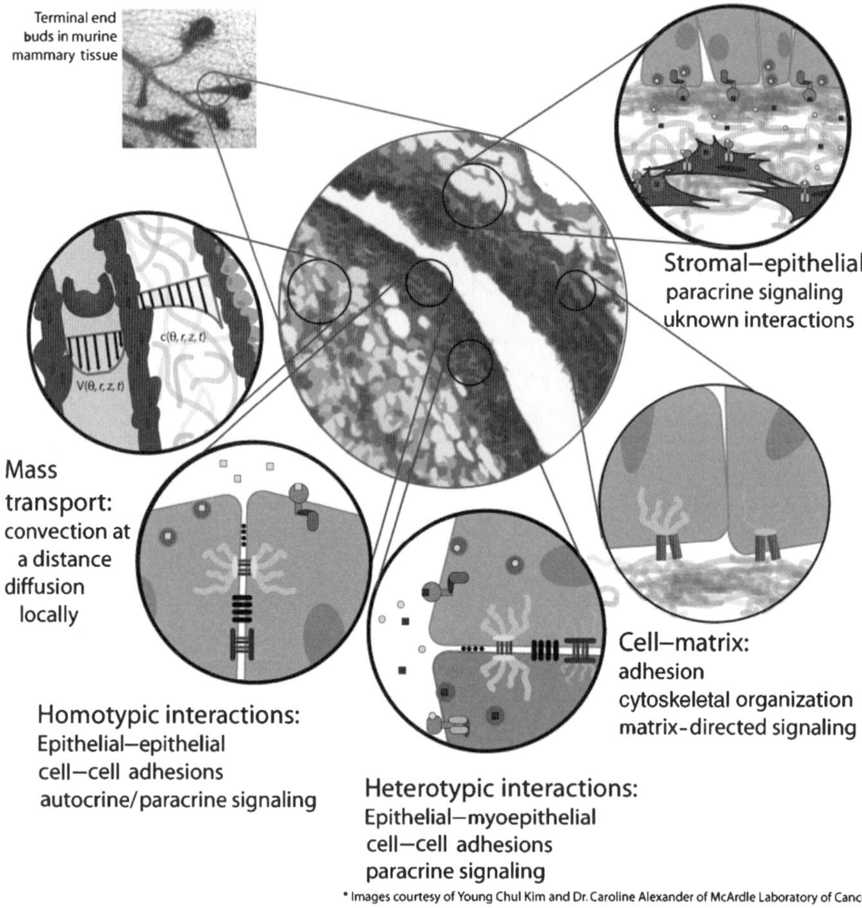

FIGURE 2.1 The *in vivo* microenvironment of mammary epithelial cells, much like that of many other tissues in the body, consists of a wide variety of interactions and stimuli. Cell–cell interactions can occur via mechanical contact either between similar cell types (homotypic, e.g., epithelial and epithelial), or dissimilar cell types (heterotypic, e.g. epithelial and myoepithelial). Soluble factors produced and received in these types of interactions along with those received from more distant cells (e.g., stromal–epithelial) can provide an infinitely diverse range of signals and instructions for proliferation, organization, polarization, migration, and numerous unknown signals and functions. Also important to cell behavior are cell–matrix interactions, which can provide both structural support, and direct signaling and expression. Lastly, mechanisms of mass transport are critical for cell survival, via providing nutrients and endocrine signals, and by removing waste products and signaling molecules produced by the cells of interest. All these interactions together are integrated by cells and all stimuli are considered when cell fate or behavior decisions are made. If *in vitro* assays are to determine accurate cellular responses to stimuli, microfluidic systems must be developed to more closely mimic the total cellular microenvironment. (See color insert)

guide tissue organization and growth via functions such as orienting cell polarization and migration, coordinating growth and apoptosis, andregulating cellular expression and behavior producing a functional and coordinated tissue. In addition, these chemical interactions can occur between cells of the same type (homotypic), adjoining cells of different types (heterotypic), cells separated by matrix (as is the case for stromal and epithelial cells), or cells in other locations in the body that communicate via endocrine (or neuroendocrine) pathways and the bloodstream.

These different sources of stimuli rely upon different mechanisms of mass transport for signal transport. *In vivo*, locally, species move via diffusion through the extracellular matrix, meanwhile tissues are continuously perfused via capillary beds, which relies on convective mass transport. Although most cells in a tissue (such as the mammary gland) are not directly exposed to convective mass transport and the shear forces that accompany it, as endothelial cells of the blood vessels and capillary beds are, they are nonetheless affected by the convective mass transport that occurs at the blood–endothelial cell interface. The convection occurring at this interface serves to regulate the local concentration gradient (established by diffusive mass transport through the matrix) of species at the boundary to the current desired concentration. Thus, the maintenance of homeostasis in a field of cells not only relies upon the diffusional mass transport locally, which allows cells to produce and maintain local gradients of soluble factors and signaling molecules, but also the continuous supply of nutrients and oxygen and the removal of waste products of metabolism by convection at a distance.

From a purely mechanical viewpoint, cells rely upon the extracellular matrix for support, but its contributions to cellular behavior extend much further than simply a passive structural component. Cell adhesions to the matrix via integrins connect to intracellular complexes that direct many functions intracellularly, including gene expression, and protein translation, modification, and secretion. Components of the extracellular matrix such as collagen and fibronectin have peptide motifs that are recognized by cell surface receptors (integrins), which when bound, link the cell's cytoskeleton's main components, actin (via focal adhesions) and intermediate filaments (via hemidesmosomes) to the extracellular matrix. In addition to proteins, proteoglycans exist in the extracellular matrix to provide important mechanical properties and act as coreceptors and function as modulators of cell signaling by binding growth factors (such as fibroblast growth factor) and by binding or acting as a reservoir for other secreted factors, proteases or other inhibitors of cell–cell signaling.

The *in vivo* microenvironment that engineers aim to mimic *in vitro* is incredibly complex, and thus if we are to attempt to understand effects of specific contributions on cellular behavior, *in vitro* assays must address potential effects (or lacktherof) of the shortcomings that may exist. The *in vitro* chemical microenvironment includes not only the soluble factors present in the fluid surrounding the cells either produced by neighboring cells or introduced by the experimenter but also the chemical characteristics of synthetic surfaces that the cells are in contact with. The *in vitro* mechanical microenvironment includes dimensionality of the environment (e.g., a 2D monolayer culture versus a 3D gel culture), mechanical properties of the materials the cells come in contact with, and components the cells actually interact with, either cellular (as in coculture experiments) or different chemical composition of gels (as in some 3D cultures).

To date, much of the work in microfluidics for biological applications has focused on testing the effects of a soluble component as compared to cell behavior in its absence. The introduction of controlled gradients of soluble factors has highlighted microfluidics' potential for expanding current techniques to include new assays or for providing a platform for simplifying and improving current assays. New tissue culture methods are being developed for applying the scalability of microfluidics to study cell proliferation and differentiation using both disperse cultures and cultures that employ surface patterning to modulate cell–cell interactions. Some promising work with cells in 3D cultures has begun, and new fabrication methods have allowed experimenters to perform cocultures of multiple cell types in a 3D culture.

Most of the microfluidic platforms currently in use provide a glimpse into only one aspect of the cellular microenvironment while lacking the complexity found in tissue. Often this type of platform is both sufficient to provide a readout in a predictive assay (such as is often the case in drug screening/toxicity assays) and also is a significant improvement over traditional techniques (either in throughput, cost or improved accuracy due to the different microenvironment). However, *in vivo*, the chemical and mechanical microenvironment are intimately linked, one influencing how a cell responds to the other. The idea of teasing apart using *in vitro* assays that responses are a result of the chemical environment versus the mechanical one assumes that these responses are not the result of an integration of all the stimuli a cell is exposed to. In some cases the characteristics that *in vitro* assays lack as compared to the *in vivo* microenvironment do not prevent a predictive assay from being performed. However, in many cases, either these aspects of the microenvironment do affect the quality of the data obtained, or not enough is known about the system to determine whether the effects were significant or not. The challenge posed to these *in vitro* assays is to provide accurate results without the need for animal studies and human trials, and thus mimicking the *in vivo* cellular microenvironment as closely as possible is an important step toward that goal.

2.2 THE CHEMICAL MICROENVIRONMENT

Engineering the chemical microenvironment is a fundamental aim of drug discovery, for which microfluidics holds immense promise (1). Preliminary drug screening can be performed using *in vitro* assays of cytotoxicity of various cell types in response to doses of molecules of interest, generally using small volumes of reagents and low cell numbers. Ideally, these *in vitro* assays could be done at a high throughput level and be fully representative of the drugs' effects *in vivo*. This would allow for rapid identification of compounds of interest based on their efficacy and low toxicity in these assays. However, current limitations in fabrication and design of engineered microenvironments also limits the usefulness of the assays performed in them.

2.2.1 Soluble — Gradients

The soluble portion of the cellular microenvironment in microfluidic devices has been engineered and the effects of these environments assayed using a variety of techniques (see Fig. 2.2 for a summary). Stimulating a field of cells with a gradient of a soluble factor can provide a unique type of microfluidic assay that can effectively produce different microenvironments on a single device. Few traditional techniques for gradient production have been able to produce as defined, controlled and repeatable gradients as those produced by the many microfluidic techniques currently in use. The use of microfluidic gradient production techniques holds great promise for creating and studying the effects of spatially defined soluble microenvironments on cellular behavior in ways not previously possible.

Complex soluble factor gradients can be created by taking advantage of diffusional mass transport across the interface between adjacently flowing fluid streams in laminar flow (2,3). These laminar flow based systems provide control over gradient profiles by modulating the flow rates or input concentration of the input streams, varying the number of input streams, or by altering the number of branches in the splitting network (see Fig. 2.2a). Recently Irimia and coworkers demonstrated a simplified methodology that uses flow obstructions of varying sizes within a microchannel to create guiding alleys that physically split the incoming stream into discrete streams. These discrete streams (each with a different concentration) mix via diffusion across their interface and create a chemical gradient downstream (4,5). The geometry and position of the flow obstructions and the input flow rate can be modified to create arbitrary gradient profiles. These laminar flow based systems are advantageous because they facilitate rapid switching of gradient characteristics and therefore provide dynamic control over the cellular microenvironment.

Defined chemical gradients have been used to stimulate migratory cells (e.g., neutrophils, bacteria, sperm cells) using chemoattractants (3,5–10), investigate cancer cells responding to a drug or growth factor (11–13), or to stimulate the differentiation of embryonic stem cells (14). Laminar flow based systems are beneficial because they facilitate quantitative correlations between environmental cues and observed cellular behavior. The ability to quantitatively compare the responses of different cell types to a defined chemical stimulus (or multiple stimuli) may provide insight into the mechanisms by which signaling cascades and expression can be altered.

While flow-based systems provide dynamic control over microenvironment cues, one disadvantage of these systems is that they require continuous perfusion to maintain the exogenous factor gradient. In these systems, nutrients and chemoattractant are continuously renewed, as waste products and intercellular signaling molecules are continuously depleted. This results in temporally uniform

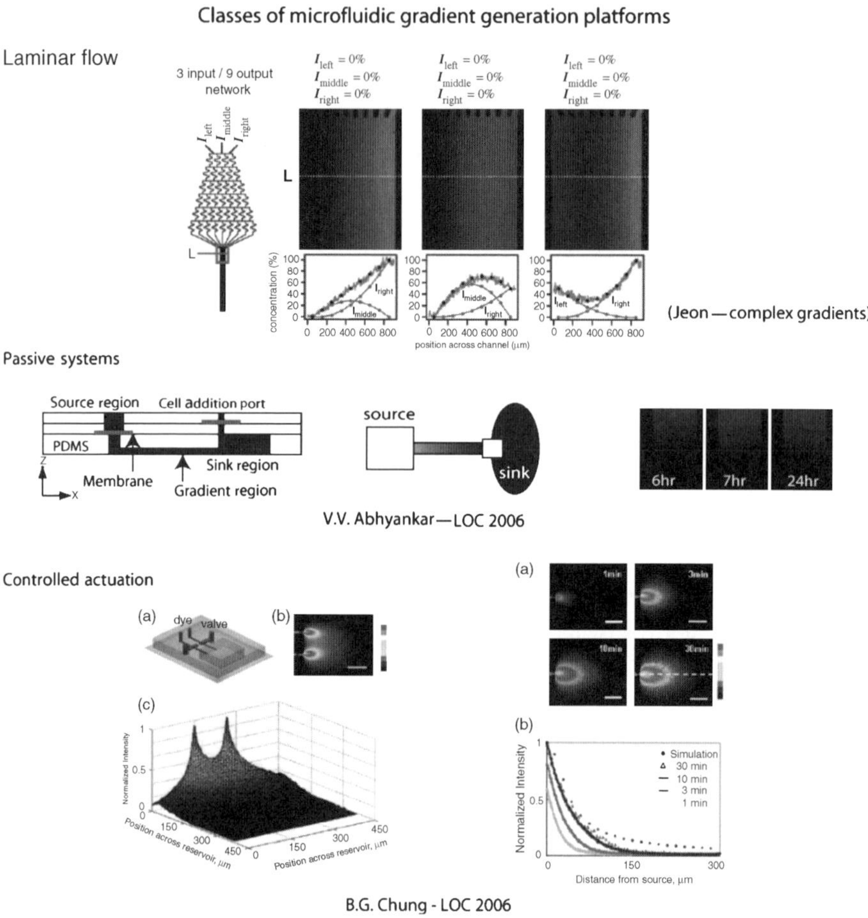

FIGURE 2.2 Schematic representation of microfluidic gradient generation techniques. (a) Laminar flow-based methods take advantage of diffusion and mixing across laminar streams. This type of method is typically driven by syringe pumps (81). Different gradient profiles can be created by altering the input flow rates. Reprinted with permission from American Chemical Society Copyright 2000. (b) Passive systems utilize porous membranes to limit convective flows and create diffusion-dominated environments (6). Reproduced by permission of The Royal Society of Chemistry. (c) Controlled actuation methods create gradients by using valves to add discrete volumes of the chemical to a media reservoir; the actuation pattern and characteristics define the gradient (1). Reproduced by permission of The Royal Society of Chemistry. (See color insert)

concentrations of media and experimental components. Although this uniformity is beneficial in that the response of cells over time to a uniform stimulus or the effects of rapid changes in the stimulus can be determined, the contribution to the response of any soluble cell–cell communication is obscured. *In vivo*, neutrophils (a common cell type used in microfluidic chemotaxis assays) participate in the immune response to inflammation. They sense and respond to the soluble microenvironment in tissue, migrate toward sites of inflammation or infection, and work cooperatively to eliminate the potential danger. In general, microfluidic chemotaxis assays include only a gradient of a single chemoattractant (or possibly a countergradient of another chemoattractant (6)), although *in vivo*, the local population of cells as a whole might be presented with multiple stimuli and thus must pick and choose which to attend to. Also, by disrupting cell–cell communication, the location and migration of nearby cells may not influence a cell's response to the stimulus. Some work has already analyzed the effects of flow alone on neutrophils and shown that they can be mechanically activated by inducing shear on them via laminar flow in microchannels (15). Walker and colleagues have also shown that the flow rate used to create gradients can affect/bias the migratory behavior of cells (16).

In situations where cell–cell communication plays an important role in modulating cellular response, the continuously flowing streams that are necessary to maintain the chemical gradients make laminar flow based methods unsuitable for probing cellular responses. An alternative approach involves using static or "passive" gradient generation schemes that do not require flowing fluid streams. Recent work by Chung and coworkers describes the use of pneumatically controlled injection pulses of a soluble factor from an orifice into a large volume reservoir to create radially evolving concentration profiles (2). This method allows the use of multiple injection sources that can develop complex overlapping gradients.

In another approach, pressurized channels are used to connect a central region with small volume interconnection channels (10). The gradient profile and concentration range is controlled with the channel pressure and input concentration. Other passive methods use porous membranes to increase the fluidic resistance of a system and create diffusion dominated environments A source–sink concept can then be used to create stable linear (as well as temporally varying) gradients along length of the channel (7). The gradient profile is controlled by adjusting the input concentration and distance from source to sink.

Takayama and colleagues demonstrate an application that uses laminar flow to simultaneously treat discrete subdomains of a single cell with different soluble factors (17); they also demonstrate a method to treat a cell population with a bolus or discrete packet of a soluble factor using laminar flow (18). These unique microscale functionalities allow a level or control the microenvironment at the cellular level that is not possible using traditional culture methods. Pinelis and coworkers has extended the use of gradients to oxygen gradients and the effects of gradients and hypoxia on cells in gels, in microfluidic devices (19). Laminar flow—using different temperature streams—can be used to create a gradient in temperature across a channel; the energy transfer is a direct analog to the mass transport described above (20).

By treating cells with soluble components (such as gradients or plugs of soluble factors), a wide range of new assays can be performed that currently are either not possible or much more challenging to perform than in microfluidic devices. The potential artifacts introduced by these devices (such as the effect of flow on the target cell population in flow-based gradient generation devices) will be important to establish for the further integration of these techniques into biological research methods. For the wealth of information, they are capable of providing to be able to be incorporated into what is already known about these biological systems from using existing technologies, a comparison between the effects of the microenvironment of the microfluidic device and those seen in traditional setups must be made.

2.2.2 Soluble — Cell Culture

Another application of microfluidics in controlling the chemical cellular microenvironment is cell culture (21,22). Traditional cell culture has relied upon media components that have been found to promote survival with the addition of both defined (e.g., insulin, EGF) and undefined (e.g., serum) components and growth factors. Typical volume densities in these types of cultures can be quite small (Fig. 2.3). The effects of volume density alone can be substantial (23,24) but have not been addressed for many biological cell culture systems. Static microfluidic culture systems can provide much larger volume densities than traditional culture, while also keeping total volumes low when defined media component costs are high or when rare or expensive cell types are to be studied. In static cultures, the lack of flow can result in less disturbance of the soluble cellular microenvironment than traditional macroscale cultures where turbulent flows result in convective mass transport being dominant (and thus eliminate local concentration gradients set up via diffusion to or from a cell). Meanwhile, continuously perfused microfluidic cultures can provide continuous supply of nutrients and allow for longer term cultures as compared to static cultures, but at the expense of effective volume density and soluble cell–cell communication. These culture conditions employ laminar flow for continuous transport of solutes and thus are convection-based devices that limit diffusive transport of solutes to and from cells in a culture. Finally, membrane-based culture devices employ a membrane to allow only diffusional mass transport between the static fluids in a channel, with flowing media in source channels (convective flow), which allow longer term cultures to be performed without sacrificing soluble cell–cell communication. Many of the phenomena that occur in microfluidic devices are unique and by leveraging either the scalability (or lack thereof) of specific designs, more control and flexibility of the microenvironments the cells are exposed to can be obtained allowing new and different assays to be performed. Stimulating cultures of cells with drugs (11,25–27) or other components that induce differentiation (28) in microfluidic platforms allow the replication of traditional tissue culture analyses in smaller volumes with fewer cells allow for expensive assays to be performed using minimal resources.

Perfusion culture of murine embryonic stem cells in microchannels at flow rates orders of magnitude apart showed improved morphology after several days with higher flow (29). However, perfusion systems have the potential to impart a range

(a)

	Surface area per culture area (cm^2)	Media volume per culture area (mL)	Surface density ($cells/cm^2$)	Volume density ($cells/cm^3$)
6-well plate	9.5	8	40,000	47,500
Microchannel	0.08	0.004	40,000	800,000
Fold difference (of 6 well over microchannel)	120x	2,000x	1x	1/17x

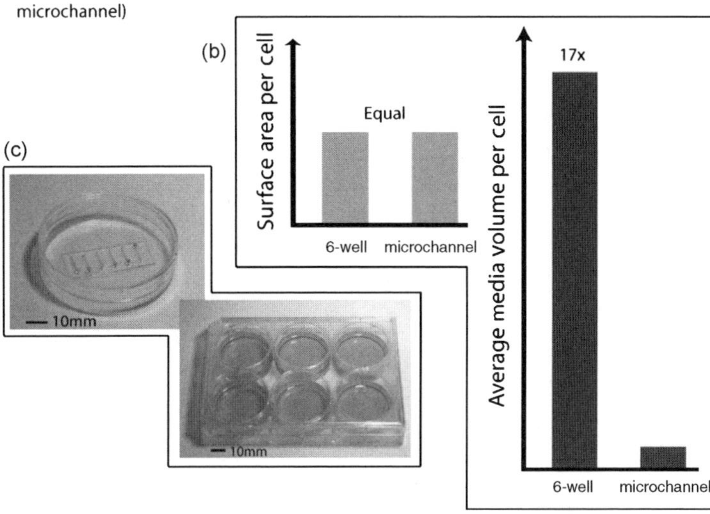

FIGURE 2.3 A comparison between volume densities of culture conditions in traditional, macroscale culture in 6-well plates and in microscale, microchannel culture. Given the same cell surface density, even a rather large microchannel (1 mm wide, 8 mm long, and 250 um tall) can provide a volume density 17 times that of a traditional well in a 6-well plate. In addition to this significantly different volume density, the presence of convection in 6 well plates results in homogenization of the media each time the plate is moved (e.g., for observation). However, in microchannels designed as shown, due to laminar flow profiles, any disturbance of the fluid in the channels would be reversed when the fluid returned to equilibrium (due to surface tension-based passive pumping). Cells in microchannels witness a significantly different microenvironment than in traditional macroscale cultures due to these issues, and the effects of these differences can be significant. (See color insert)

of artifacts to the culture, due to aspects such as continuous flow (e.g., elimination of soluble cell–cell communication, constant concentration of all components in the media) and low effective volume densities (effective volume density would be the total volume perfused rather than the channel volume divided by the number of cells in the culture area). One group even has "eliminated the incubator" by customizing media via addition of buffers (30). Although mammalian cells were cultured (myoblasts and osteoblasts) for weeks, the whole range of cellular effects of altered enzyme kinetics due to the temperature change have not been analyzed yet. A notable study done to analyze the artifacts imparted by a microfluidic culture chamber via the analysis of cellular expression profiles by DNA microarray (31,32) showed significant differences between the profiles of macro-and microscale cultures.

2.2.3 Surface Interactions — Patterning and Controlling Cell Adhesion

Folch et al., have produced an extensive review of methods of using microtechnology for cell–surface interactions (33). Of the many methods presented in this chapter, microcontact printing and stencil-based methods are still among the most commonly used, although the use of self-assembled monolayers as a basis for a patterning scheme has also become widely used.

Spatial micropatterning provides a convenient way to investigate cell–cell communication; the separation distance between cell populations can be used to probe the degree of soluble factor interaction between the cells in culture. Spatial micropatterning also facilitates the creation of organized cocultures by positioning multiple physiologically relevant populations in proximity with the hope that the increased organization and structure of the culture results in more *in vivo*-like behavior from the target cell population. In randomly distributed cultures, it is difficult to determine critical cell–cell spacing and observe localized effects of soluble factor interactions in a repeatable fashion. Numerous micropatterning techniques have been developed to precisely position proteins and small molecules to promote defined cell patterns in a culture (21,34,35).

Microcontact printing for cellular applications uses molded polymeric stamps that are first coated with a cell adhesion promoting film and then applied to the desired substrate to transfer the pattern (36). Cell-repellent films can be similarly patterned to define regions where cells are not to attach. Complex cell patterns can also be developed using combinations of adhesion promoting and cell repelling regions. While microcontact printing uses films to define the location of cell populations, stencil-based methods physically define where cells are allowed to adhere using geometrically defined holes in the stencil. The application of the stencil to a substrate physically blocks the surface and allows cells to adhere only in the unblocked regions (37–39). Adjacently placed microchannels can similarly be used to pattern cells; the geometry of the patterned regions is defined by the channel width and the separation distance between the cell populations is defined by the distance between the microchannels. Microchannels are often aligned over the cell (and protein) patterned substrates to introduce microfluidic functionality to the culture (40,41).

Other patterning methods often employ ECM components for making cell-repellent surfaces cell-adhesive in specific locations. Using a robotic liquid handling system, Chien and Bhatia looked at the differentiation of murine embryonic stem cells and maintenance of hepatocyte phenotype on combinatorial matrix mixtures (42). Seeding cells on gradients of extracellular matrix materials (42,43) or other biomolecules (44) has also been achieved using a variety of devices. One group has analyzed changes in protein expression and cell migration as a result of seeding three different cell types on countergradients of surface immobilized laminin and collagen I (45,46). These counter gradients can be formed using the microfluidic gradient production methods discussed previously for soluble factor gradient production. Electrochemical mechanisms of patterning cell adhesive areas can provide not only spatial but temporal patterning for bioassays (47) but could be extended to cell culture on various matrix components. Cells generally adhere to more hydrophilic regions, allowing surfaces to be patterned via patterning hydrophobic and hydrophilic regions as well (48). By using plasma etching, patterning hydrophilic areas on glass or polystyrene allowed cells to selectively adhere to only hydrophilic areas (49). One group has approached this issue from a cell modification, rather than surface modification method by biotinylating a cell population and then using a patterned avidin surface to pattern the cells (50).

Not only have surface chemistries been patterned and manipulated for use for monolayer cultures in microfluidics, but new work has expanded this idea to patterning surface topography as well. By using microcontact printing and hot embossing polystyrene, one group has created a selectively coated, textured surface with functionalized polymers or proteins that allows the combination of 3D topology with cell–surface interactions (51). Acoustic machining bypasses photolithographic methods to form 3D topographically controlled environments (52). Other work has shown the importance of nanoscale surface structure for cellular processes and by integration of this work into microfluidic systems, additional analyses of cellular responses could be performed (53,54).

2.3 THE MECHANICAL MICROENVIRONMENT

Tissue engineering is often interested in engineering the mechanical microenvironment (40,55,56). Mimicking the three-dimensional structure and multicellular organization found *in vivo* is an important aim when trying to create engineered tissue or doing tissue culture for basic biological research. Dimensionality is critical for the ultimate goal of creating organs for transplantation and for cell types whose function directly relies upon polarization or organized cell–cell junctions. Engineering tissues such as bone relies upon engineered scaffolds for support, and directionality for processes such as bone deposition or secretion. Three-dimensional organization is important as all tissue develops and functions in the context of the extracellular matrix, stroma, and other cells differentiating along other lineages. This aspect of the mechanical microenvironment requires technologies that allow for multiple cell types to be cultured in either fluidic or mechanical contact in a controlled manner.

2.3.1 Cell–Matrix Interactions — 3D Culture

Although three-dimensional cultures and the development of cell scaffolds for various tissue engineering purposes has been performed using a wide variety of matrix materials both synthetic or naturally occurring (57,58), few have been incorporated into microfluidic devices (Figure 2.4). Synthetic hydrogel matrices

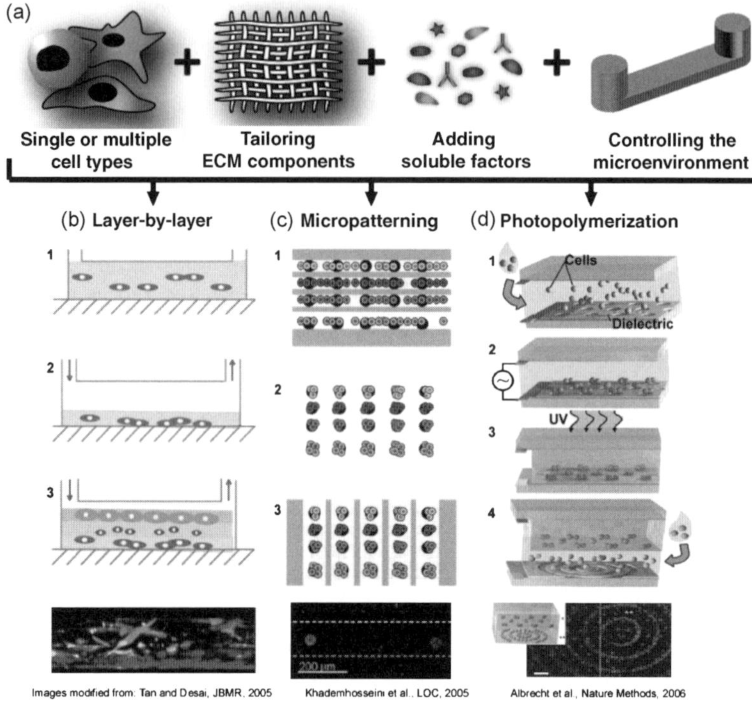

FIGURE 2.4 Mimicking the three-dimensional *in vivo* microenvironment involves the combination of four essential components that have been modeled *in vitro* using various methods by many research groups. (**a**) Single or multiple cell types, various ECM and scaffold components, soluble growth factors and culture conditions, along with a controlled microenvironment using microfluidics comprise four major elements in designing microfluidic systems. (**b**) The layer-by-layer method uses flow to introduce cell containing matrix that is fixed to the bottom surface (b1), when the cells contract a void is created (b2) to add additional layers (b3) (69). *Reproduced by permission of The Royal Society of Chemistry.* (**c**) Micropatterning cell adhesive islands under PDMS channels (c1) allows the introduction of various cell types into a device. When the channels are removed (c2) and turned 90° a coculture system is established (c4) (40). Reproduced by permission of The Royal Society of Chemistry. (**d**) Photopatterning also allows 3D control. By using dielectric patterning of cells (d2) and photopolymerization of the gel, cells can be encapsulated (d3) and these structures can be layered for coculture (d4) (60). The lower panels in b, c, and d show an application of the methods. These methods are not discrete can be combined with each other or other methods. Adapted by permission from Macmillan Publishers Ltd: Nature Methods 2006, 3(5), 369–375, copyright 2006. (See color insert)

have been formed in microchannels and cells encapsulated in them using photolithography (59). Multiple layers can be formed in these gels by using laminar flow to introduce the prepolymer solutions into a channel, and photopolymerizing the gels *in situ* (60). Also, Albrecht et al. have employed dielectrophoresis to pattern cells in three dimensions prior to photopolymerization of the gel allowing additional control over cell location and densities (60). Collagen gels have been patterned in a variety of ways for cell culture, such as using pyramidal wells in silicon chips as a substrate (61) or using laser patterning to locally gel collagen into strips or other patterns (62). More commonly used tissue engineering scaffolds for bone regeneration have been incorporated into 3D microchannel networks as well (63). Work in our lab has employed a naturally occurring enzyme to cross-link gelatin to form microfluidic channels. These biocompatible gels can then be used for cell culture by seeding cells into the channels allowing them to migrate and expand into the gel as they proliferate, resulting in three-dimensional cultures (64).

Layer-by-layer deposition is a commonly used technique that allows multiple components (such as collagen and hyaluronic acid (65) or with fibronectin as well (66)) to be layered and patterned electrostatically on a surface. Another method of doing layer-by-layer deposition in microfludics takes advantage of the natural contraction of matrices by cells (67–69). Here, cells are mixed with various matrix components, flowed into a channel and allowed to contract the matrix, allowing space for the next layer above it. These layer-by-layer techniques allow for multiple cell types to be patterned in the same 3D gel and multiple gel types to be patterned in the same microfluidic channel as well. Cell viability in systems like these are much more likely to be higher than in other techniques that require more processing or photolithography due to the lack of exogenous toxic components or processes. These techniques can be expanded upon to provide the ability to culture multiple cell types in multiple mechanical microenvironments in the same device, allowing cell–cell communication between them (either chemically or mechanically). More work incorporating three-dimensional microenvironments into microfluidic assays will be important for improving and expanding the readouts from these types of assays.

2.3.2 Cell–Cell Interactions — Coculture

Mimicking the cellular microenvironment includes not only the soluble factors, gradients, and/or extracellular matrix constructs but also the interaction between multiple cell types. The body is composed of numerous cell types in organized structures working synergistically. Microfluidics has been used to better elucidate the communication mechanisms employed by cells. The precise low Reynolds number control of microfluidics has been utilized to study gap junctions, by introducing flow in the x-direction and applying suction from small side channels in the $(+)$ y-direction for one cell type and then from the $(-)$ y-direction for another, cells are trapped in pairs along the walls of microchannels (70). Others have used the microscale phenomena to pattern multiple cell types through an array of channels into microwells, then by removing the channels and applying a second set of channels orthogonal to the first a coculture is established (40). In addition, gradient systems described above from Jeon et al. have been used with two cell types perfused with media as the source of soluble factors creating gradients of cellular expressed proteins from two cell types delivered to wells of other cells (71).

Spatial control of the cell types has also been demonstrated using various techniques of micropatterning (66,72–75). By patterning, each cell type can be separated into cellular compartments but yet remain in close contact for cell–cell interactions. One group used natural ECM components to pattern cell adhesive islands. They coated glass slides with hyaluronic acid (HA) and patterned regions in the HA using capillary force lithography. These bare regions were coated with cell adhesive fibronectin and seeded with cells. The remaining HA region was then coated with collagen to promote cell adhesion and then seeded with a second cell type giving rise to more natural cocultures (66). They also used similar methods to pattern 3D spheroids (72). Many coculture platforms have been created to recapitulate the function of the liver, culturing hepatocytes with fibroblasts for drug, and cytotoxicity prescreening prior to *in vivo* models by photopatterning techniques (73,74,76). Bhatia et al. defined regions and patterns that were subsequently coated with collagen for hepatocyte cell adhesion, the unpatterned areas then were seeded with fibroblasts for versatile designed cocultures (73). Toner's group patterned cell adhesive collagen islands in microfluidic gas and media perfusion wells (74). Electron beam lithography has been employed to create nanopore membranes to act as a divider for coculture in studying the blood-brain barrier where one cell type is seeded on top of the membrane and the other cell type on bottom thus allowing direct communication through the pores (75). Other devices have been presented that also use membranes to perfuse cultures, these have been used for primary cell cultures (of a single type) and recently been extended to cocultures of embryos and endometrium (77,78).

To better mimic the *in vivo* conditions, 3D layered structures have been demonstrated to model the organization of the body (67–69). These cultures combine advantages of cell–cell interactions with that of 3D culture and the cell–ECM interaction further mimicking the *in vivo* conditions. Here Tan et al. immobilized matrix adhesion molecules in microchannels and flowed in matrix containing cells and allowed one cell type to contract the matrix subsequently the process can be repeated creating multiple layers. One group used optical tweezers to precisely position single cells in coculture arrays both in 2D and in 3D (76). As described above with the work by Albrecht et al., 3D cocultures can be created by layering cell encapsulated photopolymerized gels (60). These cocultures ultimately have applications in tissue engineering by, for example, combining microfabricated cocultures with biodegradable scaffolds (79,80).

2.4 CONCLUSION

How cells interact with their environment is the sum of how they interact with both the chemical and mechanical components of the microenvironment. Although much of biology and engineering focus on dissecting out which response is due to which stimulus, it is becoming increasingly clear that stripping a cell of its microenvironment can drastically affect the response one sees in *in vitro* assays to stimuli. New microfluidic systems have the potential to not only increase the number and types of assays we are capable of performing but also provide the opportunity to include more aspects of the cells' *in vivo* microenvironment, while still allowing for controlled, repeatable assays to be performed. Expanding our use of patterning of soluble factors, examining mass transport and its effects on cellular behavior, and improving our abilities to use three-dimensional culture devices will allow higher quality and more relevant results to be obtained using microfluidic assays.

With new assays come new experimental issues such as designing and performing the correct controls, choosing appropriate timepoints and relevant readouts. For future microfluidic culture data to be integrated into the extensive body of existing biological data that has been collected over the past several decades that has relied upon traditional macroscale culture methods, the sources of potential artifacts must be addressed. Although these traditional methods impart their own set of artifacts, they are currently the state of the art, and thus it is the duty of the experimenter using microfluidic culture methods to illustrate its validity in the context of current methods. Before any new and unique data obtained will be utilized and integrated into what the community accepts as valid, a basis for understanding how cells respond to these environments will be critical. Additional controls, relating the microfluidic data obtained to that obtained using traditional methods will be critical for establishing a baseline for comparison.

REFERENCES

1. Dittrich PS, Manz A. Nat Rev Drug Discovery 2006;5:210–218.
2. Chung BG, Lin F, Jeon NL. Lab Chip 2006;6:764–768.
3. Jeon NL, Baskaran HSKW, Dertinger SKW, Whitesides GM, Van de Water L, Toner M. Nat Biotechnol 2002;20:826–830.
4. Irimia D, Geba DA, Toner M. Anal Chem 2006;78:472–3477.
5. Irimia D, Liu SY, Tharp WG, Samadani A, Toner M, Poznansky MC. Microfluidic Lab Chip 2006;6:191–198.
6. Lin F, Nguyen CMC, Wang SJ, Saadi W, Gross SP, Jeon NL. Ann Biomed Engi 2005;33:475–482.
7. Abhyankar VV, Lokuta MA, Huttenlocher A, Beebe DJ. Lab Chip 2006;6:389–393.
8. Koyama S, Amarie D, Soini HA, Novotny MV, Jacobson SC. Anal Chem 2006;78: 3354–3359.
9. Diao JP, Young LS, Kim LS, Fogarty EA, Heilman SM, Zhou P, Shuler ML, Wu MM, DeLisa MP. Lab Chip 2006;6:381–388.
10. Frevert CW, Boggy G, Keenan TM, Folch A. Lab Chip 2006;6:849–856.
11. Fujii S, Uematsu M, Yabuki S, Abo M, Yoshimura E, Sato K. Anal Sci 2006;22:87–90.
12. Wang SJ, Saadi W, Lin F, Nguyen CMC, Jeon NL. Exp Cell Res 2004;300:180–189.
13. Saadi W, Wang SJ, Lin F, Jeon NL. Biomed Microdev 2006;8:109–118.
14. Chung BG, Flanagan LA, Rhee SW, Schwartz PH, Lee AP, Monuki ES, Jeon NL. Lab Chip 2005;5:401–406.

15. Yap B, Kamm RD. J Appl Phys 2005;98:1930–1939.
16. Walker GM, Sai JQ, Richmond A, Stremler M, Chung CY, Wikswo JP. Lab Chip 2005;5:611–618.
17. Takayama S, Ostuni E, LeDuc P, Naruse K, Ingber DE, Whitesides GM. Chem Biol, 2003;10:123–130.
18. Adamson DN, Mustafi D, Zhang JXJ, Zheng B, Ismagilov RF. Lab Chip 2006;6: 1178–1186.
19. Pinelis M, Kasinskas R, Borno R, Park J, Chu E, Forbes N, Maharbiz MM. Microfluidic devices for the assembly and culture of three-dimensional multi-cellular constructs with diffusion-limited microenvironments in micro total analysis systems 2006. Tokyo Japan. Micro Total Anal Syst 2006.
20. Lucchetta EM, Munson MS, Ismagilov RF. Lab Chip 2006;6:85–190.
21. Chen CS, Jiang XY, Whitesides GM. MRS Bull 2005;30:194–201.
22. El-Ali J, Sorger PK, Jensen KF. Nature 2006;442:43–411.
23. Yu HM, Meyvantsson I, Shkel IA, Beebe DJ. Lab Chip 2005;5:1089–1095.
24. Raty S, Walters EM, Davis J, Zeringue H, Beebe DJ, Rodriguez-Zas SL, Wheeler MB. Lab Chip 2004;4:86–190.
25. Linder V, Koster S, Franks W, Kraus T, Verpoorte E, Heer F, Hierlemann A, de Rooij NF. Biomed Microdev 2006;8:159–166.
26. Schwartz MP, Derfus AM, Alvarez SD, Bhatia SN, Sailor MJ. Langmuir 2006;22:7084–7090.
27. Yamamura S, Kishi H, Tokimitsu Y, Kondo S, Honda R, Rao S.R, Omori M, Tamiya E, Muraguchi A, Anal Chem 2005;77:8050–8056.
28. Tourovskaia A, Figueroa-Masot X, Folch A. Lab Chip 2005;5:14–19.
29. Kim L, Vahey MD, Lee HY, Voldman J. Lab Chip 2006;6:394–406.
30. Futai N, Gu W, Song JW, Takayama S. Lab Chip 2006;6:149–154.
31. Stangegaard M, Wang Z, Kutter J, Dufva M, Wolff A. Mol Biosyst 2006;2:421–428.
32. Stangegaard M, Petronis S, Jorgensen AM, Christensen CBV, Dufva M. Lab Chip 2006;6:1045–1051.
33. Folch A, Toner M. Ann Rev Biomed Eng 2000;2:27.
34. Sia SK, Whitesides GM. Electrophoresis 2003;23:3563–3576.
35. Whitesides GM, Ostuni E, Takayama S, Jiang X, Iber DE. Ann Rev Biomed Eng 2003;3:335–373.
36. Tien J, Nelson CM, Chen CS. Proc Natl Acad Sci U S A 2002;99:1758–1762
37. Folch A, Toner M. Ann Rev Biomed Eng 2000;2:227–256.
38. Folch, A, Ayon A, Hurtado O, Schmidt MZ, Toner M. J Biomech Eng 1999;121:28–34.
39. Ostuni E, Kane RS, Chen CS, Ingber DE, Whitesides GM. Langmuir 2000;16: 7811–7819.
40. Khademhosseini A, Yeh J, Eng G, Karp J, Kaji H, Borenstein J, Farokhzad OC, Langer R. Lab Chip 2005;5:1380–1386.
41. Khademhosseini A, Suh KY, Jon SG, Eng GJ, Yeh J, Chen GJ, Langer R. Anal Chem 2004;76:3675–3681.
42. Flaim CJ, Chien S, Bhatia SN. Nat Methods 2005;2:119–125.

43. Dertinger SKW, Jiang XY, Li ZY, Murthy VN, Whitesides GM. Proc Nat Acad Sci U S A 2002;9:12542–12547.
44. Jiang XY, Xu QB, Dertinger SKW, Stroock AD, Fu TM, Whitesides GM. Anal Chem 2005;77:2338–2347.
45. Gunawan RC, Silvestre J, Gaskins HR, Kenis PJA, Leckband DE. Langmuir 2006;22:4250–4258.
46. Gunawan RC, Choban ER, Conour JE, Silvestre J, Schook LB, Gaskins HR, Leckband DE, Kenis PJA. Langmuir 2005;21:3061–3068.
47. Kaji H, Hashimoto M, Nishizawa M. Anal Chem 2006;78:5469–5473.
48. Peterson SL, McDonald A, Gourley PL, Sasaki DY. J Biomed Mater Res Part A 2005;72A:10–18.
49. Rhee SW, Taylor AM, Tu CH, Cribbs DH, Cotman CW, Jeon NL. Lab Chip 2005;5:102–107.
50. Sinclair J, Salem AK. Biomaterials 2006;27:2090–2094.
51. Dusseiller MR, Schlaepfer D, Koch M, Kroschewski R, Textor M. Biomaterials 2005;26:5917–5925.
52. Entcheva E, Bien H. Lab Chip 2005;5:179–183.
53. Karuri NW, Porri TJ, Albrecht RM, Murphy CJ, Nealey PF. IEEE Trans Nanobiosci 2006;5:273–280.
54. Liliensiek SJ, Campbell SP, Nealey F, Murphy CJ. J Biomed Mater Res Part A 2006;79A:185–192.
55. Tsang VL, Bhatia SN. Adv Drug Delivery Rev 2004;56:1635–1647.
56. Khademhosseini A, Langer R, Borenstein J, Vacanti JP. Proc Natl Acad Sci U S A 2006;103:2480–2487.
57. Shin H, Jo S, Mikos AG. Biomaterials 2003;24:4353–4364.
58. Schwarz US, Bischofs IB. Med Eng Phys 2005;27:763–772.
59. Liu VA, Bhatia SN. Biomed Microdev 2002;4:257–266.
60. Albrecht DR, Underhill GH, Wassermann TB, Sah RL, Bhatia SN. Nat Methods 2006;3:369–375.
61. Torisawa Y, Shiku H, Yasukawa T, Nishizawa M, Matsue T. Biomaterials 2005;26:2165–2172.
62. Liu YM, Sun S, Singha SMR, Cho MR, Gordon RJ. Biomaterials 2005;26: 4597–4605.
63. Leclerc E, David B, Griscom L, Lepioufle B, Fujii T, Layrolle P, Legallaisa C. Biomaterials 2006;27:586–595.
64. Paguirigan A, Beebe DJ. Lab Chip 2006;6:407–413.
65. Zhang J, Senger B, Vautier D, Picart C, Schaaf P, Voegel JC, Lavalle P. Biomaterials 2005;26:3353–3361.
66. Fukuda J, Khademhosseini A, Yeh J, Eng G, Cheng JJ, Farokhzad OC, Langer R. Biomaterials 2006;27:1479–1486.
67. Tan W, Desai TA. Biomaterials 2004;25:1355–1364.
68. Tan W, Desai TA. Biomed Microdev 2003;5:235–244.
69. Tan W, Desai TA. J Biomed Mater Res Part A 2005;72A:146–160.
70. Lee PJ, Hung PJ, Shaw R, Jan L, Lee LP. Appl Phys Lett 2005:86.
71. Wei CW, Cheng JY, Young TH. Biomed Microdev 2006;8:65–71.

72. Fukuda J, Khademhosseini A, Yeo Y, Yang XY, Yeh J, Eng G, Blumling J, Wang CF, Kohane DS, Langer R. Biomaterials 2006;27:5259–5267.
73. Bhatia SN, Balis UJ, Yarmush ML, Toner M. FASEB J 1999;13:1883–1900.
74. Kane BJ, Zinner MJ, Yarmush ML, Toner M. Anal Chem 2006;78:4291–4298.
75. Ma SH, Lepak LA, Hussain RJ, Shain W, Shuler ML. Lab Chip 2005;5:74–85.
76. Birkbeck AL, Flynn RA, Ozkan M, Song DQ, Gross M, Esener SC. Biomed Microdev 2003;5:47–54.
77. Ostrovidov S, Jiang JL, Sakai Y, Fujii T. Biomed Microdev 2004;6:279–287.
78. Mizuno J, Ostrovidov S, Nakamura H, Inui H, Sakai Y, Fujii T. Fertil Steril 2005;84: S403
79. Bettinger CJ, Weinberg EJ, Kulig KM, Vacanti JP, Wang YD, Borenstein JT, Langer R. Adv Mater 2006;18:165–169.
80. Vozzi G, Flaim C, Ahluwalia A, Bhatia S. Biomaterials 2003;24:2533–2540.
81. Jeon NL, Dertinger SKW, Chui DT, Choi IS, Strook AD, Whitesides GM. Langmuir 2000;16:8311–8316.

3

MICROFABRICATED DEVICES FOR CELL SORTING

Christopher E. Sims and Nancy L. Allbritton

3.1 INTRODUCTION

3.1.1 Why Sort Cells?

Cells, particularly mammalian cells, are known to be highly heterogeneous in their genetic, proteomic, and behavioral characteristics, even within clonal cell lines (1–6). Techniques such as microscopy, polymerase chain reaction (PCR), patch-clamp, and microanalytical chemical separations allow assay of individual cells and benefit from sorting of target cells prior to analysis. Many biochemical, pharmaceutical, and clinical studies still demand samples composed of large numbers of cells where the admixture of cells with different characteristics is a source of bias. For accurate results, these studies require highly enriched or clonal populations to reduce cellular heterogeneity, and thus significant effort is put forth to sort cells into homogeneous populations for further study or expansion (1,3,7–10). Clinical diagnostics, genomic and proteomic studies, genetic engineering protocols, cell transformation studies, cell-based assays, and stem cell isolation are but a few examples where sample preparation techniques for this purpose are of value (11–15). Microfabricated lab-on-chip (LOC) technologies are showing exceptional promise for integrating cell sorting for sample preparation with genetic and other analyses in compact, monolithic devices (14).

Biological Applications of Microfluidics, Edited by Frank A. Gomez

3.1.2 Sorting Strategies

Sorting strategies generally are of two formats, bulk sorting and single-cell sorting. In bulk sorting, cells are collected *en masse* based on a shared characteristic, whereas in single-cell sorting each cell is analyzed, then individually selected. The most frequently used conventional approaches for bulk sorting are gradient centrifugation based on cell density or size and affinity capture based on antibody or ligand binding to a cell-surface target (16). Bulk sorting tends to provide a sample enriched in target cells rather than a homogenous population. Cell sorting by flow cytometry has been a highly successful approach by virtue of the ability to perform high throughput, multiparameter analysis of single cells followed by sorting of the individual target cells (17). Of the conventional methods, flow sorting provides the most pure sample, although it is often necessary to subject the cells to two or more sorts in order to attain the desired purity (16,17). These traditional techniques have in common the need for large sample sizes wherein the cells are placed in suspension for selection and sorting. However, the majority of cell types grow adherent to a surface or to one another, and it is often desirable to screen and select cells in their native state rather than after they are stripped from their growth surface and placed in suspension. Sorting by cloning ring isolation and more recently laser microdissection enables this to be accomplished, but suffer from very low throughput and extremely high cell death rates (18). There are additional caveats to sorting cells by conventional means, chief of which involves exposure of personnel to infectious agents. Droplet-based sorting by flow cytometry has enabled sorting at rates up to 40,000 cells/s, but creates aerosols that may contain particles of infectious agents such as human immunodeficiency virus (HIV) or the hepatitis viruses (19). Limitations of traditional cell sorting schemes have motivated much of the development in microfabricated sorting systems, the topic of this chapter.

3.1.3 Steps in Sorting

3.1.3.1 Identification The initial step in any sorting protocol, whether using conventional means or using LOC technologies, is to screen the cells for one or more characteristics on which to base the sort. Biologists now have an extensive array of techniques for characterizing cells. Cells can be selected based on physical properties such as size, density, behavior in physical fields (electrical, magnetic, or optical), or resistance to chemical lysis (14,16). The advent of optical labeling techniques, particularly fluorescence, combined with flow cytometry has been an important advance for increasing selection criteria for cell sorting (17). The use of highly specific monoclonal antibodies conjugated to fluorophores or magnetic particles has further expanded the number of cellular properties that can be used as a basis for sorting (17). When the probe or labeling procedure is nontoxic, viable cells can be sorted and then expanded into a clonal population for further studies.

3.1.3.2 Separation After a cell is identified, it must be segregated from other cells in the sample. In some instances, the identification and separation occur simultaneously as when differences in physical characteristics or affinity capture are used to separate cells. In single-cell sorting, each cell is analyzed, typically for the presence of one or more fluorescent tags, and is then sorted based on a characteristic signature. Sorting may employ positive or negative selection. In positive selection, the target cell is identified by its signature and collected for further use. Positive selection is typically employed in flow sorting. In negative selection, unwanted cells possess the targeted characteristic and are retained on a capture bed permitting the unretained cells to be collected. Negative selection is frequently used in affinity capture protocols where the release of cells from the affinity trap can be difficult.

3.1.3.3 Collection Once cells are sorted, they must be collected for further use. As alluded to above, bulk sorting procedures generally use a negative selection based on differences in a single characteristic, such as the presence or absence of a cell-surface ligand. Cells are collected as an enriched population with some contaminating cells escaping entrapment. Positive selection by single-cell sorting in flow might be expected to provide a very pure population of cells, since each cell is analyzed before a sort decision is made. For reasons of practicality and throughput, the collected population is usually highly enriched in the target cell population since coincidence (the presence of an unwanted cell in the same detection volume as the target cell) and sorting errors result in contamination of the sorted population (17). This issue is particularly pertinent to isolation of rare cells where contaminating nontarget cells in the sorted population may still be in the majority. Typically in single-cell sorts, sorted cells are pooled during collection, but some sorting schemes enable collection of individual cells. Isolation of individual cells is important for studies of cell-to-cell differences, for example, genetic studies performed by single-cell PCR, and for establishing clonal cell lines, for example, hybridomas or stem cell lines. New approaches for single-cell isolation are thus an important focus of LOC devices.

3.1.4 Lab-on-a-Chip Technologies

Microfabrication is expected to bring a number of advances to cell sorting. Conventional macroscopic sorting methods generally require sample sizes in the range of 10^5-10^6 cells. Lower numbers of cells cannot typically be sorted due to cell losses during the setup and performance of the sort. The ability to use samples of only a few hundred to a few thousand cells has been verified with a variety of LOC-based sorting strategies. Moreover, a number of examples integrating cell sorting with downstream analyses have also been demonstrated. The ability to position individual cells on the micron scale and engineer the cellular microenvironment present new methods to sort cells, making possible unique approaches for cellular characterization that further broaden sort criteria. The enclosed design of LOC technologies based on fluidic switching and other sorting formats reduce the risk of handling infectious materials. Additionally, the cost-effective manufacture of disposable devices is of high value in reducing the risk of sample carry over, an important consideration for diagnostic assays. Given the pervasive enthusiasm for miniaturized devices and their potential as point-of-care diagnostic tools, the above advantages are being diligently pursued in developing LOC cell sorters. Illustrations of advances in these many areas are presented in this chapter.

This chapter summarizes recent advances in the development and application of miniaturized systems capable of sorting bulk and single cells. Emphasis is placed on LOC devices that demonstrate actual separation of cells into one or more distinct populations. Although work utilizing both prokaryotic and eukaryotic cells is covered, we wish to highlight techniques for mammalian cells as these cells are the most fragile, and therefore challenging, to sort into viable populations. We will provide measures of sorting performance such as purity, recovery (yield), and sorting speed when possible, but it is noted that many publications do not proffer a full data set describing the performance of the sorting scheme. Sorting speed (rate at which cells are sorted) will be differentiated from the throughput (speed at which cells are detected) where available. The chapter is divided into sections based on three predominant microfabrication schemes in which the sort is accomplished: (a) flow sorting in a fluidic channel, (b) sorting via a microarray format, and (c) sieving.

3.2 MICROFABRICATED FORMATS FOR CELL SORTING

3.2.1 Sorts in Fluidic Channels

3.2.1.1 Switching Hydrodynamic Flow This year marks the 40th anniversary of cell sorting on a microfluidic device. Kamentsky and Melamed reported the first example of a flow sorter based on a simple cross-channel design utilizing hydrodynamic switching in 1967 (20). Their device was created with a sealed pair of quartz microscope slides, the $100 \times 100\,\mu m$ channels having been etched with an ultrasonic cutter on one slide before bonding. The sample leg of the chip was connected via polyethylene tubing to inlet and outlet reservoirs while the sorting channel was connected to a syringe pump. Pressurizing the inlet reservoir with compressed air initiated flow of a cell suspension through the device and a four-channel optical detection scheme enabled detection at the rate of 5000 cells/s. Sensing of a cell with the desired optical properties actuated the syringe pump causing fluid flow across the "T" junction. By matching the delay time of the actuation to that between detection and displacement of the cell to the sort junction, the desired cell could be withdrawn from the flow channel into the sort side-channel. Sorted cells remained in this channel until the end of the sample processing and were then expelled for analysis. With this scheme, the investigators could achieve a 2000-fold enrichment from 1 : 10,000 to 1 : 5 desired cells. Interestingly, this hardwired system's performance compares favorably to the modern microfluidic systems discussed below (Table 3.1).

Given the intense interest in ever-increasing throughput, the hydrodynamic switching technique of Kamentsky and Melamed was displaced by droplet-based sorting for flow cytometry, but the potential created by modern microfabrication has renewed interest in fluidic switching. Flow switching may provide advantages in cost and aerosol-free sorting of biohazardous samples despite suffering from lower

TABLE 3.1 Sorting Performance of Fluidic Sorting Systems

Reference	Enrichment Method	Switch Time (ms)	Sort Rate (cells/s)	Throughput (cells/s)	Fold
Kamentsky (20)	Hydro	5	—	1000	2000
Wolff (23)	Hydro	2.5	—	12,000	100
Fu (24)	Hydro	5	—	30–40	90
Studer (26)	Hydro	—	—	1500	—
Lancaster (22)	Hydro	1	4	—	—
Yang (27)	Hydro	60	2	—	—
Krüger (28)	Hydro	200	—	—	—
Bang (21)	Hydro	1	—	—	—
Chen (29)	Bubble jet	0.02	—	—	—
Shirasaki (30)	Sol–gel	120	—	25	13
Fu (32)	Eo	—	—	20	30
Johann (34)	Hydro +Eo	90	—	—	—
Dittrich (31)	Hydro + Eo	100	1	—	10
Takahashi (35)	Eo	33	—	—	—
Yao (36)	Eo	—	—	2	—
Xia (49)	Mag	—	—	10,000	—
Wang (56)	Optical	2–4	—	105	—
MacDonald (58)	Optical	—	—	25	—
Hu (95)	Dielectric	—	—	10,000	200

sorting rates (19). In the past 5 years, a number of groups have described microfabricated hydrodynamic cell sorters whose flows have been mechanically controlled either with off-chip pumps and valves (21–23) or utilizing integrated microcomponents (24–27). Many of the systems incorporate a hydrodynamic focusing scheme to create a sheath flow topography. A cell suspension is introduced through the central of three channels. At the confluence of these channels, the cells are focused into a single-file stream for optical detection under laminar flow conditions. Detection of a cell streaming by the interrogation point triggers the sort decision. Sorting is based on redirecting the flow stream from a waste channel to a collection channel.

Flow switching has been accomplished in a variety of ways. Wolff et al. used an external valve that when opened allowed the flow stream carrying the cell of interest to enter the collection channel (23). The valve possessed a response time of 2.5 ms and was connected to the hybrid silicon/glass chip via the collection channel. The investigators were able to separate fluorescent beads from red blood cells achieving a 100-fold enrichment with a throughput of 12,000 cells/s. Quake's group described a flow sorter using multilayer soft lithography to create integrated pumps and valves for flow control (24). An upper layer contained channels controllably pressurized with nitrogen or placed under a vacuum to actuate the on-chip microvalves and micropumps. The micropumps were used to move a stream of bacterial cells at velocities of up to 10 mm/s in small channels in the lower layer of the device. The sorting region was composed of "T" junction and flow in each of the three channels leading into the junction was regulated by a microvalve. The opening and closing of the valves was used to trap the bacteria in the junction and then steer them into the appropriate channel for collection or disposal. Throughput was in the range of 30–40 cells/s with a 20%–40% rate of recovery and an enrichment of up to 90-fold. Studer et al. utilized a similar approach to sort cells by first trapping cells in a "U" shaped sorting channel using microvalves (26). After analysis, selected cells were moved into an accessory collection channel. The approach required precision manipulation of a target cell within the loop into the collection channel. Throughput was 1500 cells/s, but the system was demonstrated by manual actuation of the valves and no data on sorting speed or degree of enrichment were reported.

Lancaster et al. have described the use of a commercial benchtop instrument (Lab-Card MicroFlow System, Micronics, Inc.) to sort cells (22). The instrument employed sheath flow in the detection/sorting zone and flow was regulated by pressurized air applied to the channels. The integrated system enabled cells to be fluorescently labeled on chip and then transferred to a sorting region. To perform the sort, a small number of cells (~4 cells/group) were moved into a detection region, fluid flow was stopped, and the cells interrogated. A fluorescent signal from one or more cells triggered a flush of the entire group of cells into a collection area while the absence of a signal resulted in transfer of the cells to waste. This approach should yield higher sensitivity detection compared to that achievable under continuous flow conditions since the data collection times from the cells can be long. Throughput was approximately 4 cells (one group)/s. Unfortunately, the inability to access the cells after the sort precluded any evaluation of the efficiency of sorting. In another example of a compact benchtop instrument, Yang et al. described a prototype sorting apparatus centered on a microfluidic chip and incorporating an optical detection system, data analysis and control modules in a stand-alone benchtop instrument (Fig. 3.1a and b) (27). The chip manufactured with soft lithography possessed three integrated pneumatic micropumps to provide hydrodynamic sheath and sample flow along with three downstream microvalves to direct cells into any one of three sort channels. Preliminary testing with fluorescently labeled and unlabeled tumor cells resulted in a sorting error of less than 2%. On the basis of the response time of the valves, the authors predicted that a sorting rate of up to 120 cells/min might be achievable.

Pulsed flow has been investigated as a means of briefly deflecting a sample stream into a collection channel. Krüger et al. described a unique sorting design that incorporated hydrodynamic focusing with fluid flow controlled by syringe pumps (28). After interrogation of the cells, the cells entered into the central portion of an "X" junction through a third channel carrying the sample of cells. A brief increase in flow rate through one of the upper legs of the "X" temporarily shifted the laminar flow to one or the other of the downstream legs. Switching could be performed with a response time of 200 ms, but suffered from turbulence created by the differential pressures of the syringe pumps during switching of the flows. Bang et al. have recently described a modification to this approach (Fig. 3.1c) (21). In this case, an accessory channel is positioned between the detection point and the split of a "Y" junction. Under nonsort conditions, the sample flow is biased to flow through the "Y" junction into the arm leading to the waste reservoir. When a cell with the desired

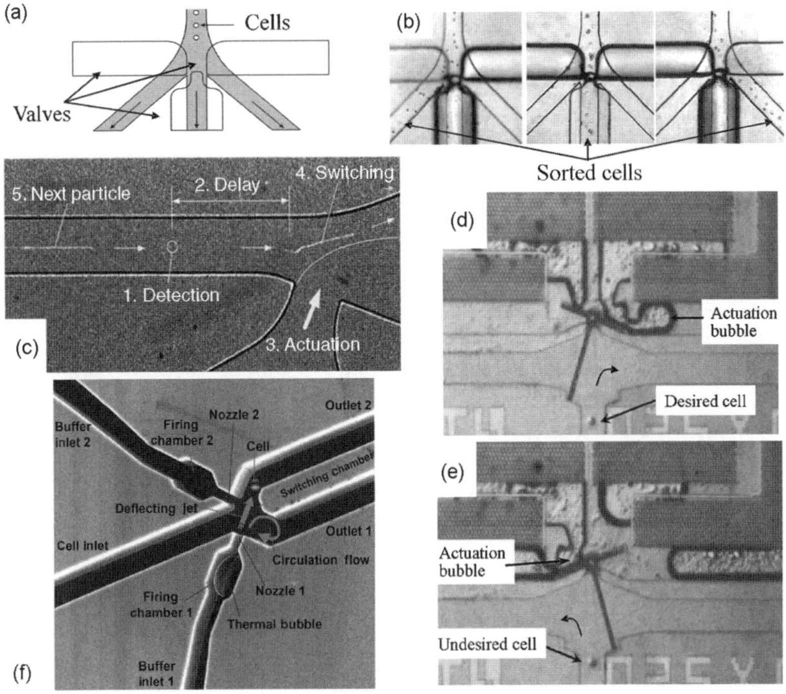

FIGURE 3.1 Sorting by hydrodynamic flow switching. (**a**) Schematic of integrated pneumatic valves and sorting channels described by Yang et al. (2006). (**b**) Photomicrograph of cell sorting in the device schematized in "A." A prefocused flow of cells is directed into one of three downstream channels by opening the valve at the entrance of the selected channel. (**c**) Cell sorting by pulsed flow from Bang et al. (2006). A cell is detected upstream of a "Y" junction (not shown) and is deflected by a pulse of fluid from the actuation channel enabled by an off-chip check valve. The deflected cell is carried into the upper collection channel. Undesired cells are continuously carried into the waste channel by biasing of the laminar flow until another target cell triggers a fluid pulse. (**d** and **e**) Integrated gating switch for control of fluid flow demonstrated by Ho et al. (2005). The "T" switch is actuated by bubbles formed from electrolysis directing cells into the collection (right) or waste channel (left). (**f**) Schematic of the bubble-jet cell sorter described by Chen et al. (2006). A bubble formed in a firing chamber generates a fluid pulse that deflects a cell passing the intersection into one of two outlet channels. (**a** and **b**) Reprinted with permission from Reference 27. (**c**) Reprinted with permission from Reference 21. (**d** and **e**) From Reference 25, reproduced by permission of The Royal Society of Chemistry. (**f**) Reprinted from Reference 29, copyright 2006, with permission from Elsevier. (See color insert)

properties is detected, a brief opening of a check valve injects a pulse of fluid through the accessory channel deflecting the sample stream and cell into the opposite arm of the "Y" junction leading to the collection reservoir. Although data on sorting rate and enrichment have yet to be published, this intriguing technique has the potential for increased sort speed if very short pulses and rapid response times can be achieved.

Three recent reports have described unusual switching schemes for cell sorting (25,29,30). Ho et al. developed a movable gate-like, bubble-actuated microelectromechanical system (MEMS) microswitch to direct hydrodynamic flow between two legs of a "T"-shaped microchannel (Fig. 3.1d and e) (25). Reversible bubble formation via electrolysis was used to move the switch within the intersection of a "T" channel. Movement of the switch was used to direct fluid flow into one of two outlet channels. Target cells flowing in the sample stream were optically identified and then directed to the desired outflow tract. The investigators used a homogenous population from a tumor cell line to test their device, thus enrichment and sorting efficiency data were not obtained. After a mock sort, more than 70% of cells remained viable suggesting that the procedure would be appropriate for live-cell sorting. Chen et al. designed a microchannel geometry with integrated heater elements to sort flowing cells using the principle of bubble-jet printing in hopes of achieving faster sort speeds (Fig. 3.1f) (29). Microheaters with a 5 μs working period were integrated into a pyrex/silicon chip to generate a bubble in a side channel connected to the sort junction. Fluid displacement caused by the bubble deflected microspheres flowing through the sort region into a collection channel in a manner similar to the pulsed flow scheme described above. The bubble was shown to collapse in 5–30 μs depending on pulse width and the energy used in its formation. The system shows promise for high speed sorting if the potential working frequency of 50 kHz, which is comparable to conventional droplet sorting systems, can be exploited. Shirasaki et al. demonstrated the use of a temperature-sensitive sol–gel block copolymer to rapidly switch flow between two channels in order to sort cells (30). Heating from laser irradiation resulted local polymerization of the polymer-doped media within the collection channel, thus blocking flow and directing the sample stream into the waste channel. Cooling upon moving the laser to the waste channel allowed reversal of the polymerization and flow into the collection channel, whereas flow was then blocked by polymerization in the waste channel. The switching and polymer reversal could be accomplished in a 120 ms time period. By switching the laser from channel-to-channel, the investigators were able to attain a 13-fold enrichment in target cells at a throughput of up to 25 cells/s. Cell damage from heat generation did not appear to be a significant problem for the bacterial species used in these experiments, but the high viscosity of the biocompatible polymer along with the need for a high power laser (3 W) and scanning optics may prove to be limitations for mammalian cell applications.

3.2.1.2 Electrophoresis/Electroosmosis The principle of electrokinetic force generation to sort cells has been implemented by a number of groups. Since the response of the electroosmotic flow to the direction of the applied field is brisk, voltage-induced fluid manipulation provides the potential for more rapid sorting speeds compared with hydrodynamic methods and allows for exceptional control over flow direction and velocity (31,32). These advantages are offset, however, by the variations in potential difference over time due to ion depletion and other factors when the electric field is applied continuously. Furthermore, when the desire is to collect live mammalian cells, the negative impact of an electric field on cell physiology is a significant weakness. Cell sorting techniques based on electrokinetic manipulation of flows generally fall into one of two categories—continuous electrokinetic driving of the sample flow (32–34) or intermittent generation of electrokinetic forces to steer cells into collection channels (31,35,36). Fu et al. described the first successful microfluidic cell sorter based on electrokinetic switching of the sample stream (32). These investigators used a poly(dimethylsiloxane) (PDMS) substrate with a three-channel "T" design. Electrodes placed in the inlet and two outlet reservoirs enabled electrokinetic steering of the sample stream into one or the other reservoir. By using computerized signal detection and amplifier switching, a throughput of 20 cells/s and a 30-fold enrichment of target cells were achieved. The device was demonstrated with bacteria expressing a fluorescent protein, some of which remained viable after the sorting. The applicability of this algorithm for the sorting of living mammalian cells is limited due to the increased susceptibility of eukaryotic cells to damage by electric fields. Another example of sorting under the influence of a continuous electric field was provided by Johann et al. who sorted yeast cells under the influence of balanced electroosmotic and pressure-driven flows (34). Similar to Fu's design, the sample stream was brought to a sort junction by electroosmotic flow. Careful balancing of electroosmotic and hydrodynamic flows was used to drive cells to a waste channel. When a target cell was identified, the electric field was switched off, temporarily unbalancing the flow and driving the target cell into a collection channel. While an interesting concept, experimental verification of the sorting efficiency remains to be shown. Pertinent to the discussion here is that the authors noted disintegration of the cells and problems with channel clogging when the device was operated continuously.

To enhance cell viability, sorting approaches using intermittent field effect have been developed. Dittrich and Schwille described hydrodynamic transport and focusing of a sample stream while sorting was performed under the influence of an electrokinetically driven cross flow through an accessory channel downstream of the detection region (Fig. 3.2a) (31). As cells were detected, the direction of electroosmotic flow in the accessory channels was manipulated to deflect cells into waste or collection channels. The authors reported a 10-fold enrichment in target cells with the ability to sort 1 cell/s. The sorting speed was limited by the binning time for detection and electrical capacitance of the switching electrodes. Improvements in the electronics of their system should enable higher sorting rates. Although this strategy reduced the time that the cells were exposed to an electric field, the scheme still required all cells to pass through such a field. It has long been known that exposing cells to electric fields of even microseconds duration can negatively impact cell physiology (37–39). Therefore, a sorting scheme in which live target cells did not experience an electric field would be preferable. Yao et al. accomplished this by utilizing an electric field to deflect unwanted cells only (36). During the run, a voltage of 500 V/cm was applied over a cross channel below the detection region that continually deflected cells passing this intersection into the waste channel. Upon optical detection of a target cell, the voltage was switched off permitting the cell to pass directly into the collection channel without exposure to the electric field. The rate of analysis in this instance was low (1–2 cells/s), but the

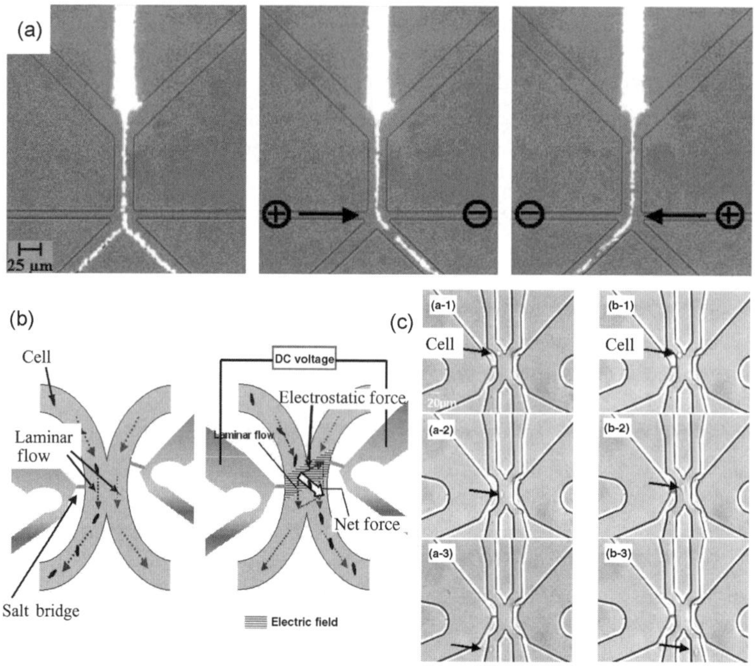

FIGURE 3.2 Electroosmotic flow switching. (**a**) Sorting scheme described by Dittrich and Schwille (2003). After hydrodynamic focusing, cells flow equally into the two output channels (left image). Initiation of an electroosmotic flow in the perpendicular channel directs the sample stream into one or the other output channels depending on the polarity applied to the accessory channel (middle and right image). (**b**) Schematic of cell sorting design of Takahashi et al. (2004). Cells flow under laminar flow conditions through an "X" intersection where two electrodes connect via salt bridges. A DC voltage is continually applied to generate a localized field deflecting cells to waste. Detection of a target cell entering the intersection switches the electric field off and the target cell enters the collection channel without exposure to the electric potential. (**c**) Photomicrographs of the sorting scheme in "B." Panels a-1,2,3 show a target cell passing into the collection channel. Panels b-1,2,3 demonstrate the deflection of an unwanted cell under the influence of a DC field. (**a**) Reprinted with permission from Reference 31, copyright 2003 American Chemical Society. (**b** and **c**) Reprinted with permission from Reference 35.

electrokinetic switch-off scheme should be capable of higher throughput. This example was one of the few published reports that specified the sample sizes required for analyses and sorts. The chip used 20 μL of a cell suspension containing only 400 cells to generate the data presented. Unfortunately, data on cell viability or sorting efficiency were not presented in order to fully assess the technique. A variation of this approach was recently described by Takahashi et al. (Fig. 3.2b and c) (35). These investigators used gel electrodes adjacent to the "X" junction of two microfluidic channels that formed a sorting region. The electrode pair was composed of 1 M NaCl in agar connected by a salt bridge positioned at the intersection of the two channels.

The pressure-driven sample stream through an upper leg of the "X" junction formed a laminar flow with a buffer stream from the other leg. Application of a DC voltage across the electrodes created a local electric field that deflected the cells from the sample stream to the waste stream on a continuous basis. Detection of a target cell entering the sorting region switched off the voltage allowing the cell to pass unperturbed down the sample stream into the collection region before the voltage was reapplied. Limited experimental data published revealed a 100% deflection of unwanted cells into the waste stream at a voltage intensity greater than 5 VDC. Data on cell viability after sorting and sorting efficiency with this technique have yet to be published.

3.2.1.3 Magnetic Magnetic sorting of cells has gained increasing acceptance in biology as a bulk purification method due to low cost and high cell viabilities, especially when used for negative selection (40). The most common technique utilizes ferromagnetic particles of nanometer to micrometer scales that are conjugated with an antibody against a cell-surface antigen. Incubation of the cells with the antibody-coated particles followed by passage of the cells/particles through a magnetic field results in entrapment of cells bound to the magnetic particles while cells lacking the surface antigen pass through the magnetic region. The technique may be used for positive or negative selection. In positive selection antibodies are chosen that will bind the desired cell, such as for purifying lymphocytes or circulating tumor cells from a whole blood sample. A limitation to this approach is that the target cells remain bound to the magnetic particles unless stringent measures, which may be detrimental to the cell, are used to reverse antibody binding. For this reason, the technique is often used for negative selection, that is, an antibody or suite of antibodies is used that will bind unwanted cells, but not the cell type of interest. Negative selection is often used to sort lymphocyte subsets from the buffy coat of blood or stem cells from a mixed cell sample.

Microfluidic systems that have incorporated magnetic-based sorting have typically utilized a positive selection scheme via a stationary magnetic capture region within the device or by sorting in a continuous flow regime. Batch capture on the microscale holds the advantage over conventional magnetic sorting by enabling efficient sorting of very small numbers of cells and with subsequent precise manipulation of the separated cells (14). Both Harrison's and Grodzinski's groups have described a straightforward capture technique for purifying cells for PCR analysis (41–43). Furdui et al. captured immunomagnetic beads coated with a CD-3 antibody in a channel using a localized magnetic field generated by an off-chip magnet (42). A blood sample was then passed over the captured beads and lymphocytes possessing the CD3 cell surface bound to the antibody-coated beads. Captured cells could be thoroughly washed, released by removal of the magnet, and collected for PCR. The investigators achieved a capture efficiency approaching 40% from a 2-μL starting sample (41). Liu et al. have demonstrated a fully integrated device for the immunomagnetic capture and sorting of *Escherichia coli* cells followed by concentration and amplification of their DNA (43). A stationary magnet was incorporated into the PCR chamber of the device to capture target cells preincubated with immunomagnetic beads. The magnetic bead-based target cell capture enabled efficient preconcentration and purification of a 1.4-μL sample, followed by washing, cell lysis, and subsequent on-chip PCR.

A variety of other magnetic-based sorting devices also demonstrate the potential for high throughput cell separation, improved purity, and high recovery rate. Austin's group has described a number of designs for magnetic separation utilizing magnetic stripes placed at an angle to the hydrodynamic flow in a microchannel (44–46). The basic concept is that the magnetic field from each stripe creates a trap that alters the movement of cells bound to immunomagnetic beads. After incubation of a mixed sample with the magnetic beads, the sample is flowed through the channel. As target cells adhere to the stripe, they flow diagonally along the stripe due to the combined force vector created by the hydrodynamic flow and magnetic force. The magnetically labeled cells are thus separated from the unlabelled cells that continue undeflected through the channel. While elegant, the major drawback with this technique was that bead-bound cells tended to permanently stick to the stripes making subsequent recovery problematic (46). This problem has been overcome in other designs by utilizing multiple laminar flow streams passing through a magnetic field created by an external magnet (47–49). With these techniques, cells bound to immunomagnetic beads are deflected laterally by a magnetic field into an adjacent flow stream. The flows separate downstream to carry the bead-bound cells to the outlet reservoir and the remaining sample to waste. Xia et al. used a dual laminar flow scheme to sort immunomagnetically labeled *E. coli* from a mixture of the bacteria with red blood cells (49). The investigators were able to achieve a throughput of 10,000 cells/s with a sorting efficiency approaching 80% (sample size not specified). Pamme et al. used a broad channel design through which multiple laminar streams flowed in parallel (47,48). Mammalian cells were induced to uptake magnetic nanoparticles by endocytosis and were then flowed through the device. Mixtures of magnetic and nonmagnet cells could readily be separated; moreover, the magnetic cells could be segregated based on their magnetic moment (amount of endocytosed nanoparticles) (Fig. 3.3a–c). The results of this study suggest that it may be possible to sort cells based on the expression level of a surface antigen by utilizing the very small nanoparticles now available (<10 nm) to label the cell surface in direct proportion to the density of antigen.

It is possible to sort cells based on their intrinsic magnetic properties, but feasibility has only been demonstrated for the separation of red blood cells from other cells. In the presence of high magnetic fields and gradients, the weak paramagnetic behavior of deoxygenated red blood cells can be exploited to separate them from leukocytes (14,46). Frazier's group has utilized this phenomenon to separate red and white blood cells in a microfluidic device (50). Moving to the microscale enabled the investigators to integrate a 0.2 T, 50 μm-thick magnetized nickel stripe into 2.5 cm channel to produce a 200 T/m field. Red blood cells were concentrated into the high magnetic flux regions over a 20 min period whereas leukocytes were unaffected by the magnetic field. These investigators also developed an optimized version of this device with hydrodynamic flow to carry the separated cells into downstream collection channels (Fig. 3.3d–f) (51). The optimized device

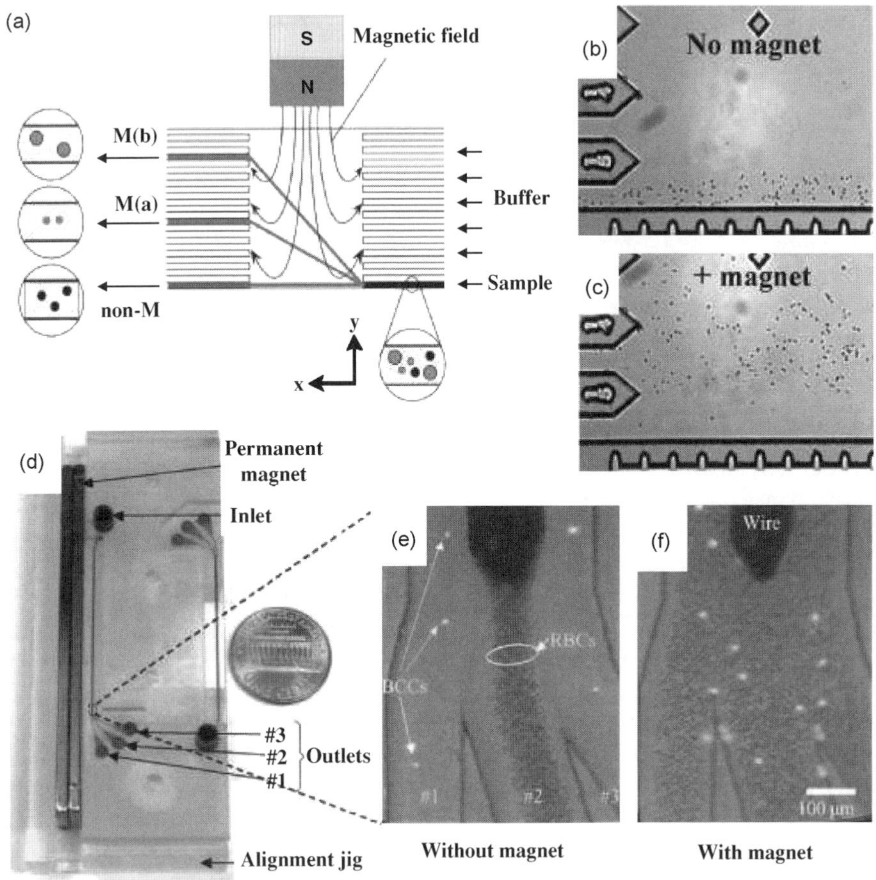

FIGURE 3.3 Flow sorting in a magnetic field. (**a**) Schematic of the magnetic field sorter described by Pamme et al. (2006). (**b** and **c**) Photomicrographs of cells being sorted under the influence of the applied field. (**b**) In the absence of a magnetic field, all cells pass into the lower left outlet channel. (**c**) Upon application of the field, cells labeled with magnetic particles are deflected from the sample stream and collected into one of various outflow tracts depending upon the amount of magnetic label on the cell. (**d**) Paramagnetic microseparator of Han et al. (2006). (**e**) Under the influence of a magnetic field, RBCs (ellipse) are focused into a central stream and separated from breast cancer cells (arrows). (**f**) With the magnetic field removed, cells flow with equal probability into each of the three outlet channels. (**a–c**) From Reference 48, reproduced by permission of The Royal Society of Chemistry. (**d–f**) Reprinted from Reference 51, copyright 2006, with permission from Elsevier. (See color insert)

was used to sort a mixture of breast cancer cells spiked in a sample of purified red blood cells with a separation efficiency of 95% at a sample flow rate of 0.05 mm/s. Use of this label-free scheme to isolate rare cancer cells circulating in raw peripheral blood would be complicated by the admixture of nucleated white blood cells that would be expected to far outnumber the tumor cells.

3.2.1.4 Optical (Trapping and Switching) Two decades have passed since cell sorting by light trapping was first reported (52). The mating of this technique with microfluidic devices has demonstrated a number of the advantages that optical manipulation of cells has to offer. Chief among these are the high selectivity and precision with which single cells can be manipulated in a noncontact, sterile manner. Optical-based sorting also permits the physical decoupling of microscale sorters with the macroscopic world, albeit requiring a transparent device of optically high quality and an expensive off-chip optical system. In a manner analogous to the magnetic flow-based sorters described above, Oakey et al. combined an optical trap with simple paired laminar flow streams to provide the force necessary to deflect a target cell from the sample stream into an adjacent collection stream (53). The simple PDMS–glass device was coupled with a single laser beam focused through a 100× oil-immersion objective and scanned with a piezoelectrically controlled mirror. *E. coli* cells could be readily trapped as they flowed along the sample stream. Moving the trap normal to the direction of fluid flow by only a few microns carried the trapped cell into the collection stream where it was released and carried downstream into the collection reservoir. Data on throughput and sorting efficiency were not published. Enger et al. used optical trapping to displace cells over long distances (54). Yeast cells were sorted from a mixture with bacteria in a sample stream carried to the optical trap region by electroosmotic flow. Individual target cells could be trapped and transported many millimeters through a side channel into a collection reservoir and released to generate a small sample of high purity. Although trapping and displacing individual cells is low in throughput, there are many applications, such as single-cell studies, which would benefit from procedures capable of specifically sorting small numbers of cells on-chip (18). In this regard, Arai et al. described an impressive technical *tour de force* integrating an optical trap with a microfluidic chip for single-cell sorting (55). The cells were suspended in a biocompatible, thermosensitive hydrogel solution, and an optical trap was employed to capture and transfer a single yeast cell adjacent to an integrated microheater. Switching on the microheater caused a local sol–gel transformation to occur at 32°C effectively immobilizing the cell. Undesired cells were washed away and the selected cell could be maintained in the gel trap or released by switching off the heater. Impressively, the yeast cells could grow and divide within the gel trap indicating their viability. Demonstrating successful application of this technique with mammalian cells that are more sensitive to mechanical and thermal stresses would significantly extend its utility.

Recent efforts have begun to address the throughput issues of optically based sorting techniques. Wang et al. reported high throughput fluorescence activated sorting using an "optical switch" which, rather than trapping the target cell, merely deflected it with enough force to move it into a collection stream flowing parallel to the sample stream (56). A laser with an "on–off" switching time of 2–4 ms was used to deflect the cell. The technology was demonstrated with mammalian cell lines and sample sizes as small as 1000 cells. Throughputs up to 105 cells/s and purities of the sorted sample as high as 98% were achieved. MacDonald et al. reported high throughput sorting by using five interfering laser beams to form a 3D optical lattice within a microfluidic device (57). As cells flow through the continuous lattice, they underwent a deflective force determined by the optical properties of the cell, including refractive index, size, and shape. By employing a paired laminar flow scheme, cells were deflected into a collection stream, and thus bulk sorted based on their optical properties. The approach, which achieved a throughput up to 25 cells/s and purities of the sorted population in excess of 95%, has been demonstrated by separating lymphocytes from red blood cells (57,58).

3.2.1.5 Size Separation Using Fluid Flow In this section, schemes for sorting cells based on their size and behavior in a flow stream will be described. Seki's group has published a number of reports using the principle of "pinched-flow fractionation" (59,60). Three streams with one of the outer streams containing a sample of peripheral blood were merged under laminar flow conditions in a microchannel. The channel was then constricted or pinched resulting in the alignment of the cells along one of the channel walls. As the cells moved through the pinched region, the geometric centers of the smaller cells were closer to the sidewall than that of the larger cells whose geometric centers were primarily within the central stream. Thus, when the channel diameter was increased after the pinched region, the small cells segregated into the original outer stream and larger cells remained aligned in the center stream. Branching of the channel then sorted cells into various reservoirs based on their size. Enrichments on the order of 20-fold could be accomplished. Recent advances reported by this group involving splitting and recombining the streams in a multiply branched system have enabled enrichments up to 80-fold (61,62).

3.2.1.6 Affinity Capture An element shared among several techniques for cell sorting is the application of a label to the cells in order to discriminate among cell types. This label is frequently based on binding of an antibody to a cell-surface antigen. To accomplish cell tagging, one or more preprocessing steps must be undertaken to attach the label (fluorescent dye, magnetic bead, etc.) to the cells. Cell affinity chromatography (CAC) bypasses these processing steps by immobilizing an affinity tag on a surface that can interact with and trap targeted cells (63,64). To maximize the amount of antibody available, polymer beads are conjugated to an antibody and cells passed through a packed bed of the beads. However, cell transit times through the bed can be as long as 1–2 h due to the low flow rates employed. To overcome this limitation, free-flow microfluidic chambers have been implemented to take advantage of the high surface-area-to-volume ratios inherent in microfluidic devices. Murthy et al. used a flow chamber with antibody-coated walls to separate T and B lymphocytes based on the presence of CD5 (T cell) or CD19 (B cell) surface markers (65). The population of cells captured on the wall was up to 97% pure, but neither the efficiency of capture nor the time required to complete the separation were described. Subsequent elution and collection of the highly purified population retained in the device was also not described. This group has used this capture device as an inexpensive tool to count CD4+ T lymphocytes in whole blood (66). These investigators have also demonstrated an improved flow chamber design incorporating multiple parallel serpentine channels that enhances capture efficiency for negative selection (67). In this work depletion of the B-cell population to produce an enriched T-cell population was demonstrated. With a starting sample a 50 : 50 mixture of B and T cells containing ~1 million cells, a near 100% pure population of T cells was obtained in less than 3 min. Arrays of these channels might provide significant time savings over conventional bead-based CAC for sorting by negative selection.

3.2.2 Array-Based Sorting

3.2.2.1 Dielectrophoresis Separating cells based on their dielectric properties has been an area of intense research for more than a decade, and the use of microfabricated systems for this purpose has been a focus of much of this investigation. Space constraints limit the scope of the chapter, so that readers are encouraged to seek out one of the recent books or reviews for an in-depth treatment of this topic (68–74). Dielectrophoresis (DEP) refers to the motion of particles resulting from the effects of nonuniform electric fields on the dielectric polarization of cells (68). The DEP force impinging on a cell under the influence of the field is determined by the magnitude and polarity of the charges induced in the cell. Since a cell consists of structures (cell membrane, mitochondria, nucleus, etc.) with very different electrical properties, large dipole moments that are dependent on the frequency of the applied field can be induced. Differences in cell size, shape, cell-surface charge, membrane conductivity, and the presence and conductivity of membrane-bound vesicles contribute to the forces created by the electric field. Therefore, cells displaying differences in morphology, stage of differentiation, and physiologic state generally show differences in their polarization characteristics, and hence dielectrophoretic behavior (70). A study of the physics involved in this process yields interesting findings. The DEP force varies with the square of the electric field, thus reversing the polarity of the applied voltage does not reverse the DEP force. AC voltages can therefore be employed, typically in the 500 Hz to 50 MHz range (68). Usually cells will be less conductive than the extracellular buffer in which they are suspended. In this situation, cells typically move away from regions of high electric field, a characteristic known as negative DEP. If, on the contrary, the cell's conductivity is greater than its suspending buffer, the cell will display positive DEP and move toward the region of high electric field (74). Sorting and separation schemes on microdevices have taken advantage of both negative and positive DEP characteristics.

Electric fields have a number of profound effects on living cells, from activation of cellular signaling pathways to cell lysis (37–39). It is generally believed that the magnitude of the applied AC fields used to affect dielectrophoretic manipulation is below that which leads to irreversible damage to cells, and studies of cell viability and proliferation support this contention (68,75,76). Nevertheless, other studies have clearly shown physiologic effects including upregulation of gene transcription as a result of dielectrophoretic manipulation (77). It is conceivable that the use of microfabricated DEP systems may be advantageous in reducing untoward effects on cells. There are significant advantages from miniaturizing the electrodes as a 100-fold reduction in size will yield a 1000-fold reduction in the operating voltage required to produce a given DEP force (68). The effect of buffer composition on cell viability must also be borne in mind when using DEP for cell sorting. Bacteria and yeast can often tolerate extremes of buffer pH and osmolarity that may be employed to provide media of appropriate conductivity for DEP separations (78). Mammalian cells have very narrow tolerances for the buffer composition in which they are suspended, and buffers with a pH or ionic strength that is not in this physiologic range can have profound and detrimental effects on cellular physiology (79).

A large range of construction designs and sorting methods has been utilized to develop microfabricated devices employing DEP for cell sorting. Most of these designs incorporate arrays of electrodes, although a few schemes utilize a single electrode pair in a flow channel to perform dielectrophoretic switching (see below). The typical device uses thin-film metallic electrodes arrayed on a glass or silicon substrate, although recent advances in polymer devices have increased the attractiveness of polymeric materials for the device substrate (73). Hybrid devices using plastic microfluidic components combined with CMOS silicon chips show promise in enabling the creation of large arrays of programmable electrodes for DEP-based sorting (80,81). A novel approach was recently reported to create DEP electrodes using a standard indium–tin oxide (ITO) coated surface as one electrode and a photoconductive hydrogenated amorphous silicon layer over an ITO layer to create the second (82). "Virtual electrodes" could be generated by exposure of the photosensitive layer to light even at low light intensity. By patterning the light exposure using a laser or mask, electrodes could be created in desired locations. This approach was used to create cell traps operating by positive DEP to attract and hold cells while cells not trapped under the DEP conditions were washed away. With this approach, flexibility in cell manipulation by DEP could be accomplished by selectively patterning the electrodes in real time.

An AC field applied across a single pair of electrodes to generate positive DEP forces is one of the most commonly used sorting schemes for the bulk separation of cells from suspension (69). Generally, conditions are chosen so that desired cells display positive DEP resulting in their capture, while undesired cells experience negative DEP forces and are repelled from the traps and washed away. The trapped cells can then be released and collected (Fig. 3.4a–d). This technique has been used in microfabricated devices to separate *E. coli* from red blood cells, transformed from nontransformed cells, and cultured neuronal cell types (83,84). Investigators at Nanogen have used this approach to capture cells from a mixed population, whereupon the trapped cells were immobilized by antibodies to form a cell array (85). Hughes' group has used positive DEP trapping to separate viable and nonviable yeast and to study apoptosis in leukemic cell lines (86,87). Wang et al. used an in-channel trap to separate and collect viable from nonviable yeast cells from a continuous flow sample stream with 90% collection efficiency and 90% purity (78). Bulk separation by these trapping methods allows concentrating a desired cell type, but the ability to discriminate among more than two cell types is limited (69).

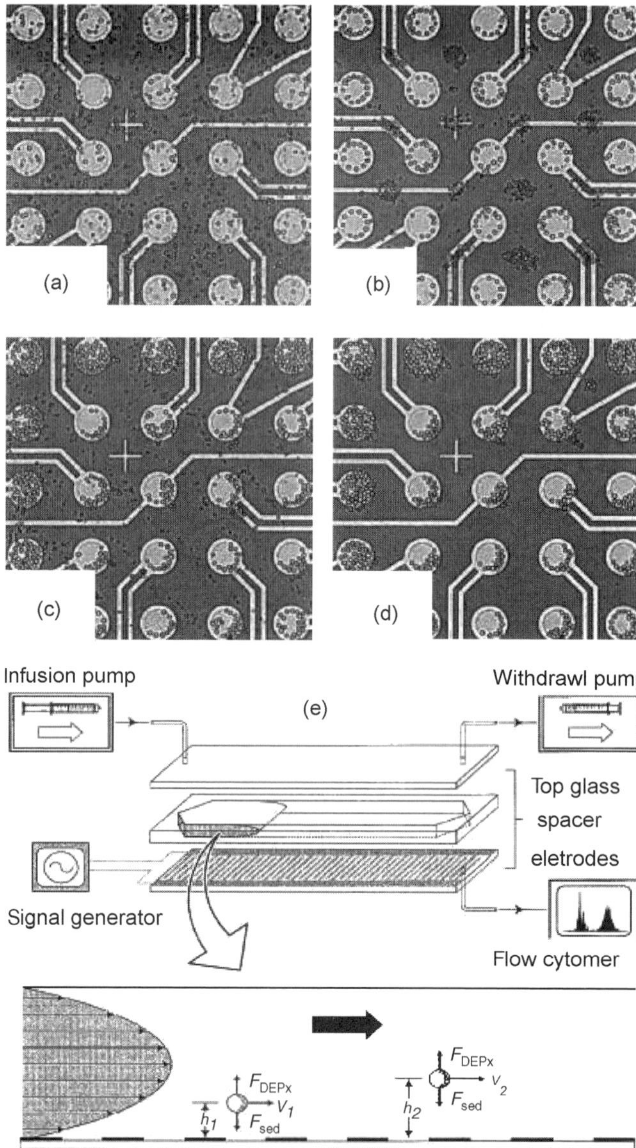

FIGURE 3.4 Dielectrophoretic cell separation. (**a**–**d**) DEP cell separation array as demonstrated by Huang et al. (2002). (**a**) A mixed population of cells is introduced. No separation is present. (**b**) Target cells are trapped by positive DEP on the circular electrodes while unwanted cells collect in the regions between the electrodes due to negative DEP forces. (**c**) Unwanted cells are washed away by fluid flow while maintaining voltage on the electrodes to hold trapped target cells. (**d**) Purified population of target cells. (**e**) Principle of cell separation by DEP field-flow fractionation from Wang et al. (2000). See text for discussion. (**a**–**d**) Reprinted with permission from Reference 83, copyright 2002 American Chemical Society. (**e**) Reprinted with permission from Reference 92, copyright 2000 American Chemical Society.

Multiple-phase AC sources can be used to trap and sequentially transport cells in traveling-wave DEP (74). In traveling-wave DEP a series of parallel planar electrodes are used to levitate cells above the electrode bed by negative DEP (88). Addressing of the electrodes with a series of sequentially phase-shifted voltages produces a traveling electric field that propels particles along the electrode array. Morgan's group has used this principle to separate cells from beads based on differences in the velocities of the respective particles as they moved along the electrode bed (89). These investigators have also described integration of this approach with optical detection schemes that may lend themselves to improvements in cell analysis for higher throughput separations (90,91). A variant of this cell separation approach is field-flow fractionation, a family of methods that uses negative DEP to position cells at varying heights in a hydrodynamic flow stream (69). Cells are carried in the stream's parabolic flow profile at various velocities dependent on their position in the stream. Since cells are positioned at varying distances from the electrodes depending on their characteristic dielectric properties, they are separated by the different velocities they achieve depending on their position in the channel. Gascoyne's group used this principle in the bulk enrichment of leukocytes from blood, a mixture of breast cancer cells and normal T-lymphocytes, and leukocyte subpopulations (Fig. 3.4e) (92,93). Enrichments up to 35-fold were reported. In a variation of field-flow fractionation, Park et al. reported the use of asymmetric electrodes to generate dielectric forces of varying magnitude in hopes of enhancing separation efficiencies (94). The investigators performed a head-to-head comparison of their method and field-flow fractionation with a symmetric electrode array. Employing a mixture composed of equal numbers of mouse tumor cells and red blood cells revealed a slightly better postsort cell purity using the asymmetric design (82% versus 79%). Notably, while both devices performed less well than the 95% purity obtained with traditional MACS used as a control, the DEP technique did not require the use of antibody labeling. Although not commented upon in this comparison, the microfabricated systems should also efficiently separate small sample sizes. In other studies, DEP has been used as the driving force to deflect cells traveling in a laminar flow streams in order to carry out sorting (95–97). To enhance the deflecting force imposed by DEP, Hu et al. labeled bacterial cells possessing a cell-surface marker with polymer beads conjugated to an antibody directed at the surface antigen (95). An electrode region angled to the axis of the cell flow was used to deflect target cells by negative DEP into a collection stream (Fig. 3.5a and b). The enhanced dielectrophoretic amplitude response of the bead attached to the cell enabled the investigators to achieve an enrichment of labeled versus nonlabeled cells of greater than 200-fold with a throughput of 10,000 cells/s.

The DEP designs described above have been for the most part directed at bulk separations. Single-cell sorting, on the contrary, has largely been performed using multipole traps operating by negative DEP. Fuhr's group detailed the use of negative DEP to spatially confine cells using an inhomogeneous electric field created by multiple electrodes (98). This trapping technique was combined with a microfluidic channel to capture individual cells from the flow stream for single-cell analysis (99). After microscopic analysis, the cell was released from the trap and directed by a

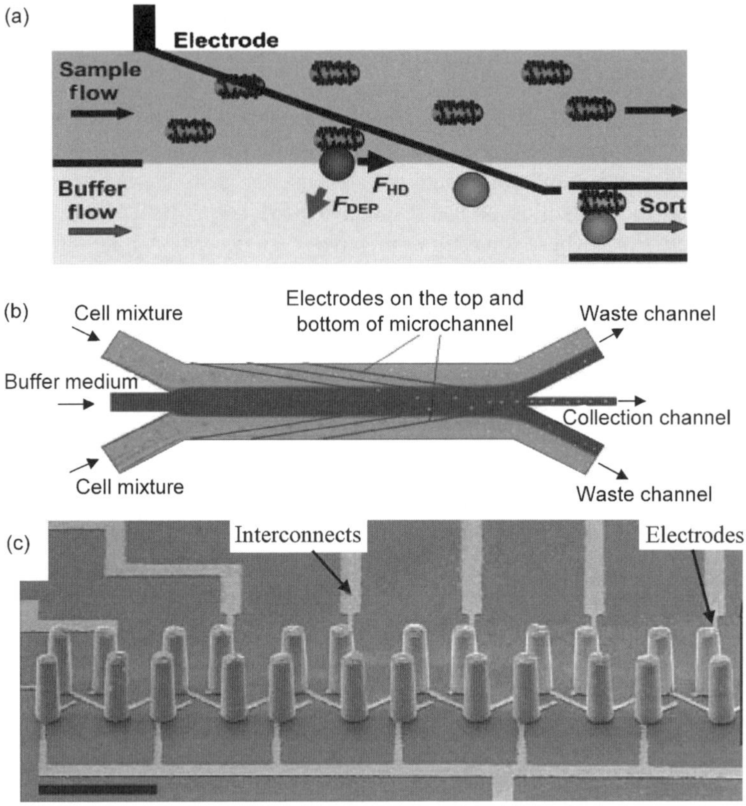

FIGURE 3.5 Dielectrophoretic cell separation. (**a** and **b**) Schematics of the DEP sorting scheme described by Hu et al. (2005). (**a**) Flowing cells of a given DEP character are deflected into a collection stream by a dielectrophoretic force (F_{DEP}) generated by electrodes embedded in the flow channel. (**b**) A three-phase laminar steam is used to enhance collection of target cells. (**c**) DEP trap array demonstrated by Voldman et al. (2002). Shown are six single-cell traps each consisting of four electroplated gold electrodes in a trapezoidal arrangement. (**a** and **b**) From Reference 95, copyright 2005 National Academy of Sciences, USA. (**c**) Reprinted with permission from Reference 101, copyright 2002 American Chemical Society. (See color insert)

second set of electrodes to a collection or waste channel. While very low in throughput, the method enables high content analysis of the captured cells (100). This DEP trap and switch now serve as the heart of a commercially available system for single-cell manipulation and sorting (Evotec Technologies, Hamburg, Germany, available at http://www.evotec-technologies.com) Voldman et al. designed and tested an array of negative DEP quadrupole traps for cell capture and analysis (Fig. 3.5c) (101). The linear array was composed of a series of individually addressable traps capable of holding single cells in the presence of fluid flow. The fluid flows enabled delivery or reagents and cell washes for performing cell-based assays. The array was imaged by microscopy and individual cells could be released or held in the trap for sorting. Device performance was demonstrated by manual capture, fluorescent labeling, and sorting of a mammalian leukemia cell line using an eight-trap array. The implementation of automated control and an expanded array size would likely be required for practical application. To address this issue, Voldman's group recently described a scalable and addressable array format utilizing positive DEP traps (102). By employing a 4×4 row/column array architecture of "ring–dot" electrodes, individual electrode pairs could be addressed to selectively release a trapped cell from the array. Using the HL-60 leukemia cell line, the array was loaded with a mixture of cells containing either a red or a green fluorescent dye. Selective release of cells of one color was demonstrated. While data on release efficiency and purity were not provided, these proof-of-principle experiments hold promise for higher throughput single-cell sorting. Scaling laws suggest that the architecture demonstrated could be used to produce an array containing 10,000 individual traps using 200 electrical connections that was deemed manageable based on currently available chip-packaging technologies (102). In this regard, investigations of the use of microelectronic CMOS (complementary metal–oxide semiconductor) silicon chips for DEP trap arrays have been described for cell sorting (75,80,81). Fuchs et al. used a hybrid silicon/plastic device incorporating a 0.64 cm^2 CMOS chip composed of a planar array of over 100,000 electrodes used as negative DEP traps (75). Sample sizes of ~2000 cells were used to capture single cells in the 4500 traps of the array. Cells were identified in traps by microscopy and individually transported from trap-to-trap into a collection area by switching the AC field between electrodes in the direction of the desired motion. Using the K562 leukemia cell line, typical experiments recovered 2–10 cells with a recovery rate of selected cells between 50% and 80%. Although requiring 1–15 min to transport each cell to the collection chamber, the large array size may prove valuable for the detection and sorting of very small sample sizes.

3.2.2.2 Affinity Capture As discussed in a prior section of this chapter, a cell specific ligand such as an antibody can be used to capture a specific cell type from a mixed population. Devices utilizing the combination of an array layout and affinity capture have incorporated spotted arrays of antibodies and arrays of posts conjugated with a cell-specific reagent. Arrays of antibodies create an especially powerful format to isolate cells. By virtue of modern array spotters' ability to handle numerous separate solutions, many unique capture regions can be created on a single glass slide. Sekine et al. have demonstrated the simultaneous capture of four different types of leukocytes from a buffy coat sample with purities exceeding 97% (103). Even more impressive is the work of Belov et al. who have described the use of a protein microarray to simultaneously capture and immunophenotype leukemia cells obtained from patients (Fig. 3.6a and b) (104). Using this approach, these investigators have phenotype the leukemias of individual patients by arraying up to 82 antigens simultaneously (105). Although the array spots with subpopulations of captured cells cannot be collected individually, the cells can be further analyzed on the array using fluorescent reagents. While the spot size of the Sekine and Belov

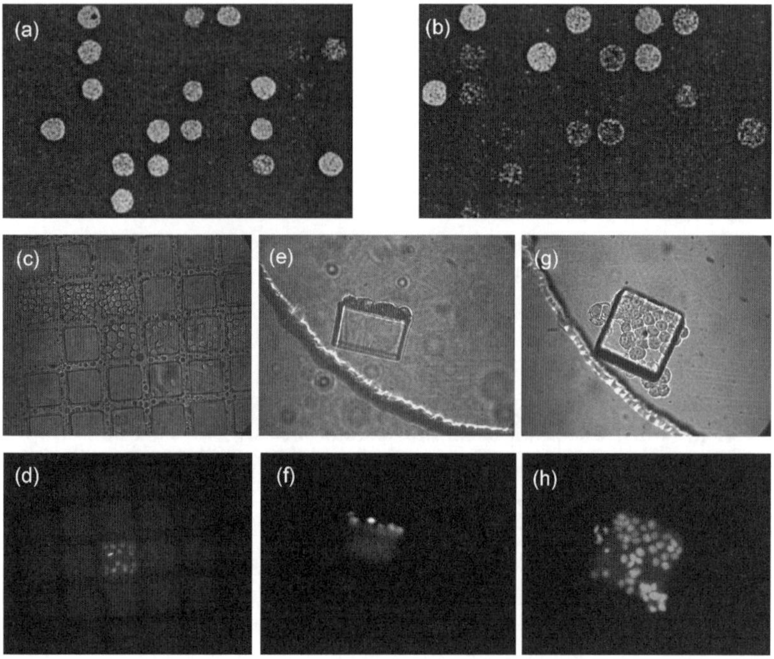

FIGURE 3.6 Array-based separations. (**a** and **b**) Affinity-mediated cell collection using an array of antibodies directed against cell-surface CD antigens from Belov et al. (2001). Each spot of the microarray contains a unique antibody. The pattern of leukocyte binding on the array varies among patients and correlates with the disease process, as demonstrated by the differential patterns seen between panels "**a**" and "**b**." (**c**–**h**) Separation of adherent cells from a cell-based array. (**c** and **d**) Transmitted light and corresponding fluorescence images of cells on a micropallet array. A single micropallet in the photomicrograph is seen to possess fluorescent cells. (**e** and **f**) Transmitted light and corresponding fluorescence images of the single released micropallet containing the target fluorescent cells immediately after collection. (**g** and **h**) Transmitted light and corresponding fluorescence images of the same pallet and collection well 2 days after collection showing growth of a fluorescent colony on the pallet. (**a** and **b**) Reprinted with permission from Reference 104.

experiments was in the millimeter scale and required sample sizes of 1–10 million cells, further miniaturization should be possible. Modern dispensing tools are now capable of generating droplets of less than 1 nL to create spots less than 100–200 μm in diameter (106). It should, therefore, be possible to generate affinity traps on the micron scale to isolate cells from suspension. By combining such arrays with microfluidic cell manipulation, smaller sample sizes should be achievable.

In a different approach for sorting by affinity arrays, Chang et al. used the adhesion protein selectin conjugated to an array of posts to retard the flow of leukocytes with the goal of separating cells based on their differential interactions with the array (107). Unlike high affinity antibodies, selectin-mediated binding tends to be weaker so that as cells transited the array, flow-induced shear forces should fractionate the cells based on the surface expression of the ligand for selectin (108). Using a mixture of leukemia and lymphoma cell lines, the investigators demonstrated that the two cell types did transit the array at different speeds, although the differential in transit speed was not great enough to achieve complete separation. This preliminary study suggested that adhesion-based cell sorting might prove feasible with optimization of device design and surface modification. Fuchs and Toner have described a similar approach for sorting circulating tumor cells from whole blood samples (109). Posts within their array were coated with one or more monoclonal antibodies directed at tumor cell markers. As the blood sample was passed through the array, target cells bound and were trapped within the latticework. The microfluidic device was shown to be capable of passing whole blood at a rate up to 20 mL/h. An asset of this device is that very large volumes of blood can be separated in reasonable times.

3.2.2.3 Laser-Based Cell Selection Microfabricated array techniques have been combined with pulsed laser methods for cell sorting. Laser microdissection has been used for over a decade to selectively isolate single or small numbers of cells from fixed tissue sections for genomic and proteomic analyses (110–112). The typical procedure involves placement of cells or tissue on a thin membrane. The desired cells are identified by microscopy and separated from adjacent cells by either adhering target cells to the membrane by thermal bonding or cutting the membrane around the cells of interest with the laser. The membrane is lifted away carrying with it the target cell(s) while leaving undesired cells behind (110). Revzin et al. utilized laser microdissection to sort lymphocytes from a microwell array (113). Poly(ethylene glycol) (PEG) photolithography was used to create microwells composed of PEG hydrogel surrounding a glass surface. The glass bottom of the wells was modified with a variety of ligands to act as affinity capture agents in order to improve trapping of the nonadherent cells used in this study. After formation of the cell array, the cells were fixed with ethanol and dried. Laser microdissection was used to remove, and thus sort, cells from selected wells. However, living cells could not be sorted using this method. Lee et al. used a similar approach to attain isolated hepatocytes for gene expression analysis (114). A microarray printed with collagen spots was used to culture hepatocytes that adhered selectively to the collagen. Fibroblasts were then added to form a mixed coculture. Individual colonies of hepatocytes were removed by laser microdissection and subjected to polymerase chain reaction. The mRNA expression patterns confirmed the cellular identity of sorted colonies as hepatocellular in origin. While living cells could not be sorted, the technique is particularly exciting since it supports the possibility of selectively collecting cells from an engineered microenvironment. Cell-to-cell interactions could thus be manipulated by tailoring the surface properties in microscale regions followed by capture of the cells for direct analysis of the genome or proteome.

Laser microdissection approaches are powerful tools for sorting of cells when maintenance of cell viability is not at issue. Unfortunately, the required cell manipulation, dehydration, and direct effects of the laser tend to reduce the survival of sorted cells (18). To circumvent this problem, the Allbritton and Li groups have collaborated in the development of a cell array technology with removable elements termed pallets (Fig. 3.6c–h) (115–117). The technology has been specifically designed to permit the analysis and sorting of adherent cells or colonies. The pallets are fabricated from SU-8 on a glass surface using standard photolithography. Covalent modification of the glass surface between the pallets is used to create stable regions of air between the pallets. These virtual walls of air restrict cell attachment and growth to the top surfaces of the pallets (116). A pulsed laser focused at the base of the pallet is used to disrupt the adhesion of the pallet to its glass substrate. Cell viability is maintained as the pallet protects the overlying cell from the direct effects of the laser. In addition, the analysis, release and collection procedures are performed in cell media or buffer so that cell dehydration does not occur (117). The laser-based techniques discussed in this section share a common and significant advantage to other sorting methods. Adherent cells can be analyzed and selected based on their morphology or their dynamic properties to significantly broaden the range of sorting criteria.

3.2.2.4 Magnetic Magnetic manipulation of cells has been accomplished in an array format using patterned microelectromagnetic matrices and by patterning of magnetic beads. Lee et al. demonstrated a crosshatched array of wires fabricated on a Si/SiO_2 substrate to create the electromagnetic matrix (118). Each layer of conducting wire was insulated to prevent electrical shorting. By varying the current in each independent wire, dynamic magnetic field patterns were created on the array. The magnetic field was then used to manipulate the motion of individual cells attached to magnetic beads. The investigators demonstrated single-cell sorting of viable yeast using a fluorescent viability stain. Recently, this group extended their work to incorporate CMOS integrated circuit technology containing a programmable microcoil array to produce the spatially patterned magnetic fields (119). In this demonstration, single bovine capillary endothelial cells with endocytosed magnetic nanoparticles were manipulated to sort a small group of cells. Although the prototype device possessed only a small 8×8 array, modern electrical engineering capabilities should enable much larger arrays capable of manipulating many individual cells simultaneously. Pregibon et al. patterned magnetic microbeads in a polymer scaffold for bulk cell separations (120). The bead patterns were created using UV photopolymerization of a hydrogel containing the micron-scale beads. A suspension of beads in the hydrogel was placed in a microfluidic chamber. The beads were evenly dispersed within the hydrogel by application of a uniform magnetic field, and a photomask was used to pattern the polymerization of the hydrogel. After washing away unpolymerized hydrogel, an organized array of bead clusters trapped in the polymer was produced in the base of the flow chamber. Notably, the entire procedure could be accomplished within an enclosed microfluidic device. An external circular electromagnet was used to reversibly magnetize the paramagnetic bead array. A sample of T and B lymphocytes was used to test the array as a sorting method. The B lymphocytes were conjugated via their CD19 surface antigen to immunomagnetic particles; therefore, the B cells were captured on the array spots formed by the embedded magnetic beads as the mixture was flowed through the chamber. After the capture step, the B cells were released into the flow channel by switching off the electromagnetic field, but statistics on enrichment and collection efficiency were not provided. The data suggest an interesting and flexible approach to creating patterned cell capture and release traps within a microfluidic device for cellular assays and sorting.

3.2.3 Arrayed Sieves, Weirs, and Posts for Sized-Based Cell Sorting

Size-based sorting by means of sieves has been demonstrated by a number of investigators as an effective bulk separation scheme. Kricka's group has experimented with a simple, but effective, method using a single-file closely spaced array of either posts or weirs constructed in silicon to create sieves for isolation of white cells from whole blood (121–123). The approach is conceptually simple with white blood cells being retained along the array due to their larger size while the smaller and more easily deformed red blood cells pass through the interstices of arrayed posts or between the roof of the flow channel and weir. Initial studies using 5–7 μm spacing whether between posts or of the weir showed a surprisingly low capture efficiency due to the ready deformability of the white cells (122). Subsequent iterations of the cell separation module found that a post array with 3.5 μm spacing improved capture and collection results for on-chip PCR, although capture efficiency based on cell counts were not described (121,123). This technique is appropriate for small samples and was demonstrated with white cells isolated from 1 μL of whole blood. At the same time, the capacity for cell isolation is limited by the width of the array. Using a similar concept, Mohamed et al. designed and tested a microfluidic device containing an array of posts characterized by sequential regions of decreasing interpost spacing and height in order to trap cells of differing size at different points within the array (124). The device was used to sort white cells and tumor cells from whole blood, but data on sample size, capture efficiency, and purity were not published. An important caveat to this system is that the target cells are embedded within the arrays after the sort, and thus subsequent collection of the cells for further use may prove a challenge.

Although not in array format, two groups have demonstrated the use of weirs in flow channels for cell sorting. Chen et al. recently published a weir-type structure similar to that described above in order to capture circulating tumor cells from blood (125). The weir structures used a 5- or 10-μm spacing to pass red cells while retaining the larger nucleated cells in a chamber upstream of the weir. The purity of the tumor or white cell population captured using the 10 μm weir was only 23%, but by incorporating two weirs in series this could be improved to 70%. The 5 μm weir, on the contrary, yielded a purity of greater than 99%. The nucleated cells were retained in an open reservoir that would enable easy collection. A key issue for capture of rare circulating tumor cells from whole blood with this technique is that the much more numerous white blood cells in the sample would also be retained by the weir. Vankrunkelsven et al. described the use of a weir or microstep that formed a constriction in the channel (126). Using a mixture of bacteria and yeast to demonstrate separation of cells by size, the investigators showed that as shear flow carried cells against the microstep, the smaller bacterial cells flowed over the step while the larger yeast cells were retained in a backwash at the leading edge of the step. Whether this approach will prove feasible with more malleable and fragile mammalian cells remains to be determined.

Austin's group has employed parallel arrays of posts to study the separation of blood cells by physical means (127–129). In earlier work, an extensive array of posts of 5 μm spacing and possessing a polar polyurethane coating was used to emulate the flow of cells in a capillary bed. While red cells readily flowed through the array, white blood cells were forced to deform under the hydrodynamic flow in order to pass through the closely spaced latticework. Interestingly, granulocytes and lymphocytes were seen to become adherent to the posts at different distances into the array resulting in an *in situ* sort. The posts lacked an affinity ligand suggesting a nonspecific mechanism underlying cell adherence. It was hypothesized that intrinsic differences in cell adhesiveness or other physical properties brought about the differential cell trapping. Recently, this group described a system for size-based separations resulting from asymmetric bifurcation of laminar flow around a continuous series of offset posts within an array (Fig. 3.7) (128,129). Beads or cells below a certain critical diameter follow streamlines through the array, tending to deviate laterally by a magnitude that depends on their diameter. Thus, cells of different size exit the array in different locations and white cells, red cells, and platelets could be fractionated using this method. The authors report a 110-fold enrichment in the white-to-red cell ratio from the original whole-blood specimen. Although this enrichment falls short of what may be accomplished by simple centrifugation, the procedure was carried out with only a drop of blood from a fingerprick.

A microfluidic device has recently been demonstrated that acts to segregate a mixture of cells based on size by employing two linear arrays of posts parallel to the long axis of a sample flow stream (130,131). As the cell sample flows through the channel, smaller cells pass through this bordering sieve and are carried by an ever-widening flow channel to a pair of collection reservoirs. The central channel becomes enriched in the larger cell type. The device has been used to separate a mixture of myocyte and nonmyocyte cells from a homogenate of neonatal rat myocardium (130). The device performs best when small cells need to be isolated from much larger cells (131).

3.3 OUTLOOK FOR THE FUTURE

The past decade has seen the growth of an extensive body of literature relating to the design and development of cell sorting technologies based on microfabricated systems. Once fully developed, these LOC devices will offer distinct advantages over conventional methods for sorting cells. Chief among the opportunities for advancing cell sorting methods is the ability to manipulate and sort very small numbers of cells. This function is crucial for enabling performance of biological experiments that at present cannot be accomplished because only small and precious samples are

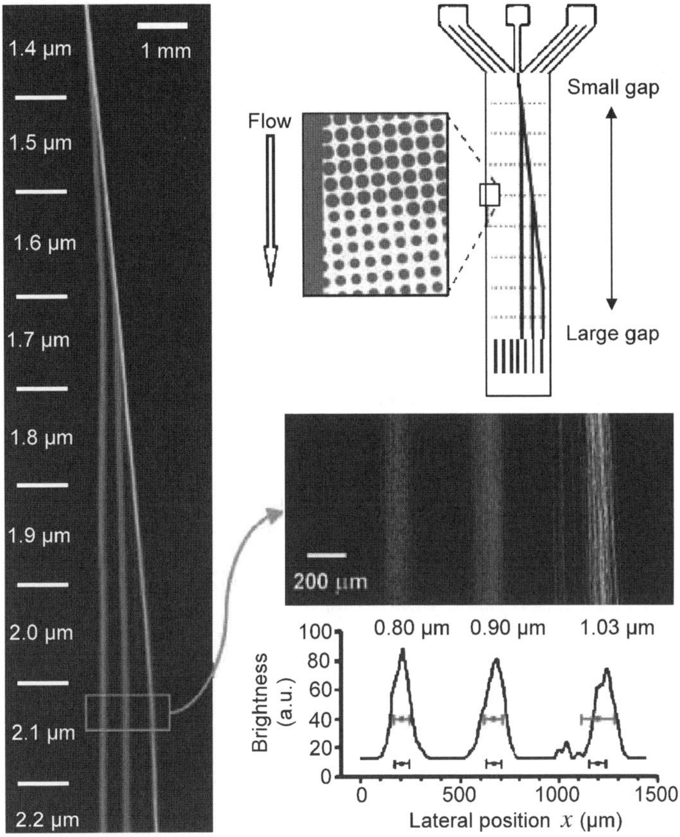

FIGURE 3.7 Bead separation by deterministic lateral displacement as demonstrated by Huang et al. (2004). A matrix composed of offset posts of progressively smaller diameter and an increasing distance between the posts was used to separate fluorescent particles (green, red, and yellow beads with diameters of 0.8, 0.9, and 1.03 μm, respectively). The distance between the centers of the posts is shown on the far left side of the figure. Beads of a given size follow equivalent paths through the array and are segregated into populations based on size. Reprinted with permission from Reference 129. (See color insert)

obtainable. Continued progress in microscale formats that benefit from the low loss sample handling capabilities of these devices is thus a crucial focus for the near future. The ability to engineer and control the cellular microenvironment within a sorting platform will increase the type of criteria by which cells can be sorted. Furthermore, integration of cell labeling protocols with the analysis and sorting of target cells offers yet another opportunity for providing improved cell sorting tools for the biologist. Although the goal of achieving throughputs and sort rates rivaling flow cytometry presents a tremendous challenge, the massively parallel designs possible with microfabricated systems may make high speed sorting possible.

Moreover, the prospect of sorting potentially infectious samples in enclosed devices offers significant advantages over droplet-based sorting for clinical use. Additional value exists in the development of cost-effective, disposable formats to enable diagnostic and therapeutic applications without concern for sample carryover. This aspect is particularly pertinent with the increasing prevalence of genetic studies based on PCR that demand highly purified cell populations for diagnostic and prognostic accuracy.

ACKNOWLEDGMENTS

This work was supported by NIH grants CA105514 to C.E.S. and EB004597 & EB004436 to N.L.A.

REFERENCES

1. Davis JE, Eberwine JH, Hinkle DA, Marciano PG, Meaney DF, McIntosh TK. Methodological considerations regarding single-cell gene expression profiling for brain injury. Neurochem Res 2004;29:1113–1121.
2. Dovichi NJ, Hu S. Chemical cytometry. Curr Opin Chem Biol 2003;7:603–608.
3. Lefievre L, Barratt CL, Harper CV, Conner SJ, Flesch FM, Deeks E, Moseley FL, Pixton KL, Brewis IA, Publicover SJ. Physiological and proteomic approaches to studying prefertilization events in the human. Reprod Med 2003;7:419–427.
4. Li H, Sims CE, Kaluzova M, Stanbridge EJ, Allbritton NL. A quantitative single-cell assay for protein kinase B reveals important insights into the biochemical behavior of an intracellular substrate peptide. Biochemistry 2004;43:1599–1608.
5. Negulescu PA, Shastri N, Cahalan MD. Intracellular calcium dependence of gene expression in single T lymphocytes. Proc Nat Acad Sci, USA 1994;91: 2873–2877.
6. Schneggenburger R, Sakaba T, Neher E. Vesicle pools and short-term synaptic depression: lessons from a large synapse. Trends Neurosci 2002;25:206–212.
7. Fend F, Raffeld M. Laser capture microdissection in pathology. J Clin Pathol 2000; 53:666–672.
8. Hack CJ. Integrated transcriptome and proteome data: the challenges ahead. Brief Funct Genomic Proteomic 2004;3:212–219.
9. Kaern M, Elston TC, Blake WJ, Collins JJ. Stochasticity in gene expression: from theories to phenotypes. Nat Rev Genet 2005;6: 451-64.
10. Taylor TB, Nambiar PR, Raja R, Cheung E, Rosenberg DW, Anderegg B. Microgenomics: Identification of new expression profiles via small and single-cell sample analyses. Cytometry A 2004;59:254–261.
11. Fernandes PB. Technological advances in high-throughput screening. Curr Opin Chem Biol 1998;2 (5):597–603.
12. Hahn WC, Weinberg RA. Rules for making human tumor cells. N Eng J Med 2002;347:1593–1603.

13. Mousses S, Caplen NJ, Cornelison R, Weaver D, Basik M, Hautaniemi S, Elkahloun AG, Lotufo RA, Choudary A, Dougherty ER, Suh E, Kallioniemi O. RNAi microarray analysis in cultured mammalian cells. Genome Res 2003;13: 2341-7.
14. Toner M, Irimia D. Blood-on-a-chip. Ann Rev Biomed Eng 2005;7:77–103.
15. Welm B, Behbod F, Goodell MA, Rosen JM. Isolation and characterization of functional mammary gland stem cells. Cell Prolif 2003;36 Suppl 1:17–32.
16. Freshney RI. Culture of Animal Cells. Hoboken: Wiley-Liss;2005.
17. Shapiro HM. Practical Flow Cytometry. Hoboken: Wiley-Liss;2003.
18. Sims CE, Bachman M, Li GP, Allbritton NL. Choosing one from the many: selection and sorting strategies for single adherent cells. Anal Bioanal Chem 2006;387:5–8.
19. Leary JF. Ultra high-speed sorting. Cytometry A 2005;67A:76–85.
20. Kamentsky LA, Melamed MR. Spectrophotometric cell sorter. Science 1967;156: 1364–1365.
21. Bang H, Chung C, Kim JK, Kim SH, Chung S, Park J, Lee WG, Yun H, Lee J, Cho K, Han DC, Chang JK. Microfabricated fluorescence-activated cell sorter through hydrodynamic flow manipulation. Microsys Technol 2006;12:746–753.
22. Lancaster C, Kokoris M, Nabavi M, Clemmens J, Maloney P, Capadanno J, Gerdes J, Battrell CF. Rare cancer cell analyzer for whole blood applications: Microcytometer cell counting and sorting subcircuits. Methods 2005;37:120–127.
23. Wolff A, Perch-Nielsen IR, Larsen UD, Friis P, Goranovic G, Poulsen CR, Kuttera JP, Telleman P. Integrating advanced functionality in a microfabricated high-throughput fluorescent-activated cell sorter. Lab Chip 2003;3:22–27.
24. Fu AY, Chou H-P, Spence C, Arnold FH, Quake SR. An integrated microfabricated cell sorter. Anal Chem 2002;74:2451–2457.
25. Ho C-T, Lin R-Z, Chang H-Y, Liu C-H. Micromachined electrochemical T-switches for cell sorting applications. Lab Chip 2005;5:1248–1258.
26. Studer V, Jameson R, Pellereau E, Pepin A, Chen Y. A microfluidic mammalian cell sorter based on fluorescence detection. Microelectron Eng 2004;73–74:852–857.
27. Yang S-Y, Hsiung S-K, Hung Y-C, Chang C-M, Liao T-L, Lee G-B. A cell counting/sorting system incorporated with a microfabricated flow cytometer chip. Meas Sci Technol 2006;17:2001–2009.
28. Kruger J, Singh K, O'Neill A, Jackson C, Morrison A, O'Brien P. Development of a microfluidic device for fluorescence activated cell sorting. J. Micromech Microeng 2002;12:486–494.
29. Chen CC, Wang JS, Solgaard O. Micromachined bubble-jet cell sorter with multiple operation modes. Senso Actuators B 2006;117:523–529.
30. Shirasaki Y, Tanaka J, Makazu H, Tashiro K, Shoji S, Tsukita S, Funatsu T. On-chip cell sorting system using laser-induced heating of a thermoreversible gelation polymer to control flow. Anal Chem 2006;78:695–701.
31. Dittrich PS, Schwille P. An integrated microfluidic system for reaction, high-sensitivity detection, and sorting of fluorescent cells and particles. Anal Chem 2003; 75:5767–5774.
32. Fu AY, Spence C, Scherer A, Arnold FH, Quake SR. A microfabricated fluorescence-activated cell sorter. Nat Biotechnol 1999;17:1109–1111.

33. Fu L-M, Yang R-J, Lin C-H, Pan Y-J, Lee G-B. Electrokinetically driven micro flow cytometers with integrated fiber optics for on-line cell/particle detection. Anal Chim Acta 2004;507:163–169.

34. Johann R, Renaud P. A simple mechanism for reliable particle sorting in a microdevice with combined electroosmotic and pressure-driven flow. Electrophoresis 2004;25: 3720–3729.

35. Takahashi K, Hattori A, Suzuki I, Ichiki T, Yasuda K. Non-destructive on-chip cell sorting system with real-time microscopic image processing. J Nanobiotechnology 2004;2:1–8.

36. Yao B, Luo G, Feng X, Wang W, Chen L-X, Wang Y. A microfluidic device based on gravity and electric force driving for flow cytometry and fluorescence activated cell sorting. Lab Chip 2004;4:603–607.

37. Chen C, Smye SW, Robinson MP, Evans JA. Membrane electroporation theories: a review. Med Biol Comput 2006;44:5–14.

38. Teruel MN, Meyer T. Electroporation-induced formation of individual calcium entry sites in the cell body and processes of adherent cells. Biophys J 1997;73:1785–1796.

39. Bilska AO, DeBruin KA, Krassowska W. Theoretical modeling of the effects of shock duration, frequency, and strength on the degree of electroporation. Bioelectrochemistry 2000;51:133–143.

40. Safarik I, Safarikova M. Use of magnetic techniques for the isolation of cells. J Chromatography B 1999;722: 33–53.

41. Furdui VI, Harrison DJ. Immunomagnetic T cell capture from blood for PCR analysis using microfluidic systems. Lab Chip 2004;4:614–618.

42. Furdui VI, Kariuki JK, Harrison DJ. Microfabricated electrolysis pump system for isolating rare cells in blood. J Micromech Microeng 2003;13:S164–S170.

43. Liu RH, Yang J, Lenigk R, Bonanno J, Grodzinski P. Self-contained, fully integrated biochip for sample preparation, polymerase chain reaction amplification, and DNA microarray detection. Anal Chem 2004;76:1824–1831.

44. Berger M, Castelino J, Huang R, Shah M, Austin RH. Design of a microfabricated magnetic cell separator. Electrophoresis 2001;22:3883–3892.

45. Inglis DW, Riehn R, Austin RH, Sturm JC. Continuous microfluidic immunomagnetic cell separation. Appl Phys Lett 2004;85:5093–5095.

46. Inglis DW, Riehn R, Sturm JC, Austin RH. Microfluidic high gradient magnetic cell separation. J Appl Phys 2006;99.

47. Pamme N, Eijkel JCT, Manz A. On-chip free-flow magnetophoresis: separation and detection of mixtures of magnetic particles in continuous flow. J Magne Magn Mater 2006;307:237–244.

48. Pamme N, Wilhelm C. Continuous sorting of magnetic cells *via* on-chip free-flow magnetophoresis. Lab Chip 2006;6:974–980.

49. Xia N, Hunt TP, Mayers BT, Aslberg E, Whitesides GM, Westervelt RM, Ingber DE. Combined microfluidic-micromagnetic separation of living cells in continuous flow. Biomedic Microdevi 2006;8:299–308.

50. Han K-H, Frazier AB. Continuous magnetophoretic separation of blood cells in microdevice format. J Appl Phys 2004;96:5797–5802.

51. Han K-H, Han A, Frazier AB. Microsystems for isolation and electrophysiological analysis of breast cancer cells from blood. Biosens Bioelectron 2006;21:1907–1914.
52. Buican TN, Smyth MJ, Crissman HA, Salzman GC, Stewart CC, Martin JC. Automated single-cell manipulation and sorting by light trapping. Appl Opti 1987;26:5311–5316.
53. Oakey J, Allely J, Marr DWM. Laminar-flow-based separations at the microscale. Biotechnol Prog 2002;18:1439–1442.
54. Enger J, Goksor M, Ramser K, Hagberg P, Hanstorp D. Optical tweezers applied to a microfluidic system. Lab Chip 2004;4:196–200.
55. Arai F, Ng C, Maruyama H, Ichikawa A, El-Shimy H, Fukuda T. On-chip single-cell separation and immobilization using optical tweezers and thermosensitive hydrogel. Lab Chip 2005;5:1399–1403.
56. Wang MM, Tu E, Raymond DE, Yang JM, Zhang H, Hagen N, Dees B, Mercer EM, Forster AH, Kariv I, Marchand PJ, Butler WF. Microfluidic sorting of mammalian cells by optical force switching. Nat Biotechnol 2005;23:83–87.
57. MacDonald MP, Spalding GC, Dholakia K. Microfluidic sorting in an optical lattice. Nature 2003;426:421–424.
58. MacDonald MP, Paterson L, Richies A, Dholakia K, Spalding GC. Cell cytometry with a light touch: sorting microscopic matter with an optical lattice. J Biol Regul Homeosta Agents 2004;18:200–205.
59. Takagi J, Yamada M, Yasuda M, Seki M. Continuous particle separation in a microchannel having asymmetrically arranged multiple branches. Lab Chip 2005;5:778–784.
60. Yamada M, Nakashima M, Seki M. Pinched flow fractionation: continuous size separation of particles utilizing a laminar flow profile in a pinched mode microchannel. Anal Chem 2004;76:5465–5471.
61. Yamada M, Seki M. Hydrodynamic filtration for on-chip particle concentration and classification utilizing microfluidics. Lab Chip 2005;5:1233–1239.
62. Yamada M, Seki M. Microfluidic particle sorter employing flow splitting and recombining. Anal Chem 2006;78:1357–1362.
63. Baran MM, Allen DM, Russell SR, Scheetz ME, Monthony JF. Cell sorting using a universally applicable chromatography matrix. J Immunol Methods 1982;53:321–334.
64. Mandrusov E, Houng A, Klein E, Leonard EF. Membrane-based cell affinity chromatography to retrieve viable cells. Biotechnol Prog 1995;11:208–213.
65. Murthy SK, Sin A, Tompkins RG, Toner M. Effect of flow and surface conditions on human lymphocyte isolation using microfluidic chambers. Langmuir 2004;20: 11649–11655.
66. Cheng X, Irimia D, Dixon M, Sekine K, Demirci U, Zamir L, Tompkins RG, Rodriguez W, Toner M. A microfluidic device for practical label-free CD4+ T cell counting of HIV-infected subjects. Lab Chip 2007;7:170–178.
67. Sin A, Murthy SK, Revzin A, Tompkins RG, Toner M. Enrichment using antibody-coated microfluidic chambers in shear flow: model mixtures of human lymphocytes. Biotechnol Bioeng 2005;91:816–826.
68. Pethig R, Markx GH. Applications of dielectrophoresis in biotechnology. TIBTECH 1997;15:426–432.

69. Gascoyne PRC, Vykoukal JV. Dielectrophoresis-based sample handling in general-purpose programmable diagnostic instruments. Proc IEEE 2004;92:22–42.
70. Gonzalez CF, Remcho VT. Harnessing dielectric forces for separations of cells, fine particles and macromolecules. J Chromatog A 2005;1079:59–68.
71. Hughes MP. Nanoelectromechnics in Engineering and Biology. Boca Raton: CRC;2002.
72. Morgan H. AC Electrokinetics: Colloids and Nanoparticles. London: Research Studies Press;2003.
73. Simmons BA, McGaw GJ, Davalos RV, Fiechtner GJ, Fintschenko Y, Cummings EB. The development of polymeric devices as dielectrophoretic separators and concentrators. MRS Bull 2006;31:120–124.
74. Voldman J. Electrical forces for microscale cell manipulation. Ann Rev Biomed Eng 2006;8:425–454.
75. Fuchs AB, Romani A, Freida D, Medoro G, Abonnenc M, Altomare L, Chartier I, Guergour D, Villiers C, Marche PN, Tartagni M, Guerrieri R, Chatelain F, Manaresi N. Electronic sorting and recovery of single live cells from microlitre sized samples. Lab Chip 2006;6:121–126.
76. Wang X, Yang J, Gascoyne PRC. Role of peroxide in AC electrical field exposure effects on Friend murine erythroleukemia cells during dielectrophoretic manipulations. Biochim Biophys Acta 1999;1426:53–68.
77. Archer S, Li T-T, Evans AT, Britland ST, Morgan H,Cell reactions to dielectrophoretic manipulation. Biochem Biophys Res Commun 1999;257:687–698.
78. Wang Z, Hansen O, Petersen PK, Rogeberg A, Kutter JP, Bang DD, Wolff A. Dielectrophoresis microsystem with integrated flow cytometers for on-line monitoring of sorting efficiency. Electrophoresis 2006;27:5081–5092.
79. Sperelakis N, editor. Cell Physiology Sourcebook: A Molecular Approach. 3rd ed. San Diego: Academic Press;2001.
80. Altomare L, Borgatti M, Medoro G, Manaresi N, Tartagni M, Guerrieri R, Gambari R. Levitation and movement of human tumor cells using printed circuit board device based on software-controlled dielectrophoresis. Biotechnol Bioeng 2003;82:474–479.
81. Chartier I, Bory C, Fuchs A, Freida D, Manaresi N, Ruty M, Bablet J, Gilbert K, Sarrut N, Baleras F, Villiers C, Fulbert L. Fabrication of hybrid plastic-silicon micro-fluidic devices for individual cell manipulation by dielectrophoresis. Proc SPIE 2004;5345: 7–16.
82. Ohta AT, Chiou P-Y, Wu MC. Optically-controlled manipulation of live cells using optoelectronic tweezers. Proc SPIE 2006;6326:17–28.
83. Huang Y, Joo S, Duhon M, Heller M, Wallace B, Xu X. Dielectrophoretic cell separation and gene expression profiling on microelectronic chip arrays. Anal Chem 2002;74:3362–3371.
84. Huang Y, Yang JM, Hopkins PJ, Kassegne S, Tirado M, Forster AH, Reese H. Separation of simulants of biological warfare agents from blood by a miniaturized dielectrophoresis device. Biomedi Microdev 2003;5:217–225.
85. Cheng J, Sheldon EL, Wu L, O'Connell JP; Nanogen, assignee. Channel-less separation of bioparticles on a bioelectronic chip by dielectrophoresis. USA patent 6,280,590.2001.

86. Fatoyinbo HO, Kamchis D, Whattingham R, Ogin SL, Hughes MP. A high-throughput 3-D composite dielectrophoretic separator. IEEE Trans Biomed Eng 2005;52: 1347–1349.

87. Labeed FH, Coley HM, Hughes MP. Differences in the biophysical properties of membrane and cytoplasm of apoptotic cells revealed using dielectrophoresis. Biochim Biophys Acta 2006;1760:922–929.

88. Cui L, Morgan H. Design and fabrication of travelling wave dielectrophoresis structures. J Micromecha Microeng 2000;10:72–79.

89. Cui L, Holmes D, Morgan H. The dielectrophoretic levitation and separation of latex beads in microchips. Electrophoresis 2001;22:3893–3901.

90. Cui L, Zhang T, Morgan H. Optical particle detection integrated in a dielectrophoretic lab-on-a-chip. J Micromecha Microeng 2002;12:7–12.

91. Holmes D, Morgan H, Green NG. High throughput particle analysis: combining dielectrophoretic particle focussing with confocal optical detection. Biosens and Bioelectron 2006;21:1621–1630.

92. Wang X-B, Yang J, Huang Y, Vykoukal J, Becker FF, Gascoyne PRC. Cell separation by dielectrophoretic field-flow-fractionation. Anal Chem 2000;72:832–839.

93. Yang J, Huang Y, Wang X, Becker FF, Gascoyne PRC. Dielectric properties of human leukocyte subpopulations determined by electrorotation as a cell separation criterion. Biophys J 1999;76:3307–3314.

94. Park J, Kim B, Choi SK, Hong S, Lee SH, Lee K-I. An efficient cell separation system using 3D-asymmetric microelectrodes. Lab Chip 2005;5:1264–1270.

95. Hu X, Bessette PH, Qian J, Meinhart CD, Daugherty PS, Soh HT. Marker-specific sorting of rare cells using dielectrophoresis. Proc Natl Acad Sci 2005;102: 15757–15761.

96. Lee G-B, Lin C-H, Chang S-C. Micromachine-based multi-channel flow cytometers for cell/particle counting and sorting. J Micromecha Microeng 2005;15: 447–454.

97. Li Y, Kaler KVIS. Dielectrophoretic fluidic cell fractionation system. Anal Chim Acta 2004;507:151–161.

98. Schnelle T, Muller T, Fuhr G. Trapping in AC octode field cages. J Electrostatics 2000;50:17–29.

99. Fiedler S, Shirley SG, Schnelle T, Fuhr G. Dielectrophoretic sorting of particles and cells in a microsystem. Anal Chem 1998;70:1909–1915.

100. Muller T, Pfennig A, Klein P, Gradl G, Jager M, Schnelle T. The potential of dielectrophoresis for single-cell experiments. IEEE Eng Med Biol Mag 2003; 22:51–61.

101. Voldman J, Gray ML, Toner M, Schmidt MA. A microfabrication-based dynamic array cytometer. Anal Chem 2002;74:3984–3990.

102. Taff BM, Voldman J. A scalable addressable positive-dielectrophoretic cell-sorting array. Anal Chem 2005;77:7976–7983.

103. Sekine K, Revzin A, Tompkins RG, Toner M. Panning of multiple subsets of leukocytes on antibody-decorated poly(ethylene) glycol-coated glass slides. J Immunol Methods 2006;313:96–109.

104. Belov L, de la Vega O, dos Remedios CG, Mulligan SP, Christopherson RI. Immunophenotyping of leukemias using a cluster of differentiation antibody microarray. Cancer Res 2001;61:4483–4489.

105. Belov L, Mulligan SP, Barber N, Woolfson A, Scott M, Stoner K, Chrisp JS, Sewell WA, Bradstock KF, Bendall L, Pascovici DS, Thomas M, Erber W, Huang P, Sartor M, Young GAR, Wiley JS, Juneja S, Wierda WG, Green AR, Keating MJ, Christopherson RI. Analysis of human leukemias and lymphomas using extensive immunophenotypes from an antibody microarray. Br J Hematol 2006;135:184–197.

106. Aerni H-R, Cornett DS, Caprioli RM. Automated acoustic matrix deposition for MALDI sample preparation. Anal Chem 2006;78:827–834.

107. Chang WC, Lee LL, Liepman D. Biomimetic technique for adhesion-based collection and separation of cells in a microfluidic channel. Lab Chip 2005;5:64–73.

108. Reinhardt PH, Kubes P. Differential leukocyte recruitment from whole blood via endothelial adhesion molecules under shear conditions. Blood 1998;92:4691–4699.

109. Fuchs M, Toner M. Devices and methods for enrichment and alteration of circulating tumor cells and other particles. US Patent #20070026414.2006.

110. Emmert-Buck MR, Bonner RF, Smith PD, Chuaqui RF, Zhuang Z, Goldstein SR, Weiss RA, Liotta LA. Laser capture microdissection. Science 1996;274:998–1001.

111. Stich M, Thalhammer S, Burgemeister R, Friedemann G, Ehnle S, Luthy C, Schutze K. Live cell catapulting and recultivation. Pathol Res Pract 2003;199 (6): 405-9.

112. Ladanyi A, Sipos F, Szoke D, Galamb O, Molnar B, Tulassay Z. Laser microdissection in translational and clinical research. Cytometry A 2006;69A:947–960.

113. Revzin A, Sekine K, Sin A, Tompkins RG, Toner M. Development of a microfabricated cytometry platform for characterization and sorting of individual leukocytes. Lab Chip 2005;5:30–37.

114. Lee JY, Jones C, Zern MA, Revzin A. Analysis of local tissue-specific gene expression in cellular micropatterns. Anal Chem 2006;78:8305–8312.

115. Salazar GT, Wang Y, Young G, Bachman M, Sims CE, Li GP, Allbritton NL. Micropallet arrays for the separation of single, adherent cells. Anal Chem 2007;79:682–687.

116. Wang Y, Sims CE, Marc P, Bachman M, Li GP, Allbritton NL. Micropatterning of living cells on a heterogeneously wetted surface. Langmuir 2006;22:8257–8262.

117. Wang Y, Young G, Bachman M, Sims CE, Li GP, Allbritton NL. Collection and expansion of single cells and colonies released from a micropallet array. Anal Chem 2007;79:2359–2366.

118. Lee H, Purdon AM, Westervelt RM. Manipulation of biological cells using a microelectromagnet matrix. Appl Phys Lett 2004;85:1063–1065.

119. Lee H, Liu Y, Westervelt RM, Ham D. IC/microfluidic hybrid system for magnetic manipulation of biological cells. IEEE J Solid-State Circuits 2006;41:1471–1480.

120. Pregibon DC, Toner M, Doyle PS. Magnetically and biologically active bead-patterned hydrogels. Langmuir 2006;22:5122–5128.

121. Panaro NJ, Lou XJ, Fortina P, Kricka LJ, Wilding P. Micropillar array chip for integrated white blood cell isolation and PCR. Biomol Eng 2005;21:157–162.

122. Wilding P, Kricka LJ, Cheng J, Hvichia G, Shoffner MA, Fortina P. Integrated cell isolation and polymerase chain reaction analysis using silicon microfilter chambers. Anal Biochem 1998;257:95–100.

123. Yuen PK, Kricka LJ, Fortina P, Panaro NJ, Sakazume T, Wilding P. Microchip module for blood sample preparation and nucleic acid amplification reactions. Genome Res 2001;11:405–412.

124. Mohamed H, McCurdy LD, Szarowski DH, Duva S, Turner JN, Caggana M. Development of a rare cell fractionation device: application for cancer detection. IEEE Trans Nanobiosci 2004;3:251–256.

125. Chen Z, Zhang S, Tang Z, Xiao P, Guo X, Lu Z. Pool-dam structure based microfluidic devices for filtering tumor cells from blood mixtures. Surf Interface Anal 2006;38:996–1003.

126. Vankrunkelsven S, Clicq D, Pappaert K, Ranson W, De Tandt C, Ottevaere H, Thienpont H, Baron GV, Desmet G. A novel microstep device for the size separation of cells. Electrophoresis 2004;25:1714–1722.

127. Carlson RH, Gabel CV, Chan SS, Austin RH, Brody JP, Winkelman JW. Self-sorting of white blood cells in a lattice. Phys Rev Lett 1997;79:2149–2152.

128. Davis JA, Inglis DW, Morton KJ, Lawrence DA, Huang LR, Chou SY, Sturm JC, Austin RH. Deterministic hydrodynamics: taking blood apart. Proc Natl Acad Sci 2006;103:14779–14784.

129. Huang LR, Cox EC, Austin RH, Sturm JC. Continuous particle separation through deterministic lateral displacement. Science 2004;304:987–990.

130. Murthy SK, Sethu P, Vunjak-Novakovic G, Toner M, Radisic M. Size-based microfluidic enrichment of neonatal rat cardiac cell populations. Biomed Microdev 2006;8:231–237.

131. Sethu P, Sin A, Toner M. Microfluidic diffusive filter for apheresis (leukapheresis). Lab Chip 2006;6:83–89.

4

ADVANCED MICROFLUIDIC TOOLS FOR SINGLE-CELL MANIPULATION AND ANALYSIS

Erik S. Douglas, Nicholas M. Toriello, Chung N. Liu and Richard A. Mathies

4.1 INTRODUCTION

Microfluidic devices present a powerful platform for working with living cells (1). The length and volume scales of these devices are approaching the size and volume of single cells, providing a convenient tool for manipulating cellular microenvironment. While traditional cell culture, performed on the scale of millions of cells, can provide valuable information about the average activity and properties of a population, the interesting stochastic behavior of individual cells is lost (2). Using an integrated microfluidic device, it is possible to perform detailed analyses of the activity of many individual cells in parallel, while still sampling a sufficient number to characterize the group dynamics.

The small size scale of microfluidic devices confers several advantages relevant to working with cells. Laminar flow in microsystems enables reproducible, diffusion-limited mixing, which can be used to tailor the fluid exposure over the surface a single cell (3). The ability to manipulate very small, encapsulated volumes with pumps and valves increases the efficiency of reagent usage while minimizing the opportunity for contamination (4). The small volume facilitates response detection by effectively increasing the local concentration. Additionally, photolithography can readily create highly parallel structures that provide detailed data on a large number of individual cells.

Biological Applications of Microfluidics, Edited by Frank A. Gomez

Here we provide a brief review of the elements necessary for manipulating and analyzing single cells in a microfluidic device. These elements include (a) channels, valves, and pumps for fluidic control, (b) heaters and temperature sensors for thermal control of the microfluidic environment, (c) cell or surface modification for single-cell manipulation and capture, and (d) detection to quantify the intracellular analyte of interest. We conclude by presenting some examples of microdevices that integrate at least three of these necessary elements and point to the development of fully integrated single-cell manipulation and analysis microdevices.

4.2 FLUIDIC CONTROL

4.2.1 Channels

The first design consideration in a microfluidic system is the substrate material. Many different materials including silicon (5–7), glass (8–11), plastic, and elastomer (12–14) have been employed for microsystems. Each substrate has its own technical and chemical merits and drawbacks.

Silicon is the most widely accepted substrate and leverages commercially available equipment for fabrication from the well-established semiconductor and microelectronics fields. Advanced fabrication techniques with anisotropic etching can yield high aspect ratio silicon microstructures that are not possible with other substrates (Fig. 4.1a). However, silicon is seldom used for cellular analysis due to a lack of surface modification protocols, low breakdown voltage, and optical opacity. While silicon can be modified with a thermal oxide or silanization, the passivated surface has been shown to degrade rather rapidly due to protein adsorption, limiting its long-term utility in cell-based assays (15). Since silicon is a semiconductor, applications employing high voltage such as capillary electrophoresis (CE) and dielectrophoresis are challenging. Although oxide modification can minimize this

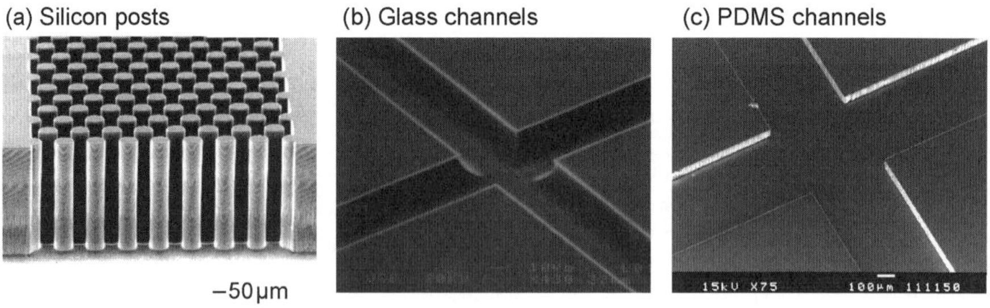

FIGURE 4.1 Materials for fabricating microsystems. (**a**) Scanning electron micrograph of high aspect ratio 50 μm diameter columns fabricated on silicon wafer (courtesy Losey et al. (74)) (**b**) SEM micrograph of an intersection of two 25 μm deep channels etched in borofloat glass (courtesy Simpson et al. (75)). (**c**) SEM micrograph of a cross-shaped microchannel fabricated with polydimethyl siloxane. PDMS can replicate fine structures down to a submicron feature size, and produce smooth and vertical sidewalls (courtesy Fujii et al. (76)).

problem to permit microgel electrophoresis on chip, separation field strengths have been limited to ~25 V/cm, resulting in longer analysis time and low resolution separations (16). To enable optical detection, the opaque silicon substrate can be anodically bonded to a glass wafer to form a silicon–glass microchannel.

Glass is a useful platform because is offers chemical surface modification for compatibility with many different assays, including CE. Advantages of glass microsystems include fabrication with conventional wet etching techniques, high dielectric strength, low background fluorescence, and optical transparency in the visible wavelengths. Because wet etching of glass is isotropic, the critical dimension attained is larger than other substrates. A typical glass microchannel is shown in Fig. 4.1b. Thermal bonding at the glass transition temperature is employed to fuse two glass substrates together and create enclosed structures such as channels and reaction chambers. Disadvantages of glass microsystems include larger critical dimensions over other substrates, thermal bonding time of ~6 h to form glass–glass channels, and lack of disposability due to its high unit cost.

Plastic and elastomer have the key advantage of simple and rapid fabrication using either injection molding or soft lithography techniques. Soft lithography is the technique most often employed for creating microfluidic systems. Devices are cast from a mold, so high aspect ratio features can be easily generated (Fig. 4.1c). Elastomer can either bond to another elastomer to form an all-elastomer microchannel or bond to glass using UV–ozone cleaning or oxygen–plasma to form hybrid elastomer–glass microchannels (17). The primary drawbacks to using a polymer-based substrate are the native fluorescence and porosity. Additionally, elastomeric compounds do not readily permit surface modification. However, using perfluoro-based monomer subunits increases compatibility with organic chemicals, limiting the need for postfabrication modification (18,19).

4.2.2 Microvalves and Pumps

Many groups have developed microvalves, micromixers, and micropumps to manipulate liquids in a microfluidic environment. In general, microvalves can be grouped into four subcategories: (a) active mechanical, (b) active nonmechanical, (c) passive mechanical, and (d) passive nonmechanical types. Active mechanical microvalves are actuated when mechanical movable membranes are coupled with electromagnetic (20), electrostatic (21), piezoelectric (22), and pneumatic (13,17) actuation methods. Active nonmechanical microvalves are actuated when the valving materials undergo a phase change upon activation. Representative materials used for nonmechanical microvalves are sol–gel (23), hydrogel (24), and paraffin (25). Passive mechanical microvalves allow flow in only one direction showing diode-like characteristics and are generally check valves (26). Passive nonmechanical microvalves are those without mechanical moving parts and are based on the geometries or surface properties of the microchannels for passive valving (27). The applications and substrate will dictate which of the four types of microvalves discussed is most suitable for an integrated microsystem. Active microvalves are very versatile and the most commonly used in microsystems. In this section, four representative examples of active microvalves will be discussed for different substrates.

Mathies and coworkers have developed monolithic elastomer membrane valves suitable for large-scale integration into glass microsystems (17). Microvalves are formed by sandwiching an elastomer membrane between etched glass fluidic channel and manifold wafers. Two different architectures have been developed for different applications. The three-layer valve structure allows easy fabrication for a hybrid glass–PDMS fluidic channel with simple non–thermal device bonding; the four-layer structure presents an all glass fluidic system to minimize fluid-elastomer contact for improved chemical and biochemical biocompatibility (Fig. 4.2a). With an integrated manifold, the microvalves can be fabricated in dense arrays, and can be addressed in parallel with <10 nL dead volumes. The normally closed membrane valves allow flow rates up to 380 nL/s at 30 kPa driving pressure and can seal reliably against fluid pressure as high as 75 kPa. When three valves are arranged in series, it forms a self-priming diaphragm pump capable of pumping from a few nanoliters to a few microliters per cycle at overall rates from 1 to over 100 nL/s and can reliably pump against a 42 kPa pressure head. Microliter-volume pumps have actuation rates greater than 1 s due to the longer filling time.

Quake and coworkers utilize soft lithography to develop monolithic elastomeric microvalves and pumps (13). The microvalves are fabricated using a crossed-channel

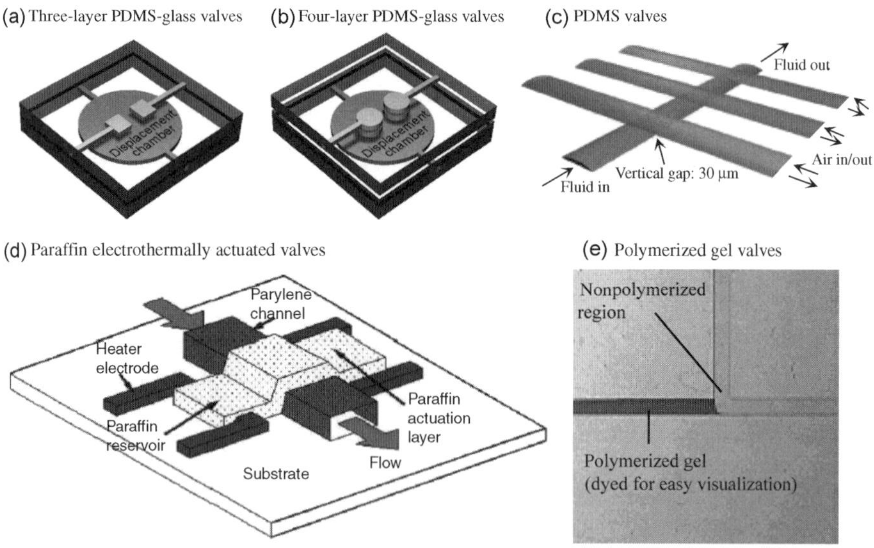

FIGURE 4.2 Representative examples of microfluidic valves and pumps. 3D perspective views of three-layer (**a**) and four-layer (**b**) normally closed monolithic PDMS valves for glass microsystems (adapted from Grover et al. (17)). (**c**) A 3D scale diagram of an elastomeric peristaltic pump for PDMS microsystems (courtesy Unger et al. (77)). (**d**) A 3D schematic view of a normally open electrothermally actuated inline microvalve (courtesy Selvaganapathy et al. (25)). (**e**) A micrograph of the *in situ* gel polymerization valve for plastic microsystems (courtesy Koh et al. (28)).

architecture (Fig. 4.2b). Typical channels are 100 μm wide and 10 μm high, forming a 100 μm by 100 μm active valve area. The polymer membrane between the channels is roughly 30 μm thick. The normally open microvalves close when the membrane deflects down with pressure applied to the upper (control) channels. The microvalves allow quick response time of 1 ms and have a packing density of 30 mm^{-2}. Building upon this platform, Quake and coworkers have also demonstrated simple on–off switching valves as well as a peristaltic pump. Three valves arranged in series forming a peristaltic pump on a single channel achieve maximum pumping rate of 2.35 nL/s at ~75 Hz.

Selvaganapathy et al. have developed a normally open, inline microvalve based the thermally triggered phase change of paraffin (Fig. 4.2c) (25). These actuators employ the high volumetric expansion of a sealed paraffin patch when heated above its melting point to provide large displacements and forces while minimizing power consumption. Complete closure of sealed microchannels is obtained with power as low as 40 mW and response time of 15 ms. Compared to the elastomeric microvalves, the response time of this microvalve is rather slow, making it appropriate for applications that do not require rapid response but demand inline valving capability.

Koh et al. have developed a gel valve for plastic devices made from poly(cyclic olefin) (28). The gel valve is based on the concept of using an in-channel plug of gel as a closed valve (Fig. 4.2d). Acrylamide monomer containing 1-hydroxycyclohexylphenyl ketone (HCPK) is first filled into all channels to form the gel valve. Localized UV exposure allows UV-initiated photopolymerization to occur in the exposed area, while unexposed, unpolymerized solution is flushed away from the channels. The exposed solution is polymerized to form a plug of gel, thus forming the gel valve. The performance of the gel valve is evaluated by the amount of material diffused through or leaked into the gel. A diffusional leak of ~2 mm with 30 PCR cycles, and <1 mm after 3 h at room temperature is observed. After PCR is completed, the amplicons are injected electokinetically through the gel valve. Performance of the gel valves increases with the degree of gel cross-linking, as the degree of leakage in 4% cross-linked gel is about twice that in 10% cross-linked gel. With proper polymerization, the gel valve can consistently withstand hydrostatic pressures up to 100 psi. In-line polymer valves are typically limited to disposable plastic-based microsystems due to the inability to remove the gel after photopolymerization.

4.3 TEMPERATURE CONTROL

Temperature control of fluids is critical, particularly when reactions and separation are performed. Temperature gradients in capillaries for electrophoresis are responsible for band broadening and subsequent reduction in separation efficiency (29). The importance of temperature control in micrototal analysis systems has been demonstrated for enzymatic reactions and polymerase chain reaction (PCR) amplifications of DNA as enzymes have very narrow optimal operating windows. In addition, cell viability and activity are typically optimal at temperatures between

32 and 37°C. To achieve precise temperature control, many groups have developed methods both to sense temperature and to heat/cool microsystems. Methods of temperature sensing based on fluorescence (30), infrared (IR) spectra (31), reflectance spectra (32), ultrasonic transducers (33), and resistivities of metallic thin films (9) and ionic liquid (4) have been reported. Noncontact heating has been demonstrated with an IR source (34) and ultrasonic methods (33). Thin-film heaters made of Ti/Pt (9), Cr/Au (35), or transparent indium-tin oxide (ITO) (36) have also been used, as well as Joule heating of ionic liquids (4) and various other physical processes for heating and cooling (37). In this section, we will briefly discuss representative examples of contact and noncontact heating methods.

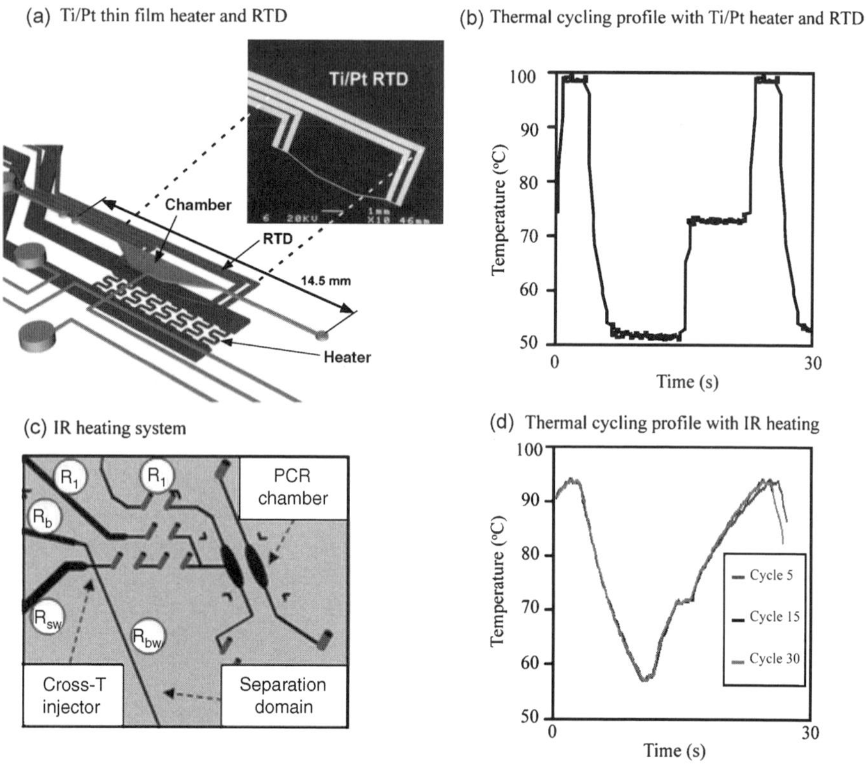

FIGURE 4.3 Temperature control for microsystems. (**a**) Schematic of an integrated glass microsystem showing relative positions of the PCR chamber, Ti/Pt RTD, and Ti/Pt heater with gold contact pads. *Inset*: SEM showing details of the RTD structure (adapted from Lagally et al. (9)). (**b**) Thermal cycling profile obtained from the integrated microsystem, showing very fast heating and cooling (~20°C/s); 20 cycles are completed in 10 min. (**c**) Schematic of relative positions of the PCR chamber and separation domain for a non-contact infrared light cycler (courtesy Easley et al. (34)). Thermal cycling is performed using noncontact IR heating and a thermocouple. (**d**) A representative thermal cycling profile for this microdevice. Thirty cycles are completed in only 11.8 min, with heating rates ranging from 4.8 to 7.8°C/s, and a cooling rate of 5°C/s.

Lagally et al. reported the integration of microfabricated heaters and resistance temperature detectors (RTDs) in PCR microsystems (9). The integrated components improve the overall thermal transfer from the heating element to the PCR chamber, increasing heating and cooling rates, and increase the accuracy of the temperature measurement. The heating elements are made of Ti/Pt and the heater contact pad is electroplated with gold to minimize heating in the contact pad and to ensure localized heating at the PCR chamber (Fig. 4.3a). The RTDs are made of Ti/Pt, and are placed parallel to the PCR chamber to permit accurate temperature measurements in the thermal plane of interest. Heating and (active convective) cooling rates are 20°C/s, and 20 PCR cycles are completed in only 10 min to achieve successful and efficient PCR. The representative thermal cycling profile is shown in Fig. 4.3b.

Easley et al. employ IR light as a heating mechanism to thermal cycle a PCR chamber (34). Instead of integrating a temperature sensor, a miniature commercially available thermocouple is inserted into a parallel reference chamber filled with buffer for temperature feedback control during PCR (Fig. 4.3c). Heating rates of 4.8 and 7.8°C/s (for the two heating regions) and −5.0°C/s cooling resulting in 30 PCR cycles in 11.8 min and representative thermal cycling profiles are shown in Fig. 4.4. Although infrared radiation provides very fast heating while eliminating substrate heating, it inherently requires a laser or other external source of focused infrared radiation, making its use in portable microsystems more difficult than the integrated on-chip resistive thin film heaters.

4.4 CELL MANIPULATION

4.4.1 Cell and Surface Modification

Precise control over cell captures and patterning on a microfluidic device is essential to enabling single-cell assays. This procedure often includes both chemical modification of the substrate or cell, along with manipulation of the cell for capture. Surface modification for cell capture can be subdivided into substrate modification and cell-surface modification. Cell manipulation can take the form of chemical capture, physical traps, and energetic forces, such as optical tweezers and electrophoresis.

Microcontact printing, pioneered by Whitesides and colleagues (Fig. 4.4a), has seen widespread use in cell patterning (38,39). This technique uses a molded polymer stamp, "inked" with a cell adhesion molecule such as laminin or fibronectin, to create an adhesive substrate on which cells will grow. The area outside the pattern is chemically blocked to prevent nonspecific adhesion. This technique has been popular because of its simplicity, and can be performed without expensive photolithography equipment once the mold is produced. Microcontact printing is essentially a technique for producing a patterned cell culture. Because of the initial low cell-seeding density, this technique requires an incubation period of several hours to days before the cell pattern fully emerges. Also, because the adhesion molecules are typically not cell-specific, the patterning is generally limited to one cell type.

The substrate-treatment technique can be modified to permit sequential patterning of multiple cell types. Sequential cell patterning is achieved by employing an electroactive surface, whose binding capability can be switched by applying a potential to the selected areas (40). A nonspecific cell-binding molecule is first patterned on an electrically addressable substrate using microcontact printing (Fig. 4.4b). The remaining area is then backfilled with an initially nonbinding electroactive monolayer that is electrically switched to promote cell adhesion. The first cell

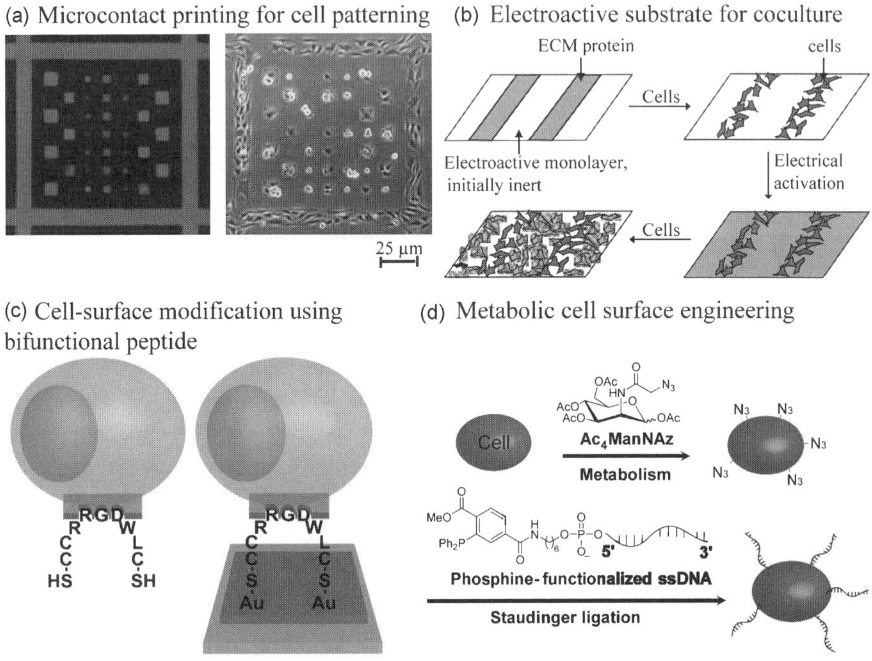

FIGURE 4.4 Strategies for surface modification with cells. (**a**) Microcontact printing for creating cell-adhesive substrates. *Left*: Extracellular matrix protein fibronectin patterned with microcontact printing and labeled with a fluorescent antibody complex. *Right*: Cells growing in culture on adhesive-patterned areas. Bar = 40 μm (courtesy Kane et al. (39)). (**b**) Heterotypic coculture patterned on an electroactive substrate. Clockwise from top-left: Lines of adhesive protein are patterned on the surface, surrounded by inert area. Cells are seeded on the device, and grow only in the patterned area. Electrical activation of the surface allows the chemical modification of the previously inert area, making it cell adhesive. The second cell type is seeded and grown, leading to the desired coculture system (courtesy Yousaf et al. (40)). (**c**) Cell-surface modification using adhesion peptide. Cells are treated with a bifunctional peptide containing, the R–G–D sequence, and thiol-containing cysteine residues. Once the peptide is bound, the thiolated cell will readily bind to a gold substrate. Capture efficiency can be increased using a driving electric field. (adapted from Toriello et al. (41)). (**d**) Cell-surface modification using metabolic engineering. Cells are treated with an artificial sugar, $Ac_4ManNAz$, which is metabolized to produce azide groups on the surface. Phosphine-terminated DNA is then reacted with these groups via the Staudinger ligation, leading to DNA-functionalized cells that can then be readily patterned (adapted from Douglas et al. (44)).

type is seeded on the device, rinsed to remove unbound cells, and then cultured for 48 h to fill in the adhesive area. At this point the inert area is electrically switched to permit the binding of the second cell type, resulting in a heterotypic coculture. While this provides a useful technique for patterning a coculture, the timescale required limits its extension into more complex systems.

Chemical modification can also be used to treat the cell surface, thereby removing the biomolecular recognition step from the cell capture. We have previously demonstrated the use of a bifunctional peptide, containing the RGD sequence commonly recognized by cell-surface integrin proteins, and terminal cysteine residues that contain thiol groups (Fig. 4.4c). This surface modification effectively coated the cell surface with thiols, which readily bind to a gold substrate. This pretreatment of the cell enabled rapid capture when the cell was electrically driven to the attachment site (41).

4.4.2 Cell Capture

Cell and substrate modification have been successfully combined to rapidly produce complex cell patterns. Using single-stranded DNA artificially attached to the cell surface, we have patterned cells on a substrate containing the complementary ssDNA in a sequence-specific manner. DNA is attached to the cell surface by feeding the cell an unnatural sugar ($Ac_4ManNAz$), which the cell processes to produce azide groups on its exterior, serving as a reactive chemical handle (42). DNA with an attached phosphine group can then be reacted with the azide, via the Staudinger ligation, resulting in the cell surface being decorated with ssDNA (43), shown in Fig. 4.4d. With this chemical handle, DNA-barcode directed patterning can be used with both adherent and naturally nonadherent cells, like T and B cells. A wide array of standard techniques can be employed to pattern the complementary DNA on the substrate, such as a microarray spotting system, microcontact printing, or simple manual deposition. In Fig. 4.5a, we used a micropipette and a micromanipulator to directly write the capture DNA, forming a script "Cal." Upon the addition of Jurkat T cells bearing surface DNA, the self-assembled pattern with single-cell resolution was formed within minutes (44). The ability to produce high resolution cell patterns with an arbitrary design could be especially valuable in studies of cell communication. Multiple cell types can potentially be captured in parallel from a mixed population, due to the specificity of the DNA-mediated capture, enabling the formation of complex cell arrays.

In addition to the chemical modification approach, a variety of physical techniques have also been demonstrated for the capture and patterning of cells in a microfluidic device. One simple and effective technique for passive capture uses molded PDMS enclosures and a fluidic weir. The weir is blocked upon the entry of a single cell, preventing further cells from entering the trap (Fig. 4.5c) (45,46). This provides a platform on which data can be collected from tens to hundreds of discrete cells simultaneously. Since the same individual cells can be continuously observed, a cell's unique response to stimuli can be traced over the course of time.

Cell culture substrates have also been physically patterned to promote selective attachment and growth based on surface topography. Turner et al. used photolithography to fabricate arrays of micron-scale pillars and wells on silicon (47). They found that astroglial cells preferentially attached to the featured area, and spread across the pillars without reaching down into the space between, even when the interpillar gap was 5 μm (Fig. 4.5b). They suggested that this preference of the cells was due to the fact that sitting on top of a pillar array allowed growth media to circulate freely underneath. They also found that cells were less likely to grow in etched wells than on the featureless area. In the same way that pillars would enable increased flow, a cell isolated in a well would experience inhibited growth because

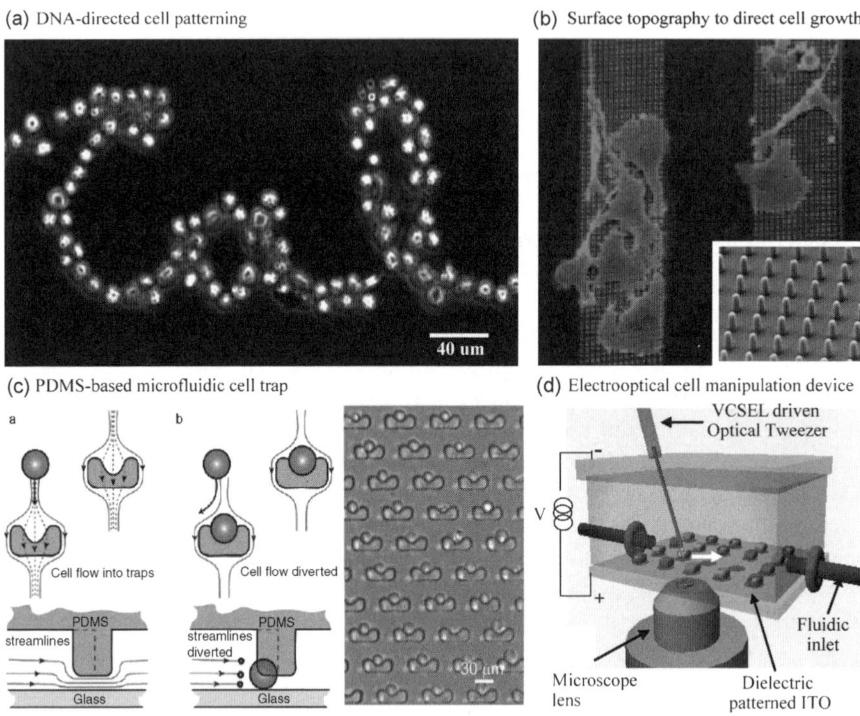

FIGURE 4.5 Cell capture and patterning. (**a**) Naturally nonadherent cells patterned using DNA-based cell-adhesion barcodes form the world's smallest living Cal fan cub (adapted from Douglas et al. (44)). (**b**) Topographical substrate modification enables preferentially cell growth on raised pillars, and limits growth in depressed wells (courtesy Turner et al. (47)). (**c**) Microfluidic cell capture system using autovalving physical traps. When empty, fluid flows under the weir, maintaining streamlines through the trap. When a cell enters the trap flow is blocked, preventing the accumulation of additional cells (courtesy Di Carlo et al. (45)). (**d**) Electrooptical platform for cell sorting and manipulation. Cells are first arrayed onto electrode capture sites in a microfluidic channel using dielectrophoresis. Once cells are initially isolated, individual cells can be manipulated using the integrated VCSEL-driven optical tweezer (courtesy Ozkan et al. (50)). (See color insert)

it is less capable of exchanging waste products with the bulk solution. Such topographical features can be used to direct cell binding and growth in a device without the need for chemical modification.

For additional control over cell capture, active manipulation is possible using non-contact methods such as dielectrophoresis and optical tweezers. Dielectrophoresis (DEP) uses a nonuniform electric field to manipulate cells based on their relative polarizability (48). This technique requires the electric field be maintained to keep the cells in position, and would need to be combined with another technique to achieve a stably patterned system. Control over cell positioning is also limited by the fixed structure of the electrodes in most devices. Chiou et al. have overcome this barrier by using a virtual electrode formed by projecting an optical image onto a semiconductive substrate (49). This approach enabled real-time modification of the cell trap, based on the pattern of light projected. Ozkan et al. demonstrated an electrooptical platform that combines DEP with optical tweezers (50). DEP was used for the initial cell sorting, and optical tweezers for independent control over single cells (Fig. 4.5d). While energetic traps such as optical tweezers have typically been labor-intensive, holographic masks for optical traps have been demonstrated to reduce this limitation. A diffractive mask can be used with a single laser to control many cells simultaneously (51).

4.5 DETECTION

Detection techniques for use with cell assays on a microdevice are typically either optical (fluorescent, chemiluminescent, or bioluminescent) or electrochemical. While optical detection can be easily integrated into both simple and complex microfluidic systems, electrochemical detection offers its own advantages in multiplexing and sensitivity.

4.5.1 Optical Detection

Optical detection is a natural fit for microfluidic devices, given its potential ease of integration. The instrumentation is typically external to the device, and based on commercially available equipment, such as fluorescence microscopy. Several developments in fluorescent probes have made them powerful tools for biological analysis. Energy-transfer dyes couple a donor and acceptor pair, to offer a variety of enhanced output wavelengths following excitation at a single laser wavelength. The dyes have been used in DNA sequencing instruments to allow for the differential detection of the four bases (A, T, C, and G) using a single excitation laser (52). These fluorescent dyes have made possible high throughput, capillary array electrophoresis for DNA sequencing to be miniaturized onto a chip as shown in Fig. 4.6a, producing the sequencing data shown in Fig. 4.6a. Energy transfer cassettes could also be used in cell labeling by coupling them to cell-reactive functional groups.

In addition to finding application in molecular analysis, fluorescent probes are a staple in cellular analysis. The variety of specific dye applications is tremendous,

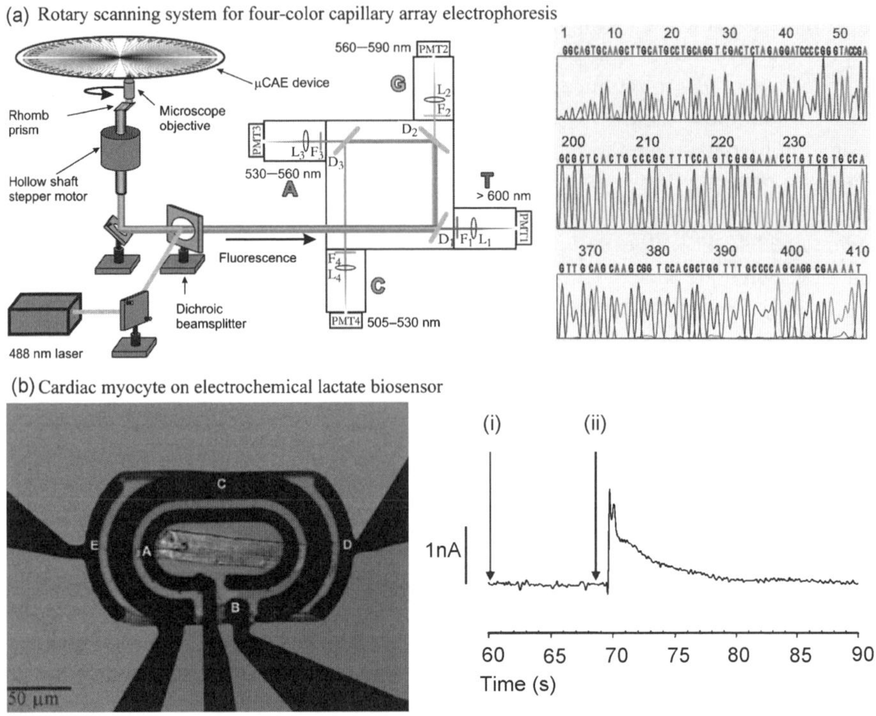

FIGURE 4.6 Detection techniques for on-chip systems. (**a**) Rotary confocal scanning system for DNA sequencing. Capillary electrophoresis is performed in each of the 96 arrayed channels on the chip. A single laser is used to excite all four fluorescent energy-transfer labels, which enables the identification of each nucleotide (courtesy Blazej et al. (78)). *Right*: DNA sequence data collected with the system (adapted from Paegel et al. (79)). (**b**) Electrochemical detection system used with a single cardiac myocyte. The cell is manually positioned on an electrode array in a 20-μm deep well. Electrodes labeled A, B, and C make up an electrochemical sensor for lactate release, while D and E are stimulating electrodes. *Right*: Typical data produced by the lactate sensor. (i) The sensor is at baseline before the cell is electrically stimulated. (ii) Electropermeabilization of the cell results in lactate release, which is detected amperormetrically by the sensor (courtesy Cheng et al. (57)).

ranging from selective labeling of cell components to cell-state indication. One type of fluorescent probe that has been especially useful in the device context is the ion indicator dyes. Pioneered by Tsien and coworkers (53), fluorescent dyes that change emission wavelength or quantum yield upon binding with calcium have been widely used in microfluidic devices. Calcium signaling is a component of many pathways, so these dyes can serve as a reporter of many different signals. These dyes are typically cell permeable, removing the need for a complicated loading protocol, and rely upon modification once inside the cell to reduce the background intensity. In addition to calcium, fluorescent indicators are available for a variety of ions, including sodium, chloride, magnesium, potassium, and pH (54).

4.5.2 Electrical Detection

While optical detection benefits from ease of integration, electrochemical sensors have also become a valuable tool for detection of cell activity in real time. Planar electrodes can be fabricated on a microfluidic device to detect a variety of cell products and analyze cell activity with high sensitivity. Metal electrodes have been used for amperometric measurements of electroactive species (55), and have been modified with pH-sensitive layers to detect changes in cell respiration rates (56). Integrated electrodes have also been modified with a variety of enzymes, to enable the detection of biological products. Cheng et al. have incorporated an electrode modified with lactate oxidase into a microfluidic analysis device for individual cardiac cells (Fig. 4.6b) (57). The enzyme catalyzed the oxidation of lactate, producing hydrogen peroxide that could be detected amperometrically.

4.6 INTEGRATION

While there have been many reports of isolated microsystems offering the ability to manipulate fluids, temperature, cells or detect a desired analyte, there have been few truly integrated microdevices for cellular analysis. This realization has been slowed in large part because integration is more than the sum of its individual parts. Rather, integration of new elements is exponentially more difficult than the previous element. Despite that, there have been numerous successful integrations of two or three of the required elements outlined here for manipulating and analyzing single cells. Each of the achievements in integration can be grouped into one of the following categories: (a) genetic analysis, (b) gene expression analysis, (c) fluorescent, and (d) electrical recordings from active whole cells.

4.6.1 Genetic Analysis of Single Cells

Integrated microsystems for genetic analysis of single cells have yielded the highest degree of device integration. This high level of device integration is because PCR, the most important genetic analysis tool, offers the ability to amplify DNA from as little as a single copy (58). Successful PCR requires thermal cycling between 50 and 95°C. This in turn requires an integrated system with fluidic control for proper sealing of the reaction chamber, temperature control for rapid heating/cooling and temperature sensing, and downstream detection of the PCR amplicons. There are numerous reviews of integrated microsystems that employ PCR amplification (59–61).

Mathies and coworkers have developed highly integrated microsystems for genetic analysis (Fig. 4.7a). This approach centers on thermally cycling an immobilized reaction volume and then performing a sized-based separation of the resulting products by capillary electrophoresis. A 10 cm diameter glass substrate is used with etched trenches that when thermally bonded to a blank glass substrate, create reactors, and separation channels (62). Building upon this platform, we have

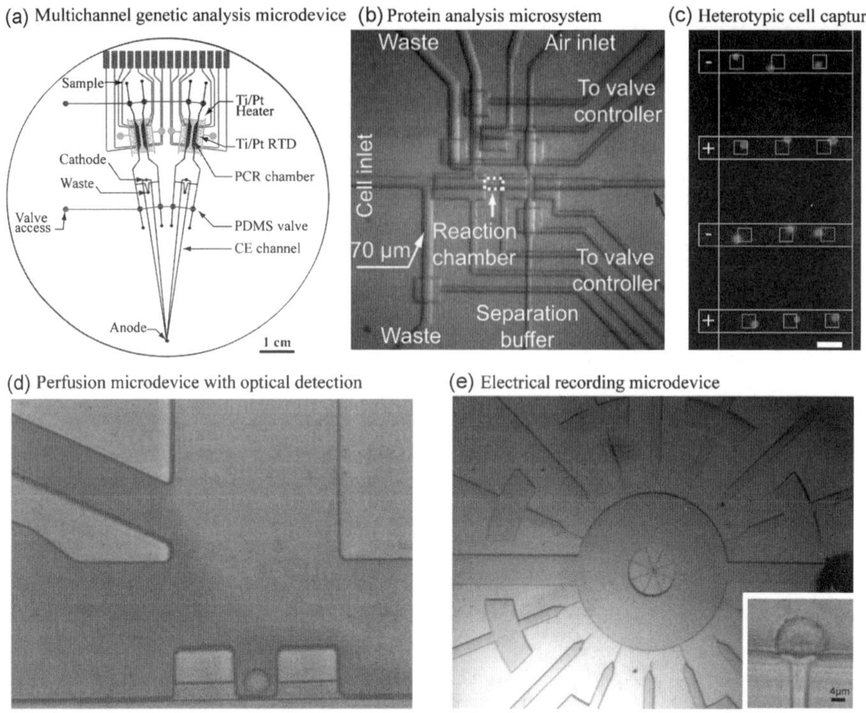

FIGURE 4.7 Integrated analysis microsystems. (**a**) Multichannel genetic analysis microdevice. The four-layer glass–PDMS hydrid structure integrates Ti/Pt temperature detectors and heaters for temperature control, PDMS microvalves for fluidic control, along with glass etched reactors and CE separation channels for detection of amplified genetic material (adapted from Toriello et al. (67)). (**b**) Protein analysis microsystem. The PDMS microsystem offers the ability to capture, lyse, label and analyze protein content from single cells (courtesy Huang et al. (19)). (**c**) An electrically driven, heterotypic cell capture and activation microdevice. The glass–PDMS microsystem offers cell capture in an arrayable format, fluid control and optical readout of the cells activity (adapted from Toriello et al. (41)). (**d**) Single cell perfusion microdevice with optical detection. The PDMS microsystem combines size-limiting wells for mechanically trapping individual cells, fluidic control for administering metered amounts of stimuli directly to the immobilized cell and optical readout of the cells response (courtesy Wheeler et al. (46)). (**e**) Electrical recording microdevice. A single-cell electroporation microsystem offers fluidic control, rapid cell capture, and the ability to perform electrical and fluorescent measurements from individual cells (courtesy Khine et al. (73)).

integrated RTDs for temperature sensing (9), resistive heaters for rapid thermal cycling (9,63,64), microvalves and pumps for fluidic control and to physically isolate the reaction chamber (17,63,65), and a-Si fluorescent detectors (66). Systems employing this approach have been arrayed for highly parallel analysis (63,67), compartmentalized into a laptop sized portable system (64,65), and have been utilized for infectious disease detection (65), forensic analysis (64), DNA sequencing (68), gene expression analysis, and biomarker analysis (67).

Other approaches, that have been used to perform PCR amplification, have utilized an immobilized reaction system or a flow-through PCR scheme. Landers and coworkers employ a similar PCR scheme as mentioned above, but utilize infrared heating instead of integrating resistive heaters onto the substrates surface (34). Burns and coworkers employ a postamplification enzyme digestion to further characterize their amplicons and reduce the demands on their separation resolution (16). Many researchers including Quake and coworkers use soft lithography over glass for rapid device prototyping and highly parallel analysis (69). However, polymer-based systems typically exhibit poor thermal transfer leading to longer amplification times. In addition, they are not easily coupled to separation systems due to their permeability, so they rely on real-time detection of product.

An alternative to stationary amplification is flow-through PCR, where the reactants are physically moved from one temperature zone to another. This allows the temperatures in the device to remain fixed over the course of the reaction, reducing the burden of creating a system capable of a heating and cooling (70). This amplification scheme has generated many intriguing designs but continues to be hampered by poor amplification efficiency. Possible explanations for this reduced efficiency are the adsorption of the active PCR amplification enzyme, *Taq*, on the sidewalls of the channels over the course of the amplification as well as the parabolic profile of the pressure-driven flow resulting in a nonuniform temperature zone. Overall, the use of PCR as a means for interrogating the genetic makeup of cells has led to a very high level of microdevice integration.

4.6.2 Gene Expression Analysis of Single Cells

While understanding the genetic composition of a cell is highly desirable for deciphering its functionality, gene expression analysis aids in understanding what actions are actually being performed. There have been two primary methods developed for analyzing the expression patterns in cells under microfluidic environment, (a) whole cell lysis and protein detection and (b) cells lysis followed by RNA content analysis.

Analyzing the protein content directly is the optimal platform for understanding the expression pattern within a single cell. The fundamental limitation associated with measuring proteins directly is there are no tools for amplifying them analogous to PCR. Typically, proteins are present in very wide dynamic range from a few copies to a thousand copies, far below the detection limits of standard laboratory equipment. However, by combining a cell lysis microchamber with CE separation and detection of fluorescently labeled proteins, researchers have begun to probe the protein content of single cells (71) (Fig. 4.7b). These devices integrate fluidic control, single cell capture and downstream detection without the need for temperature control because there is no analyte amplification. Highly specific detection schemes in concert with low volume reaction chambers enable very sensitive analysis. Currently observed limits-of-detection are on the order of

100 copies, and multiplex analysis has been limited to only a few proteins, but this direct analysis without amplification shows great promise for understanding single cells.

Cell lysis followed by reverse transcriptase PCR (RT-PCR) offers the ability to investigate gene expression at all mRNA transcript levels (67,72). This technique leverages the same microfluidic device outlined above for genetic analysis of cellular material. By incorporating an isothermal reverse transcription step at the beginning of a standard PCR reaction, mRNA can be targeted for amplification over DNA. The primary advantage of technique over other gene expression analysis microsystems is the ability to amplify more than one target of interest. Thus, multiplex low transcript level analysis is possible.

4.6.3 Fluorescence Imaging of Whole Cells

Fluorescence recording from discrete cells in an integrated microdevice offers the ability to measure their activity and viability in response to various stimuli. Our group has created a microfluidic platform for the manipulation of cells coupled to fluorescence readout (Fig. 4.7c) (41). By electrically driving thiol-functionalized cells to size-limiting gold pads, we are able to sequentially pattern multiple types of individual cells in a single channel. The cells are loaded with a calcium-sensitive fluorescent indicator and stimulated with carbachol. Cells that possess the membrane-bound receptor for the agonist fluoresce upon stimulation, providing a platform for agonist or cell screening. Another example of an integrated system capable of capturing a single cell, controlling its fluidic environment, and measuring its fluorescent response to reagents was presented by Wheeler et al. (Fig. 4.7d) (46). This cell perfusion microdevice utilized soft lithography to mechanically trap individual cells in size-limiting wells and then exposed the cells to metered volumes of reagent. By loading the cells with a membrane permeable calcium-sensitive fluorescent dye and using a standard epifluorescent microscope, the researchers were able to measure the cell's activity and eventual apoptosis in real time. While whole cell fluorescent recording from viable and active cells offers many advantages, it is often limited by fluorescent overlap and detection sensitivity. Thus, only a couple of cellular markers can be probed from a given cell and only at high levels.

4.6.4 Electrical Recording from Whole Cells

The desire to perform electrical recordings from single cells was largely driven by neurobiologists looking to investigate the activity of individual neurons. Electrical measurements from intact cells can yield information on viability and subcellular activity. An electroporation and a multielectrode microdevice have demonstrated the highest level of device integration. The electroporation microdevice created by Lee and coworkers combines fluidic control, cell capture along with electrical and optical detection for rapid, and parallel recording from individual cells (73) (Fig. 4.7e). In addition, because the cells are reversibly

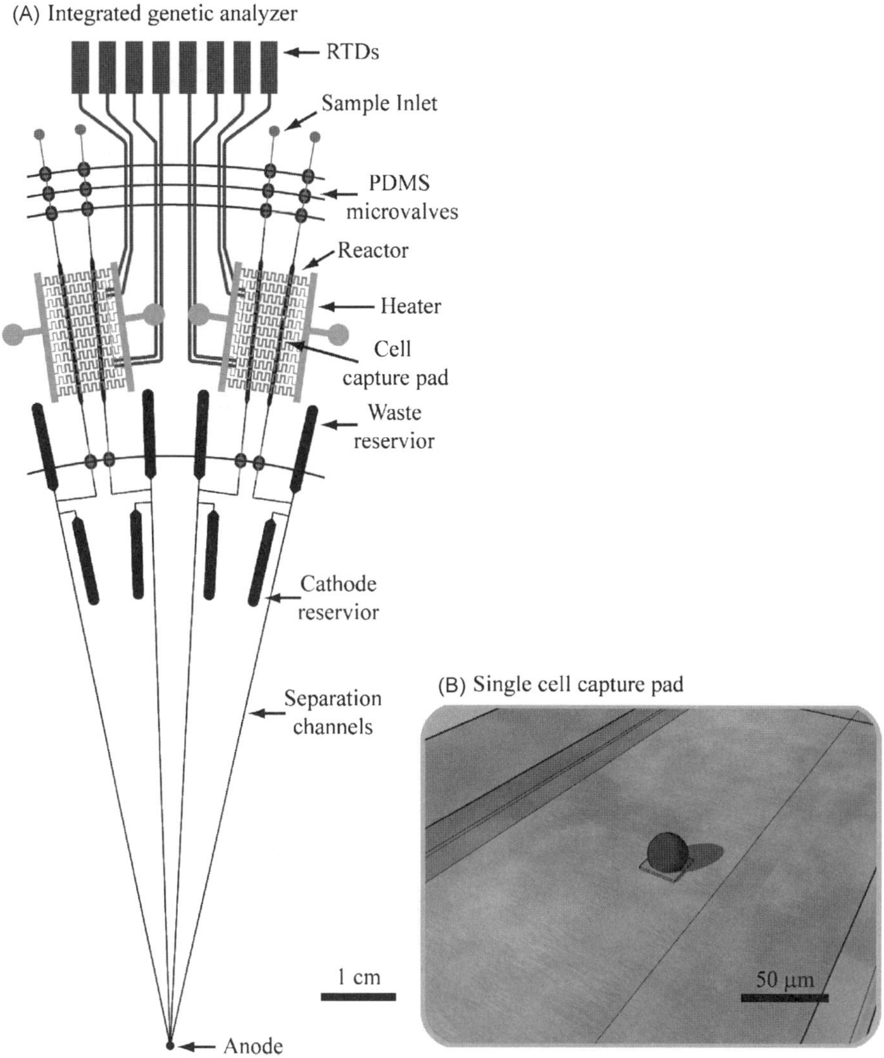

FIGURE 4.8 Fully integrated single cell genetic analyzer. (**a**) The proposed device integrates all the elements to perform rapid and highly-parallel genetic analysis from individual cells. The 150 mm device integrates (1) fluidic control through independently addressable microvalves, (2) temperature control with RTDs and microfabricated heaters, (3) cell isolation on size-limiting capture pads, and (4) analysis of amplified material through microcapillary electrophoresis. (**b**) An expanded depiction of the cell capture area inside the microreactor. A cell is captured on a size-limiting metal pad inside the reactor. The cell can be analyzed with fluorescent measurements, electrical measurement, or lysed for amplification of its subcellular genetic material. (See color insert)

immobilized with suction, a cell of interest can be easily removed from the system for further interrogation.

A multielectrode microdevice offers the ability to capture a cell over an array of independently addressable electrodes in a fluidically controlled environment. The electrodes can be utilized for stimulation or recording from the immobilized cell. Systems with valves and pumps allow for nutrient regeneration and stimuli perfusion.

4.6.5 Fully Integrated Microdevice for Cellular Analysis

As noted, the highest level of microdevice integration has combined three of the desired four elements outlined above. A schematic of a highly parallel microsystem with all the necessary elements to analyze single cells is presented in Fig. 4.8. The device has four individually addressable systems with (a) microvalves and pumps for the controlled movement of cells and fluids, (b) microfabricated heaters and RTDs for temperature control, (c) size-limiting gold pads in the reaction chamber to capture cells and perform electrical measurements, and (d) coupled downstream separation channels to detect the subcellular target of interest. The integrated microsystem allows for cell capture, stimulation, and analysis in a fluidically and thermally controlled environment. While integrating all of these necessary elements into a single device is an engineering challenge, recent progress suggests that it is only a matter of time before it is fully realized.

4.7 CONCLUSIONS

We have presented a survey of the fundamental elements necessary for a fully integrated microsystem to perform single-cell manipulation and analysis. Control over fluid motion, volumes, and content provides a means to tailor the micro-environment to suit the assay. Precise control of cell placement enables highly ordered, heterotypic cell systems to be cultured in the device. Finally, the variety of optical and electrochemical tools available for detection complete the toolkit. While to date on-chip integrated systems have lacked sophistication in at least one of these elements, current developments in this rapidly advancing field hold great promise.

ACKNOWLEDGMENTS

We thank Mary Hammond for assistance in preparing the manuscript, and Charlie Emrich for helpful discussions. E.S.D., N.M.T., and C.N.L. were supported by an NSF Graduate Research Fellowship, an NIH Molecular Biophysics Training Grant (T32GM08295) and Chevron-Texaco Graduate Fellowship, respectively. This work was supported by NIH grants R01HG01399 and HG003329 and by the Chemical Sciences Division of the U.S. Department of Energy under contract DE-AC03-76SF00098.

REFERENCES

1. El-Ali J, Sorger PK, Jensen KF. Nature 2006;442:403–411.
2. Di Carlo D, Lee LP. Anal Chem 2006;78:7918–7925.
3. Takayama S, McDonald JC, Ostuni E, Liang MN, Kenis PJA, Ismagilov RF, Whitesides GM. Proc Acad Sci U S A 1999;96:5545–5548.
4. deMello AJ. Nature 2006;442:394–402.
5. Daniel JH, Iqbal S, Millington RB, Moore DF, Lowe CR, Leslie DL, Lee. MA, Pearce MJ. Sens Actuators A Physical 1998;71:81–88.
6. Northrup MA, Benett B, Hadley D, Landre P, Lehew S, Richards J, Stratton P. Anal Chem 1998;70:918–922.
7. Wilding P, Kricka LJ, Cheng J, Hvichia G, Shoffner MA, Fortina P. Anal Biochem 1998;257:95–100.
8. Ferrance JP, Wu Q, Giordano B, Hernandez C, Kwok Y, Snow K, Thibodeau S, Landers JP. Anal Chim Acta 2003;500:223–236.
9. Lagally ET, Emrich CA, Mathies RA. Lab Chip 2001;1:102–107.
10. Waters LC, Jacobson SC, Kroutchinina N, Khandurina J. Anal Chem 1998;70: 5172–5176.
11. Zhang NY, Tan HD, Yeung ES. Anal Chem 1999;71:1138–1145.
12. Kricka LJ, Fortina P, Panaro NJ, Wilding P, Alonso-Amigo G, Becker H. Lab Chip 2002;2:1–4.
13. Quake SR, Scherer A. Science 2000;290:1536–1540.
14. Wang J, Pumera M, Chatrathi MP, Escarpa A, Konrad R, Griebel A, Dorner W, Lowe H. Electrophoresis 2002;23:596–601.
15. Felbel J, Bieber I, Pipper J, Kohler JM. Chem Eng J 2004;101:333–338.
16. Pal R, Yang M, Lin R, Johnson BN, Srivastava N, Razzacki SZ, Chomistek KJ, Heldsinger DC, Haque RM, Ugaz VM, Thwar PK, Chen Z, Alfano K, Yim MB, Krishnan M, Fuller AO, Larson RG, Burke DT, Burns MA. Lab Chip 2005;5: 1024–1032.
17. Grover WH, Skelley AM, Liu CN, Lagally ET, Mathies RA. Sens Actuators B Chemical 2003;89:315–323.
18. Huang B, Wu HK, Bhaya D, Grossman A, Granier S, Kobilka BK, Zare RN. Science 2007;315:81–84.
19. Rolland JP, Van Dam RM, Schorzman DA, Quake SR, DeSimone JM. J Amer Chem Soc 2004;126:2322–2323.
20. Capanu M, Boyd JGIV, Hesketh PJ. J Microelectromech Syst 2000;9:181–189.
21. Yobas L, Huff MA, Lisy FJ, Durand DM. J Microelectromech Syst 2001;10:187–196.
22. Li HQ, Roberts DC, Steyn JL, Turner KT, Yaglioglu O, Hagood NW, Spearing SM, Schmidt MA. Sens Actuators A Physical 2004;111:51–56.
23. Zhang ZJ, He DY, Liu W, Lv Y. Luminescence 2005;20:377–381.
24. Liu RH, Yu Q, Beebe DJ. J Microelectromech Syst 2002;11:45–53.
25. Selvaganapathy P, Carlen ET, Mastrangelo CH. Sens Act A: Physical 2003;104: 275–282.
26. Wang XQ, Lin Q, Tai YC. Proceedings of 1999 IEEE International MEMS Conference 1999. 177–182

27. Feng Y, Zhou Z, Ye X, Xiong J. Sen Actuators A Physical 2003;108:138–143.
28. Koh CG, Tan W, Zhao MQ, Ricco AJ, Fan ZH. Anal Chem 2003;75:4591–4598.
29. Knox JH, McCormack KA. Chromatographia 1994;38:207–214.
30. Ross D, Gaitan M, Locascio LE. Anal Chem 2001;73:4117–4123.
31. Floyd TM, Schmidt MA, Jensen KF. Ind Eng Chem Res 2005;44:2351–2358.
32. Iles A, Fortt R, de Mello AJ. Lab Chip 2005;5:540–544.
33. Jagannathan H, Yaralioglu GG, Ergun AS, Khuri-Yakub BT. Proc SPIE 2003;4982: 243–247.
34. Easley CJ, Karlinsey JM, Landers JP. Lab Chip 2006;6:601–610.
35. Schwab K, Arlett JL, Worlock JM, Roukes ML. Physica E Low-Dimensional Syst Nanostruct 2001;9:60–68.
36. Sun K, Yamaguchi A, Ishida Y, Matsuo S, Misawa H. Sens Actuators B Chemical 2002; 84:283–289.
37. Guijt RM, Dodge A, van Dedem GWK, de Rooij NF, Verpoorte E. Lab Chip 2003; 3:1–4.
38. Folch A, Toner M. Ann Rev Biomed Eng 2000;2:227–256.
39. Kane RS, Takayama S, Ostuni E, Ingber DE, Whitesides GM. Biomaterials 1999;20:2363–2376.
40. Yousaf MN, Houseman BT, Mrksich M. Proc Natl Acad Sci U S A 2001;98: 5992–5996.
41. Toriello NM, Douglas ES, Mathies RA. Anal Chem 2005;77:6935–6941.
42. Prescher JA, Bertozzi CR. Nat Chem Biol 2005;1:13–21.
43. Chandra RA, Douglas ES, Mathies RA, Bertozzi CR, Francis MB. Angew Chem Intl Ed 2006;45:896–901.
44. Douglas ES, Chandra RA, Bertozzi CR, Mathies RA, Francis MB. Lab Chip 2007;10.
45. Di Carlo D, Aghdam N, Lee LP. Anal Chem 2006;78:4925–4930.
46. Wheeler AR, Throndset WR, Whelan RJ, Leach AM, Zare RN, Liao YH, Farrell K, Manger ID, Daridon A. Anal Chem 2003;75:3581–3586.
47. Turner AMP, Dowell N, Turner SWP, Kam L, Isaacson M, Turner JN, Craighead HG, Shain W. J Biomed Mater Res 2000;51:430–441.
48. Voldman J. Ann Rev Biomed Engin 2006;8:425–454.
49. Chiou PY, Ohta AT, Wu MC. Nature 2005;436:370–372.
50. Ozkan M, Pisanic T, Scheel J, Barlow C, Esener S, Bhatia SN. Langmuir 2003;19:1532–1538.
51. Grier DG. Nature 2003;424:810–816.
52. Glazer AN, Mathies RA. Curr Opin Biotechnol 1997;8:94–102.
53. Tsien RY. Biochemistry 1980;19:2396–2404.
54. Tsien RY. Ann Rev Neurosci 1989;12:227–253.
55. Huang WH, Cheng W, Zhang Z, Pang DW, Wang ZL, Cheng JK, Cui DF. Anal Chem 2004;76:483–488.
56. Ges IA, Ivanov BL, Schaffer DK, Lima EA, Werdich AA, Baudenbacher FJ. Biosens Bioelectron 2005;21:248–256.
57. Cheng W, Klauke N, Sedgwick H, Smith GL, Cooper JM. Lab Chip 2006;6:1424–1431.

58. Saiki RK, Gelfand DH, Stoffel S, Scharf SJ, Higuchi R, Horn GT, Mullis KB, Erlich HA. Science 1988;239:487–491.
59. Auroux PA, Koc Y, deMello A, Manz A, Day PJR. Lab Chip 2004;4:534–546.
60. Lagally ET, Mathies RA. J Phys D Appl Phys 2004;37:R245–R261.
61. Roper MG, Easley CJ, Landers JP. Anal Chem 2005;77:3887–3893.
62. Lagally ET, Simpson PC, Mathies RA. Sens Actuators B Chemical 2000;63:138–146.
63. Liu CN, Toriello NM, Mathies RA. Anal Chem 2006;78:5474–5479.
64. Liu P, Seo TS, Beyor N, Shin KJ, Scherer JR, Mathies RA. Anal Chem 2007;79:1881–1889.
65. Lagally ET, Scherer JR, Blazej RG, Toriello NM, Diep BA, Ramchandani M, Sensabaugh GF, Riley LW, Mathies RA. Anal Chem 2004;76:3162–3170.
66. Kamei T, Toriello NM, Lagally ET, Blazej RG, Scherer JR, Street RA, Mathies RA. Biomed Microdev 2005;7:147–152.
67. Toriello NM, Liu CN, Mathies RA. Anal Chem 2006;78:7997–8003.
68. Blazej RG, Kumaresan P, Mathies RA. Proc Natl Acad Sci U S A 2006;103:7240–7245.
69. Ottesen EA, Hong JW, Quake SR, Leadbetter JR. Science 2006;314:1464–1467.
70. Obeid PJ, Christopoulos TK, Crabtree HJ, Backhouse CJ. Anal Chem 2003;75: 288–295.
71. Huang Y, Castrataro P, Lee C-C, Quake SR. Lab Chip 2007;7:24–26.
72. Marcus JS, Anderson WF, Quake SR. Anal Chem 2006;78:956–958.
73. Khine M, Lau A, Ionescu-Zanetti C, Seo J, Lee LP. Lab Chip 2005;5:38–43.
74. Losey MW, Jackman RJ, Firebaugh SL, Schmidt MA, Jensen KF. J Microelectromech Syst 2002;11:709–717.
75. Simpson PC, Woolley AT, Mathies RA. Biomed Microdev 1998;1:7–26.
76. Fujii T. Microelectron Eng 2002;61/62:907–914.
77. Unger MA, Chou H-P, Thorsen T, Scherer A, Quake SR. Science 2000;288:113–116.
78. Blazej RG, Paegel BM, Emrich CA, Mathies RA. In: van den Berg A, Oosterbroek RE, editors. Lab Chip: Miniaturized Systems for (Bio) Chemical Analysis and Synthesis. Amsterdam: Elsevier Science;2003. p.231–248.
79. Paegel BM, Emrich CA, Weyemayer GJ, Scherer JR, Mathies RA. Proc Natl Acad Sci U S A 2002;99:574–579.

5

ENGINEERING CELLULAR MICROENVIRONMENTS WITH MICROFLUIDICS

GEETA MEHTA, YU-SUKE TORISAWA AND SHUICHI TAKAYAMA

5.1 INTRODUCTION

5.1.1 The Microenvironment Gap

There are many biomedical applications where it is necessary to manipulate mammalian cells outside of the body. The most common of such applications are (a) Basic cell biological research, (b) drug testing, and (c) Cell-based therapies. A frequent problem in these types of *in vitro* cell manipulations, which typically occur in culture dishes, is that one can study, test, or use the same cells but get very different responses from them because the microenvironment they are subjected to *in vitro* is very different from what the cells experience physiologically *in vivo*. This "microenvironment gap" may cause the cell biologist to make scientific conclusions about cells that are contrary to their physiological behavior, or the drug developer to identify false positives and false negatives in drug testing, or the clinician to have to settle for a suboptimal therapeutic outcome. Biologists have identified and made available many bimolecular components that comprise the physiologic microenvironment for *in vitro* culture, and the list of available components continues to increase. However, the typical culture process of adding excessive volumes of culture media that contain a variety of these biomolecules together in a dish with a thin layer of cells attached often only goes so far in closing the microenvironment gap to elicit physiological cellular responses. Why is this? What comprises the physiological microenvironment and what are the ways in which the microenvironment gap

Biological Applications of Microfluidics, Edited by Frank A. Gomez

between *in vivo* and *in vitro* can be narrowed? This chapter will address these questions with a particular focus on resolving the microfluidic aspects of this problem.

5.1.2 What Specifically is the Cellular Microenvironment?

For discussions in this chapter, we define "microenvironment" as a specific set of physical, chemical, and biological conditions in the vicinity of cells within a distance where these conditions can have an effect on or be sensed by the cells. The microenvironment provides cells with physical architecture, fluid and solid mechanical stimulation, adhesive signals, growth factors and cytokines, nutrients, inhibitor chemicals, and cell–cell interactions. The *in vivo* cellular microenvironment is composed of an intricate blend of extracellular matrix proteins, immobilized protein factors, proteoglycans, mineralized tissue, soluble protein factors, small molecule signals, and various adjacent cell types, which vary in space and time due to material deformation, fluid flow, chemical reactions, or cell proliferation, differentiation, death, or migration (1–3). Cells continually sense these inputs, process the information through signal transduction and genetic regulation, and execute a cell behavior or determine a cell fate. Therefore, to study cell biology under well-controlled and versatile yet physiologically relevant conditions *in vitro* or to manipulate cells for clinical applications to elicit their maximum therapeutic potential, it is important not only to have the correct mixture of biomolecules but also to recreate the physical characteristics of the tissue.

5.1.3 Opportunities for Microenvironment Engineering Using Microfluidics

Use of microfluidics makes intuitive sense for cell microenvironment engineering. The physiological microenvironments in living systems are largely microfluidic in nature. Just a glance at the vascular network, pulmonary system, liver sinusoids, reproductive tracts, other tissues and organs of animals and humans reveals a wealth of physiologic microfluidic structures. The structure is very different from the architecture of the typical cell culture dish. An example of the physiologic microenvironment in the human liver is presented in Fig. 5.1.

More substantially, there are differences in the chemical and the mechanical environment cells experience in a microfluidic environment, even if one cultures the same cells in the same medium. Some of the major microenvironment differences are discussed conceptually in the next few paragraphs, and examples of how microfluidics can be used to engineer more physiologic conditions are described in the subsequent section.

5.1.3.1 The Fluid Volume to Cell Volume Ratios Leads to Chemical Changes In a typical culture dish, the volume of culture media is vastly greater compared to the volume of cells adhered to the bottom of the dish. In contrast, physiologically, a majority of the volume within tissues is taken up by cells rather

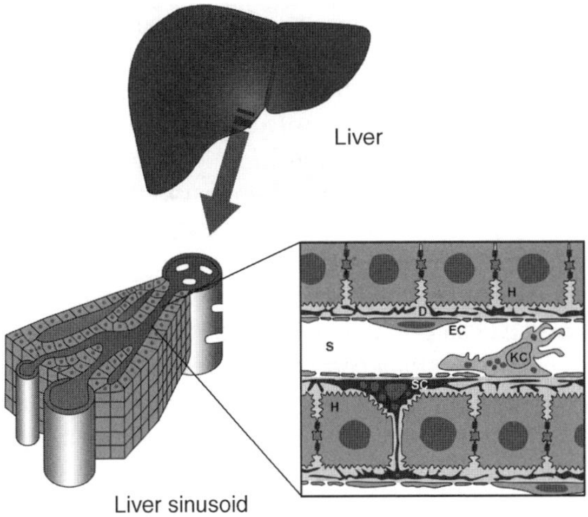

FIGURE 5.1 Physiology is microfluidic. An example of a microfluidic environment presented in human liver. The liver is composed of parenchymal cells, hepatocytes (H) (epithelial cells), and at least four major types of nonparenchymal cells (mesenchymal cells), Kupffer cells (KC) (resident macrophages), perisinusoidal or stellate cells (SC), sinusoidal epithelial cells, and large granular lymphocytes (natural killer cells). Blood flows through the sinusoids (S) formed by fenestrated endothelial cells (EC). Kupffer cells and large granular lymphocytes bulge into the sinusoidal lumen from the endothelial cells. The space between the endothelial cell lining and the hepatocyte plates is the space of Disse (D) in which perisinusoidal or stellate cells and the extracellular matrix are located. Adjacent hepatocytes form bile canaliculi with their apical membranes. (See color insert)

than the fluid surrounding them. This difference in fluid volume to cell volume ratio can lead to differences in the chemical microenvironment of cells. Nanoliter-scale microfluidic bioreactors have an advantage over conventional bioreactors or other macroscopic culture systems in that the small volumes allow even one or a few cells to condition the culture environment significantly. By having physiological ratios of cells to liquid volumes, better *in vitro* model systems can be developed in microfluidic devices in terms of autocrine and paracrine signaling, for example, and pave way for biological studies that may otherwise not be possible. The small volumes also facilitate the formation of significant cell-generated chemical gradients because cells can consume and secrete at appreciable levels where compounds are depleted or accumulated along the channel length.

5.1.3.2 Flow Effects on Chemical Environment Another difference between *in vitro* and *in vivo* cellular environments is fluid flow. The dish is static whereas the physiologic environment is dynamic. In the past 40 years, many studies have made it clear that the fluid flow and biology are inextricably intertwined. On a macroscopic scale, endocrine effects are exerted through transport by blood flow. On a smaller scale, convection can affect chemical gradients and chemical sensibility of cells secreted by themselves (autocrine effects) as well as by other cells (paracrine effects). The static nature of conventional cell culture also poses limitations for studying the effects of steady-state and intermittent stimulation with chemical and mitogenic factors.

In conventional cell cultures, spontaneously occurring spatial variability in temperature, solute concentration, or dissolved gas concentration can lead to surface tension differences at the gas–solution interface. These fluctuations cause variable convection and mass transfer making precise control of mitogens difficult. On the contrary, there is no defined fluid flow as is present in the blood stream, lymph, liver, or other physiologic conditions. Convection is an important aspect of the *in vivo* microenvironment, particularly with respect to autocrine or paracrine signaling, which can be recreated to an extent in microfluidic bioreactors with controlled flow. The generation of gradients during perfusion is relevant to a variety of physiological cellular microenvironments in developing tissues where morphogenetic gradients are important in specifying cell fate, as well as in mature tissue for regulating and maintaining function.

5.1.3.3 Flow Effects on Mechanical Stimuli Fluid flow also exerts mechanical effects. Cells exist *in vivo* in a dynamic environment where mechanical forces such as shear stress, tension, and compression are important. In fact, cells respond to fluid shear stress through a complex process involving several transduction elements and multiple signaling pathways. For example, fluid shear forces acting on endothelial cells (ECs) produce marked, time-dependent changes in their morphology (4,5). Fluid-dynamic stresses have been observed to influence the adhesion of leukocytes to the endothelium and the tendency of the blood to clot. Cellular level fluid mechanics are complicated by the three-dimensional nature of the flow, the compliance of the interacting surfaces, and the receptor–ligand dynamics.

Mechanical forces are also involved in cell–cell interactions. L-selectin, an adhesion molecule, mediates lymphocyte binding to, and rolling on, high endothelial venules. Puri et al. manipulated the bond between selectin and mucin-like ligand by the selective and reversible chemical modification of the ligand to show the effect of mechanical properties on the bond strength of interaction between adhesion molecules on two different cells (6). Additionally, to promote and maintain rolling interactions between leukocytes and vascular vessel wall through L-selectin, shear above a critical threshold is required but not through E-selectin, P-selectin, or VCAM-1. The shear threshold requirement for L-selectin may be physiologically important in low shear to prevent inappropriate aggregation of leukocytes and interaction with the vessel wall (7,8). It is important to have not only applied forces in cell culture but also temporal variations in the forces applied. The endothelium has a completely different response to turbulent shear stress than to a laminar shear stress of the same time-mean magnitude (9,10). Temporal shear also produces a sustained response in intracellular calcium, whereas steady shear leads to an unrepeated single impulse. Endothelial cells also respond to vascular region-specific flow patterns by changing phenotypes as evidenced from their differential patterns of gene expression. They have distinct phenotypic modulation in response to the wall shear stresses present in prototypic human arterial waveforms (atherosclerosis susceptible versus atherosclerosis resistant) (11).

The comparison of biological responses of human umbilical vein ECs to arterial pulsatile flow for 24 h at equivalent time-averaged steady laminar shear stress and at no-flow (static) culture conditions showed changes in cell shape and alignment in case of flow stimuli. However, different patterns of flow stimuli gave rise to distinct patterns of responses in the distribution of actin stress fibers and vinculin-associated adhesion complexes, intrinsic migratory characteristics, and the expression of eNOS mRNA and protein (12). Human endothelial cells exposed to two well-defined biomechanical stimuli—a steady laminar shear stress and a turbulent shear stress of equivalent spatial and temporal average intensity—have the capacity to discriminate among these specific biomechanical forces. They even translate these input stimuli into distinctive phenotypes and also change gene expression as a result (13).

5.1.3.4 Opportunities for Microfluidics Because of the microenvironmental requirements described above, efforts to understand and control cellular signaling, proliferation, and differentiation would benefit from the development of artificial cell microenvironment where cells can be cultured in nanoliter volumes of fluids under investigator-defined conditions. A summary of the difference between conventional 2D cultures and static or dynamic microfluidic cultures are depicted in Fig. 5.2.

In the next part of this chapter, we give specific examples of how microfluidic cell culture aids in each of the three above-mentioned categories. In the first section of the examples of microfluidic cultures, we take a look at the reports of static microfluidic cell culture and the knowledge gained from them. In the second section, we follow the progress made by engineering chemical environments with active fluid flow. In the third section, we discuss the fluid mechanical effects of microfluidic cell culture. After discussing these three topics of how microfluidics can contribute to cellular microenvironment engineering, we describe additional useful capabilities that microfluidic devices offer including culture of precious cells, integrated biochemical analyses, and 3D culture. We end the chapter with future prospects of microfluidic cell biological studies.

5.2 MICROFLUIDIC CULTURES CAN SIMULATE *IN VIVO* MICROENVIRONMENTS

The following are specific examples of how microfluidic cultures can be used to simulate *in vivo* conditions.

5.2.1 Use of Substantially Static Microfluidic Devices to Engineer the Chemical Microenvironment

Simply culturing cells in small volumes of fluids can be a beneficial and more physiological approach. An early account of "microfluidic" mammalian cell culture is from 1948 when Sanford et al. described successful growth of isolated cells in a glass pipette that did not grow well in the typical culture dish (14). A small number of cells grew in the capillary (but not in a dish) due to their ability to self-condition the

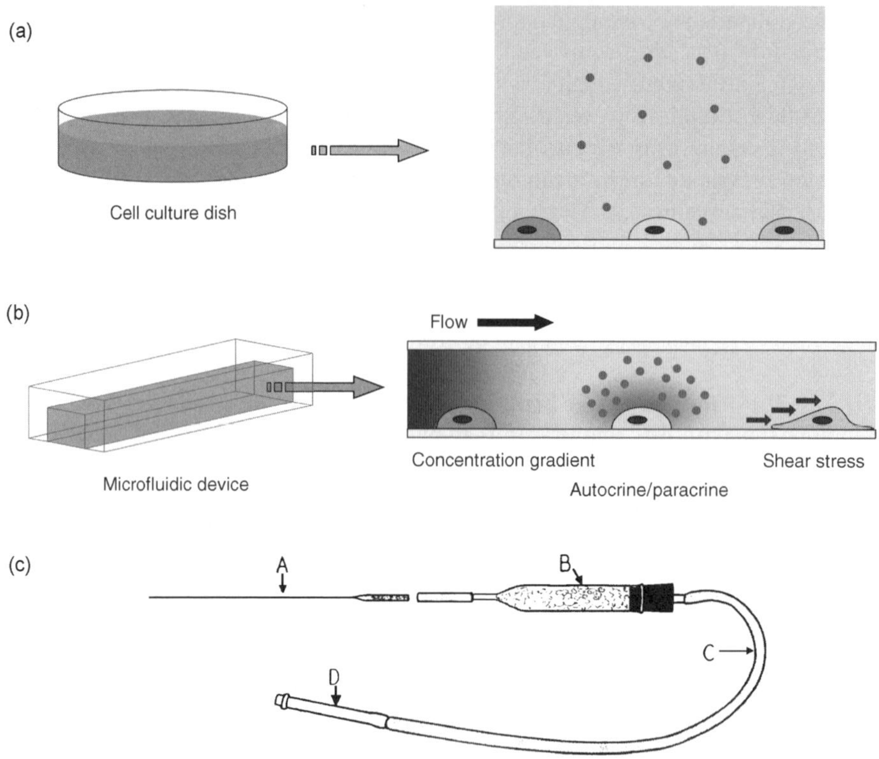

FIGURE 5.2 Cellular microenvironment in a microfluidic culture compared to a conventional 2D culture. (**a**) Cells cultured in a dish grow in a large volume of medium under static condition. Biomolecules secreted from cells are dissipated in bulk medium by diffusion and some uncontrolled convection. (**b**) Cells cultured in a microfluidic device grow in small volumes of media under static or dynamic conditions. Under static conditions, biomolecules secreted from cells distribute around cells (autocrine/paracrine) and into the surrounding medium by diffusion but in a limited way compared to in a dish. Cell consumption and secretion as well as fluid flow can create concentration gradients, and fluid flow can generate shear stress. (**c**) Cell culture in a micropipette from a report in 1948. A = micropipette, B = tube filled with cotton that acts as a bacterial air filter, C = rubber tube, and D = mouthpiece. (See color insert)

small volumes of media surrounding them. The cells, however, eventually died because of nutrient depletion and waste buildup that eventually prevailed, and also caused by the small volumes of media. Microfluidics has come a long way since then, thanks to the semiconductor industry, and now 60 years later, many sophisticated microfluidic tools are being developed with enhanced structures, materials, and flow manipulation to manipulate mammalian cells. There are still many beneficial uses, however, of static microfluidic cell cultures.

Poly (dimethylsiloxane) (PDMS) microfluidic devices, for example, have been used for the static culture of mammalian and fruit fly embryos (15–20). The diffusion dominant microchannel environment has also been characterized by static culture of fall armyworm ovarian cells (Sf9) in different-sized microchannels (height, width, length) at different seeding densities. The cell proliferation was found to be dependent on seeding density and channel height (21).

An alternative to the conventional macroscopic passive gradient generation systems, such as the Boyden chamber and Zigmond chamber, has been developed in a microfluidic format. Chemotaxis of neutrophils to gradients of formyl-Met-Leu-Phe (fMLP) was shown to cause polarization and migration in 6 h in this passive flow microfluidic device (22).

Substantially static microfluidic devices can be useful for the formation of microscale tissue-like constructs such as spheroids. Multicellular tumor spheroids, for example, have been widely used in 3D culture methods for solid tumor model. Tumor spheroids exhibit a characteristic *in vivo*-like morphology and cellular microenvironment that can be used to determine gene expression and biological behavior of cells. A tumor spheroid is composed of proliferating, quiescent, and necrotic cells. Cells located at the periphery of the spheroid are actively proliferating, whereas those located at the center are necrotic. These characteristics make tumor spheroids a useful tool for a variety of experimental studies on radiotherapy, chemotherapy, and photodynamic therapy. A microchip has been used to perform long-term culture (2 weeks) of HepG2 and MCF-7 spheroids and to evaluate viability and response to different concentrations of the cytotoxic agent sodium azide (23).

A straightforward method to control the formation of embryoid bodies (EBs) using microfluidic techniques has also been recently developed using static culture conditions. The device consists of two microchannels separated by a semiporous membrane treated to be resistant to cell adhesion. Mouse embryonic stem (ES) cells are introduced into the upper channel where they self-aggregate to form uniformlysized EBs. The size of EBs is regulated by adjusting the cross-sectional size of the microchannels. The semiporous membrane allows the subsequent treatment of the nonattached EBs with different reagents without the need for EB washout. This method is compatible with on-chip EB culture, differentiation, and manipulation opening the way for more comprehensive ES cell processing on a chip (24).

5.2.2 Use of Microfluidic Devices with Active Fluid Flow to Engineer the Chemical Microenvironment

Active fluid flow and recirculation give rise to many physiologic chemical phenomena including, but not limited to, generation of long-range biochemical gradients, formation of spatial and temporal heterogeneity, autocrine/paracrine/endocrine effects, and metabolic interactions between different cells and tissues. Fluid flow also enables the use of microfluidic phenomena such as patterned laminar flows not possible in static systems. Examples of studies made possible by microfluidic devices with active fluid flow are showcased in Fig. 5.3.

5.2.2.1 Cell Culture with Flow Cell types vary in their need or tolerance for static or dynamic flow environments. For example, Hickman et al. demonstrated that static

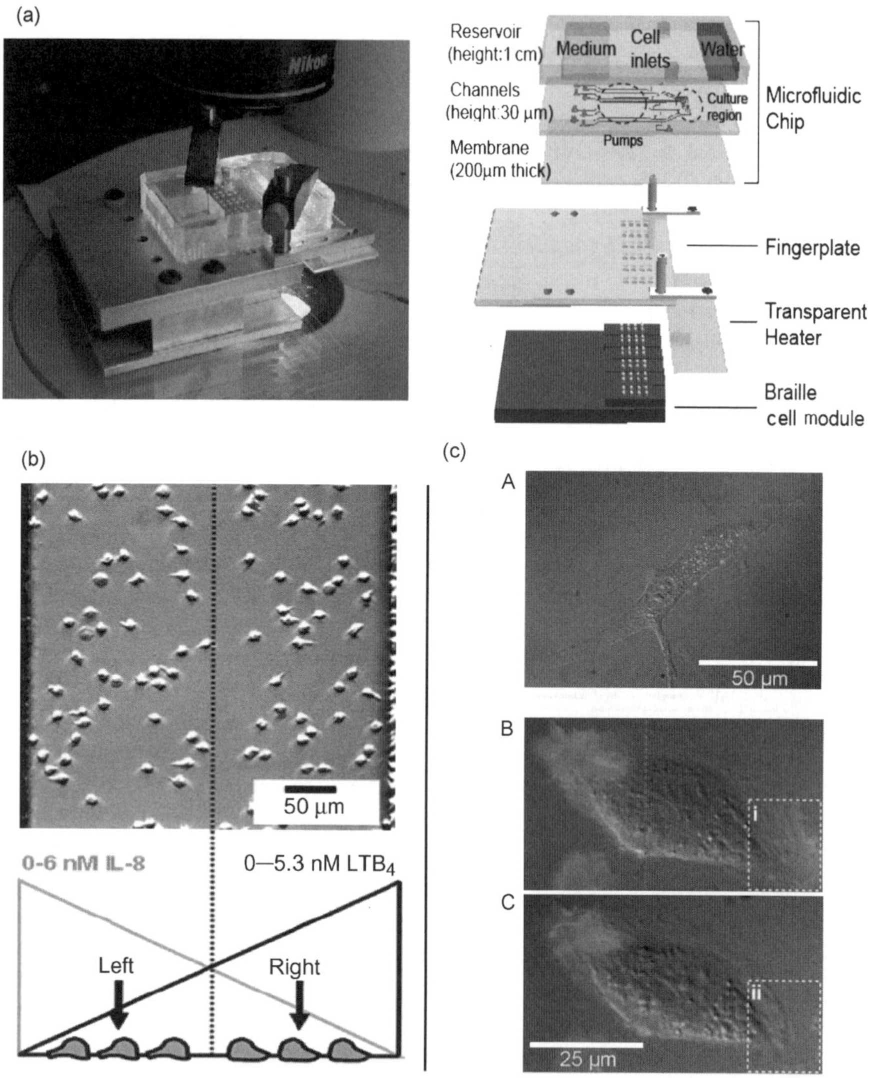

FIGURE 5.3 (**a**) A Braille display-based microfluidic cell culture system (166). (**b**) Chemotaxis of neutrophils in opposing gradients of IL-8 (0–6 nM) and LTB4 (0–5.3 nM). A selected frame from the time-lapse images of a representative experiment, where most of the cells of the "left" population polarized to the right and most of cells of the "right" population polarized to the left (Reference (57)). (**c**) PARTCELL on bovine endothelial cells. (A) Area-selective delivery of fluorescently labeled, acetylated, low density lipoprotein (DiI Ac-LDL) to a portion of the cell surface. The green dotted line indicates the position of the interface between the flows, with and without DiI Ac-LDL. (B) Intracellular transport of cell-surface glycoproteins labeled with wheat germ agglutinin (WGA) conjugated to green (Alexa 488-WGA) and red (TRITC-WGA) fluorescent markers. (C) Cell after 1 h of WGA treatment (47). (See color insert)

culture of embryos developing from the two-cell stage to the blastocyst stage produced higher numbers of blastocysts and morulas compared to a dynamic flow culture condition at two different flow rates (0.1 and 0.5 μ/L) (25). Static culture conditions also had lower numbers of abnormal and eight-cell embryos compared to constant flow. Thus, in this case, static culture was better for embryo development compared to dynamic flow conditions. Allen et al. were able to create an *in vitro* model of liver zonation with three sinusoidal zones by forming oxygen gradients along the length of the bioreactor by perfusion (26,27). Here, the oxygen concentration decreased along the length of the channel because the oxygen uptake by hepatocytes depleted oxygen in the downstream regions. Primary rat hepatocytes and their cocultures with nonparenchymal fibroblasts showed regionally heterogeneous expression of CYP2B and CYP3A proteins in a manner that mimics their distribution in the zonated liver *in vivo*, thus demonstrating the importance of perfusion in creating *in vitro* models of liver physiology. In another study using HFF11 HeLa cells, the signal transduction resulting in the expression of the reporter protein *Photinus* luciferase was shown in microfluidic channels to have no unspecific activation, which is caused by stress to cells, and common in microtiter plate assays (28).

For some applications, a delicate balance of substantially static and dynamic environments needs to coexist. Taylor et al. (29) created microdevices with two large microfluidic compartments that are separated by a physical barrier of very small micron-sized grooves. Neurites but not the cell body can extend from one microfluidic compartment across the small grooves into the adjacent microfluidic compartment. The authors plated primary rat E18 cortical neurons into the somal (cell body) compartment, and after 3–4 days neurites had extended into the neuritic compartment. When combined with surface micropatterning, neuronal attachment and orientation of neurite outgrowth could be further regulated. The system opens many opportunities for engineering the neuronal microenvironments. For example, the same authors demonstrated the ability to perform localized physical or chemical treatment of just the axons or soma independently (30).

Whereas many dynamic microfluidic cell culture systems are flow-through devices, some physiological applications require fluid recirculation. The Shuler group has developed a cell culture analog reactor (CCA) for culture of mammalian cells together with a corresponding physiologic pharmacokinetic model (PBPK) to study drug toxicology. The devices have a liver compartment (containing rat H4IIE hepatoma cells), a lung compartment (with rat lung L2 cells), and other tissue compartments interconnected to each other with fluid recirculating among the compartments with physiologic residence times. The systems have been used, for example, to test the toxicity of naphthalene and the bioaccumulation of naphthalene in adipocytes (31–33). They have also used a four-chamber CCA device (with compartments for lung, liver, fat, and other tissue) to screen drugs for colon cancer (34,35).

A challenge for microfluidic recirculation systems where submicroliter volumes of fluid are used is evaporation. This problem is particularly problematic when using permeable polymeric microdevices such as PDMS. Heo et al. characterized the evaporation-mediated osmolality shifts of cell culture media through thin

PDMS membranes in microfluidic devices (36). They also demonstrated that a PDMS–parylene–PDMS hybrid membrane can slow down evaporation and osmolality shifts for static culture of single-cell mouse embryos into blastocysts as well as dynamic culture of human endothelial cells with recirculation of submicroliter volumes of media.

Microfluidic architecture can be used to enhance mass transport. The Fujii lab fabricated microfluidic bioreactors by stacking 10 PDMS layers to create four cell culture chambers and one chamber for oxygen supply, and then used the system to grow a large number of HepG2 cells at high cell densities (37). Glucose consumption and albumin production measures showed that cellular activity gradually increased and eventually became saturated. The authors also cultured rat liver cells within the microfluidic bioreactors for 15 days and demonstrated that albumin secretion and ammonia removal by these cells were increased by seven times compared to static culture (38). In a simpler device, hepatocytes and 3T3-J2 fibroblasts were cocultured in a microgrooved substrate where perfusion provided oxygen for stable liver-specific function and the grooves protected cells from shear stress due to the perfusion (39).

Recently, a microfluidic device for stem cell culture with both logarithmically varying perfusion rates and concentration gradients has been developed, making it possible to explore a wide range of biological perfusion conditions (including effect of shear) simultaneously for mouse embryonic stem cells (40). Human embryonic stem cells have been cocultured with a feeder layer of human foreskin fibroblasts in microfluidic channels and culture conditions have been established as a function of flow (41).

In other examples, cell arrays of multiple phenotypes (hepatocytes, fibroblasts, and embryonic stem cells) have been fabricated by sequential delivery of fluids or cells into microwells on substrate using reversibly sealable PDMS molds (42). Astrocyte and glioblastoma cells have been cultured in PDMS microdevices modified with silane coupling agents to study cell adhesion and growth (43). An integrated microfluidic bioreactor with integrated indium-tin-oxide heater and miniature temperature probe and uniform media perfusion system has been developed to culture HeLa cells for up to 2 weeks (44).

5.2.2.2 Patterned Microfluidic Flows for Spatio-Temporal Patterning of Biochemicals The signal transduction in cells occurs through cytoskeleton, various scaffolding proteins, other subcellular components. Such structural elements are not randomly distributed in cells but are present heterogeneously throughout the cell with micron-scale spatial variations. The structural order of these components is necessary for the propagation and processing of cellular signals. Taking inspiration from *in vivo* microenvironments where molecular and spatial discrimination can be at the micrometer scale, patterned microfluidic flows have been used to perform localized cell and tissue stimulation with spatial precision that allows even subcellular biochemical delivery. Partial treatment of cells using laminar flows (PARTCELL) uses multiple laminar streams in a microfluidic channel to deliver molecules to selected subcellular microdomains on the surface or inside the single mammalian cells in the case of cell-permeable molecules. The low Reynolds number flows in these microfluidic channels ensure that there is no turbulent mixing of chemicals between the adjacent laminar streams, although diffusive mixing still occurs.

Using this technique, rapidly diffusing, membrane-permeable molecules are directed to subcellular regions of a single cell. Two subcellular processes of mitochondrial movement and changes in cytoskeletal structure of bovine capillary endothelial cells were studied by using PARTCELL. Localization of molecules is achieved by a combination of rapid influx and efflux of molecules in distinct regions of the cell created by using multiple laminar streams, and the process is simple and requires no special equipment (45–47).

Using the PARTCELL approach, the Miyawaki lab took a closer look at the epidermal growth factor receptor (EGFR) signaling in single live COS cells (a monkey cell line) to find if the activation of signaling is localized or propagated over the entire cell. The EGFR signaling pathway is one of the most important pathways that regulate growth, survival, proliferation, and differentiation in mammalian cells. The receptor is often upregulated in tumor cells and the difference in EGF signaling between normal and malignant cells is of great biological and medical interest. Through the combined use of microfluidic subcellular EGF delivery and real-time visualization of intracellular signaling based of fluorescent protein reporters, cells were shown to exhibit localized activation of receptor phosphorylation and Ras activation, or ligand-independent lateral propagation of signals depending on the cell surface receptor density and receptor internalization. The results provide unique insights into intracellular events in ways that would be very difficult to obtain without the use of microfluidics (48).

Using a similar microfluidic approach, embryonic patterning of Drosophila with warm (27°C) and cool (17°C) media under time-specific conditions showed that anterior and posterior halves of the embryo, which are exposed to different temperatures, develop at different rates (20). In another related use of microfluidics, Tourovskaia et al. (49) performed localized delivery of agrin to myotubes formed inside microchannels to induce localized acetylcholine receptor clustering in the stimulated portions of the myotubes (50,51).

Using a different microfluidic approach, Piston group developed a microfluidic bioreactor capable of partially stimulating islet beta cells to create glucose gradients across the islet and observed cellular NAD(P)H and $[Ca^{2+}]_i$ responses (52). They demonstrated that the extent of Ca^{2+} propagation across the islet depends on a delicate interaction between the degree of metabolic coupling and the extent of ATP-sensitive K^+ channel activation.

5.2.2.3 Biochemical Gradients The two common methods for using perfused microfluidic systems to create chemical gradients are (a) use of patterned laminar flows, where adjacent streams contain different concentrations of the chemical of interest, to create chemical gradient in a direction lateral to the direction of fluid flow and (b) use of the ability of cells to secrete and consume biochemicals in their vicinity combined with slow perfusion to create gradients in a direction parallel to the direction of fluid flow.

A useful category of devices to create lateral gradients of biochemicals is the branched network microchannels (53,54). Homogeneous and heterogeneous (hill gradient, cliff gradient) linear gradients have been generated and used to study the chemotaxis of human neutrophils toward IL-8 (53). The chemotaxis behavior of these cells was quantified and important observations regarding the migration of neutrophils with respect to different gradient profiles of IL-8 were made.

A microfluidic stem cell culture platform with a concentration gradient has been used to study the effect of growth factor concentration on human neuronal-stem-cell behavior. The observed proliferation rate in the device was proportional to growth factor concentration, whereas differentiation (to astrocytes) was inversely proportional. In these studies, flow in the device minimized autocrine and paracrine signaling (55). The same group also quantified the migration of the MDA-MB-231 human metastatic breast cancer cells in various conditions by simultaneously generating multiple growth factor (EGF and anti-EGFR) gradients in the microfluidic chemotaxis chamber (56). Microfluidic gradient generator was also utilized to differentiate between neutrophil chemotactic response to equal host-derived conflicting gradients of chemotactic factors IL-8 and LTB4 (57). Chemotaxis of human T cells in response to single and competing gradients (spatial and temporal) of chemokine CCL19 and CXCL12 was studied in "Y"-shaped microfluidic bioreactor (58). Concentration gradients over few hundred microns were generated by the combination of controlled fluid distribution at each intersection of a microfluidic network (due to liquid pressure) and subsequent diffusion between streams in the microchannel (59).

Chemical gradients have been created in microfluidic devices by using arrays of externally connected, horizontally oriented tubes compatible with commercialized multichannel pipettors. This horizontally oriented mini-reservoirs pumping scheme (HOMR pump) provides a steady, pulse-free flow by keeping a constant hydraulic pressure drop across a microchannel; thus the pump overcomes the problem of decreasing flow rate typical for gravity-driven pumps. The HOMR pump does not require any external power connection, is compact in size, and is compatible with generation of multiple laminar streams (up to 8 in this example) of alternating liquids, smooth gradients of various concentration profiles, as well as a step gradient of biochemicals that can be used for microfluidic drug assays (60).

Apart from the generation of lateral chemical gradients, microfluidic systems with slow perfusion can take advantage of cellular consumption or production of biochemicals to generate gradients along the length of a channel. For example, changes in oxygen concentration have been detected in a continuously perfused PDMS microfluidic bioreactor for cell culture (61–63). Steep oxygen gradients caused by uptake by cells have been observed along the length of a mammalian cell culture channel in continuously perfused microfluidic bioreactors (26,27,64).

The area of microfluidic gradient generation is a very active and fruitful area. Considering that *in vivo* microenvironments have more complex gradients, such as pulsed gradients, exponential gradients, or other nonlinear gradients of multiple factors simultaneously, one can expect many more enhancements and innovations in this area (65).

5.2.2.4 Paracrine Effects Active perfusion allows the control of directionality in terms of intracellular chemical communications. For example, interactions between macrophages and osteoblasts were studied while the cells remained physically separated in different parts of a microfluidic bioreactor. The cytokines released upstream by macrophages (IL-1β and TNF-α) flowed through the device in linear gradients and induced downstream osteoblasts to secrete prostaglandin E2 (PGE2) (66).

5.2.2.5 Perfusion Culture of Three-Dimensional Cell Aggregates Spherical cell aggregates of hepatocyte spheroids retain the 3D architecture and cell–cell contacts required for the maintenance of many liver-specific functions (67). Spheroids are generated when their cellular adhesion sites to a substrate are blocked, thereby forcing the cells to adhere to each other. Recently, microscale technologies have been applied to spheroid culture for the development of alternative tools for cellular biology. Coculture spheroid arrays have been developed using micropatterning techniques (68,69). Hepatocytes spheroids cocultured with endothelial cells or fibroblasts maintain their liver-specific functions for at least 1 month. Such 3D cell cultures have been combined with microfluidic systems. 3D culture inside PDMS microfluidic bioreactors (with uniform and nonuniform flow) was achieved by stacking two pieces of PDMS containing 3D features. The culture of HepG2 cells was demonstrated in this bioreactor and metrics of cell functionality (albumin production, glucose consumption) were evaluated (70). Kojima et al. created agar microstructures with cell-culture holes to culture primary neonatal rat cardiac myocytes and observed synchronized beating in them 90 min after they had made physical contact (71).

5.2.2.6 Microarrayed Bioreactors A number of interesting microfluidic array systems have been developed.

A 10×10 microfluidic cell culture array for long-term monitoring of HeLa cells was developed by Hung et al. (72,73). By continuous perfusion, it was possible to passage confluent cells within the microfluidic chambers, thereby creating a stable culture environment.

A bioreactor array (8×8) with culture regions disjoint from flow channels to create low shear spaces for cells and convective flow for transport of fluids was made for continuous flow in multiplexed environment (74). The device enabled uniform cell loading, low shear for cell culture units, and low pressure stresses on cultured cells for the culture of HeLa, NIH 3T3 fibroblasts, primary bovine endothelial cells, HepG2 hepatocytes, and human SY5Y neuroblastoma cells.

A 64 (8×8) array of microfluidic wells connected with two microfluidic perfusion networks for culture medium and oxygen was fabricated to support micropatterned coculture of primary rat hepatocytes and 3T3-J2 fibroblasts (75). The hepatocytes cultures in the array were able to continuously generate albumin synthesis and urea production in a steady state. Chueh et al. have recently reported fabrication of 8×12 criss-crossing microfluidic channel array by overlaying polyester and polycarbonate membranes over PDMS structures (76).

5.2.2.7 Use of Microfluidic Flow for Creating Layered Structures Defined 3D microstructural multilayered arrangements of live cells have been created by photopatterning of 3D hydrogel to culture different cell types. 3D layers of extracellular matrix biopolymers (collagen, chitosan, fibronectin) have been micropatterned on PDMS substrates for culture of human umbilical vascular endothelial cells (HUVECs) and fibroblasts (77,78). Microfluidic layer-by-layer 3D biomimetic structures were also created using collagen, chitosan, matrigel, and fibrin to culture human lung fibroblasts (IMR-90), human umbilical vein smooth muscle cells (HUVSMCs), human umbilical vein endothelial cells (HUVECs) for 5 days (79).

Microfluidic channels were utilized to create polyelectrolyte multilayers of poly (allylamine hydrochloride) (PAH) and poly(styrene sulfonic acid) (PSS) on PDMS substrate in order to culture neuron-like cells (NLCs) derived from pluripotent embryonal carcinoma P19 cells (80). Additionally, primary murine bone marrow stromal cells were grown on nanocomposite coatings deposited onto PDMS using layer-by-layer assembly in microfluidic bioreactors (81). The cells attached, spread, and proliferated well on the nanocomposite coatings, whereas the cells detached, rounded, and died on native PDMS even when coated with cell-adhesive proteins.

5.2.2.8 Long-Term Culture Using microfluidic bioreactors, long-term culture of cells can easily be carried out in controlled environments for many physiological functional assays that take place over time in the range of a few weeks. For example, differentiation of alveolar epithelial cells takes two weeks to complete and finding the frequency of long-term culture-initiating hematopoeitic cells takes over 3 weeks to complete (82–84). Long-term culture combined with imaging of cells has always been one of the important goals of microfluidic cell culture labs. Although still somewhat low resolution in terms of cell imaging, a recently developed microfluidic recirculation system with built-in transparent heater when used with customized media allowed time-lapse imaging of proliferation of C2C12 myoblasts and MC3T3-E1 osteoblasts for over 2 weeks in ambient atmosphere without medium exchange (85).

5.2.3 Use of Microfluidic Devices with Active Fluid Flow to Engineer the Fluid Mechanical Microenvironment

The most well-known and important example of the influence of fluid mechanical forces on physiology is that of endothelial cells. Vascular endothelial cells are located at the innermost layer of the blood vessel wall and are always exposed to three different mechanical forces: shear stress due to blood flow, hydrostatic pressure due to blood pressure, and cyclic stretch due to vessel deformation. It is well known that endothelial cells respond to these mechanical forces and change their shapes, cytoskeletal structures, and functions. After applying fluid shear stress, cultured endothelial cells show marked elongation and orientation in the flow direction. In addition, thick stress fibers of actin filaments appear and align along the long axis of the cell. Thus, endothelial cell morphology is closely related to the cytoskeletal structure (86).

When endothelial cells were exposed to hydrostatic pressure, they exhibited a marked elongation and orientation in a random direction, together with the development of centrally located, thick stress fibers. Pressured endothelial cells also exhibited a multilayered structure with less expression of VE-cadherin unlike under control conditions. Simultaneous hydrostatic pressure loading and shear stress inhibited endothelial cell multilayering and induced elongation and orientation of endothelial cells with well-developed VE-cadherin in a monolayer.

Wall shear stress also regulates adaptive vessel growth and angiogenesis, and might be a local risk factor in the pathogenesis of atherosclerosis. Shear stress also modulates the production of vasoactive substances such as endothelium-derived relaxing factor, prostacyclin, histamine, and endothelin, and regulates macromolecule permeability and endocytosis. More recent studies have shown that shear stress exerts an influence on the expression of mRNA such as tissue plasminogen activator mRNA (87).

Additionally, different flow patterns elicit distinct responses of Kruppel-like factor-2 (KLF2) in endothelial cells *in vitro* and *in vivo*. The pulsatile flow with a significant forward direction induces sustained expression of KLF2 in cultured ECs, but oscillatory flow with little forward direction causes prolonged suppression after a transient induction. The KLF2 protein level ECs at arterial branch points are related to local hemodynamics (88).

5.2.3.1 Two-Phase Flows Gas–liquid two-phase flow configurations have been used to study alveolar epithelial cells and also for microfluidic flow cytometer (89). To expose pulmonary epithelial cells to various fluid flows, Huh et al. developed a compartmentalized microfluidic bioreactor that mimics *in vivo* microenvironment with different flows under normal and pathological pulmonary conditions (82,83).

5.2.3.2 Shear Stress in Microfluidic Bioreactors Exposure of cultured endothelial cells to controlled levels of shear stress is possible in microfluidic culture systems using computer-controlled pulsatile fluid flows, as represented in Fig. 5.4. The high velocity flows created by these microfluidic bioreactors generated shear stress levels capable of fluid mechanical regulation of vascular endothelial cells (90).

A variable height flow cell has been developed to create an increasingly stronger laminar shear stress along a flow path. Flow rates of 5–105 ml/min created a wide range of shear stress to study adhesion characteristics of neuroblastoma X glioma cells, NG108-15, on different self-assembled monolayers of organosilicates (91).

By culturing of endothelial cells in microfluidic channels of different widths and subjecting them to steady state shear stress levels, shape and function effects were quantified (92). Narrow channels had more elongated bovine aortic endothelial cells; however, if channel width was 200 μm or larger, the elongation was similar to that in plastic dishes.

Culture of osteoblasts requires shear stress for enhanced differentiation and expression of alkaline phosphatase and osteocalcin. The ALP expression has been shown to increase three times (even with no flow due to a 3D environment) and 7.5 times (with low flow due to simulating shear stress and 3D environment) in a

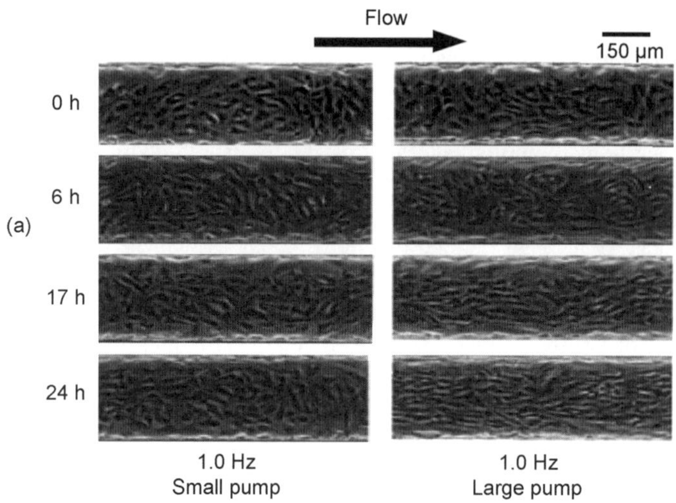

FIGURE 5.4 Study of effects of shear stress on microvascular endothelial cells inside a continuously perfused microfluidic bioreactor (90). Time-lapse images of ECs comparing changes in their elongation and alignment in response to varying levels of shear stress.

bioreactor made of 3D network of microstructures, compared to a conventional dish culture (93).

Shear stress has also been shown to induce embryonic stem cell differentiation. F1k-1-positive mouse ES cells subjected to shear stress selectively differentiated into vascular endothelial cells. The shear stressed F1k-1-positive ES cells formed tube-like structures and developed an extensive tubular network significantly faster than the static controls (94).

By creating a microfluidic bioreactor with channels whose width narrows uniformly, just as *in vivo* microcirculation, release of ATP from erythrocytes was found to be proportional to channel cross-sectional area (95). Weinberg et al. developed three-dimensional microfluidic constructs for tissue engineering that have uniform wall shear stress throughout the network (96).

Hang et al. used microfluidic devices to quantify cell adhesion as a function of shear stress by manipulating the geometry and the surface chemistry (97). The adhesion of fibroblasts to fibronectin-coated substrates was evaluated as a function of shear stress or fibronectin concentration in channels. They also confirmed that the fibroblasts cultured in the perfusion-shear device reduced their adhesion strength to the substrate in response to epidermal growth factor stimulation.

Griffith group has presented a bioreactor that enables morphogenesis of 3D tissue structures under continuous perfusion (98). Three-dimensional scaffolds were created by deep reactive ion etching of silicon wafers to create an array of channels (through hole) with cell-adhesive walls. Scaffolds were combined with a cell-retaining filter and supported across the top of the array and through the 3D tissue mass in each channel. Reactor dimensions were constructed so that media flow rates meet the estimated value of cellular oxygen demands while providing fluid shear stress at or below a physiological range (<2 dyne cm^2). Primary rat hepatocytes seeded in the bioreactor rearranged to form tissue-like structures reminiscent of liver cords. The aggregates formed bile canaliculi and tight junctions and maintained viability for at least 2 weeks (99).

The embryonic flow–structure interactions influence whether development will proceed normally or become pathogenic. Genetic, pharmacological, or surgical manipulations that alter the flow environment can thus profoundly influence morphologic and functional cardiovascular phenotypes (100).

To get a better understanding of fluid dynamics in microchannels, a three-dimensional numerical flow model incorporating mass transport has recently been developed to compute substrate concentrations and shear stresses in the fluid environment of the microfluidic bioreactors (101).

5.2.3.3 Squeezing Cells Through Microchannels Microchannels have also been used to observe healthy and malaria (*Plasmodium falciparum*) infected red blood cells (102). The normal erythrocytes were very elastic and passed through channels of varying widths (2–8 μm) and squeezed through blockages readily, while the infected cells in different stages of the disease could not do so.

5.2.3.4 Osmotic Loading Chao et al. developed a microfluidic device to apply hydrostatic pressure-driven dynamic osmotic loading to observe dynamic changes in cell sizes of cultured bovine articular chondrocytes (103). The cells showed a trend of increasing changes in cell size with decreasing loading frequency (for a given osmotic loading magnitude), with significant differences from isotonic control at 0.0125 Hz.

5.3 OTHER USEFUL CAPABILITIES OF MICROFLUIDIC CELL CULTURE DEVICES

5.3.1 Efficient Manipulation of Precious Cells or Cells Available in Only Small Numbers

Some of the cell types are so precious and rare or hard to isolate—for example, primary cells like neurons, hematopoietic stem cells, and human embryos for clinical treatment— that any study or handling involving them has to be done with great care. Moreover, the biochemical responses of such a small population of cells in culture are difficult to observe because of the dilution effects in the large pool of media surrounding them. Biological studies with such a small number of precious cells would greatly benefit from the use of microfluidic culture systems integrated with microanalysis capabilities where even miniscule quantities of secreted and soluble products can be assayed with accuracy. One example of rare cell culture is shown in Fig. 5.5, where murine embryos are grown in a microfluidic device for *in vitro* fertilization.

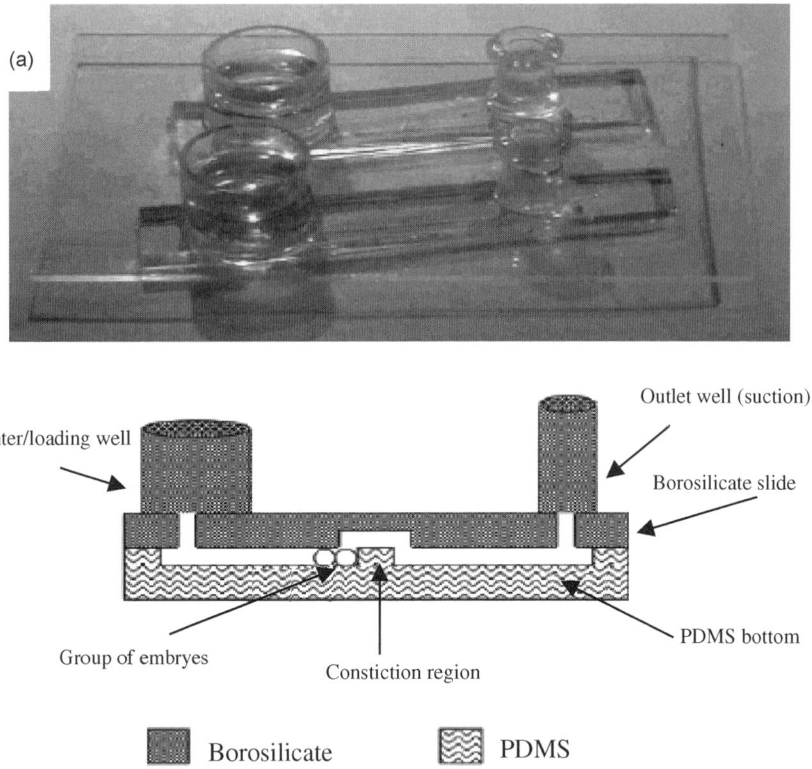

FIGURE 5.5 A PDMS/borosilicate device for embryo culture: (**a**) side view and (**b**) schematic cross-section of the PDMS/borosilicate device. The basic structure of the device is a PDMS bottom covered with a borosilicate slide that forms the top of the microchannels (16).

Stem cell biology in microfluidic bioreactors has focused on two key aspects of differentiation and self-renewal of the cells. Recently, attempt has been made to expose primary human neural precursor cells in a parallel manner to various ECM signals and perform a quantitative, multiparameter analysis of cellular responses. The ECM arrays of signals provided to cells were composed of mixtures of extracellular matrix components, morphogens, and other signaling proteins. The signals that affected developmental decisions on the differentiation of these precursor cells were quantified. Wnt and Notch costimulation maintained the cells in an undifferentiated-like proliferative state, whereas bone morphogenetic protein 4 induced an "indeterminate" differentiation phenotype characterized by simultaneous expression of glial and neuronal markers (104).

ES cell differentiation is often initiated by the generation of embryoid bodies (EBs), which are three-dimensional cell aggregates formed by culturing ES cells on a nonadherent substrate (105). EB differentiation has been shown to recapitulate many aspects of early embryogenesis, resulting in the generation of a wide variety of cell types within EBs (106). Regulating the size of EBs is crucial for controlling the differentiation of ES cells. Recently, several approaches, including microtechnological approaches, have been developed to control EB size. There have been reports of regulating EB size by using microwells to produce ES colonies of defined size and then forming EBs from these colonies (107,108). Uniformly sized EBs were formed in microfluidic device with semiporous membrane between two microchannels. This method is compatible with on-chip EB culture and differentiation, opening the way for more comprehensive ES cell processing on a chip (24).

5.3.2 Integrated Assays

Rather than monitoring only simple phenotypic changes, integrated microfluidic bioreactors can gather precise biochemical and mechanistic data from cells and tissues, as shown in Fig. 5.6, in which secretion of a soluble factor by living cells is continuously monitored over time. The rationales for integration of sensing and analysis in the microdevice include greater accuracy and reproducibility, smaller sample sizes, and higher throughput (110).

Controlled perturbations followed by monitoring cellular responses can provide insight into cellular pathways that govern phenotype and behavior. Thus, a fundamental study of biochemical pathways, cell-fate decisions, and tissue morphogenesis is underway in microfluidic bioreactors. These "integrated microfluidic bioreactors" are capable of fast and reproducible measurements with as little as femtomolar sample concentrations and nanoliter volumes. Combining these aspects with the integration of sample preparation and separation steps on the

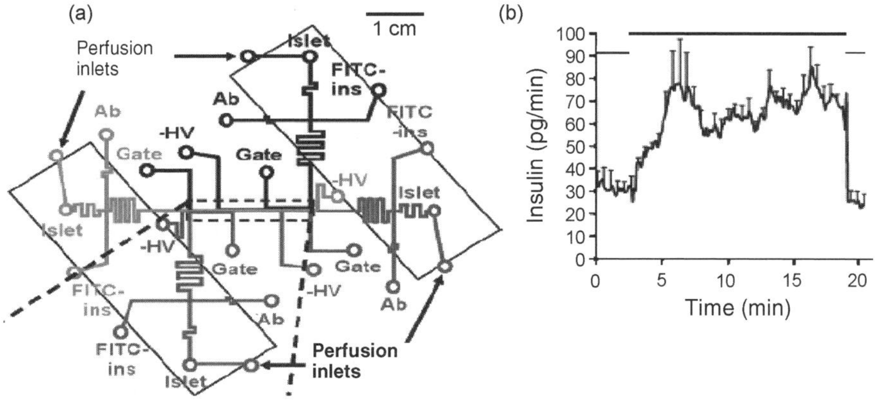

FIGURE 5.6 Integration of cell culture and analysis in a single microfluidic bioreactor. (128). (**a**) Channel layout of a microfluidic device for monitoring insulin secretion from four islets of Langerhans. (**b**) Average detected insulin secretion from four islets on a single chip.

same microfluidic bioreactor has resulted in these systems finding considerable commercial applications in the fields of genomics and proteomics (111).

Immunoassays can be readily analyzed by absorption techniques and fluorescent markers. Quantitative protein analysis and detection can be performed with antibody capture in microdevices as they have high selectivity and affinity of antibody–antigen binding to achieve specificity in analysis. Enzyme-linked immunosorbent assays (ELISA) that amplify the signal from each labeled molecule are now routinely used to detect analytes with high sensitivity in microfluidic bioreactors (112–127).

Microfluidic bioreactors that can continuously monitor insulin secretion from islets of Langerhans with four integrated electrophoresis-based immunoassay channels have been recently developed. The detection limit of the bioassay was 10-nM insulin and insulin release was evaluated every 6.25 s from four islets (128). A microfluidic bioreactor with an online electrophoresis immunoassay and continuous perfusion was developed by same group to monitor insulin levels in live islet culture. The detection limit of the assay was 0.8 nM, and it allowed quantitative monitoring of classical biphasic and oscillatory insulin secretion with 6 s sampling frequency following step changes in glucose from 3 to 11 mM (129). Electrophoretic separation of anti-insulin antibody (Ab), fluorescein isothiocyanate-labeled insulin (FITC-insulin), and insulin from different reservoirs was also carried out in a microfluidic bioreactor. The detection limits for insulin were 3 nM and assays could be collected at 15-s intervals with continuous sampling (130).

In a glass microdevice with integrated cell culture, chemical and enzymatic reactions, and detection capabilities, nitric oxide released from macrophage-like cells stimulated by lipopolysaccharide was successfully monitored. The micro-immunoassay was much faster (assay time 4 h compared to 24 h), detection limit of NO was improved (from 1×10^{-6} to 7×10^{-8} M), and the system was also able to monitor a time course of the release (131). Moreover, microfluidic bioreactor with an integrated assay modality was able to test potency of different concentrations of 5-Fluorouracil (in gradients) as a cancer drug against live stomach cancer cells (132).

Real-time characterization of gene expression in living hepatocytes in a device has been carried out by King et al. (133). High density matrixes of stimulus-response experiments were performed with the help of microvalve arrays (control row seeding and column stimulation) in 256 nL-scale bioreactors to obtain quantitative live cell imaging of fluorescent protein transcriptional reporters. Weider et al. fabricated a microfluidic device to monitor GFP-based gene expression in a high throughput manner and found that transcription factor NF-kappa B response of live reporter cells on stimulation TNF-alpha shows the same fluorescence kinetics as observed in standard tissue culture (134). In another report of on-chip immunoassay, Matsubara et al. fabricated a microfluidic device for on-chip detection of allergic response of rat basophilic leukemia cell line RBL-2H3 to antigen stimulus by fluorescence (135).

Impedance sensors, which monitor morphological changes and motility of adherent cells and pH sensors that evaluate functional responses of cells upon receptor stimulation by measuring the extracellular H+ have also been developed in microfluidic bioreactors (110,136). There are also various reports of measuring oxygen concentrations inside microfluidic devices using optical as well as polarographic measurements (137–141,61–63,81). A PDMS bioreactor coupled with pH and glucose sensors has also been developed to characterize culture of fibroblast cell line, CHO cells, hepatocytes, and hybridoma cell lines (142).

5.3.3 Three-Dimensional Culture

There are many cells where 3D culture would be more physiological compared to attachment on a 2D surface. There are many accounts of microfluidic cell culture with defined three-dimensional cellular matrixes (70,71,77,98). Some examples are discussed below.

5.3.3.1 Cells in Hydrogels Inside Microfluidic Devices Cells have been cultured in hydrogels inside the microfluidic devices. For example, macrophages and 3T3 fibroblasts were encapsulated in poly(ethylene glycol)(PEG) hydrogel microstructures created by photolithographic polymerization inside PDMS microdevices (143). HepG2 cells immobilized in channels by hydrodynamic focusing of biocompatible self-assembling peptide hydrogel have been cultured with linear concentration gradients by injecting different reagents on the sides of the scaffold (144). Primary rat hepatocytes were immobilized *in situ* in a localized matrix of methylated collagen and a HEMA–MMA–MAA in a microfluidic bioreactor (145). Cells have also been grown in 3D layers of extracellular matrix biopolymers (collagen, chitosan, fibronectin, matrigel, and fibrin) that were micropatterned in PDMS devices (78,79). Anticancer drug sensitivity test was performed using PDMS microfluidic devices in which collagen gel-embedded cells were entrapped. The proliferation rate of collagen gel-embedded cells was close to that *in vivo* and their drug sensitivity was less than that of 2D cells, suggesting that the 3D culture conditions provide a more *in vivo*-like environment (109).

5.3.3.2 Cells in Gel Channels Microfluidic devices were even made by molding gelatin to produce microchannels for culture of adherent epithelial cells (146). Microfluidic bioreactors have also been fabricated from biocompatible photopolymerized poly(ethylene) glycol diacrylate (PEGDA) (147). Recently, Leclerc et al. used a photosensitive biodegradable polymer (pCLLA acrylate) to create single and multistep microstructures, as well as microstructure with simple fluidic channels. These microstructures are able to support the culture of mammalian cells to grow and spread on it with HepG2 cells, human umbilical blood vessel endothelial cells (HUVEC), 3T3-L1 mouse fibroblasts (148).

5.4 MICROFLUIDIC DEVICES USEFUL FOR CELL APPLICATIONS OTHER THAN CULTURE

Microfluidic devices are useful for more than cell culture. Devices have been used in cell sorting, for example, in the separation of nonmotile and motile cells such as sperm, in the separation of the contents of single cells, in the separation coupled to mass spectroscopy, high throughput screening in drug development bioanalyses, and examination and manipulation of samples consisting of a single cell or a single molecule (149–157).

Microfluidic devices have been used to isolate small numbers of target cells (or even single cells) from a large population of cells on the basis of physical and chemical properties such as electrical characteristics, chromatography, or fluorescent markers (158–164). A comprehensive review of recent developments in microfabricated flow cytometers and related microfluidic devices that can detect, analyze, and sort cells or particles is provided by Huh et al. (165).

5.5 FUTURE PROSPECTS FOR BIOLOGICAL STUDIES IN MICROFLUIDIC BIOREACTORS

One of the useful applications of microfluidics is the culture of cells under engineered physiologic conditions. In microfluidic bioreactors, well-defined physiologically relevant chemical and fluid mechanical microenvironments can be created for answering fundamental questions in cell biology. These *in vitro* microenvironments can be powerful tools for studying disease mechanisms, testing drug safety and efficacy, engineering microscale tissues, and developing cell-based biosensors. In these systems, one can systematically perturb cell culture conditions for specific experiments, test many different conditions on the same or different cells or even different parts of the same cell. On-chip integration of quantitative analysis and sensing greatly expands the biological applications of these devices.

Just as the electronic industry has integrated more and more components onto smaller and smaller an area, the trend in the bioMEMS community is to integrate more and more types of cells and analysis capabilities on a chip, with a possible eventual establishment of an animal-on-a-chip or a human-on-a-chip. Having all tissues growing in one single device could be a great boon to drug screening and clinical research. However, in order to achieve an organotypic culture of such complexity, we must first be able to engineer the microenvironment of each cell type with high quality and precision in terms of recapitulating physiologic conditions. This type of research requires in-depth knowledge of not only microfluidic engineering and materials science but also the underlying biology. Collaborations between bioengineers and biologists or clinicians need to be further enhanced to unleash the true potential of microfluidics for the study of cell biology and the application to the clinic.

REFERENCES

1. Adams GM, Scadden DT. Nat Immunol 2006;7:333–337.
2. Jang JH, Schaffer DV. Mol Syst Biol 2006;2:39.
3. Shim J, Bersano-Begey TF, Zhu X, Tkaczyk AH, Linderman JJ, Takayama S. Curr Top Med Chem 2003;3:687–703.

4. Dewey CF, Bussolari SR, Gimbrone MA, Davies PF. J Biomed Eng 1981;103:177–185.
5. Flaherty JT, Ferrans VJ, Tucker WK, Fry DL, Patel DJ, Prierce JE. Circ Res 1972;30:23–33.
6. Puri KD, Chen S, Springer TA. Nature 1998;392:930–933.
7. Finger EB, Purl KD, Alon R, Lawrence MB, von Andrian UH, Spinger TA. Nature 1996;379:266–269.
8. Alon R, Hammer DA, Springer TA. Nature 1995;374:539–542.
9. Davies PF, Dewey CF, Bussolari SR, Gordon EJ, Gimbrone MA. J Clin Invest 1984;73:1121–1129.
10. Davies PF, Remuzzi A, Gordon EJ, Dewey CF, Gimbrone MA. Proc Natl Acad Sci U S A 1986;83:2114–2117.
11. Dai GH, Kaazempur-Mofrad MR, Natarajan S, Zhang YZ, Vaughn S, Blackman BR, Kamm RD, Garcia-Cardena G, Gimbrone MA. Proc Natl Acad Sci U S A 2004;101: 14871–14876.
12. Blackman BR, Garcia-Cardena G, Gimbrone MA. J Biomech Eng 2002;124:397–407.
13. Garcia-Cardena G, Comander J, Anderson KR, Blackman BR, Gimbrone MA. Proc Natl Acad Sci U S A 2001;98:4478–4485.
14. Sanford KK, Earle WR, Likely GD. J Natl Cancer Inst 1948;9:229–246.
15. Walker GM, Zeringue HC, Beebe DJ. Lab Chip 2004;4:91–97.
16. Raty S, Walters EM, Davis J, Zeringue H, Beebe DJ, Rodriguez-Zas SL, Wheeler MB. Lab Chip 2004;4:186–190.
17. Walters EM, Clark SG, Beebe DJ, Wheeler MB. Methods Mol Biol 2004;254:375–382.
18. Clark SG, Haubert K, Beebe DJ, Ferguson CE, Wheeler MB. Lab Chip 2005;5:1229–1232.
19. Glasgow IK, Zeringue HC, Beebe DJ, Choi SJ, Lyman JT, Chan NG, Wheeler MB. IEEE Trans Biomed Eng 2001;48:570–578.
20. Lucchetta EM, Lee JH, Fu LA, Patel NH, Ismagilov RF. Nature 2005;434:1134–1138.
21. Yu H, Meyvantsson I, Shkel IA, Beebe DJ. Lab Chip 2005;5:1089–1095.
22. Abhyankar VV, Lokuta MA, Huttenlocher A, Beebe DJ. Lab Chip 2006;6:389–393.
23. Torisawa Y, Takagi A, Nashimoto Y, Yasukawa T, Shiku H, Matsue T. Biomaterials 2007;28:559–566.
24. Torisawa Y, Chueh B, Huh D, Ramamurthy P, Roth TM, Barald KF, Takayama S, Lab Chip 2007;7:770–776.
25. Hickman DL, Beebe DJ, Rodriguez-Zas SL, Wheeler MB. Comp Med 2002;52:122–126.
26. Allen JW, Bhatia SN. Biotechnol Bioeng 2003;82:253–262.
27. Allen JW, Khetani SR, Bhatia SN. Toxicol Sci 2005;84:110–119.
28. Davidsson R, Boketoft A, Bristulf J, Kotarsky K, Olde B, Owman C, Bengtsson M, Laurell T, Emneus J. Anal Chem 2004;76:4715–4720.
29. Taylor AM, Rhee SW, Tu CH, Cribbs DH, Cotman CW, Jeon NL. Langmuir 2003;19:1551–1556.

30. Taylor AM, Blurton-Jones, Rhee SW, Cribbs DH, Cotman CW, Jeon NL. Nat Methods 2005;2:599–605.
31. Ghanem A, Shuler ML. Biotechnol Prog 2000;16:334–345.
32. Ghanem A, Shuler ML. Biotechnol Prog 2000;16:471–479.
33. Viravaidya K, Shuler ML. Biotechnol Prog 2002;18:174–181.
34. Viravaidya K, Shuler ML. Biotechnol Prog 2004;20:590–597.
35. Sung *et al.* 2006 Francisco, CA, United States, Sept. 10–14, BIOT-393.
36. Heo YS, Cabrera LM, Song JW, Futai N, Tung Y–C, Smith GD, Takayama S. Anal Chem 2007;79:1126–1134.
37. Leclerc E, Sakai Y, Fujii T. Biotechnol Prog 2004;20:750–755.
38. Ostrovidov S, Jiang J, Sakai Y, Fujii T. Biomed Microdev 2004;6:279–287.
39. Park J, Berthiaume F, Toner M, Yarmush ML, Tilles AW. Biotechnol Bioeng 2005;90:632–644.
40. Kim L, Vahey MD, Lee HY, Voldman J. Lab Chip 2006;6:394–406.
41. Korin N, Bransky A, Dinnar U, Levenberg S. Proc SPIE—Int Soc Optical Eng 2007; 6416 (Biomed Appl Micro Nanoeng), 64160N/1-64160N/8.
42. Khademhosseini A, Ferreira L, Blumling J, Yeh J, Karp JM, Fukuda J, Langer R. Biomaterials 2006;27:5968–5977.
43. Cox JD, Curry MS, Skirboll SK, Gourley PL, Sasaki DY. Biomaterials 2002;23: 929–935.
44. Petronis S, Stangegaard M, Christensen CBV, Dufva M. Biotechniques, 2006;40: 368–376.
45. Takayama S, McDonald JC, Ostuni E, Liang MN, Kenis PJA, Ismagilov RF, Whitesides GM. Proc Natl Acad Sci U S A 1999;96:5545–5548.
46. Takayama S, Ostuni E, LeDuc P, Naruse K, Ingber DE, Whitesides GM. Nature 2001;411:1016.
47. Takayama S, Ostuni E, LeDuc P, Naruse K, Ingber DE, Whitesides GM. Chem Biol 2003;10:123–130.
48. Sawano A, Takayama S, Matsuda M, Miyawaki A. Dev Cell 2002;3:245–257.
49. Tourovskaia A, Figueroa-Masot X, Folch A. Lab Chip 2005;5:14–19.
50. Tourovskaia A, Kosar TF, Folch A. Biophys J 2006;90:2192–2198.
51. Tourovskaia A, Figueroa-Masot X, Folch A. Nat Protoc 2006;1:1092–1104.
52. Rocheleau JV, Walker GM, Head WS, McGuinness OP, Piston DW. Proc Natl Acad Sci U S A 2004;101:12899–12903.
53. Jeon NL, Baskaran H, Dertinger S KW, Whitesides GM, Water LVD, Toner M. Nat Biotechnol 2002;20:826–830.
54. Lin F, Saadi W, Rhee SW, Wang S–J, Mittal S, Jeon NL. Lab Chip 2004;4:164–167.
55. Chung BG, Flanagan LA, Rhee SW, Schwartz PH, Lee AP, Monuki ES, Jeon NL. Lab Chip 2005;5:401–406.
56. Saadi W, Wang S–J, Lin F, Jeon NL. Biomed Microdev 2006;8:109–118.
57. Lin F, Nguyen CM, Wang SJ, Saadi W, Gross SP, Jeon NL. Ann Biomed Eng 2005;33:475–482.
58. Lin F, Butcher EC. Lab Chip 2006;6:1462–1469.

59. Yang M, Yang J, Li C–W, Zhao J. Lab Chip 2002;2:158–163.
60. Zhu X, Chu LY, Chueh B, Shen M, Hazarika B, Phadke N, Takayama S. Analyst 2004;129:1026–1031.
61. Park J, Bansal T, Pinelis M, Maharbiz MM. Lab Chip 2006;6:611–622.
62. Szita N, Boccazzi P, Zhang ZY, Boyle P, Sinskey AJ, Jensen KF. Lab Chip 2005;5:819–826.
63. Zanzotto A, Szita N, Boccazzi P, Lessard P, Sinskey AJ, Jensen KF. Biotechnol Bioeng 2004;20:243–254.
64. Mehta G, Mehta K, Sud D, Song JW, Bersano-Begey T, Futai N, Heo YS, Mycek M–A, Linderman JJ, Takayama S. Biomed Microdev 2006, Epub ahead of print.
65. Katz MJ, Lasek RJ. Cell Motil 1980;1:141–157.
66. Wei C–W, Cheng J–Y, Young T–H. Biomed Microdev 2006;8:65–71.
67. Koide N, Sakaguchi K, Koide Y, Asano K, Kawaguchi M, Matsushima H, Takenami T, Shinji T, Mori M, Tsuji T. Exp Cell Res 1990;186:227–235.
68. Otsuka H, Hirano A, Nagasaki Y, Okano T, Horiike Y, Kataoka K. Chem Bio Chem 2004;5:850–855.
69. Fukuda J, Khademhosseini A, Yeo Y, Yang X, Yeh J, Eng G, Blumlimg J, Wang C–F, Kohane DS, Langer R. Biomaterials 2006;27:5259–5267.
70. Leclerc E, Sakai Y, Fujii T. Biomed Microdev 2003;5:109–114.
71. Kojima K, Moriguchi H, Hattori A, Kaneko T, Yasuda K. Lab Chip 2003;3:292–296.
72. Hung PJ, Lee PJ, Sabounchi P, Lin R, Lee LP. Biotechnol Bioeng 2005;89:1–8.
73. Hung PJ, Lee PJ, Sabounchi P, Aghdam N, Lin R, Lee LP. Lab Chip 2005;5:44–48.
74. Lee PJ, Hung PJ, Rao VM, Lee LP. Biotechnol Bioeng 2006;94:5–14.
75. Kane BJ, Zinner MJ, Yarmush ML, Toner M. Anal Chem 2006;78:4291–4298.
76. Chueh B, Huh D, Kyrtsos CR, Houssin T, Futai N, Takayama S. Anal Chem 2007;79(9):3504–3508.
77. Liu VA, Bhatia SN. Biomed Microdev 2002;4:257–266.
78. Tan W, Desai TA. Tissue Eng 2003;9:255–267.
79. Tan W, Desai TA. Biomaterials 2004;25:1355–1364.
80. Forry SP, Reyes DR, Gaitan M, Locascio LE. Langmuir 2006;22:5770–5775.
81. Mehta G, Kiel MJ, Lee JW, Kotov N, Linderman JJ, Takayama S. Adv funct Mater. Published online: 28 August 2007. DOI: 10.1002/adfm.200700016.
82. Huh D, Kamotani Y, Grotberg JB, Takayama S. Proc Micro Total Anal Syst, 2(297), Royal Society of Chemistry Special Publications, Cambridge, UK; 2004:282–284.
83. Huh D, Fujioka H, Tung Y–C, Futai N, Paine R, Grotberg JB, Takayama S. Proc Natl Acad Sci U S A forthcoming.
84. Bock TA. Stem Cells 1997;15:185–195.
85. Futai N, Gu W, Song JW, Takayama S. Lab Chip 2006;6:149–154.
86. Sato M, Ohashi T. Biorheology 2005;42:421–441.
87. Ando J, Kamiya A. Front Med Biol Eng 1993;5:245–264.
88. Wang N, Miao H, Li YS, Zhang P, Haga JH, Hu Y, Young A, Yuan S, Nguyen P, Wu CC, Chien S. Biochem Biophys Res Commun 2006;341:1244–1251.

89. Huh D, Tkaczyk AH, Bahng JH, Chang Y, Wei HH, Grotberg JB, Kim CJ, Kurabayashi K, Takayama S. J Am Chem Soc 2003;125:14678–14679.
90. Song JW, Gu W, Futai N, Warner KA, Nor JE, Takayama S. Anal Chem 2005;77:3993–3999.
91. Schneider TW, Schessler HM, Shaffer KM, Dumm JM, Yonce LA. Biomed Microdev 2001;3:315–322.
92. Gray BL, Lieu DK, Collins SD, Smith RL, Barakat AI. Biomed Microdev 2002;4:9–16.
93. Leclerc E, David B, Griscom L, Lepioufle B, Fujii T, Layrolle P, Legallaisa C. Biomaterials 2006;27:586–595.
94. Yamamoto K, Sokabe T, Watabe T, Miyazono K, Yamashita JK, Obi S, Ohura N, Matsushita A, Kamiya A, Ando J. Am J Physiol Heart Circ Physiol 2005;288: H1915–H1920
95. Price AK, Martin RS, Spence DM. J Chromatogr A 2006;1111:220–227.
96. Lu H, Koo LY, Wang WM, Lauffenburger DA, Griffith LG, Jensen KF. Anal Chem 2004;76:5257–5264.
97. Power MJ, Domansky K, Kaazempur-Mofrad MR, Kalezi A, Capitano A, Upadhyaya A, Kurzawski P, Wack KE, Stolz DB, Kamm R, Griffith LG. Biotechnol Bioeng 2002;78:257–269.
98. Weinberg EJ, Borenstein JT, Kaazempur-Mofrad MR, Orrick B, Vacanti JP. Materials Reserach Society Symposium Proceedings, Vol. 823, Warrendale, PA: MRS Press; 2004, 157–162.
99. Power MJ, Janigian DM, Wack KE, Baker CS, Stolz DB, Griffith LG. Tissue Eng 2002;8:499–513.
100. Hove JR. Pediatr Res 2006;60:6–13.
101. Zeng Y, Lee TS, Yu P, Roy P, Low HT. J Biomech Eng 2006;128:185–193.
102. Shelby JP, White J, Ganesan K, Rathod PK, Chiu DT. Proc Natl Acad Sci U S A 2003;100:14618–14622.
103. Chao PG, Tang Z, Anqelini E, West AC, Costa KD, Hung CT. J Biomech 2005;38:1273–1281.
104. Soen Y, Mori A, Palmer TD, Brown PO. Mol Syst Biol 2006;2:37.
105. Desbaillets I, Ziegler U, Groscurth P, Gassmann M. Exp Physiol 2000;85:645–651.
106. Choi D, Lee H–J, Jee S, Jin S, Koo SK, Paik SS, Jung SC, Hwang S–Y, Lee KS, Oh B. Stem Cells 2005;23:817–827.
107. Khademhosseini A, Yeh J, Eng G, Karp J, Kaji H, Borenstein J, Farokhzad OC, Langer R. Lab Chip 2005;5:1380–1386.
108. Mohr JC, de Pablo JJ, Palecek SP. Biomaterials 2006;27:6032–6042.
109. Torisawa Y, Shiku H, Yasukawa T, Nishizawa M, Matsue T. Biomaterials 2005;26:2165–2172.
110. Park TH, Shuler ML. Biotechnol Prog 2003;19:243–253.
111. Clayton J. Nat Methods 2005;2:621–627.
112. Weigl BH, Bardell RL. Lab Med 2004;35:233–237.
113. Rossier J, Girault HH. Lab Chip 2001;1:153–157.
114. Roberts MA, Durst RA. Anal Chem 1995;67:482–491.

115. Diamandis EP, Yu H, Melegos DN. Clin Chem 1996;42:853–857.
116. Hatch A, Kamholz AE, Hawkins KR, Munson MS, Schilling EA, Weigl BH, Yager P. Nat Biotechnol 2001;19:461–465.
117. Bernard A, Michel B, Delamarche E. Anal Chem 2001;73:8–12.
118. Schweitzer B, Wiltshire S, Lambert J, O'Malley S, Kukanskis K, Zhu ZR, Kingsmore SF, Lizardi PM, Ward DC. Proc Natl Acad Sci U S A 2000;97:10113–10119.
119. Schweitzer B, Roberts S, Grimwade B, Shao W, Wang M, Fu Q, Shu Q, Laroche I, Zhou Z, Tchernev VT, Christiansen J, Velleca M, Kingsmore SF. Nat Biotechnol 2002;20:359–365.
120. Carroll A, Scampavia L, Luo D, Lernmark A, Ruzicka J. Analyst 2003;128: 1157–1162.
121. Hayes MA, Polson TN, Phayre AN, Garcia AA. Anal Chem 2001;73:5896–5902.
122. Roos P, Skinner CD. Analyst 2003;128:527–531.
123. von Heeren F, Verpoorte E, Manz A, Thormann W. Anal Chem 1996;68:2044–2053.
124. Koutny LB, Schmalzing D, Taylor TA, Fuchs M. Anal Chem 1996;68:18–22.
125. German I, Kennedy RT. Anal Chem 2000;72:5365–5372.
126. Chiem N, Harrison DJ. Anal Chem 1997;69:373–378.
127. Jiang XY, Ng JM, Stroock AD, Dertinger SK, Whitesides GM. J Am Chem Soc 2003;125:5294–5295.
128. Dishinger JF, Kennedy RT. Anal Chem 2007;79:947–954.
129. Shackman JG, Dahlgren GM, Peters JL, Kennedy RT. Lab Chip 2005;5:56–63.
130. Roper MG, Shackman JG, Dahlgren GM, Kennedy RT. Anal Chem 2003;75: 4711–4717.
131. Goto M, Sato K, Murakami A, Tokeshi M, Kitamori T. Anal Chem 2005;77:2125–2131.
132. Fujii S, Uematsu M, Yabuki S, Abo M, Yoshimura E, Sato K. Anal Sci 2006;22:87–90.
133. King KM, Wang SIrimia D, Jayaraman A, Toner A, Yarmush ML. Lab Chip 2007;7: 77–85.
134. Wieder KJ, King KR, Thompson DM, Zia C, Yarmush ML, Jayaraman A. Biomed Microdev 2005;7:213–222.
135. Matsubara Y, Murakami Y, Kobayashi M, Morita Y, Tamiya E. Biosens Bioelectron 2004;19:741–747.
136. Giaever I, Keese CR. Nature 1993;366:591–592.
137. Benninger RKP, Hofmann O, McGinty J, Requejo-Isidro J, Munro I, Neil MAA, deMello AJ, French PMW. Opt Express 2005;13:6275–6285.
138. Chang-Yen DA, Gale BK. Lab Chip 2003;3:297–301.
139. Mitrovski SM, Nuzzo RG. Lab Chip 2005;5:634–645.
140. Sin A, Chin KC, Jamil MF, Kostov Y, Rao G, Shuler ML. Biotechnol Prog 2004;20:338–345.
141. Vollmer AP, Probstein RF, Gilbert R, Thorsen T. Lab Chip 2005;5:1059–1066.
142. Prokop A, Prokop Z, Schaffer D, Kozlov E, Wikswo J, Cliffel D, Baudenbacher F. Biomed Microdev 2004;6:325–339.
143. Koh WG, Pishko MV. Anal Bioanal Chem 2006;385:1389–1397.

144. Kim MS, Yeon JH, Park J–K. Biomed Microdev 2007;9:25–34.
145. Toh YC, Ng S, Khong YM, Samper V, Yu H. Assay Drug Dev Technol 2005;3:169–176.
146. Paguirigan A, Beebe DJ. Lab Chip 2006;6:407–413.
147. Butterworth Amy, Lopez Garcia Maria del Carmen, Beebe D. Proc Micro Total Anal Systems 2004;2:4–6.
148. Leclerc E, Furukawa KS, Miyata F, Sakai Y, Ushida T, Fujii T. Biomaterials 2004;25:4683–4690.
149. Cho BS, Schuster TG, Zhu X, Chang D, Smith D, Takayama S. Anal Chem 2003;75:1671–1675.
150. McClain MA, Culbertson CT, Jacobson SC, Allbritton NL, Sims CE, Ramsey JM. Anal Chem 2003;75:5646–5655.
151. Vilkner T, Janasek D, Manz A. Anal Chem 2004;76:3373–3385.
152. Lion N, Rohner TC, Dayon L, Arnaud IL, Damoc E, Youhnovski N, Wu ZY, Roussel C, Josserand J, Jensen H, Rossier JS, Przybylski M, Girault HH. Electrophoresis 2003;24:3533–3562.
153. Andersson H, van den Berg A. Sens Actuators B Chem 2003;92:315–325.
154. Sia SK, Whitesides GM. Electrophoresis 2003;24:3563–3576.
155. Verpoorte E. Electrophoresis 2002;23:677–712.
156. Helmke BP, Minerick AR. Proc Natl Acad Sci U S A 2006;103:6419–6424.
157. Stavis SM, Edel JB, Samiee KT, Craighead HG. Lab Chip 2005;5:337–343.
158. Brehm-Stecher BF, Johnson EA. Microbiol Mol Biol Rev 2004;68:538–559.
159. Becker FF, Wang XB, Huang Y, Pethig R, Vykoukal J, Gascoyne PR. Proc Natl Acad Sci U S A 1995;92:860–864.
160. Hu X, Bessette PH, Qian J, Meinhart CD, Daugherty PS, Soh HT. Proc Natl Acad Sci U S A 2005;102:15757–15761.
161. Huang Y, Joo S, Duhon M, Heller M, Wallace B, Xu X. Anal Chem 2002;74: 3362–3371.
162. Fu AY, Chou HP, Spence C, Arnold FH, Quake SR. Anal Chem 2002;74:2451–2457.
163. Chang WC, Lee LP, Liepmann D. Lab Chip 2005;5:64–73.
164. Revzin A, Sekine K, Sin A, Tompkins RG, Toner M. Lab Chip 2005;5:30–37.
165. Huh D, Gu W, Kamotani Y, Grotberg JB, Takayama S. Physiol Meas 2005;26: R73–R78
166. Futai N, Naruse K, Smith GD, Takayama S. Igaku no Ayumi (in Japanese)2006;218: 4523–5452

6

MICROFLUIDIC CULTURE PLATFORMS FOR STEM CELL AND NEUROSCIENCE RESEARCH

Bong Geun Chung, Behrad Vahidi, Jeong Won Park, Jia Sheng Hu, Joseph Harris, Edwin S. Monuki and Noo Li Jeon

6.1 INTRODUCTION

Stem cells have the unique ability of self-renewal as well as being able to give rise to different cell types (1). Embryonic stem cells (ESCs) are derived from the inner cell mass of the blastocyst and are capable of generating all differentiated cell types in the body (pluripotent stem cells). ESCs in turn can generate organ- or tissue-specific stem cells (multipotent stem cells) (2). They have the potential to develop into mature cells that have specialized shapes and functions such as nerve cells (neurons), blood cells (lymphocytes), and muscle cells (myocytes) (3). Because stem cells can differentiate into diverse cell types, they hold tremendous promise for fundamental biological studies and cell-based therapies in human disorders (4–6). The ability to control differentiation of stem cells can generate a renewable source of cells for regenerative medicine and be utilized in a wide array of applications including use as models of human disease.

Neurons are among the most polarized animal cell types. The complex nature in which neurons connect to each other in the brain gives us unique cognitive abilities. Highly polarized neurons generally transmit and receive information via axons and dendrites. Neurons are connected and communicated through electrochemical junctions between axons and dendrites, known as synapses. Losses of synaptic connectivity and deficits in axonal transport have been linked to neurodegenerative

Biological Applications of Microfluidics, Edited by Frank A. Gomez

diseases such as Alzheimer's disease, Parkinson's disease, and multiple sclerosis. Therefore, advance in stem cell and neuroscience research will increase our understanding of the central nervous system (CNS) and may allow us to develop treatments for neurodegenerative diseases.

Although widely used, there are some limitations for stem cell and neuroscience research using conventional culture system. First, a major problem is the inability to precisely control cellular behaviors (e.g., stem cell differentiation, purity, and axonal isolation from soma and dendrites) in culture. For example, precise and thorough combinatorial studies with multiple growth factors are difficult using traditional cultures (5,6). In addition, therapeutic strategies for the axonal regeneration in CNS injury are limited due to difficulty in isolating axons. The *in vitro* experimental manipulation of synaptic environments is also difficult. Second, traditional cell culture methods lack the ability to generate stable single or complex gradients over time and require large volumes of medium and cells.

To overcome these limitations imposed by conventional culture methods, microfluidic culture platforms have been developed for cell biology study (7–10). Microfluidic culture platforms are capable of precisely controlled fluid flow (7), generation of dynamically controlled temporal and spatial gradients (11), and generation of compartmental cultures (12). Extracellular microenvironments play an important role in regulating cellular behaviors such as proliferation, differentiation, axon guidance, and synaptogenesis (13,14). Microfluidic culture platforms can manipulate extracellular microenvironments such as cell–cell, cell–extracellular matrix, and cell–soluble factor interactions (15). This chapter reviews recent developments in microfluidic culture platforms for stem cell and neuroscience research. The first part of the chapter describes applications for stem cell research such as cell growth and differentiation. The second part of the chapter reviews applications for neuroscience research including axon guidance, synaptogenesis, and patch-clamp applications.

6.2 APPLICATIONS FOR STEM CELL RESEARCH

Stem cells hold tremendous promise for fundamental biological studies and cell-based therapies in human disorders. Microfluidic culture platforms have some advantages over conventional culture methods. They can precisely manipulate fluid flow, generate stable gradients over times, and require small volume of medium. Simple yet highly adaptable microfluidic culture platforms will be needed because stem cells are difficult to culture in microfluidic devices. Here, we provide a detailed review of growth and differentiation of stem cells in microfluidic devices.

6.2.1 Microfluidic Devices for Culturing Stem Cells

Extracellular microenvironments (e.g., soluble gradients, extracellular matrix, and shear stress) play a significant role in controlling cell growth and survival. Proliferation may be further qualified as symmetric or asymmetric cell division.

The asymmetric cell division can give rise to one daughter cell for differentiation while the other remains a stem cell. In general, dissociated cells from neuronal tissues are cultured with mitogens such as epidermal growth factor (EGF) and fibroblast growth factor (FGF) (16). Proliferation of stem cells in culture is controlled by the addition of FGF and EGF. The length of cell cycle is another parameter for manipulating stem cell growth. In some cases, proliferation can be blocked by using mitotic inhibitors. Cells exposed to mitotic inhibitors can be differentiated into other cell phenotypes.

To precisely regulate and optimize cellular behaviors of stem cells, microfluidic culture platforms have been recently developed. Microfluidic culture platform can precisely manipulate fluid flows and generate stable gradients in well-defined microenvironments. Chung et al. developed the gradient-generating microfluidic device for studying proliferation and differentiation of human neural stem cells (NSCs) (17). Figure 6.1a shows schematic of the microfluidic culture device that consists of control and gradient chambers. Gradient profile of growth factor mixtures was confirmed by fluorescence imaging from fluorescein isothiocyanate (FITC)-dextran as shown in Fig. 6.1b. Human NSCs exposed to gradients of growth factor mixtures were cultured for 7 days *in vitro*. Human NSCs proliferated and differentiated into astrocytes in a graded and proportional fashion that varied

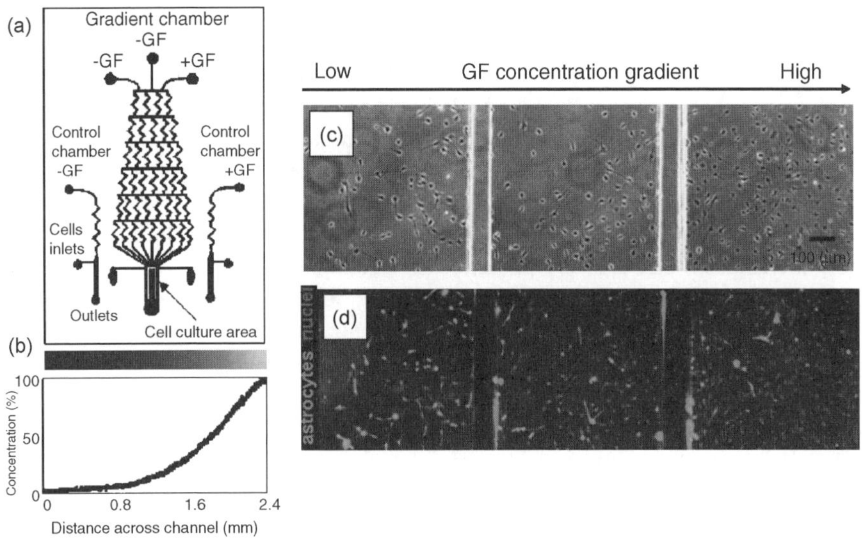

FIGURE 6.1 A gradient-generating microfluidic device for proliferation and differentiation of human neural stem cells. (**a**) Schematic of the microfluidic culture device consisted of control and gradient chambers. (**b**) Gradient profile is confirmed by fluorescence imaging from fluorescein isothiocyanate-dextran. (**c** and **d**) Proliferation and differentiation of human NSCs exposed to gradients of the growth factor mixtures in the microfluidic device. (**c**) Phase-contrast images of cell growth and (**d**) fluorescence images of the human NSCs stained by an astrocytes marker (green) and cell nuclei (blue). (Reference 17 — reproduced by permission of The Royal Society of Chemistry.) (See color insert)

directly with concentration of growth factors in Fig. 6.1c and d. Kim et al. described the proliferation of mouse ESCs in the logarithmical perfusing microfluidic device (18). This device consisted of four interconnected microfluidic layers: mixer, delivery layer and via, chamber layer. The logarithmic concentration gradients were generated across the microfluidic device. The relationship between mean colony area of ESC and flow rate revealed that the colonies exposed to higher flow rates ($>10^{-2}$ μL/min) showed healthy round shape, while colonies at the lowest flow rate formed small average colony areas. Futai et al. also reported the use of the microfluidic recirculation culture chamber for studying the proliferation of C2C12 myoblasts and MC3T3-E1 osteoblasts for 2 weeks (19). This recirculation culture system was integrated with 48-pin Braille display modules for fluidic actuations in addition to a transparent heater. They demonstrated the proliferation at the different cell densities as well as myotube formation in the microfluidic recirculation system. At high density, C2C12 myoblasts and MC3T3-E1 osteoblasts exposed to growth medium reached confluence by day 2. Therefore, concentration gradients of growth factors, flow rates, and cell density strongly affected stem cell growth in the microfluidic perfusion device.

Human ESCs have the potential ability to proliferate and differentiated into different embryonic cell lineages. There is a need to maintain undifferentiated human ESCs as well as the ability to differentiate into other cell types for the cell replacement therapy. The proliferation and differentiation of human ESCs can be controlled by cell–cell contact, cell–extracellular matrix interaction, and cell–soluble factor interaction. Recently, cell–cell interaction studies have been performed using surface patterning techniques and coculture systems. Cell growth and adhesion could be precisely controlled by high throughput microarray platforms without the continuous perfusion of medium. Khademhosseini et al. described the patterning cellular cocultures of ESCs and NIH-3T3 fibroblast on fibronectin circular islands using layer-by-layer deposition of hyaluronic acid and poly-L-lysine (20). ESCs were patterned in specific islands, while a monolayer of fibroblasts was cultured. In general, human ESCs were cultured on top of the feeder layer of mitotically inactivated murine embryonic fibroblasts (MEFs) as well as under feeder-free environments. For feeder-free conditions, human ESCs were cultured with conditioned medium obtained from MEFs that were cultured on the matrigel-coated substrates (21). To study human ESCs on a feeder layer in the microwell-patterned assay, Khademhosseini et al. cocultured human ESCs on top of a feeder layer of MEFs to maintain undifferentiated human ESCs (22). The microwell array can control the cluster size and uniformity of cell clusters. MEF monolayer cells were seeded on surfaces treated with fibronection and then human ESCs were loaded inside the microwells. Humane ESCs were stained by red Vybrant DiD cell-labeling solution and MEFs were stained by green membrane dye carboxyfluorescein diacetate succinimidyl ester (CFSE). The aggregate sizes and numbers of human ESCs in the microwells were quantified. Like the previous microwell array system, Mohr et al. reported a 3D microwell-based method to maintain undifferentiated human ESC cultures and spatially localized cells for long periods of time (23). The polyurethane microwells were coated with gold to allow a triethylene glycol-terminated alkanethiol self-assembled monolayers (SAMs) to form an Au to defer cell attachment. Cells were only localized and cultured in the matrigel-coated wells because the bottoms of the wells were shielded from gold. Human ESCs on the microwell array were identified by the transcription factor Oct-4. Colony sizes were controlled and the pluripotency of the human ESCs was maintained. Orner et al. also developed a method to pattern SAMs for cell adhesion and signaling studies of human ESCs (24). The pluripotent human ESCs were seeded on the microarrays to study proliferation and cell adhesion. The microarray was patterned with different percentages of poly(ethylene glycol) (PEG) acid-alkane thiol (AT) and polyol-AT on the gold squares surrounded by a fluoro-AT SAM. Human ESCs were not differentiated for 2 days, and this method could be useful for understanding and controlling cell adhesion and proliferation.

To dissociate human ESCs using a microfludic device, Park et al. developed the novel ESC divider with square and hexagonal patterns (25). Human ESC colonies were dissociated into regular ESC clump sizes without any enzymatic treatment. Undifferentiated human ESCs were maintained for several passages in the microfluidic device. In addition, Zeringue et al. described the embryo development in the microfluidic device (26). Cumulus removal at the zygote stage requires most mammalian *in vitro* production. To remove cumulus in the traditional method, vortexing and enzymatic treatment have been used. The microfluidic device not only allowed cumulus removal without any damages but also significantly increased early embryo developmental rates.

6.2.2 Differentiation of Stem Cells in the Microfluidic Device

Human ESCs are the potential source of differentiated cells and they are generally differentiated from embryoid bodies (15,23). Specifically, NSCs can be differentiated into the three different cell types in the CNS system—namely, neurons, astrocytes, and oligodendrocytes. Like proliferation processes, extracellular microenvironments play an important role in regulating cell fate and function (15). Growth factors have been used extensively for stem cell differentiation. Cells are either induced to differentiate by withdrawing mitogens (e.g., EGF, FGF) or by exposing the cells to another factor that induces some of the cells to develop into different lineages (16). For example, NSCs cultured using medium with growth factor mixtures can proliferate while NSCs cultured using medium without growth factor mixtures can differentiate into other cell lineages (5,6,17). Differentiation may be brought about by inhibition of the cell cycle or by the induction of differentiation without cell cycle inhibition (27). However, precise and homogeneous control of NSC differentiation is difficult. It is one of the critical challenges in making stem cell-based regenerative research.

A microfluidic culture platform can be used for investigating proliferation as well as differentiation. Tourovskaia et al. described a microfluidic perfusion system for a long-term (2 weeks) culture of murine skeletal muscle cells (28). This system manipulated accurate perfusion rates and controlled differentiation from myoblast to myotubes. The device was fabricated by two-layer photolithography: the first layer had the three converging microchannels and 16 parallel long-term perfusion channels; the second layer defined the main channel. Cells were selectively patterned on the substrate by oxygen plasma etching technique. Micropatterned cells under long-term continuous perfusion differentiated into myotubes. Lee et al. described the differentiation of mouse P19 embryonic carcinoma stem cells injected into the Zebrafish egg chorion in a simple microfluidic perfusion device (29). The chamber consisted of four polymer layers: cover, microwell culture, microhole, and medium perfusion layer. Embryonic carcinoma stem cells were injected into the empty chorion on the microwell layer. They demonstrated neural cells in the outer layer of the cell mass and beating cardiomayocytes inside the cell mass on day 15.

Differentiation was precisely controlled by high throughput microarray platforms without the continuous perfusion of medium. Chin et al. reported cellular behaviors of adult hippocampal progenitor cells infected with a retrovirus to express green fluorescent protein (GFP) (30). Figure 6.2a shows a low magnification image of a microarray chip. The platform was made by two glass coverslips sandwiching the silicone rubber gasket. The bottom coverslip was the microwell culture array treated with polyornithine and laminin. Cells were injected through the silicone gasket using a syringe and were tracked with time-lapse microscopy to follow their proliferation and differentiation. To study the neuronal differentiation from NSCs, cell culture medium was changed from with FGF2 to with retinoic acid and forskolin. Figure 6.2b demonstrates that GFP positive cells were differentiated along the neuronal lineage in the microarray platform. These cells were stained by Tuj1 and glial fibrillary acidic protein (GFAP). Bani-Yaghoub et al. developed poly (dimethylsiloxane) (PDMS)-based micropatterned neurochips for studying neurogenesis and synthetic neural communication networks (31). The micropatterned neurochip precisely controlled cell position, and regulated proliferation and differentiation. N2a neuroblast differentiated into neurons within a week in differentiation medium. F-actin immunostaining analysis demonstrated that N2a cells extended neurites along the edge of the channel. Neurons in the neurochip developed parallel neurites along the length of the microchannels. It was similar with well-structured neuronal alignment in the cortex. McBeath et al. described that the differentiation of human mesenchymal stem cells were regulated on the fibronectin-patterned substrate (32). Cell shapes on the patterned substrates controlled the adipocyte and osteoblast fates from human mesenchymal stem cells. At high cell density, cell adhesion decreased, but cell–cell contact and paracrine signaling increased. Adipogenesis of human mesenchymal stem cells appeared on small island patterns (1024 μm^2), while osteogenesis occurred on large size patterns (10,000 μm^2). Anderson et al. developed a high throughput polymer microarray chip integrated with multiple combinations of biomaterials for studying proliferation and differentiation of human ESCs (13). This chip consisted of 576 different combinations of acrylate monomers and had the potential to understand cell–biomaterial interactions and biomaterial-based therapies. The total number of cells decreased in the presence of retinoic acid while the percentage of cytokeratin-positive cells increased. Similar with previous cell-compatible polymer chip

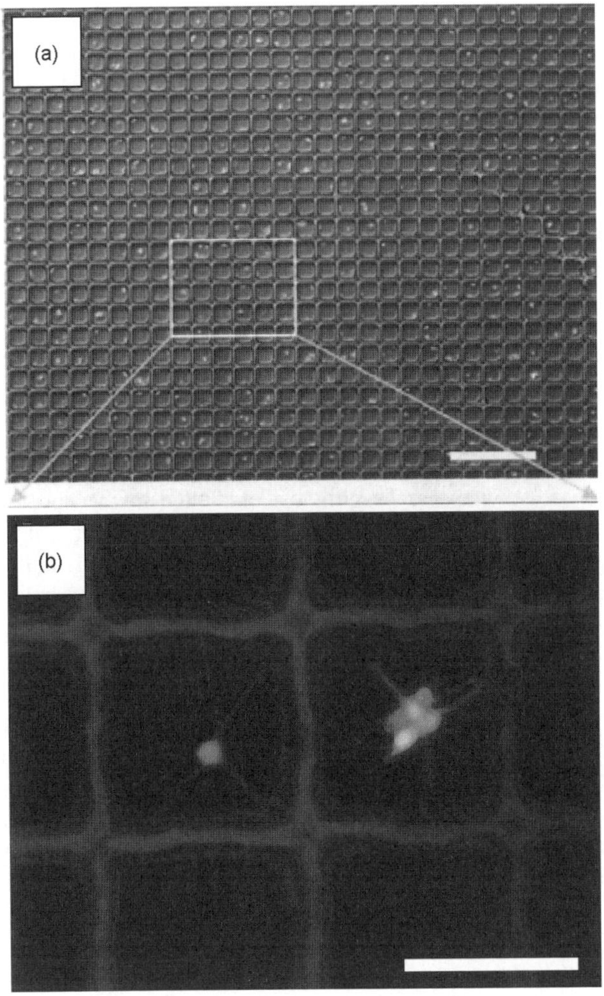

FIGURE 6.2 Differentiation of adult NSCs in the microfabricated array. (**a**) A low magnification image of a microarray chip. Scale bar is 500 μm. (**b**) GFP-positive NSCs were differentiated into neuron and were stained by neuronal marker Tuj1 (red) and glial fibrillary acidic protein (GFAP, blue). Scale bar is 100 μm. (Reprinted with permission from Chin et al., copyright 2004 Wiley-InterScience.) (See color insert)

developed by Anderson et al., Flaim et al. developed an extracellular matrix microarray for mouse ESC differentiation into hepatic fate (15). This platform consisted of 32 different combinations of five extracellular matrix molecules such as collagen I, III, IV, laminin, and fibronectin. Human ESCs exposed to retinoic acid induced differentiation and reduced proliferation. Warren et al. developed the digital RT-PCR microfluidic chip to study the transcription factor profiling of hematopoietic progenitors (33). The digital PCR chip performing 14,400 PCRs at once was thermocycled and pneumatically partitioned into 1200 isolated chambers. The transcription factor profiling in developmental hamatopoietic progenitors was quantitatively analyzed in a digital RT-PCR chip.

6.3 APPLICATIONS FOR NEUROSCIENCE RESEARCH

6.3.1 Axon Guidance in the Microfluidic Device

Neurons are among the most polarized animal cell types. Their unique elongated structure is typically composed of three functionally different compartments: the cell body or soma, containing the nucleus and the cytoplasmic machinery for biosynthesis of proteins; dendrites, highly branched process that receive signals from other neurons; and a single axon, that carries nerve signals away from the soma. The neurons are connected and communicated through electrochemical junctions, known as synapses. The biggest challenge in the development of the nervous system is the question of how axons extend over considerable long distances and through varying extracellular microenvironments, find their synaptic target cells, and form these precise synaptic connections. The signaling cues that guide the axons to their synaptic targets, critical in neuronal development, have been categorized into two different groups based on the effective distance at which they act. Short-range cues, such as extracellular matrix, and cell adhesion molecule contact (e.g., CAMs, and NCAMs) dependently affect the behavior of the elongating axon, while the long-range cues, mediated by gradients of soluble signaling molecules, induce their effect at a distance from the source. More specifically, at the growing tip of an axon, there is a dynamic, actin-supported enlargement known as the growth cone. The growth cone, originally observed by the Spanish histologist Ramón Cajal in the late nineteenth century, is the engine and steering apparatus that guides the axons to their proper targets, by sensing a variety of attractive and repulsive cues.

It was only after the discovery and isolation of these signaling cues in the past two decades that scientists have been able to conduct axon guidance studies *in vitro*. The first demonstration that the growth cones can be attracted by a diffusible signaling cue was obtained in late 1990s, by Gundersen and Barret (34). They observed that nerve growth factor (NGF), which could mediate neuronal survival, showed chemoattractive properties on dorsal root ganglion (DRG) neurons *in vitro* (34,35). It was shown that growth cones responded and reoriented toward the source of NGF ejected from a tip of a glass pipette. Many other studies demonstrated a variety of other signaling molecules and their role in guiding growth cones *in vitro* (35,36).

With the emergence of fields such as microfluidics, a multidisciplinary field that combined microfabrication technologies with chemistry and biology, in the last few decades the study of growth cones and axon guidance has taken a new turn. There were many methods for modeling *in vivo* environments in the lab, including the use of microfluidic devices (37). The reusable molds were created using soft lithography techniques. The potential microfluidic devices were nearly limitless, given the indefinite number of shapes and design possibilities (38).

Taylor et al. have developed a microfluidic-based culture system that can be used to precisely control the fluidic environment around single cells (12,25). The device, which truly exemplifies the evolution of traditional "Campenot chambers" through microfabrication technology, allows selective manipulation of neuronal micro-environments. Figure 6.3a illustrates the schematic of the microfluidic device, which is composed of two fluidically isolated compartments separated by a physical barrier. Microfabricated barriers with embedded microgrooves are used to guide the growth of axons into a fluidically isolated microenvironment as shown in Fig. 6.3b. Neurons were immunostained by Tau (green) and microtubule-associated protein 5 (MAP5, red). Utilizing this device, Taylor and colleagues have successfully demonstrated long-term culture of CNS neurons without the use neurotrophins and other targeting molecules. In addition to compatibility with live-cell imaging techniques, these chambers have several advantages over other culture methods including the ability to (a) isolate CNS axons without somas or dendrites and (b) apply localized physical and chemical treatments to CNS axons.

The study of how axons choose their growth paths was in its infancy. Although with the latest surge in microfluidics and microfabrication technologies, several

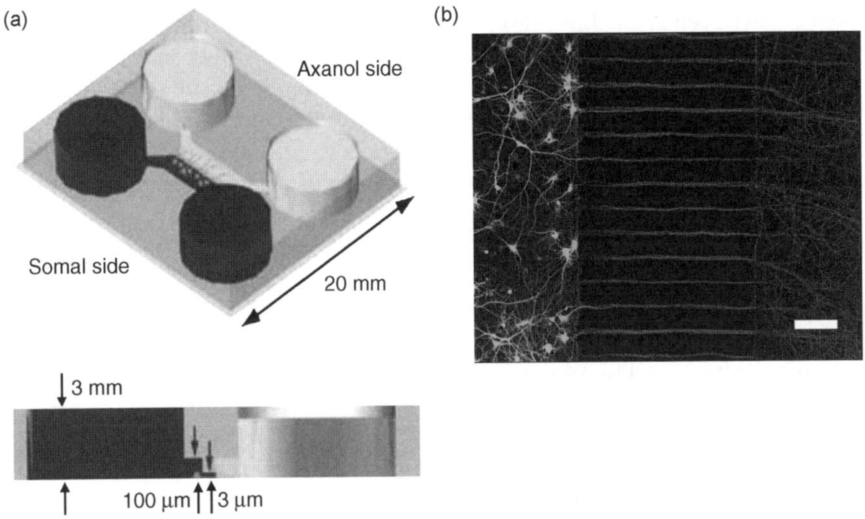

FIGURE 6.3 A microfluidic culture platform for axonal regeneration and transport. (**a**) A microfluidic-based culture platform consisted of somal and axonal compartment connected by microgroove channels. A volume difference between the somal and axonal compartment allowed axonal isolation through microgrooves. The axonal growth was guided into the axonal compartment. (Reprinted by permission from Macmillan Publishers Ltd, Reference 12, copyright 2005.) (**b**) Confocal micrograph of neurons stained by Tau (green) and MAP5 (red). Axons were isolated and guided through microgroove channels. Scale bar is 100 µm. (Reprinted by permission from Macmillan Publishers Ltd, Reference 25, copyright 2006.) (See color insert)

years of research has been devoted to the 2D modeling of axon growth, there were still many unknowns (37). However, it has generally been proven that microfluidic devices were just as accurate as larger scale (>1 mL) *in vitro* experiments (38). Experiments have been carried out using only one or two independent variables in order to break down the several factors known to axonal growth. Utilizing microfabrication technologies, standard cell culture plates have been transformed to reusable highly integrated platforms suitable for conducting a variety of neuroscience experiments. Wittig et al. have reported the design of a chamber allowing the migrating cells or growing cell processes to freely choose between two chemically distinct environments. In a typical experiment, the explants were placed in the designated tissue culture wells as close as possible to the microchannel opening. Following the attachment of the explant to the petri dish surface, appropriate growth medium were fed through the microfluidic networks, presenting the extending process an opportunity to choose between two chemically distinct environments. Utilizing this design, that exemplified the fusion of standard cell culture environments with microfluidics, neurons from spiral ganglion explants preferentially grew in medium containing neurotrophin-3 (NT-3) as opposed to media without NT-3 (38).

Since gradients, as opposed to sharp boundaries, allowed for more accurate representations of *in vivo* environments, several gradient-generating microfluidic devices have been created (37–40). Gradients could be established in these devices using calculated flow rates based on channel size, fluid density, channel design, and bonding characteristics (37). The successful creation of a known microfluidic gradient has only recently been proven, and the establishment has led to creation of substrate-bound gradients based on laminar flow of fluids in microchannels. Dertinger and colleagues reported the development of a generally applicable technique for the fabrication of substrate-bound gradients in Fig. 6.4a. By generating gradients of naturally occurring extracellular protein (e.g., laminin) adsorbed on the surface, they have created an *in vitro* platform suitable for investigating the role of immobilized gradients in neuronal development and axonal path finding. Gradients were created using a network of several mixing channels that branched out in a factorial arrangement. As fluidic streams were allowed to divide and mix by diffusion in a network of serpentine channels, concentration gradients were created. Neurons were then plated on the floor of the channel in the region where the gradient had formed. By generating surface-bound gradients that ranged from pure laminin to pure BSA, they successfully demonstrated that axon specification of rat hippocampal neurons was oriented in the direction of increasing concentration of laminin. Figure 6.4b illustrates the potentiality of this technique for the screening of molecules that are involved in cell differentiation, cell migration, and axon guidance.

Although *in vitro* models have been established as experimentally convenient methods for the investigation of neuronal responses to a variety of guidance cues, quantification methods that require manual tracing and measuring the axonal growth are becoming more obsolete. The measurement and quantification of axonal growth, developed by Poo and coworkers (39), was normally carried out by

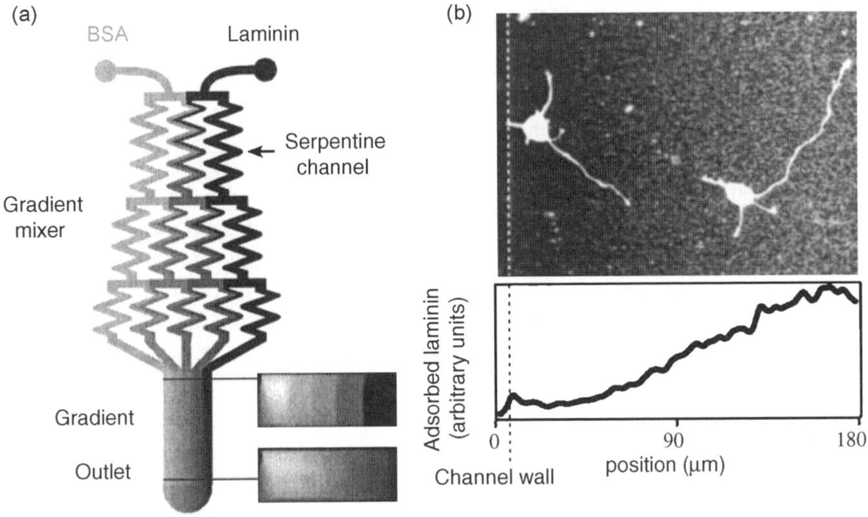

FIGURE 6.4 A microfluidic device for fabrication of immobilized gradients. (**a**) Schematic of the design of a typical microfluidic network in PDMS for fabrication of immobilized gradients. Solutions of laminin and BSA were injected into the inlets. As the fluid streams were allowed to divide and mix by diffusion in a network of serpentine channels, several concentrations were created. When reunited at the bottom, a very specific and known gradient was established and adsorbed on the floor of the channel. Neurons were then plated on the floor of the channel in the region where the gradient had formed. (**b**) A fluorescence micrograph of neurons that were fixed after 24 h in culture and immunostained for laminin and neurons with antilaminin and antitubulin antibodies, respectively. The shape of the immobilized gradient in laminin was shown in the graph below the micrograph. (Reprinted with permission from Dertinger et al., copyright 2002 National Academy of Sciences, USA.)

manually tracing and measuring the axonal growth on phase-contrast or fluorescence micrographs. However, the method had impractical limitations for statistically significant experiments, where a large number of processes need to be analyzed. Growth cones were first identified via hand, then the advanced software program monitored the axon for growth, following changes and calculating the length and area of the axon. The output was then reviewed and could be subjected to several changes, including thresholds for movement. Automated identification had the potential to vastly reduce experiment times and allow for more axons to be measured given the speed and efficiency of computers over manual analysis (40). Future experiments in axon guidance show several new variations, such as the formation of 3D substrates. Several layers of PDMS are carefully setup on top of one another to create a three-dimensional scaffold, where neurons are loaded. Axons can be challenged by several factors in all dimensions. This setup most resembles *in vivo* environments by allowing cells to grow in any dimension, just like in real tissue (41).

6.3.2 Synapse Microfluidic Device

A synapse is a specialized junction between two electrically active cells that allows for electrochemical communication. A synapse is typically 10–20 nm wide, and is typically composed of an axon that releases chemical signals and dendrites from the postsynaptic cells that receive the signals. The tip of the axon is equipped with voltage-gated ion channels that release neurotransmitters into the synaptic cleft. Neurotransmitter receptors on the postsynaptic terminals pick up the signals based on their bonding characteristics, allowing the signal to propagate.

A device that can release chemical compounds in small, controlled volumes at several, well-defined areas would prove itself to be an invaluable tool in studies on neurons and neurotransmitters released in synaptic junctions. The ability to create and control precise chemical gradients can further enhance the study of cell behavior at all levels, especially in the field of neurons (42). Current research includes several methods involving microfluidic devices. One such example developed by Peterman et al. is a device that uses electroosmosis as the driving force that drives fluid through a network of microfluidic channels and allows repeatable delivery of chemical compounds through multiple apertures (5 μm in diameter) (42). The device, built for application in retinal prosthesis, is labeled the "artificial synapse chip" as it mimics synapses with timed chemical releases.

The apertures were operated by applying a sinusoidal voltage range of −2.5 to +2.5 V. Electrodes placed next to the microfluidic channel received this input and initiated the electroosmotic flow, where the bulk flow of fluids resulted from the electrolytes in the solution responding to the applied electric field (42). Using the artificial synapse chip, it was successfully demonstrated that PC12 cells were stimulated by spatially and temporally controlled release of bradykinin, a physiologically active peptide. Although the objective for the chip was for application in retinal prosthesis, the technology could be effectively translated to studies on neurons and neurotransmitter release in synaptic junctions. This mechanism showed promise in the research field due to its lack of moving parts and relatively low power requirements (on the order of 20 μW) (42).

Synapses were highly specialized junctions through which neurons communicated to one another and to nonneuronal cells such as muscles. The synapse or junction between the axon terminal and the highly excitable region of the muscle cell plasma membrane was termed the neuromuscular junction. The signal passed through this junction via the release of neurotransmitter acetylcholine and was responsible for initiation of action potentials across the muscle's surface, ultimately causing the muscle to contract. The neuromuscular synaptogenesis was initiated by the exchange of a variety of spatially localized signals between the nerve terminal and muscle. The release of proteoglycan agrin from the axonal tip, for example, mediated the focal accumulation of acetocholine receptors on the postsynaptic membrane of the muscle fibers (43). Since neuromuscular synapse formation played a critical role in the correct development of the organism, the *in vivo* experimental manipulation of the synaptic environment has been highly hindered. This selective and focal administration of such compounds was difficult to replicate *in vitro*, limiting the majority of studies to the application of agrin baths, where acetylcholine receptors appeared randomly across the cell body. However, current research shows the promise. One device attempted to create an environment similar to the *in vivo* synaptic microenvironment, where focused and selective exposure of agrin was possible (43). They used a microfluidic device to set up a laminar flow of agrin over a precise location onto muscle cells, myotubes. Figure 6.5a illustrates the schematic of the device and the procedure, where the agrin flow is concentrated into a 100–150-μm wide stream in the center of the device. The myotubes were spatially organized on a microengineered substrate as to be parallel and equally spaced relative to each other. Following the attachment of myoblasts to microtracks of adhesive proteins, agrin was applied perpendicular across a small part of the

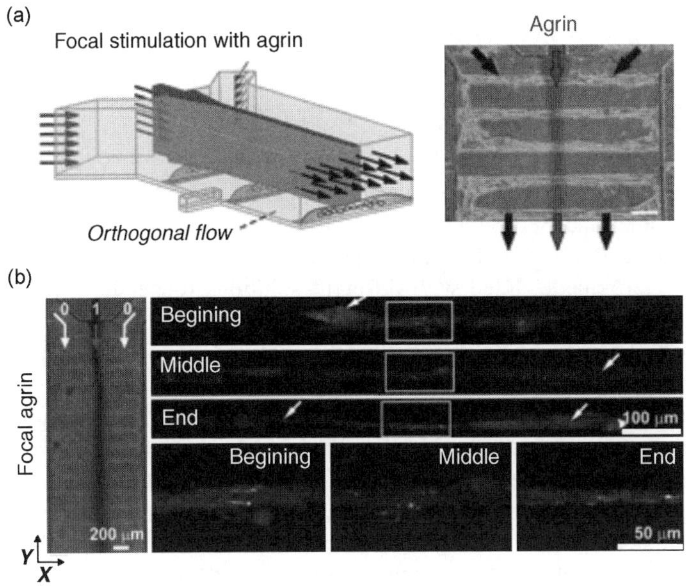

FIGURE 6.5 Local induction of acetylcholin receptor clustering in myotubes cultures using microfluidic application of agrin. (**a**) Three-dimensional schematic of the experimental design (*left*) and the corresponding phase-contrast image of the micropatterned myotubes inside the device (*right*). Cells were cultured on 30–40 μm tracks of adhesive proteins (polylysine/matrigel). The orthogonal flow was then started to stimulate the cells with a heterogeneous stream containing agrin at the center (*red arrow*) and no agrin on either side (*black arrows*). A red dye was added to the agrin solution for better visualization (*right*). (**b**) Focal exposure of central regions of myotubes to agrin induced localized AChR clustering. Phase-contrast micrograph taken during perfusion in the *y* direction, where the agrin-containing stream appeared darker (*left*). Fluorescence micrographs of AChR cluster staining (fluorescently conjugated α-BTX) corresponding to different locations along the channel. White arrows pointed to the microclusters. (Reprinted with permission from Tourovskaia et al., copyright 2006 The Biophysical Society.)

myotubes, resulting in exclusive accumulation of acetylcholine receptors at the agrin-stimulated areas in Fig. 6.5b (43).

The release of these synaptogenic factors by the nerve terminal played a central role in the accumulation of neurotransmitter receptors, such as the acetocholine receptors (43). Further research by Kosar et al. have resulted in another microfluidic-based device with the intent to further specify the delivery of agrin to a single spot rather than longitudinally across the myotubes. The device incorporated nanofabricated apertures on a transparent membrane through which soluble molecules could be focally delivered to individual cells. These apertures, ranging from 2 to 8 μm in diameter, were big enough to allow agrin to flow into the cell but small enough to prevent cells from growing into the agrin solution (44). The areas that were subject to the agrin were measured for acetycholine receptor formation using fluorescence and digital imaging calculations. Utilizing this technique, the results demonstrated that the accumulation of the acetylcholine receptors lay directly above the apertures, implicating agrin as an important factor in their formation (44).

6.3.3 Patch Clamp in the Microfluidic Device

Patch-clamp recording, a major tool in electrophysiology, allows the study of ion channels in excitable cells (e.g., neurons, muscle fibers). The technique, developed by Erwin Neher and Bert Sakmann in the late 1970s and early 1980s, involves the use of a conductive fluid-filled micropipette to form a high resistance seal with a cell. The micropipette, filled with different solutions (depending on the specific variation of the technique), can be used to record from a single ion channel.

The ion channel current measurements obtained by patch-clamp recordings have proven invaluable in numerous biological studies, such as drug target identifications. Patch clamping flux into and out of the cell can be measured as well, since voltage potential is directly related to the change in ions between the two boundaries (45). However, traditional micropipetting is extremely slow, labor-intensive, and requires expensive equipments such as micromanipulators and vibration-free environments (e.g., antivibration tables) (45). Through the use of microfluidics and lab-on-a-chip technologies, high throughput electrophysiology has become more tangible than ever before.

Microfabrication technologies, for instance, have led to development of planar patch-clamp arrays of silicon, silicone nitrite, glass, polyimide, and silicone elastomer that offer a variety of advantages over the traditional micropipette handling methods. In the planar patch configuration, instead of positioning a pipette on an adherent cell, cell suspension is pipetted on a chip containing microstructured arrays of apertures. Cells are positioned by suction and the relative movements between the cell and aperture can be excluded after sealing (eliminating the need for micromanipulators and vibration-free environment). Due to overall thickness of the device, however, the planar arrays have proven to be inefficient in single-cell level studies, where simultaneous high resolution imaging is essential (45).

Using soft lithography techniques, a novel lateral patch-clamp setup has been recently introduced by Lau et al. The innovative approach has shown to be advantageous mainly due to the transparent property of the material and low cost fabrication steps. In this configuration, which highly resembles the traditional micropipette clamping techniques, the cells are trapped in lateral microchannels raised above the bottom plane of the chip. In addition to microscale lateral patch channels, the device is equipped with a macroscale open access compartment, where cells or drug samples can be easily pipetted in. Larger manipulation channels allow a better seal by bringing cells closer to the patch channels, leading to superior measurement capabilities. The benefits here include higher throughput, easy chip creation, and compatibility with high resolution microscopy (45). Upon uptake of a cell, the patch-clamp forms a seal and applies a rapid negative pressure to the syringe, effectively breaking the cell membrane. Upon rupture, a step voltage is applied and the current responses are recorded. With the cell membrane ruptured, one could easily introduce drugs and other chemicals into the cell and measure its current changes in response.

However, the macroscale polymer mold does have a few drawbacks. First, the vibrations from normal handling are transmitted directly to cells, causing difficulty in high resolution imaging. Ionescu-Zanetti et al. have expanded on the idea of patch clamping based on microfluidics and reported the fabrication of a device that took advantage of microfluidic junctions between a main chamber and lateral recording capillaries. The elastic qualities of PDMS effectively absorb all vibrations, allowing for much sharper images to be acquired (46). PDMS construction allows for wells to be placed as little as 20 μm apart and to only require as little as 0.5 nL of fluid. The microfluidic-based patch-clamp design allows for efficient seals between cells and the device. Using this device, successful whole-cell recordings of mammalian cells were demonstrated.

To analyze neuronal networks and synaptic plasticity *in vitro*, several groups have expanded on the patch-clamp recordings in brain slices where planar microelectrode arrays are used to perform the recordings between synoptically connected cells. Egert et al., for example, has developed a system in which simultaneous stimulation and recording from rat hippocampus neurons are made possible. The system used rat hippocampal slices that were brought to close contact with a planar microelectrode array (MEA) (47). Fromherz et al. have introduced a similar design that utilizes transistor array with an organotypic brain slice (48). In a typical experiment hippocampal slices were cultured for 14 days on a linear array of silicon transistor chips and the recordings of the field potentials provided an image of the current-source density, significant for a variety of neurophysiological studies. Nam et al. also developed a single-use planar MEA using a PDMS microstencil for studying brain tissue slices (49). MEA was fabricated by microfabrication processes such as photolithography and metal deposition. PDMS microstencil insulator was aligned and attached on microelectrodes. Hippocampal neurons on microelectrodes were stimulated. A single-use MEA can be useful for high throughput cell-based biosensor research.

To study single synaptic contacts, however, considerable decrease in network complexity, inherent to use of brain slices, is essential. Vogt et al. have successfully implemented a system that combines low density neuronal cultures with microcontact printing, where experimental manipulations of neuronal interconnections are possible (50). Rat embryonic cortical neurons were cultured on microcontact printed neuronal networks using PDMS stamps. Cells adhered to the points of the grid and chemical synapses were identified by patch-clamp measurement through a gap junction at different stimulation frequencies. Microcontact printed substrates held the potential for the creation of well-defined neuronal circuits.

6.4 SUMMARY AND FUTURE DIRECTIONS

Microfluidic culture platform is a versatile experimental method for studying cell replacement therapy, regenerative medicine, and neurodegenerative diseases. This platform can precisely manipulate fluid flow, make high throughput experiments in parallel, and generate stable gradients in well-defined microenvironments. Cellular behaviors in different microenvironments (e.g., cell–cell, cell–extracellular matrix, cell–soluble factors, and cell–biomaterial interactions) can be manipulated using microfluidic platforms. Although the microfluidic approach has several advantages, there are a number of limitations such as it is more complex than biological system sometimes, and it is difficult to interface with conventional tissue culture protocols. In order for microfluidics to gain broader acceptance by the biology community, there are several major challenges that need to be addressed.

First, we need implementation for stem cell and neuroscience studies because stem cells and neurons are sensitive to microenvironments (e.g., shear stress, gas, and medium perfusion) in microfluidic devices. These microfluidic devices can integrate with various extracellular microenvironments for directing cellular fates. To perform this challenge, we can generate stable concentration gradients on micropatterned substrates coated by various extracellular matrices and biomaterials. Different cell types on a single chip can be cocultured and exposed to gradients simultaneously for studying cell–cell, cell–extracellular matrix, and cell–soluble factor interactions.

Second, 3D culture systems will be developed to investigate the regulation of cellular behaviors in well-defined microenvironments. Advance technologies (e.g., nanofiber scaffold, new biomaterial) in tissue engineering for generating 3D models will be needed. For studying axon guidance in a 3D scaffold, several layers of PDMS will be set up on top of one another in a 3D scaffold, where neurons will be loaded. Axons can be challenged by several factors in all dimensions. In addition, gradients in 3D extracellular matrix can also be used to study stem cell differentiation and migration. However, most of all, close collaborations between engineers and neurobiologists are more significant. These efforts will open the door for stem cell and neuroscience research.

ACKNOWLEDGMENT

We thank Roman Reed Spinal Cord Injury Research Fund of California.

REFERENCES

1. Zandstra PW, Nagy A. Annu Rev Biomed Eng 2001;3:275–305.
2. Gage FH, Verma IM. Proc Natl Acad Sci USA 2003;100:11817–11818.
3. Kirschstein R, Skirboll LR. NIH stem cell report 2001.
4. Svendsen CN, Smith AG. Trends Neurosci 1999;22:357–364.
5. Palmer TD, Schwartz PH, Taupin P, Kaspar B, Stein SA, Gage FH. Nature 2001;411:42–43.
6. Schwartz PH, Bryant PJ, Fuja TJ, Su H, O'Dowd DK, Klassen H. J Neurosci Res 2003;20:838–851.
7. Whitesides GM, Ostuni E, Takayama S, Jiang X, Ingber DE. Annu Rev Biomed Eng 2001;3:335–373.
8. Beebe DJ, Mensing GA, Walker GM. Annu Rev Biomed Eng 2002;4:261–286.
9. Sia SK, Whitesides GM. Electrophoresis 2003;24:3563–3576.
10. Weibel DB, Garstecki P, Whitesides GM. Curr Opin Neurobiol 2005;15:560–567.
11. Lin F, Saadi W, Rhee SW, Wang SJ, Mittal S, Jeon NL. Lab Chip 2004;4:164–167.
12. Taylor A, Blurton-Jones M, Rhee SW, Cribbs DH, Cotman CW, Jeon NL. Nat Methods 2005;2:599–605.
13. Anderson D, Levenberg S, Langer R. Nat Biotechnol 2004;22:863–866.
14. Metallo CM, Mohr JC, Detzel CJ, de Pablo JJ, Van Wie BJ, Palecek SP. Biotechnol Prog 2007;23:18–23.
15. Flaim CJ, Chien S, Bhatia SN. Nat Methods 2005;2:119–125.
16. Gage FH. Science 2000;287:1433–1438.
17. Chung BG, Flanagan LA, Rhee SW, Schwartz PH, Lee AP, Monuki ES, Jeon NL. Lab Chip 2005;5:401–406.
18. Kim MS, Kim J, Han H, Cho YS, Han Y, Park J. Lab Chip 2007;7:513–515.
19. Futai N, Gu W, Song JW, Takayama S. Lab Chip 2006;6:149–154.
20. Khademhosseini A, Suh KY, Yang JM, Eng G, Yeh J, Levenberg S, Langer R. Biomaterials 2004;25:3583–3592.
21. Khademhosseini A, Langer R, Borenstein J, Vacanti JP. Proc Natl Acad Sci U S A 2006;103:2480–2487.
22. Khademhosseini A, Ferreira L, Blumling IIIJ, Yeh J, Karp JM, Fukuda J, Langer R. Biomaterials 2006;27:5968–5977.
23. Mohr JC, Pablo JJ, Palecek SP. Biomaterials 2006;27:6032–6042.
24. Orner BP, Derda R, Lewis RL, Thomson JA, Kiessling LL. J Am Chem Soc 2004;126:10808–10809.
25. Park JW, Vahidi B, Taylor AM, Rhee SW, Jeon NL. Nat Prot 2006;1:2128–2136.
26. Zeringue HC, Rutledge JJ, Beebe DJ. Lab Chip 2005;5:86–90.
27. Rajan R, Snyder E. Methods Enzymol 2006;419:23–52.
28. Tourovskaia A, Figueroa-Masot X, Folch A. Lab Chip 2005;5:14–19.
29. Lee J, Na D, Kang J, Lee S, Ju B. Biosci Biotechnol Biochem 2006;70:1325–1330.
30. Chin VI, Taupin P, Sanga S, Scheel J, Gage FH, Bhatia SN. Biotechnol Bioeng 2004;88:399–415.

31. Bani-Yaghoub M, Tremblay R, Voicu R, Mealing G, Monette R, Py C, Faid K, Sikorska M. Biotechnol Bioeng 2005;92:336–345.
32. McBeath R, Pirone DM, Nelson CM, Bhadriraju K, Chen CS. Dev Cell 2004;6: 483–495.
33. Warren L, Bryder D, Weissman IL, Quake SR. Proc Natl Acad Sci U S A 2006;103: 17807–17812.
34. Gundersen RW, Barrett JN. Science 1979;206:1079–1080.
35. Gundersen RW, Barrett JN. J Cell Biol 1980;87:546–554.
36. Tessier-Lavigne M, Placzek M, Lumsden AGS, Dodd J, Jessell TM. Nature 1988;336:775–778.
37. Dertinger SKW, Jiang X, Li Z, Murthy VN, Whitesides GM. Proc Natl Acad Sci U S A 2002;99:12542–12547.
38. Wittig JH, Ryan AF, Asbeck PM. J Neurosci Methods 2005;44:79–89.
39. Lohof AM, Quillan M, Dan Y, Poo M. J Neurosci 1992;12:1253–1261.
40. Keenan TM, Hooker A, Spilker ME, Li N, Boggy GJ, Vicini P, Folch A. J Neurosci Methods 2006;151:232–238.
41. Li N, Folch A. Exp Cell Res 2005;311:307–316.
42. Peterman MC, Noolandi J, Blumenkranz MS, Fishman HA. Proc Natl Acad Sci U S A 2004;101:9951–9954.
43. Tourovskaia A, Kosar TF, Folch A. Biophys J 2006;90:2192–2198.
44. Kosar TF, Tourovskaia A, Figueroa-Masot X, Adams ME, Folch A. Lab Chip 2006;6:632–638.
45. Lau AY, Hung PJ, Wu AR, Lee LP. Lab Chip 2006;6:1510–1515.
46. Ionescu-Zanetti C, Shaw RM, Seo J, Jan YN, Jan LY, Lee LP. Proc Natl Acad Sci U S A 2005;102:9112–9117.
47. Egert U, Schlosshauer B, Fennrich S, Nisch W, Fejtl M, Knott T, Muller T, Hammerle H. Brain Res Protoc 1998;2:229–242.
48. Besl B, Fromherz P. Eur J Neurosci 2002;15:999–1005.
49. Nam Y, Musick K, Wheeler BC. Biomed Microdev 2006;8:375–381.
50. Vogt AK, Wrobel G, Meyer W, Knoll W, Offenhausser A. Biomaterials 2005;26: 2549–2557.

PART II

ENZYMATIC AND NONENZYMATIC REACTIONS ON MICROCHIPS

7

MICROFLUIDICS FOR STUDYING ENZYME INHIBITION

Rattikan Chantiwas, Xiaoyan Yan and S. Douglass Gilman

7.1 ENZYME ASSAYS AND INHIBITION

7.1.1 Enzymes and Enzyme Inhibition

Enzymes are biomolecular catalysts — substances that increase the rate of biochemical reactions without themselves being consumed in the process (1–6). Almost all enzymes are proteins. Ribozymes, which are RNA molecules, also exhibit catalytic activity (1,2,5). Enzyme catalysis plays a vital role in life processes in all known organisms. The complicated, dynamic chemistry of a living organism is controlled primarily through enzyme catalysis. Enzymes facilitate the most basic functions of life such as the conversion of food to energy. Enzymes are attractive drug targets in modern pharmaceutical research. Enzymes with abnormal activities are involved in many diseases, and they are vital to the function of pathogens (2,6). Enzymes can still function as catalysts *in vitro* after extraction from a living organism, and they have long been utilized in food and beverage manufacturing processes. Enzymes are ingredients in detergents and are also used in the processing of textiles and leather (6,7). Synthetic chemists employ enzymes to catalyze a wide range of organic reactions (7–9). Enzymes are important tools for analytical chemists. The polymerase chain reaction and many DNA sequencing and analysis methods rely on enzymes as tools (10,11). Enzymes that cleave proteins in a sequence-specific manner (e.g., trypsin) are used extensively for proteomic studies (12,13), and enzymes are the central components of many biosensors (14,15).

Biological Applications of Microfluidics, Edited by Frank A. Gomez

Enzyme inhibitors are molecules that bind to enzymes and decrease their catalytic activity (1–6). Enzyme inhibitors play an important role in the natural regulation of enzyme catalysis, and enzyme inhibitors are found throughout nature. At the basic research level, enzyme inhibition studies can provide insight into the mechanisms of enzyme-catalyzed reactions (1,6). Enzyme inhibitors are also used in many commercial applications. They have been developed as herbicides (16,17), insecticides (18,19), and chemical warfare agents (19,20). Enzyme inhibitors are used extensively as drugs for clinical treatment of diseases by inhibiting specific enzymes involved in the disease process (1,2,6).

The critical roles that enzymes and enzyme inhibitors play in the fundamental biochemistry of life and as tools for humans to manipulate biochemistry lead to a demand for bioanalytical methods to measure enzyme catalysis and study enzyme inhibition. This chapter will focus on the role that microfluidics plays in the measurement of enzyme inhibition. We first will present a very brief overview of enzyme catalysis, enzyme inhibition, and enzyme assays and then provide a more in-depth treatment of analytical methods that take advantage of microfluidic devices to measure enzyme inhibition.

7.1.2 Enzyme Catalysis

Chemically an enzyme functions as a catalyst, increasing the rate of a reaction without altering the equilibrium between substrate (reactant) and product. In an enzyme-catalyzed reaction, a substrate binds to a specific region of the enzyme called the active site by multiple weak interactions. The active site typically is a cleft or crevice on the surface of an enzyme molecule, and it contains the catalytic functional groups that directly influence the kinetics of making and breaking bonds (1–6). A scheme for a typical enzyme-catalyzed reaction is

$$\mathrm{E+S} \underset{k_{-1}}{\overset{k_1}{\rightleftharpoons}} \mathrm{ES} \overset{k_2}{\rightarrow} \mathrm{E+P}$$

Enzyme (E) combines with substrate (S) to form an intermediate complex (ES) that proceeds to form product (P) or dissociates back to enzyme and substrate (1–6). This reaction scheme assumes that the rate of the reaction of product to form substrate is negligible. Moreover, the rate constant k_2 suggests a single chemical step for product formation. In many cases, the reaction of ES complex to form product actually requires a series of chemical steps. For simplicity, k_{cat} a composite rate constant, is defined, and it represents the reaction rate of ES to form P at saturating substrate conditions (1–4,6).

The kinetic characteristics of enzyme-catalyzed reactions were first described mathematically by Henri in 1903 and later by Michaelis and Menten in 1913 (2,3,6). The Henri–Michaelis–Menten model assumes that the formation of the ES complex reaches equilibrium rapidly, and the formation of product from the ES complex proceeds relatively slowly. In other words, this model assumes that $k_2 \ll k_{-1}$ in the reaction scheme (2–4,6). This model also assumes that the reverse reaction of product to form substrate is insignificant, as it would be the case just after substrate and enzyme are added together when the concentration of product is very low. In 1925, Briggs and Haldane developed a more general approach that does not require $k_2 \ll k_{-1}$ (2,3,6). Their approach is based on the steady-state assumption, which states that the steady state, at which the concentration of enzyme–substrate complex [ES] remains constant, is established shortly after the beginning of the enzyme-catalyzed reaction. The rate equation derived using the steady-state assumption is

$$v = \frac{V_{\mathrm{max}}[S]}{K_{\mathrm{m}} + [S]} \tag{7.1}$$

Equation 7.1, referred to as the Michaelis–Menten or Henri–Michaelis–Menten equation, has become the principal equation used to describe enzyme kinetics. According to the Michaelis–Menten equation, when the substrate concentration [S] is much less than the Michaelis constant K_{m}, the reaction rate v is directly proportional to [S]. Under these conditions, the majority of enzyme is free enzyme. When [S] is much greater than K_{m}, the reaction rate is at its maximum V_{max} and is independent of substrate concentration. In this case, the enzyme is fully saturated with substrate, and no free enzyme is present. An enzyme molecule immediately binds another substrate molecule once a product molecule is formed and released. Fig. 7.1 shows a Michaelis–Menten plot, which describes the reaction rate v as a function of the substrate concentration [S]. The catalytic constant k_{cat} can be obtained by dividing the maximum reaction rate by the enzyme concentration (1–4,6).

$$k_{\mathrm{cat}} = \frac{V_{\mathrm{max}}}{[E]} \tag{7.2}$$

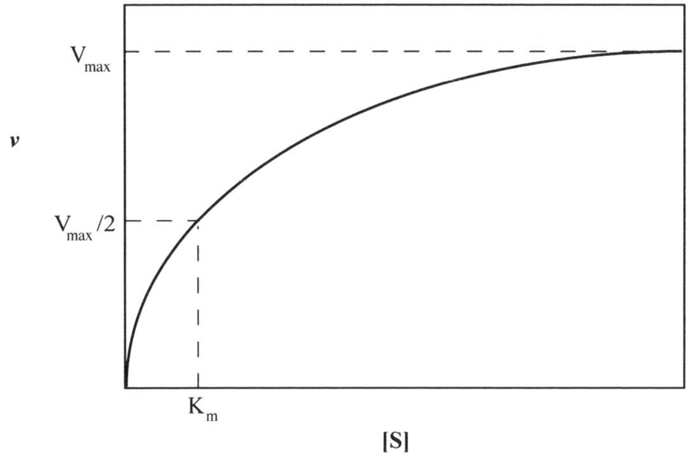

FIGURE 7.1 General plot of reaction rate, v, versus substrate concentration, [S], for an enzyme-catalyzed reaction described by Michaelis–Menten kinetics. Other terms are defined in the text.

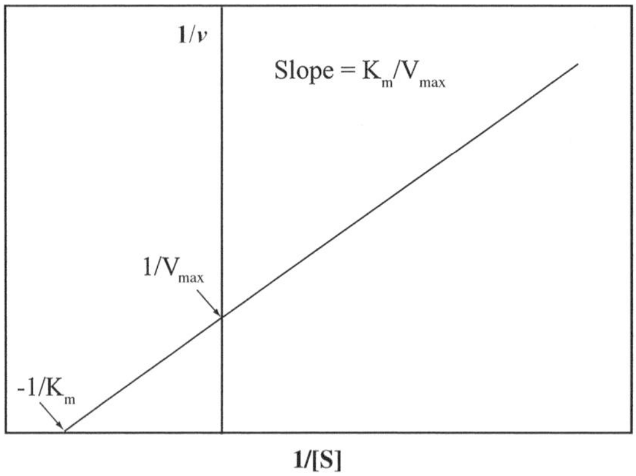

FIGURE 7.2 General Lineweaver–Burk plot 1/v versus 1/[S] for an enzyme-catalyzed reaction.

Equation 7.3 is obtained by taking the reciprocal of both sides of Equation 7.1:

$$\frac{1}{v} = \frac{1}{V_{\text{max}}} + \left[\frac{K_{\text{m}}}{V_{\text{max}}}\right]\frac{1}{[S]} \tag{7.3}$$

Equation 7.3 indicates that a plot of 1/v versus 1/[S] should result in a straight line as shown in Fig. 7.2. Such a plot is called a Lineweaver–Burk plot (1–6).

It is important to remember that the Michaelis–Menten treatment of enzyme kinetics includes several assumptions that are not true for every enzyme-catalyzed reaction. While this is a useful tool for examining many enzyme-catalyzed reactions and understanding enzyme kinetics at a basic level, it cannot be applied to all enzyme-catalyzed reactions. The reader is directed to other texts for more detailed treatments of enzyme kinetics (1–6).

7.1.3 Enzyme Assays

Enzyme assays are some of the most frequently performed analyses in biochemistry, owing to the significance of enzymes as biological catalysts for life. Fundamentally, enzyme assays are measurements of the rate of a chemical reaction and the effect of a catalyst (the enzyme) on the reaction rate. The reaction rate is determined by measuring either the decrease in reactant (substrate) as a function of time or the formation of product as a function of time (2,6).

In a typical enzyme assay, all the required components for the enzyme-catalyzed reaction are added to the reaction vessel except either the enzyme or the substrate. The reaction is then initiated by adding this missing component. The substrate concentration or the product concentration is then measured and plotted as a function of time. The slope of this plot is the reaction velocity:

$$v = -\frac{d[\text{S}]}{dt} = \frac{d[\text{P}]}{dt} \tag{7.4}$$

By varying the substrate concentration [S] in a series of separate experiments, a hyperbolic Michaelis–Menten plot of v versus [S] (Fig. 7.1) can be generated. Two important parameters in enzyme kinetics, K_m and V_{max}, can be determined from this plot or from the corresponding Lineweaver–Burk plot (Fig. 7.2) (1–6). These reactions can be carried out in vessels ranging from arrays of microliter-volume wells to standard 1-cm cuvettes or flasks, but the procedure is fundamentally the same.

To carry out an enzyme assay, it is necessary to selectively detect either the product or the substrate in the presence of the other reaction components. One common approach is to use a separation method to resolve the substrate and the product from other components. Typically, small aliquots of the reaction mixture are removed periodically and analyzed. Once separated, any detection method (selective or nonselective) can be used to quantify the substrate or the product. Alternatively, spectroscopic or electrochemical techniques are used that can selectively determine the substrate or the product. These approaches eliminate the time-consuming separation step and allow for *in situ* measurements of substrate or product concentration as a function of time. A wide variety of separation and detection methods have been used to measure the rates of enzyme-catalyzed reactions, including TLC, HPLC, gel electrophoresis, fluorescence, absorbance, radioisotope detection, mass spectrometry, and electrochemical methods.

7.1.4 Enzyme Inhibition

The rate of an enzyme-catalyzed reaction can be decreased by enzyme inhibitors. This is analogous to the fouling of a catalyst in a nonbiological chemical reaction. Generally, there are two types of enzyme inhibition: reversible inhibition and irreversible inhibition. A characteristic feature of reversible inhibition is that binding between the enzyme and the inhibitor is noncovalent and reversible so that the enzyme–inhibitor (EI) complex dissociates rapidly (1–4,6). If the inhibitor could be removed from the reaction mixture, the enzyme-catalyzed reaction would return to its initial, uninhibited rate. The binding of a reversible inhibitor and the target enzyme is typically characterized by K_i, which is the dissociation constant for the enzyme–inhibitor complex:

$$\mathrm{E} + \mathrm{I} \rightleftharpoons \mathrm{EI}$$

$$K_i = \frac{[\mathrm{E}][\mathrm{I}]}{[\mathrm{EI}]} \tag{7.5}$$

The term K_i is also called the inhibition constant, and it is an indication of the potency of a reversible inhibitor (1–4,6).The relative potency of a reversible inhibitor can also be represented by IC_{50}. The IC_{50} is defined as the concentration of an inhibitor required to reduce the enzyme activity to 50% of its uninhibited value (2). Reversible inhibitors can further be classified as competitive, noncompetitive, or uncompetitive based on their influence on the kinetics of the enzyme-catalyzed reaction. The Lineweaver–Burk plots for these three inhibitor classes are shown in Fig. 7.3 (1–6).

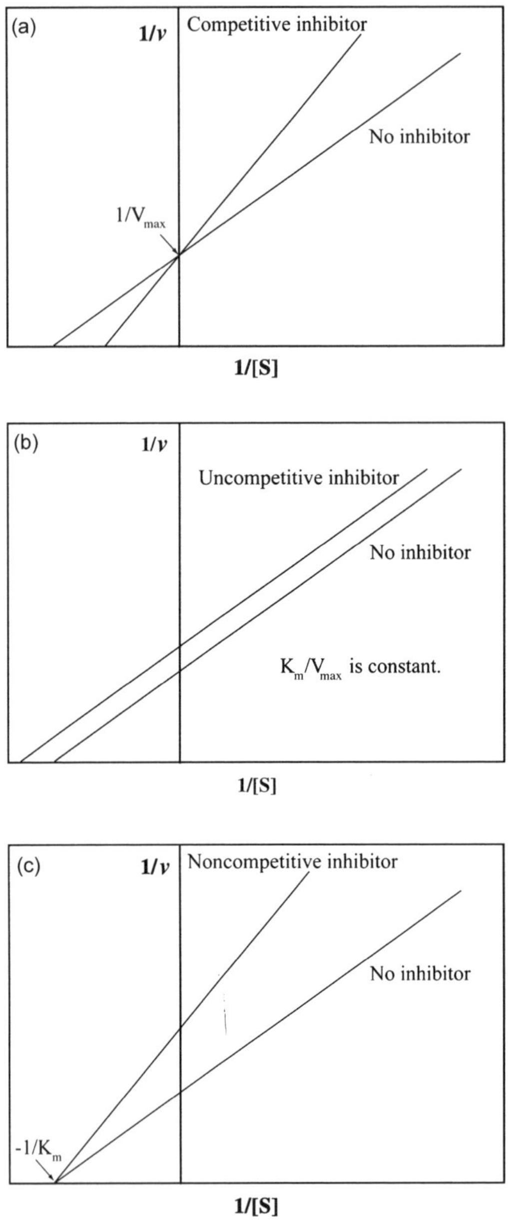

FIGURE 7.3 General Lineweaver–Burk plots illustrating the effects of three types of reversible inhibitors on the kinetics of an enzyme-catalyzed reaction: (**a**) competitive inhibitor, (**b**) uncompetitive inhibitor, and (**c**) noncompetitive inhibitor.

In the case of competitive inhibition, the inhibitor and the substrate bind to the same active site on the enzyme molecule, and their binding is mutually exclusive. In essence, the substrate and the inhibitor compete to bind at the active site of the enzyme. The kinetic effect of a competitive inhibitor is that both the Michaelis–Menten constant K_m and the slope of the Lineweaver–Burk plot K_m/V_{max} are increased, but the maximum reaction rate V_{max} remains unchanged (Fig. 7.3a) (1–6).

In uncompetitive inhibition, the inhibitor has no affinity for the free enzyme. It binds reversibly and exclusively to the ES complex, resulting in an inactive ESI complex. An uncompetitive inhibitor requires the prior formation of ES for binding and inhibition. Under the influence of an uncompetitive inhibitor, both K_m and V_{max} are decreased by the same factor, and thus the ratio K_m/V_{max} does not change (Fig. 7.3b) (2–6).

In cases involving noncompetitive inhibition, the inhibitor binds to both the free enzyme and the ES complex. The binding of the inhibitor and the substrate to the free enzyme is not competitive, since they bind reversibly and independently at different sites on the enzyme. A Lineweaver–Burk plot like that shown in Fig. 7.3c is obtained in cases where the binding affinities of the inhibitor for both the free enzyme and the ES complex are the same. The lines converge at the x-axis. When the binding affinities of the inhibitor for the two enzyme forms are finite but not equal, the lines intersect above or below the x-axis; however, there are two common features for all noncompetitive inhibition cases: V_{max} is decreased and K_m/V_{max} is increased (2–4,6).

Irreversible inhibition, in comparison to reversible inhibition, permanently reduces the ability of the enzyme to catalyze a reaction. Irreversible inhibitors typically inactivate an enzyme by attaching to it covalently (1–6). An irreversible enzyme–inhibitor complex (EI) is formed by the interaction of an enzyme with an irreversible inhibitor (I), which can be described as

$$\mathrm{E} + \mathrm{I} \xrightarrow{k_i} \mathrm{EI}$$

where k_i is the rate constant of inhibition (4). A kinetic characteristic of irreversible inhibition is that the inhibitory effect is time dependent. As time passes, more and more enzyme molecules will be inhibited irreversibly. A more detailed treatment of enzyme inhibition and kinetics is available in textbooks and primary literature (2–4,6).

The catalytic activity of enzymes can also be affected by compounds categorized as enzyme activators. Enzyme activators bind to the enzyme to increase the rate of an enzyme-catalyzed reaction. They may convert an inactive form of the enzyme to a catalytically active form, or increase the activity of an already active enzyme. More information about kinetics of enzyme activation can be found in the literature (3,5).

7.2 MICROFLUIDIC ASSAYS FOR ENZYMES AND ENZYME INHIBITION

For the purposes of this chapter, an analytical system is considered to be a microfluidic device when total channel volumes are a few microliters or less, and typical sample volumes used for analysis are in nanoliters or less. We are also limiting this chapter to microfluidic methods that include fluid flow or electrophoresis as a central part of the methodology. By this definition, a typical fused-silica capillary used for capillary electrophoresis (CE) is a microfluidic device as well as the much more sophisticated devices containing networks of interconnecting channels that can be created in glass or in plastic "chips" using microfabrication. In this section, we will briefly summarize the use of microfluidic devices for carrying out enzyme assays and then provide a more detailed overview of the application of microfluidic devices to study enzyme inhibition.

7.2.1 Microfluidic Platforms for Enzyme Assays

The application of microfluidics to study the kinetics of enzyme-catalyzed reactions offers several significant advantages compared to more traditional approaches (21–28). One advantage of performing enzyme assays using microfluidic devices is that these techniques can be capable of separating substrate from product so that they can be detected selectively. Both column-based capillary electrophoretic separations and microscale liquid chromatography have been applied to study enzyme assays (21–24,26). The first applications of capillary electrophoresis to study enzyme kinetics were reported in 1991 by Krueger et al. (29) and Lee et al. (30). These initial reports were off-line and are described in more detail in Section 7.2.2.1. Liquid chromatography and capillary electrochromatography at the scales considered in this chapter (Section 7.2) have also been used to perform kinetic studies of enzyme-catalyzed reactions (23,31–33).

Another significant advantage of carrying out enzyme assays in microfluidic devices is scale. The amount of enzyme and substrate consumed in an assay can be significantly reduced when sample volumes are only a few nanoliters or less. An early study that took advantage of small scale of microfluidic enzyme assays was the determination of lactate dehydrogenase activity in single cells by Xue and Yeung (34). The reduced scale of an enzyme assay carried out in a microfluidic device also makes the detection of enzyme catalysis by single molecules possible as shown in early studies by Xue and Yeung (35) and Craig et al. (36). Carrying out single-molecule studies in microfluidic devices reduces dilution by limiting diffusion, which allows product concentrations to increase rapidly so that single enzyme molecule catalysis can be measured.

Microfluidic devices enable multiple analytical steps in an enzyme assay to be carried out rapidly and online with minimal manual intervention from the analyst. In capillaries, a simple and elegant example is electrophoretically mediated microanalysis (EMMA) (21,22,24,26,37). This technique was first reported by Bao and Regnier in their 1992 study of glucose-6-phosphate dehydrogenase catalysis (38). Mixing of enzyme and substrate as well as separation of substrate and product for detection can be carried out by EMMA using a simple capillary and a commercial CE instrument. Section 7.2.2.2.2.1 includes a more detailed description of EMMA.

Microfabrication offers the possibility of building sophisticated, integrated microfluidic analytical devices with specialized structures for sample preparation, separation, detection, and carrying out derivatization reactions for detection. This design flexibility is being exploited to perform enzyme assays. The first enzyme assay in a microfabricated microfluidic device, as defined for this chapter, was reported by Hadd et al. (39), and this study is discussed in more detail in Section 7.3.3. Enzyme assay methods that have been developed in capillaries can be adapted to microfabricated devices, including EMMA (40–45). The most important and exciting advantages of using microfabricated devices for performing enzyme assays are realized when their flexibility is used to create online methods with multiple steps that cannot be carried out using only a simple capillary (39,46–53). The application of microfabricated devices for studying enzyme kinetics is treated in more detail in Sections 7.2.2, 7.3.3, and 7.3.4.

A significant limitation of enzyme assays carried out in microfluidic devices is the serial nature of the experiment, which results in low sample throughput despite short analysis times and small sample volumes required for single measurements. In principle, this limitation also exists for a traditional assay performed in a cuvette or in other reaction vessels. In practice, this limitation is overcome in traditional assays by performing large numbers of experiments in parallel using microtiter plates and robotics for sample handling. Parallel analysis has also been shown to be practical for microfluidic devices and can be used to overcome throughput limitations for serial experiments. The most successful example is the use of capillary array electrophoresis for DNA sequencing, (54,55) and there are several reports of parallel enzyme assays performed in capillaries and microfabricated devices (56–62).

7.2.2 Enzyme Inhibition Assays in Microfluidic Devices

The remainder of this chapter will focus on microfluidic methods used specifically to study enzyme inhibition. Both online and off-line techniques have been developed. Off-line enzyme inhibition assays are defined as assays in which the enzyme-catalyzed reaction (inhibited or uninhibited) takes place in a sample container outside of the microfluidic device. In an online assay, the enzyme-catalyzed reaction takes place in the microfluidic device. Section 7.2.2.1 will briefly summarize off-line studies since they do not take full advantage of microfluidics. In Section 7.2.2.2, online enzyme inhibition assays will be treated in more detail. The detection methods used to quantify product and/or substrate for enzyme-catalyzed reactions in microfluidic inhibition studies will be treated in Section 7.2.2.3.

7.2.2.1 Off-Line Analysis of Enzyme Inhibition For most off-line microfluidic enzyme inhibition assays, capillary electrophoresis is used simply as a technique to separate substrate and product so they can be quantified selectively in the same way that TLC or HPLC might be used for this purpose. The first example of off-line CE to measure enzyme inhibition was published by Hoffmann et al. (63). This work examined the inhibition of dipeptidyl peptidase IV (DP IV) by anti-DP IV antibodies and several peptides. There are a number of subsequent reports of off-line enzyme inhibition assays, and new methods continue to be developed (21,26,64–68). Off-line assays to study enzyme inhibition using capillary array electrophoresis have been reported (57,60,62). An off-line enzyme inhibition assay where CE is performed in a microfabricated device has been used to investigate the inhibition of a number of enzymes including β-glucuronidase by D-saccharic acid-1,4-lactone, (44) β-galactosidase by β-D-thiogalactoside and β-lactose, (41,61) and src-kinase and protein kinase A by a proprietary inhibitor and commercially available inhibitor PKI-(6–22), respectively (45).

Off-line enzyme inhibition assays using microfluidic devices do offer some advantages such as rapid, high efficiency CE separations and the use of only nanoliters to picoliters of sample for each injection from the reaction solution; however, these off-line assays do not take full advantage of the capabilities of microfluidic devices. If the enzyme-catalyzed reactions are carried out online inside the microfluidic devices, smaller total reaction volumes can be used, thus reducing reagent consumption. Delays between the time when the enzyme-catalyzed reaction takes place and when the reaction mixture is sampled and analyzed by CE are reduced or eliminated (21). The intermediate case between off-line and online assays, where the enzyme-catalyzed reaction is carried out in an open well located on a microchip device, but the reaction components (substrate, enzyme, and inhibitor) are added directly to the well from outside the device is defined as off-line for this chapter. Methods where the reaction components are added to the device in separate wells and are mixed in the microfluidic device are treated as online assays here.

7.2.2.2 Online Enzyme Inhibition Assays Online enzyme inhibition assays are classified as either heterogeneous or homogeneous for this chapter. In a heterogeneous assay, one of the reaction components (enzyme, substrate, or inhibitor) is immobilized on a solid phase, and the enzyme-catalyzed reaction takes place at the interface between the solution and the surface containing the immobilized component. In a homogeneous enzyme inhibition assay, all of the reaction components (enzyme, substrate, and inhibitor) are in solution, and the enzyme-catalyzed reaction occurs in solution. The advantages and disadvantages of both heterogeneous and homogenous approaches to studying enzyme inhibition kinetics are presented in Sections 7.2.2.2.1 and 7.2.2.2.2.

7.2.2.2.1 Heterogeneous Enzyme Inhibition Assays In most heterogeneous enzyme inhibition assays, the enzyme is immobilized inside the capillary or the microchip. The substrate and the inhibitor are then flowed past the immobilized enzyme, and the enzyme-catalyzed reaction rate is determined by measuring product formation or substrate depletion. The enzyme can be immobilized on the wall of the capillary or the microchannel, or it can be attached to particulate or monolithic solid supports immobilized inside the microfluidic device. Krenkova and Foret recently reviewed strategies for immobilizing enzymes on solid supports to make microfluidic enzyme reactors (69). Immobilizing the enzyme offers both advantages and disadvantages. A small amount of enzyme can be reused for multiple assays, and many enzymes are more stable when immobilized than when in solution. The time at which the enzyme-catalyzed reaction starts and stops can be defined by controlling when the solution containing the substrate and the inhibitor is located in the part of the microfluidic device containing the enzyme. The separation of the immobilized enzyme from the other reaction components is simple. One challenge is that the immobilized enzyme needs to be catalytically active. The kinetics of the immobilized enzyme should ideally be the same as in solution, and the kinetic effects of inhibitors on the immobilized enzyme should be the same as for the enzyme in its native environment. Also, if the enzyme becomes inactive, the immobilization procedure must be repeated or a new microfluidic device must be used. Finally, one immobilization procedure may not work well for all enzymes, so different enzymes may require different immobilization strategies.

The most common strategy for carrying out heterogeneous microfluidic enzyme inhibition studies has been to immobilize the enzyme in a short zone at the injection end of a capillary and use capillary electrophoresis to transport substrate and inhibitors through the enzyme microreactor. Substrate, product, and inhibitor are electrophoretically separated after passing through the microreactor for selective detection. Sakai-Kato et al. encapsulated microsomes containing uridine diphosphate glucuronosyltransferase (UGT) in a sol–gel monolith formed within the first 2.0 cm of a 40 cm long, 75 μm inner diameter (i.d.) capillary (70). A solution containing the substrate, acetaminophen, and the inhibitor was introduced into the microreactor part of the capillary and then was allowed to incubate with no fluid flow for 10 min. Next, a potential was applied, and the substrate was separated from the reaction product, acetaminophen glucuronide, and the inhibitors by CE. The inhibition of UGT by 4-nitrophenol, probenecid, and valproate was studied. Tang and Kang investigated the inhibition of angiotensin-converting enzyme (ACE) by captopril and cilazaprilat using an enzyme microreactor with the enzyme immobilized on the column surface in a short segment at the injection end of the capillary (71). This study is described in more detail in Section 7.3.2. This same general approach has been used in microfabricated devices to study enzyme inhibition. Park et al. immobilized β-galactosidase on a nitrocellulose membrane that was coated in a short zone on one wall in the separation channel of the device (72). The substrate and reaction product were separated by CE and detected by laser-induced fluorescence (LIF) after a zone containing substrate and inhibitor was introduced to the section of the microchannel containing the membrane and enzyme and allowed to incubate at zero potential for 0, 4, 13, and 30 s. The IC_{50} and K_i values were determined for the inhibition of β-galactosidase by 2-phenylethyl β-D-thiogalactoside using this thin-film enzyme reactor.

Heterogeneous enzyme inhibition assays have also been developed in microfluidic devices using pressure-driven flow to transport fluids and reaction components instead of electrokinetic transport (electrophoresis and electroosmotic flow) (23,50,73). Rashkovetsky et al. immobilized an HIV-protease substrate on a 2–3 mm plug of magnetic beads that were held in place inside a 75 μm i.d. fused-silica capillary by a magnet (73). Mixtures of HIV protease and the inhibitor, acetyl pepstatin, were injected as a long zone (30 s, 0.5 psi) and transported through the stationary magnetic beads by application of pressure (0.5 psi) using a commercial CE instrument. The applied pressure was stopped for 5 min to allow more reaction product to form, and it was applied again to push the reaction solution to the detector.

The enzyme-catalyzed reaction resulted in the cleavage of the immobilized substrate, and the fluorescein-labeled product was detected by a downstream LIF detector. The inhibition of the enzyme by acetyl pepstatin was measured using this method. A significant advantage of this approach is that the immobilized reaction component (substrate in this case) can be replaced by simply removing the old plug of magnetic beads using high pressure (20 psi) and injecting a fresh batch of beads at low pressure to immobilize them.

Garcia et al. described a technique that used a microfabricated device to study the kinetics of β-galactosidase inhibition by the competitive inhibitors, lactose, and galactose (50). In this study the enzyme was immobilized on one surface of the microchannel, and inhibitor and substrate were mixed in the microchip device from two intersecting channels using pressure-driven flow so that a spatial gradient of inhibitor concentration was created across the microchannel, perpendicular to the direction of fluid flow. This allowed the enzyme activity to be probed rapidly as a function of inhibitor concentration by imaging the fluorescent product formed at different positions in the microchannel as the inhibitor gradient and substrate (constant concentration) flowed by. This method was used to determine IC_{50} values for the inhibitors.

7.2.2.2.2 Homogeneous Enzyme Inhibition Assays Homogeneous enzyme inhibition assays have a fundamental advantage relative to heterogeneous assays. The enzyme-catalyzed reaction takes place with all the reaction components in solution so that the results of a kinetic study should be comparable to those obtained in a traditional enzyme assay and closer to the kinetic behavior of the enzyme in its native state for a soluble enzyme. Also, the potential complications of immobilizing the enzyme are avoided. There are, of course, disadvantages associated with homogeneous enzyme inhibition assays. There is less control over the timing of the reaction since either the enzyme and the substrate must be separated or the reaction conditions be changed in order to stop the reaction. The separation of the enzyme from the other reaction components is more difficult in a homogeneous assay. The unwanted adsorption of the enzyme to the surface of the microfluidic device can be deleterious to the measurement.

As mentioned in Section 7.2.1, microfabricated devices offer superior design flexibility relative to capillaries, and complicated analytical procedures that are not possible or practical in a capillary can be carried out in a microfabricated device. Therefore, this section starts by reviewing homogeneous enzyme inhibition assays performed in capillaries (Section 7.2.2.2.2.1), followed by homogeneous enzyme inhibition assays carried out in microfabricated microfluidic devices (Section 7.2.2.2.2.2).

7.2.2.2.2.1 HOMOGENEOUS ENZYME INHIBITION ASSAYS IN CAPILLARIES Most online methods to study the kinetics of enzyme inhibition in capillary-based microfluidic devices are based on EMMA (22,24,26,37). Electrophoretically mediated microanalysis is based on the mixing and separation of zones of reactants and products by CE to study a reaction. For enzyme inhibition assays, the reaction components that must be mixed and separated include enzyme, substrate, product, and inhibitor. There are two basic formats of EMMA: continuous engagement and transient engagement, and both have been used to study enzyme inhibition (22,24,26,37).

In transient engagement EMMA, which is also called plug–plug mode EMMA, reaction components are injected into the capillary as discrete zones. These zones then mix electrophoretically in the capillary as the faster moving zone overtakes the slower moving zone to initiate the enzyme-catalyzed reaction. When the mixed zones electrophoretically separate again, the reaction stops. Zones of product and substrate can be detected later to quantify the reaction rate. Transient engagement EMMA was the first EMMA mode applied to study enzyme inhibition kinetics by Saevels et al. to investigate the competitive inhibition of adenosine deaminase by *erythro*-9-(2-hydroxy-3-nonyl)adenine (EHNA) (74). In this study, the enzyme and the substrate were injected as separate zones in the capillary as indicated in Fig. 7.4. Note that the slower migrating zone must be injected first. The running buffer in the capillary as well as the enzyme sample solution and substrate sample solution all contained the inhibitor EHNA at nanomolar concentrations. The voltage was turned off when the zones of substrate and enzyme were overlapped for 5.0 min to allow more reaction product to form for detection (zero potential incubation). Then the electrophoretic potential was applied again to separate the zones of enzyme,

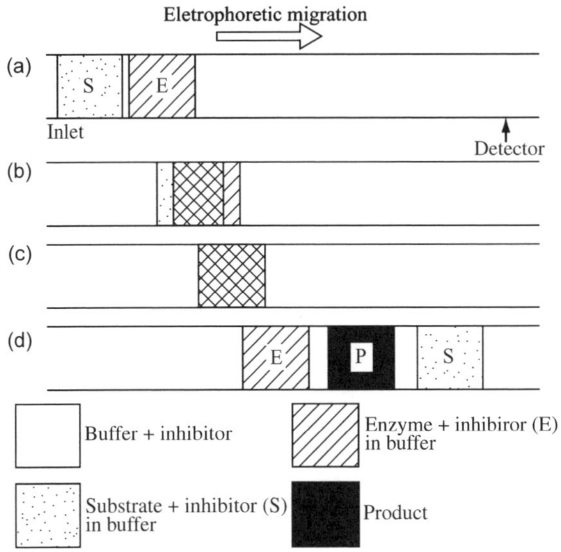

FIGURE 7.4 Schematic of transient engagement EMMA for an enzyme inhibition assay. (**a**) A solution of enzyme and inhibitor (E) and a solution of substrate and inhibitor (S) are injected into the capillary as distinct zones. Inhibitor is also present in the running buffer as indicated. (**b**) Enzyme (E) and substrate (S) zones are partly overlapped, and the enzyme-catalyzed reaction starts. (**c**) Enzyme (E) and substrate (S) zones are fully overlapped. (**d**) The zones of enzyme, substrate, and product are separated electrophoretically, and the reaction has stopped.

substrate, and product, and to deliver them to the downstream absorbance detector. It was necessary to accurately know the electrophoretic mobilities of the enzyme and substrate zones in order to stop electrophoresis at the appropriate time for zero potential incubation. Using this transient engagement EMMA method with zero potential incubation, the inhibition constant K_i of EHNA was determined. A closely related method has been used by other researchers to study the inhibition of acetylcholinesterase (75).

A significant limitation of transient engagement EMMA for performing enzyme inhibition studies is that the same buffer must be used for carrying out the enzyme-catalyzed reaction and separating the reaction components for selective detection. The optimal solution conditions for the enzyme-catalyzed reaction may not be suitable for separating the reaction components. Van Dyck et al. (76) introduced a modification of transient engagement EMMA that overcomes this limitation, and this approach was later applied to enzyme inhibition studies by Novakova et al. to study the inhibition of bovine liver rhodanese by 2-oxoglutarate (77). Fig. 7.5 shows a schematic representation of the partial filling technique. The major change relative to a standard transient engagement EMMA assay for enzyme inhibition is that the enzyme and substrate zones are in a buffer that is optimal for the enzyme-catalyzed reaction, and they are surrounded by zones of the same buffer. The rest of the capillary contains buffer that is optimized for the CE separation and the detection of the reaction components. It is critical that the zones of reaction-optimized buffer are long enough so that the enzyme and substrate zones mix before leaving this buffer and entering the separation-optimized buffer.

In the study of rhodanese inhibition by Novakova and coworkers, the enzyme-catalyzed reaction was carried out in HEPES buffer at pH 8.5, while separation and detection were carried out in a buffer containing 100 mM β-alanine at pH 3.5 (77). This separation buffer enabled the separation of cyanide and bromide (78). The inhibition of rhodanese by 2-oxoglutarate was studied, and K_i values were determined with respect to each substrate (cyanide and thiosulfate) of the enzyme.

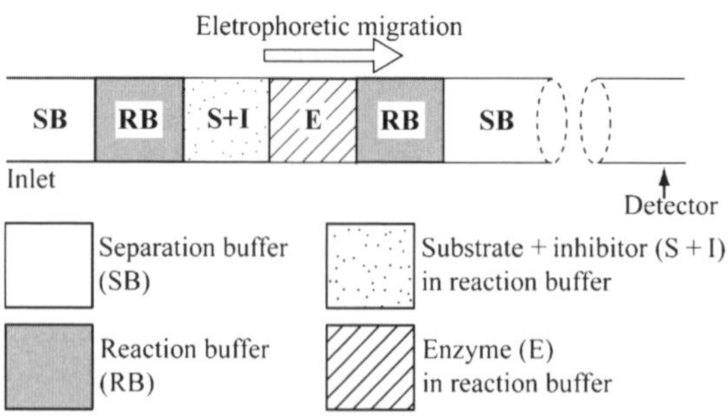

FIGURE 7.5 Illustration of the partial filling technique for an enzyme inhibition assay. The enzyme zone (E) and a zone containing a mixture of substrate and inhibitor (S + I) are shielded from the separation buffer (SB) by the reaction buffer (RB) so that the enzyme-catalyzed reaction can be carried out under optimal conditions.

This research group later used the same method to study the self-inhibition of haloalkane dehalogenase by 1,2-dibromoethane (79). This laboratory also used a similar approach to study the inhibition of haloalkane dehalogenase by a competitive inhibitor 1,2-dichloroethane using brominated substrates; however, in this work, the separation buffer was primarily optimized to facilitate the detection of the reaction product Br^- by indirect UV absorbance detection (80). The reaction buffer was 20 mM glycine at pH 8.6, and the separation and detection solution contained 10 mM chromate and 0.1 mM cetyltrimethylammonium bromide (CTAB) at pH 9.2. The K_i value for 1,2-dichloroethane was determined. Iqbal et al. used a partial filling transient engagement EMMA method with direct UV absorbance detection for screening of adenosine kinase inhibitors (65). After initial mixing of the enzyme, substrate, and inhibitor by EMMA, zero potential incubation (5 min) was used to increase the amount of product formed. An internal standard was used to improve quantification. The K_i values for two inhibitors of adenosine kinase were determined using this method.

In continuous engagement EMMA for enzyme inhibition studies, the capillary is filled with a solution of the substrate and the inhibitor. A plug of enzyme is injected into the capillary. As the enzyme plug migrates through the capillary under a constant applied potential, it interacts with the substrate and the inhibitor continuously to form the reaction product while its catalytic function is being suppressed by the inhibitor. This experiment is very similar to the original EMMA experiments reported by Bao and Regnier (38), except for the addition of an inhibitor to the substrate solution. This continuous engagement EMMA approach has been used by Belenky et al. for screening protein tyrosine phosphatase inhibitors in natural extracts (81). In this study, the authors also used zero-potential incubation in the middle of the experiment to increase the amount of product formed. Whisnant and coworkers developed a method to study enzyme inhibition that combines transient engagement EMMA and continuous engagement EMMA (82,83). This work is described in more detail in Section 7.3.1.

An approach for carrying out homogeneous enzyme inhibition assays has been developed that is similar to EMMA but does not rely on electrophoresis to mix enzyme, substrate, and inhibitor in the capillary. As shown in Fig. 7.6, a short zone of enzyme is injected into the capillary inlet followed by a zone of substrate and inhibitor. Then a second zone of enzyme is injected into the capillary inlet. The substrate and inhibitor solution is sandwiched between the two enzyme zones, and the enzyme, substrate, and inhibitor mix in the capillary inlet by diffusion during the incubation step. Before and after the sandwich injection, a small plug of water is injected. This water zone is reported to improve peak shape, presumably due to a sample stacking effect (84). After the incubation step, a potential is applied, and the reaction components and product are separated by CE so that the reaction product and substrate can be quantified. This technique for carrying out reactions in a capillary inlet was originally developed for fluorescent derivatization (85) and was later applied to study enzyme inhibition kinetics by Van Dyck et al. in 2003 (86). Van Dyck and coworkers applied this technique to study the inhibition of angiotensin-converting enzyme by five different inhibitors (captopril, lisinopril, perindoprilat,

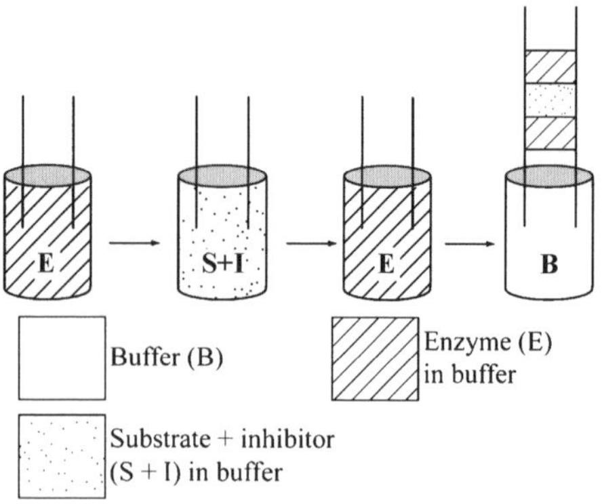

FIGURE 7.6 Schematic of an enzyme inhibition assay at the capillary inlet with mixing by diffusion. The reaction solutions are injected sequentially into the inlet of the capillary: 1, enzyme solution (E); 2, substrate solution with inhibitor (S + I); 3, enzyme solution (E). A short plug of water is injected before and after these three zones (not shown). The solutions are then allowed to mix by diffusion during incubation at zero potential while the capillary inlet is immersed in buffer.

quinaprilat, and benazeprilat), and they determined IC_{50} and K_i values for captopril. Iqbal et al. used a modified version of this technique to investigate the inhibition of three isoforms of ecto-nucleoside triphosphate diphosphohydrolase (NTPDase) by reactive blue 2, pyridoxalphosphate-6-azophenyl-2′,4′-disulfonic acid (PPADS), suramin, and N^6-diethyl-β,γ-dibromomethylene-ATP (ARL67156) (87). In their method, a short zone of enzyme was sandwiched between two zones containing substrate and inhibitor. A final zone of water was injected at the inlet before incubation (5 min) and subsequent separation by CE. Both IC_{50} and K_i values were determined for the four inhibitors.

Du et al. described an automated microfluidic sequential injection analysis system based on a capillary reactor that did not use electrokinetic transport (88). More details about sequential injection analysis can be found in recent reviews (89,90). This device was used to study the inhibition of β-galactosidase. In this work, a fused-silica capillary (6 cm long, 50 μm i.d.) was used as a reactor, and a slotted-vial rotating autosampling device was used to sequentially introduce enzyme solution, inhibitor sample, substrate solution, and carrier solution. Gravity flow was used for fluid transport. A fluorogenic substrate was used so that the reaction product could be selectively detected by LIF in the presence of other reaction components. Enzyme inhibition by diethylenetriaminepentaacetic acid (DTPA) was studied, and the IC_{50} for this inhibitor was determined. Pu and Liu reported the use of a microchip as an electroosmotic pump for sequential injection analysis for an enzyme inhibition assay (91). The β-galactosidase-catalyzed reaction was carried out in a 100 μm i.d. capillary at a flow rate of 1.5 μL/min. Plugs of enzyme, inhibitor, and substrate were injected sequentially using a selection valve and the microfabricated electroosmotic pump to generate fluid flow. The inhibition of β-galactosidase by DTPA was studied using this method.

7.2.2.2.2.2 Homogeneous Enzyme Inhibition Assays in Microchips The ability to create multiple intersecting microfluidic channels of different geometries using microfabrication means that there are many more ways possible to carry out a homogeneous enzyme inhibition assay in a microfabricated microfluidic device compared to in a capillary. In principle, all of the methods discussed in Section 7.2.2.2.2.1 could be carried out in microfabricated devices. In practice, some of the methods that require the injection of multiple short zones of different solutions, such as the partial filling transient engagement EMMA technique (Fig. 7.5), may be easier to carry out by CE in capillaries. This is due to the ability of CE to sample from different solutions by simply moving the capillary end to a different sample well and performing pressure or electrokinetic injection. Nonetheless, microfabricated devices offer superior design flexibility and room for creative departure from designs based on single columns that most analytical chemists are familiar with. Although reports of enzyme inhibition studies are not yet as common in microfabricated devices as in capillaries, it is anticipated that the exploitation of microfabricated devices for such studies will expand in the future.

Confluent mixing of solutions from two or more fluid streams intersecting at a common point has been utilized in several methods for studying the kinetics of enzyme inhibition in microfabricated microfluidic devices (39,46,47,49,50,52,92). This seemingly simple procedure is much more difficult to carry out at very small scales using capillaries due to the need to use chromatographic fittings with minimal dead volumes. A recent study by Burke and Regnier on the inhibition of β-galactosidase by phenylethyl-β-D-thiogalactoside illustrates this concept and some important related issues (52). Fig. 7.7 shows the device used in this study. To

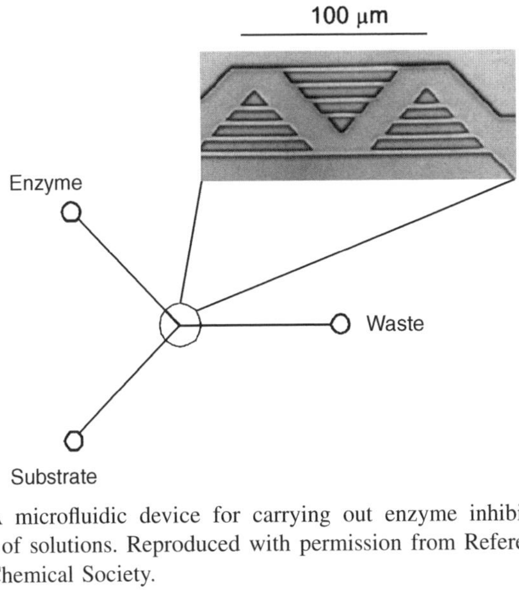

FIGURE 7.7 A microfluidic device for carrying out enzyme inhibition studies using confluent mixing of solutions. Reproduced with permission from Reference 52. Copyright 2003 American Chemical Society.

initiate the enzyme-catalyzed reaction, electrokinetic transport was used to flow substrate and enzyme from reservoirs to the Y-shaped intersection as shown in Fig. 7.7. The fluorescent reaction product was detected in the waste channel downstream from the mixing junction. Because the device is microfabricated, creating a channel intersection with near zero dead volume is trivial. Rapidly and completely mixing the two solutions meeting at the Y junction is not simple. The device includes a mixing structure just at the channel intersection, as indicated in Fig. 7.7, to enhance mixing of the two fluid streams. Thorough mixing of the two streams was complete in less than 1 s (93). Without this mixing structure, the two flow streams will remain essentially separate and parallel in the waste channel. Mixing of the two solutions will only occur by lateral diffusion and will be limiting for very fast experiments. Fluid mixing is a challenging and important issue for microfabricated microfluidic devices (94,95).

Mixing of confluent streams in a microfabricated device was employed by de Boer et al. for screening inhibitors separated by HPLC (92). Inhibitors of the protease, cathepsin B_1, were studied using *Z*-Phe-Arg-AMC as a substrate. Inhibitors were separated by HPLC, and the column effluent was combined with the enzyme solution at a three-channel intersection by confluent mixing. The substrate was added to the fluid stream containing enzyme and HPLC effluent at the second three-channel intersection on the microfabricated device. Enzyme and substrate were continuously pumped into the microchip channels using two syringe pumps. The enzyme-catalyzed reaction took place in the microfabricated device. All mixing of the different solution streams was due to lateral diffusion in the microchannels. Mass spectrometry was used to monitor product formation. The IC_{50} values were determined for *N*-acetyl-Leu-Leu-Arg-al (leupeptin), (L-3-*trans*-(propylcarbamyl)-oxirane-2-carbonyl)-L-isoleucyl-L-proline (CA-074), *N*-(*N*-α-carbonyl-Arg-Val-Arg-al)-Phe (antipain), and L-*trans*-3-carboxyoxiran-2-carbonyl-L-leucylagmatine (E-64), and inhibition by compounds separated from green tea extracts was also examined.

The first report of an enzyme inhibition study in a microfabricated microfluidic device by Hadd et al. utilized confluent mixing to adjust the concentration of substrate by mixing it with buffer at a T intersection (three channels) and also to mix enzyme and inhibitor with the diluted substrate solution at a cross intersection (four channels) downstream from the first intersection (39). The channel intersections in this device did not include specialized structures to enhance mixing, so mixing was only due to lateral diffusion. This study of β-galactosidase inhibition is described in more detail in Section 7.3.3. Cohen et al. developed a protein kinase A (PKA) inhibition assay in a microfabricated device using electrokinetic fluid transport (46). Confluent mixing was used to control the concentration of inhibitor for these experiments and to add enzyme and substrate to initiate the reaction. A short zone of the reaction solution was injected into a separation channel using a cross intersection (four intersecting channels) so the substrate and the product of the enzyme-catalyzed reaction could be separated. The K_i value was determined for the competitive inhibition of PKA by *N*-[2(*p*-bromocinnamyl)amino)ethyl]-5-isoquinoline sulfonamide using fluorescently labeled Kemptide (LeuArgArgAlaSerLeuGly) as a substrate.

In 1999, Hadd and coworkers reported the development of a more sophisticated microchip system for online studies of acetylcholinesterase AcChE inhibition (47). This device (shown in Fig. 7.8) used three microchannel intersections to carry out kinetic studies of enzyme inhibition. At the first intersection, enzyme flowed continuously through the intersection to the separation channel. Inhibitors were injected through this intersection as short zones by controlling the potentials applied to the reservoirs connected to this intersection. At the second intersection (T-1, three channels), substrate, acetylthiocholine, was added by confluent mixing with the enzyme and inhibitor stream from the separation channel. Mixing of the two streams was by lateral diffusion only, and this initiated the enzyme-catalyzed reaction. At the third intersection (T-2, three channels), a derivatization reagent was added to react with thiocholine to produce a fluorescent product for LIF detection. Basic confluent mixing was again used in this step. This multistep methodology requiring three intersections would not be practical in a capillary-based device. This device was used

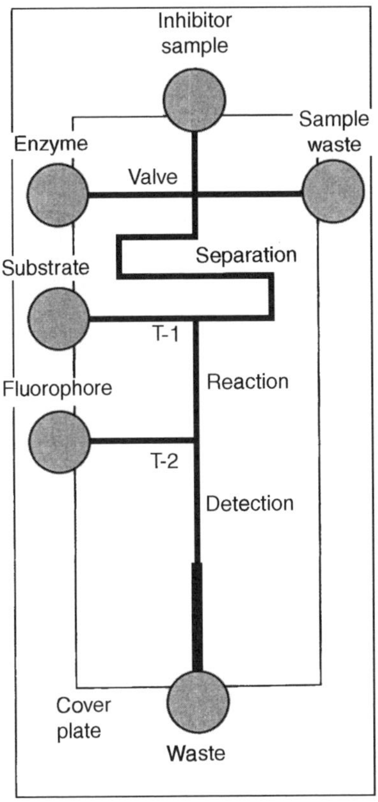

FIGURE 7.8 Schematic of a microchip with three microchannel intersections used for analysis of AcChE inhibitors. Reproduced with permission from Reference 47. Copyright 1999 American Chemical Society.

to measure K_i values for reversible inhibitors of acetylcholinesterase and to detect irreversible inhibitors. Separations of mixtures of inhibitors were also demonstrated. Another method for studying acetylcholinesterase inhibition in a microfabricated device was reported by Heleg-Shabtai et al. (49). This method included confluent mixing to dilute reagents, online derivatization of the enzyme-catalyzed reaction product, and separation of the reaction product from an internal standard and the derivatization reagent. This study is presented in more detail in Section 7.3.4.

In 1999, Duffy and coworkers reported a microfluidic study of the inhibition of alkaline phosphatase by theophylline using a microfabricated device (48). What was unique about this microfluidic method for studying enzyme inhibition was that centrifugal force was used to control and transport fluids in the device. The device was fabricated in a polydimethyl siloxane (PDMS) substrate shaped like a compact disc. Channels radiated from the center of the device toward its outer edge. Centrifugal force was applied by spinning the device. The device contained structures that acted as valves and released fluid only if a certain rotation rate and centrifugal force were exceeded. A valve could be actuated to release fluid by increasing the rotation rate of the device. The device had 48 channels to perform multiple enzyme assays in parallel. The K_i value for inhibition of alkaline phosphatase by theophylline was determined using this device.

7.2.2.3 Detection Methods for Microfluidic Enzyme Inhibition Assays Any detection method that can be used with microfluidic devices (capillaries and microfabricated devices) can potentially be used as a detection method for enzyme inhibition assays performed in a microfluidic device. The issues to consider first when selecting a detection method are what substrates are available for the enzyme of interest and what detection method provides the best combination of selectivity and sensitivity to allow either the reaction product or the substrate concentration to be measured. If the reactant and the product are separated in the assay, selectivity will be less of a concern. On the contrary, if the reactant and the product remain together as a mixture, the detection method must be capable of selectively detecting one of these two species. Selectivity for detection of the reactant or the product in the presence of the sample matrix is also a concern.

For general CE, UV–vis absorbance is the most common detection technique followed by laser-induced fluorescence. A broad range of additional detection methods has been used, including mass spectrometry and electrochemical detection. The reader is directed to other reviews for more detailed discussions about and comparisons of different detection methods for CE (96,97). Detection for microfabricated microfluidic devices is similar to detection for CE—the same issues of sensitivity, selectivity, small detection volumes, and short optical pathlengths are relevant. However, there are some important differences. The planar geometry of most microfabricated devices can make absorbance detection and interfacing with a mass spectrometer more challenging. Faster measurements compared to capillary electrophoresis are common, and channel dimensions and detection volumes are often smaller relative to CE. The design flexibility provided by microfabrication also makes some detection techniques more practical compared to CE, and detection systems with optical and electrical components integrated into the device are common. Laser-induced fluorescence and fluorescence microscopy are the most common detection methods used with microfabricated devices. The reader is directed to other reviews for more details about detection methods for microfabricated devices (98–100). The remainder of this section will discuss detection methods that have been applied to online studies of enzyme inhibition in microfluidic devices.

Absorbance detection is the most common detection method used for enzyme inhibition assays performed in capillaries because of its general applicability and the availability of commercial detectors. If the method used includes separation of the substrate and product from each other and other sample and assay components, detection selectivity is not a major concern (57, 63–66, 70, 71, 74, 75, 77, 79, 86, 87). Indirect absorbance detection has been used by Telnarova et al. to study the inhibition of haloalkane dehalogenase (80). Indirect absorbance detection was used to quantify bromide cleaved from the substrate. If the method does not include separation of the substrate from the product, it is necessary for the reactant and the product to have substantially different molar absorptivities at the wavelength used for detection. This approach was used in the initial report of EMMA by Bao and Regnier (38), but it has rarely been used for enzyme inhibition studies in microfluidic devices. Visible absorbance detection in a microfabricated device without substrate and product separation was used to study the inhibition of alkaline phosphatase by Duffy et al. (48). In this work, an LED was used as the light source for absorbance measurements.

Fluorescence detection offers lower detection limits than absorbance detection and better selectivity; however, LIF detection is not as general as UV–vis absorbance detection, and analytes often must be labeled with a fluorophore for detection. Furthermore, if both the substrate and the product for an enzyme-catalyzed reaction are fluorescent at the excitation and emission wavelengths used, they must be separated for selective detection as in the case with absorbance detection. Laser-induced fluorescence detection was used by Hadd and coworkers in their early studies of enzyme inhibition in microfabricated devices (39,47). In both cases a fluorogenic substrate was used so that separation of the substrate and product was not necessary. Whisnant and coworkers also used a fluorogenic substrate for their studies of alkaline phosphatase inhibition using EMMA in a capillary (82,83). Both Du et al. and Pu and Liu used a fluorogenic substrate and sequential injection analysis to study the inhibition of β-galactosidase using LIF and fluorescence microscopy, respectively, for detection (88,91). Fluorogenic substrates have been used in additional studies discussed elsewhere in this chapter (48,50,73). Fluorescence microscopy and LIF have also been used in capillaries and microchips for the studies of enzyme inhibition in methods where the substrate and product were separated and a fluorogenic substrate was not necessary (41,45,46,61,62,72). Derivatization of the reaction product with a fluorophore has also been used for enzyme inhibition studies in microfluidic devices (44,47,49,67).

The selectivity of mass spectrometry makes it an extremely attractive detection technique for microfluidic enzyme inhibition studies despite the expense of mass spectrometry and the challenges of interfacing a mass spectrometer to a microfluidic device. A mass spectrometer is ideally suited to selectively measuring both substrate and product in a mixture as long as these two species have different mass-to-charge ratios when ionized. Mass spectrometric detection was used by de Boer et al. with a microchip device for screening inhibitors of cathepsin B_1 (92). After the enzyme-catalyzed reaction took place in the microfluidic device, the mixture containing the inhibitor, enzyme, substrate, and product was introduced into the mass spectrometer, which was able to selectively determine reaction product concentrations. Mass spectrometry has been used with other studies of enzyme inhibition, but the column/channel volumes and flow rates used were above the upper limit for a microfluidic device as defined for this chapter (101–103).

7.3 ENZYME INHIBITION STUDIES IN MICROFLUIDIC DEVICES: SPECIFIC STUDIES

In the first two sections of this chapter, a general review of techniques for studying enzyme inhibition using microfluidic devices with an emphasis on online approaches was presented. In this final section of the chapter, we present four specific studies, which highlight the potential of microfluidic devices as platforms to study enzyme inhibition and demonstrate the diversity of possible techniques and applications.

7.3.1 Capillary Electrophoretic Analysis of Alkaline Phosphatase Inhibition Using a Combination of Continuous and Transient Engagement EMMA (82,83)

Whisnant et al. developed a technique for performing online enzyme inhibition assays in capillaries, in which the two basic formats of EMMA, continuous and transient engagement EMMA, were combined (83,82). This approach takes advantage of online electrophoretic mixing and separation of zones of enzyme and inhibitor to study the inhibition of the enzyme. This technique was first applied to study the inhibition of alkaline phosphatase by theophylline. Theophylline is a reversible, noncompetitive inhibitor of this enzyme. A fluorogenic substrate, AttoPhos (2′-[2-benzothiazoyl]-6′-hydroxybenzothiazole phosphate), was used for monitoring the enzyme-catalyzed reaction rate by detecting product fluorescence.

The alkaline phosphatase enzyme assays without inhibitor were carried out by injecting electrokinetically a zone of the enzyme (0.18 nM alkaline phosphatase, 3.0 s, 17.8 kV, 310 V/cm) into a capillary filled with the substrate at high concentration ($[S] \gg K_m$) and the running buffer. A constant potential (310 V/cm) was then applied. As the zone of enzyme–substrate complex migrated through the capillary, the fluorescent product was formed continuously and observed as a plateau by a downstream LIF detector (Fig. 7.9). For the alkaline phosphatase-catalyzed reaction of AttoPhos, the enzyme-substrate complex migrated faster than the fluorescent product during CE. Therefore, the electropherogram in Fig. 7.9 is reversed in time relative to the sequence of events taking place inside the capillary. The first product detected was produced near the end of the experiment just as the enzyme-substrate

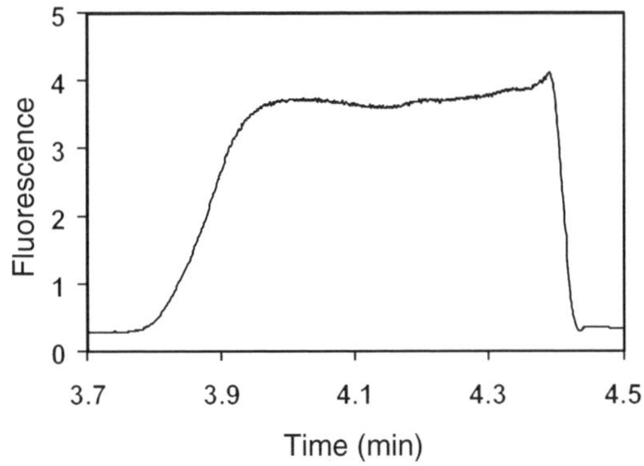

FIGURE 7.9 Electropherogram of an alkaline phosphatase enzyme assay. A zone of 0.18 nM alkaline phosphatase was injected for 3.0 s at 17.8 kV (310 V/cm) into the capillary filled with 0.10 mM AttoPhos and 50 mM diethanolamine (DEA) at pH 9.5. The separation potential was 17.8 kV. Reproduced with permission from Reference 83.

complex migrated past the detector. The last product that reached the detector was formed when the enzyme was first injected into the capillary.

For the assays of alkaline phosphatase inhibition by theophylline, the capillary was first filled with the substrate and the running buffer as in the enzyme assays without inhibitor. A zone of the inhibitor, theophylline, was injected into the capillary electrokinetically (4.0 s, 310 V/cm). Theophylline was injected first because it has a lower net electrophoretic mobility than the enzyme–substrate complex. After theophylline injection, a potential was applied for 20 s (310 V/cm) to create a zone of substrate and running buffer between the inhibitor and enzyme zones. Next, 0.18 nM alkaline phosphatase was injected for 3.0 s at 310 V/cm. Finally, a constant potential was applied, and the fluorescent product was formed continuously. When the zones of inhibitor and enzyme–substrate complex mixed electrophoretically, product formation decreased due to the inhibition and a negative peak was observed on the product plateau (Fig. 7.10). After the zone of enzyme–substrate separated from the theophylline zone again, the enzyme activity was restored since theophylline is a reversible inhibitor of alkaline phosphatase. Using Michaelis–Menten treatment of the electrophoretic data, the K_i value for theophylline was determined to be $102 \pm 4\,\mu M$, which was in agreement with literature values.

Using the same approach, several other inhibitors of alkaline phosphatase were investigated, including sodium vanadate and sodium arsenate (competitive, reversible inhibitors) and EDTA (irreversible inhibitor). Since sodium vanadate and sodium arsenate are reversible inhibitors, the resulting electropherograms resembled the electropherogram for theophylline inhibition shown in Fig. 7.10. The K_i values were calculated by analyzing the data using Michaelis–Menten theory, and they were consistent with literature values for these inhibitors. Experiments with EDTA showed that it irreversibly inhibits alkaline phosphatase at concentrations of 1.0 mM and above. Fig. 7.11 shows an electropherogram for inhibition by 1.0 mM

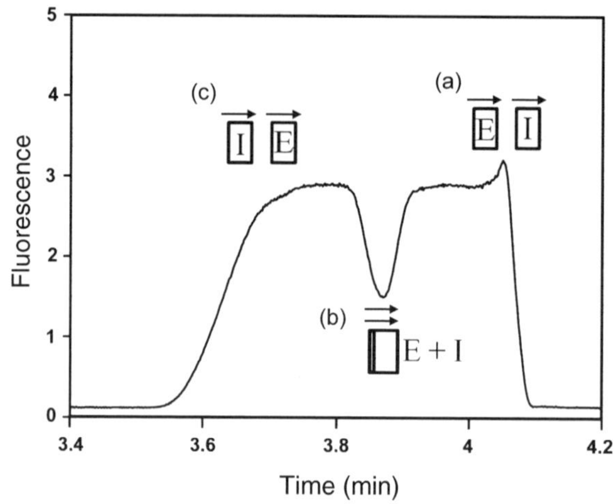

FIGURE 7.10 Electropherogram of a theophylline–alkaline phosphatase enzyme inhibitor assay. A zone of 100 μM theophylline was injected for 4.0 s at 17.8 kV into a capillary filled with 0.10 mM AttoPhos and 50 mM DEA at pH 9.5. Next, a potential of 17.8 kV was applied for 20 s. Then, a zone of 0.18 nM alkaline phosphatase was injected into the capillary for 3.0 s at 17.8 kV. Finally, the separation potential (17.8 kV) was applied. (**a**)–(**c**) The relative positions of the enzyme and inhibitor zones in the capillary at the corresponding positions in the resulting electropherogram. Reproduced with permission from Reference 83.

EDTA. Product formation decreased when the zones of EDTA and enzyme-substrate complex overlapped electrophoretically (4.2 min, Fig. 7.11). After these two zones separated again, the enzyme did not return to its original activity (4.0 min, Fig. 7.11). For EDTA injections at concentrations from 20 to 400 μM, activation for

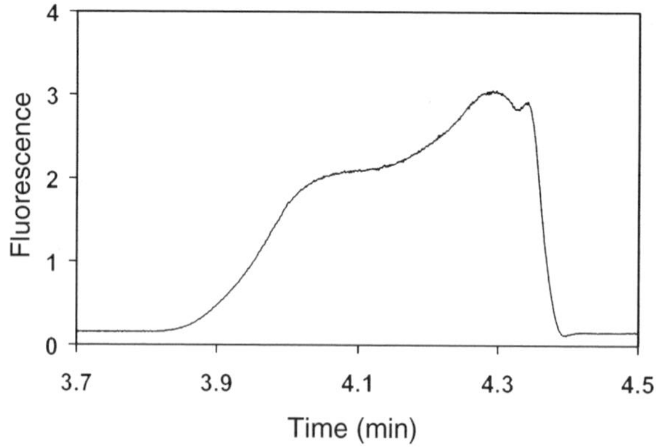

FIGURE 7.11 Electropherogram of an EDTA–alkaline phosphatase enzyme inhibition assay. A zone of 1.0 mM EDTA was injected for 55.0 s at 17.8 kV into a capillary filled with 0.10 mM AttoPhos and 50 mM DEA. Next, a voltage of 17.8 kV was applied for 180 s. Then, a zone of 0.18 nM alkaline phosphatase was injected for 3.0 s at 17.8 kV into a capillary, and the separation potential was applied (17.8 kV). Reproduced with permission from Reference 83.

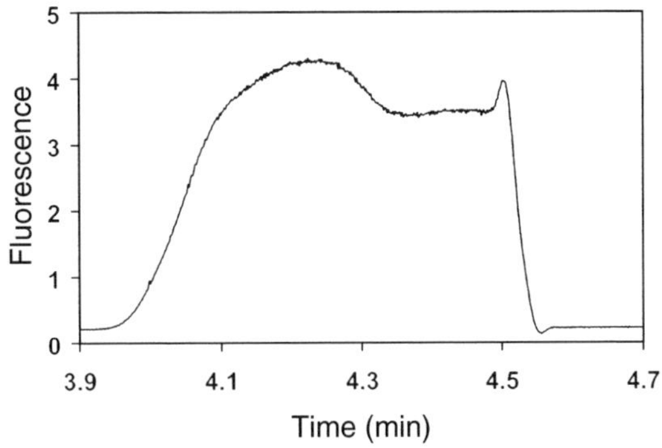

FIGURE 7.12 The experimental conditions are the same as Fig. 7.11 except the concentration of EDTA was 0.030 mM. Reproduced with permission from Reference 83.

alkaline phosphatase was observed, and Fig. 7.12 shows the electropherogram at an EDTA concentration of 30 μM.

This study highlights the wealth of information about enzyme inhibition that can be obtained using only the simplest microfluidic device, a 50 μm i.d. fused-silica capillary. This online EMMA method for studying enzyme inhibition can distinguish reversible inhibition, irreversible inhibition, and activation based on the shape of electropherograms. Competitive and noncompetitive reversible inhibitors can be quantified using Michaelis–Menten treatment of the data, and K_i values can be determined for these inhibitors. Also, much less enzyme is consumed compared to conventional microplate assays. One significant limitation of this method used in this paper is that a fluorogenic enzyme substrate is required for LIF detection so that very low concentrations of enzyme (0.18 nM) can be used.

7.3.2 Screening of Enzyme Inhibitors Using a Capillary-based Enzyme Microreactor (71)

Tang and Kang studied the inhibition of angiotensin-converting enzyme using an on-column enzyme-catalyzed reaction followed by CE separation of substrate and product by UV absorbance detection (71). Central to this work was a novel approach to immobilize enzyme in the capillary using an ionic binding technique to create the on-column microreactor. This assay resembles off-line enzyme assays using CE for separation and detection, but the enzyme-catalyzed reaction is actually carried out on-column in the first 1.5 cm of the 34.5 cm capillary. Zones of substrate were injected into the capillary and allowed to incubate to initiate the enzyme-catalyzed reaction.

The enzyme microreactor was prepared using a two-step protocol. First, the capillary surface was coated with hexadimethrine bromide (HDB) by flushing the capillary with an aqueous solution of this polycationic molecule. This coating reversed electroosmotic flow and provided a positively charged surface to which the negatively charged enzyme could adsorb. In the second step, enzyme solution was

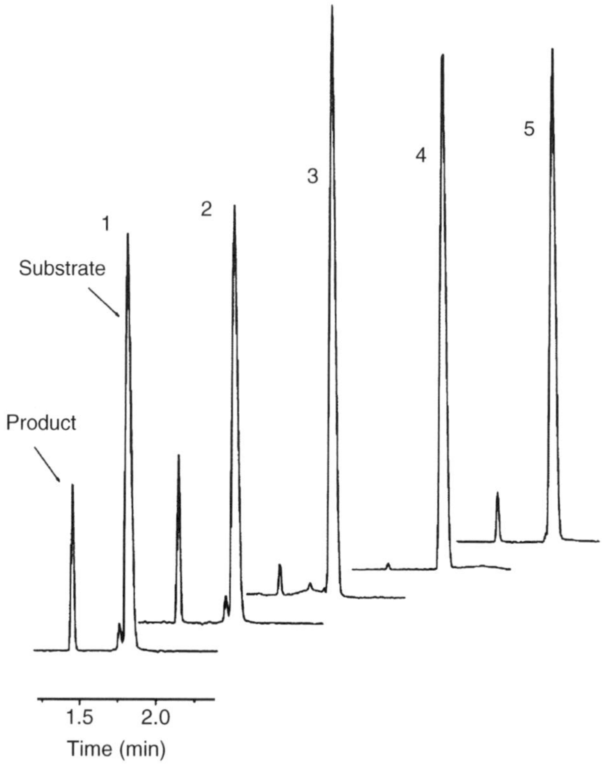

FIGURE 7.13 Electropherograms showing the separation and the detection of substrate and reaction product for ACE inhibition assays. Compounds were detected by absorbance at 230 nm. Samples: 1, blank; 2, extract of *Rhizoma coptidis* (0.5 mg/mL); 3, extract of *Rhizoma coptidis* (0.5 mg/mL) spiked with 38.9 μg of cilazaprilat; 4, cilazaprilat (500 nM); 5, captopril (500 nM). Reproduced with permission from Reference 71. Copyright 2006 American Chemical Society.

introduced into the first 1.5 cm of the capillary and was allowed to incubate for 5 min so that the enzyme could adsorb to the HDB-coated capillary surface.

Enzyme and enzyme inhibition assays were carried out by injecting a short zone (4 s, pressure) of substrate hippuryl-His-Leu (HHL) into the enzyme-coated region at the injection end of the capillary. The substrate was allowed to incubate for 1 min before high voltage (435 V/cm) was applied to perform CE and separate the substrate from the reaction product hippuric acid (HA). Fig. 7.13 shows separations of HHL and HA. These molecules were detected by UV absorbance at 230 nm, and the area of the product peak was used to determine the enzyme-catalyzed reaction rate during the 1-min incubation period. The enzyme-coating procedure was shown to produce a stable enzyme coating that preserved the catalytic activity of this enzyme. The enzyme activity dropped only 40% after 300 assays, and the activity of the coated enzyme was shown to be more stable over several days than that of the enzyme in solution. The K_m value determined for the immobilized ACE was found to be slightly lower than literature values for the same enzyme in solution.

For inhibition studies, the inhibitor was added to the substrate solution and was injected by pressure. Two reversible inhibitors of ACE, captopril and cilazaprilat, were studied in detail using this method. Fig. 7.14 shows Lineweaver–Burk plots generated for these two inhibitors. These plots were used to determine K_i values for the inhibitors and to characterize their mechanisms of inhibition. Captopril displayed competitive inhibition behavior, and cilazaprilat showed noncompetitive inhibition behavior. These results and the K_i values determined were consistent with literature reports. The assay was then used to screen for inhibition by a third known ACE inhibitor dithiothreitol, as well as 31 natural extracts. Dithiothreitol inhibited ACE

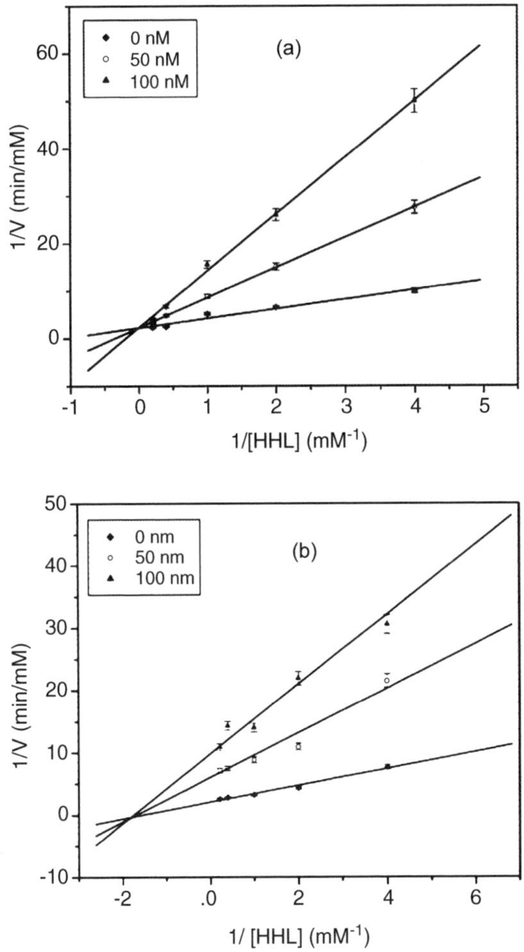

FIGURE 7.14 Lineweaver–Burk plots for captopril (**a**) and cilazaprilat (**b**). Reproduced with permission from Reference 71. Copyright 2006 American Chemical Society.

catalysis as expected, but none of the natural extracts inhibited the enzyme. Although the assay failed to identify new inhibitors in this initial study, the method is attractive because of its simplicity and the ability to perform the experiments using a commercial CE system. It is not certain that this immobilization approach will be effective for a broad range of enzymes, but there is a recent report that another research group used a similar approach successfully to study catalysis by glucose oxidase and glutamate oxidase (104).

7.3.3 Determination of β-Galactosidase Inhibition in a Microchip Device (39)

Hadd and coworkers reported the first study of enzyme inhibition in a microfabricated device in 1997 (39). A schematic of the microchip used in these studies is shown in Fig. 7.15. This device was used to measure the kinetics of a reaction catalyzed by β-galactosidase (β-Gal). Inhibition of this enzyme by three different inhibitors was also investigated. The enzyme-catalyzed reaction is carried out online and is based on the confluent mixing of substrate, enzyme, and inhibitor at a four-way intersection into the reaction channel. Electrokinetic transport was used to mix the enzyme, substrate, and inhibitor continuously in the reaction channel, and the fluorescent reaction product was continuously detected 20 mm downstream from the four-way channel intersection by LIF. This work takes advantage of the flexibility of microfabricated devices for mixing solutions to study reaction kinetics. Similar

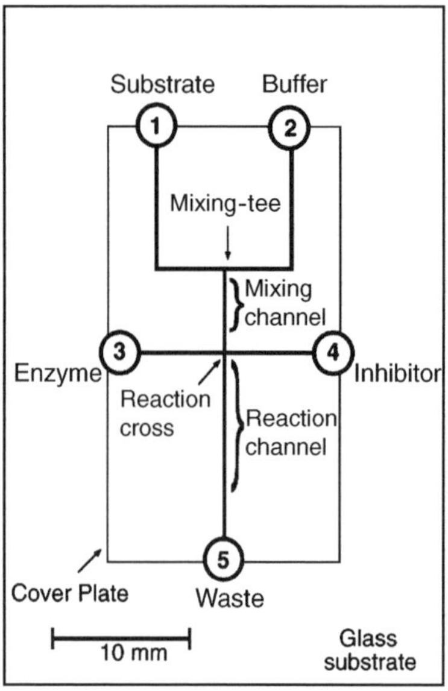

FIGURE 7.15 Microchip device with two microchannel intersections for microfluidic enzyme inhibition studies. Reproduced with permission from Reference 39. Copyright 1997 American Chemical Society.

experiments using a network of capillaries connected by unions would be impractical, but the device shown in Fig. 7.15 is fairly simple for a microfabricated microfluidic device.

Substrate and buffer were mixed at the three-way intersection in Fig. 7.15. The substrate concentration for the enzyme-catalyzed reaction could be adjusted by adjusting the potentials in reservoirs 1 and 2, which altered the dilution of the substrate. Enzyme and inhibitor were added to the diluted substrate solution from separate reservoirs at the four-way intersection. The enzyme and inhibitor concentrations were changed by replacing the solutions in the appropriate reservoirs (3 and 4, Fig. 7.15). The enzyme-catalyzed reaction then took place as these solutions moved down the reaction channel. The concentration of substrate was rapidly controlled simply by adjusting applied potentials under computer control. Kinetic studies to generate Lineweaver–Burk plots could be carried out rapidly under computer control. Calibration experiments indicated that the mixing of substrate and buffer from reservoirs 1 and 2 was accurate within the precision of this process (2% RSD). A fluorogenic substrate for β-Gal resorufin β-D-galactopyranoside was used. The fluorescent reaction product resorufin was detected selectively by LIF ($\lambda_{ex} = 543$ nm, $\lambda_{em} > 570$ nm) in the presence of excess substrate, and the fluorescent signal was proportional to the reaction rate. The microchip device was fabricated in glass with a channel depth and width at half-height of 9 μm and 35 μm, respectively. The sizes of the reservoirs were not reported. The field strength in both the mixing channel and the reaction channel was 220 V/cm.

Enzyme assays were performed using this device both with and without inhibitors present. The data for the enzyme assays without inhibitor are similar in appearance to data for enzyme inhibition measurements (Fig. 7.16). The authors determined the K_m and k_{cat} values for the enzyme by determining the reaction rate (resorufin fluorescence) at a series of substrate concentrations. The values determined with the microfluidic device were compared to values obtained in the same laboratory using a conventional assay in a cuvette with absorbance detection, and these values were similar.

Inhibition of β-Gal by three molecules was studied—phenylethyl β-D-thiogalactoside (PETG), *p*-hydroxymercuribenzoic acid (PHMB), and lactose. Fig. 7.16 presents data for the inhibition of β-Gal by PETG, which is a competitive, reversible inhibitor of the enzyme. Each plot in Fig. 7.16a shows the fluorescence at a particular inhibitor concentration as the substrate concentration is increased stepwise at regular time intervals from 28 to 115 μM. The inhibitor concentration was manually changed by changing the solution in reservoir 4. The β-Gal concentration was constant (570 μg/L) in these experiments. As the inhibitor concentration increased (0, 6, 12, and 30 μM), the fluorescent signal observed at each substrate concentration decreased, indicating that the reaction rate had decreased. Lineweaver–Burk plots for this data are presented in Fig. 7.16b, and these plots are typical for a reversible, competitive inhibitor. A K_i value of 8 μM was obtained from this data for PETG, which was significantly higher than values reported in the literature. Fig. 7.17 shows a plot of percent inhibition versus percent inhibitor concentration for PETG, PHMB, and lactose. These data were used to determine the IC_{50} for PETG and lactose. This

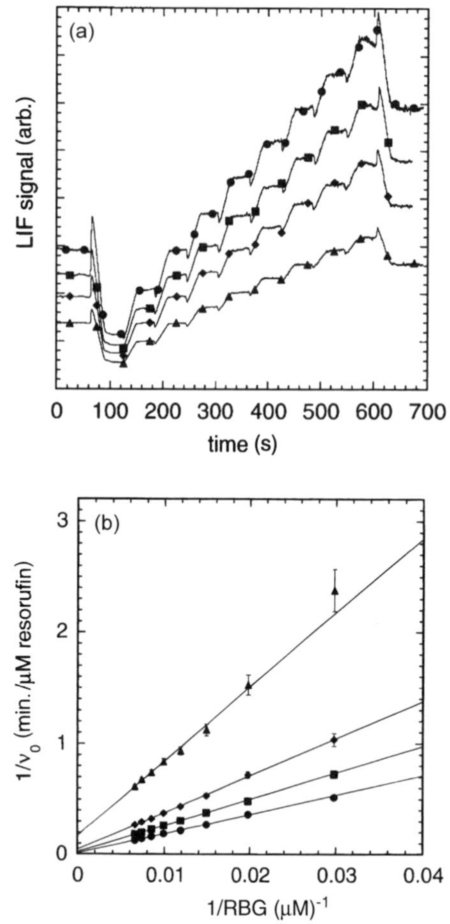

FIGURE 7.16 (**a**) Resorufin (reaction product) fluorescence at different substrate concentrations with PETG inhibition of β-galactosidase. Substrate concentrations varied with time from 28 to 115 μM with (•) no inhibitor, (■) 6 μM PETG, (♦) 12 μM PETG and (▲) 30 μM PETG. (**b**) Lineweaver–Burk plots of the initial reaction rate versus substrate concentration with (•) no inhibitor, (■) 6 μM PETG, (♦) 12 μM PETG, and (▲) 30 μM PETG. Reproduced with permission from Reference 39. Copyright 1997 American Chemical Society.

value could not be obtained for PHMB, which the authors attributed to a slow rate of inhibition.

7.3.4 Online Studies of Acetylcholinesterase Inhibition in a Microchip Device (49)

Heleg-Shabtai et al. reported the development of an on-chip assay for studying acetylcholinesterase inhibition (49). Acetylcholinesterase is a target of some

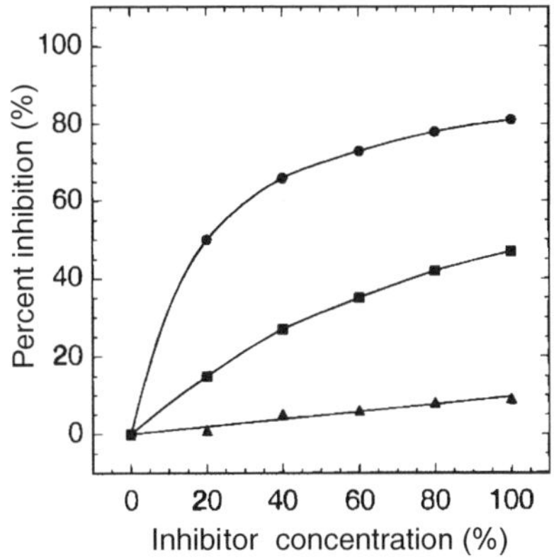

FIGURE 7.17 Plots of percent inhibition versus inhibitor concentration: PETG, 0–40 μM (●); lactose, 0–4.0 mM (■); and PHMB, 0–2 mM (▲). Reproduced with permission from Reference 39. Copyright 1997 American Chemical Society.

chemical warfare agents commonly referred to as nerve gases, and the k_i value for irreversible inhibition of acetylcholinesterase by *O*-ethyl *S*-[2-(diisopropylamino) ethyl] methylphosphonothiolate (VX) was measured in this study. The authors, approach to study acetylcholinesterase inhibition used EMMA both to fluorescently label and selectively detect the product of the enzyme-catalyzed reaction as well as to carry out the enzyme-catalyzed reaction. Hadd et al. also reported a microchip study of acetylcholinesterase inhibition (47). Their study used a fluorogenic reagent to label the product of the acetylcholinesterase-catalyzed reaction, which simplified the procedure. In the study by Heleg-Shabtai and coworkers, acetylthiocholine (AcCh) was used as an acetylcholinesterase (AcChE) substrate, and the product thiocholine (TC) was derivatized with eosinmaleimide (Eo-MI) for detection by LIF. Both the derivatization reagent and the derivatized thiocholine (Eo-TC) are fluorescent, so they were separated by CE in the microchip device for selective detection. Fig. 7.18

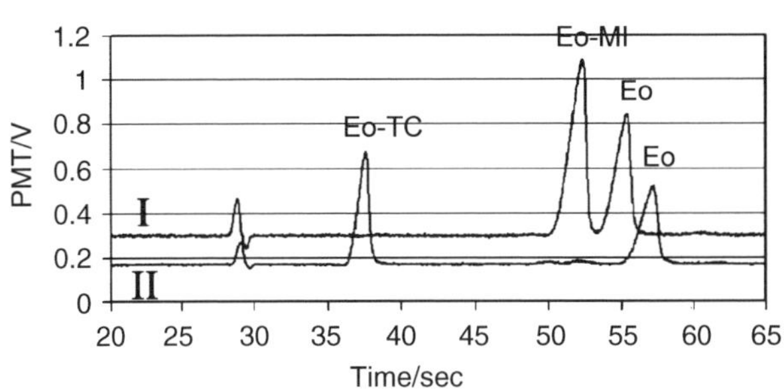

FIGURE 7.18 Electropherograms for an AcChE assay. Peak labels are defined in the text. (I) Electropherogram without enzyme; (II) in the presence of 0.5 U/mL AcChE. Reproduced with permission from Reference 49.

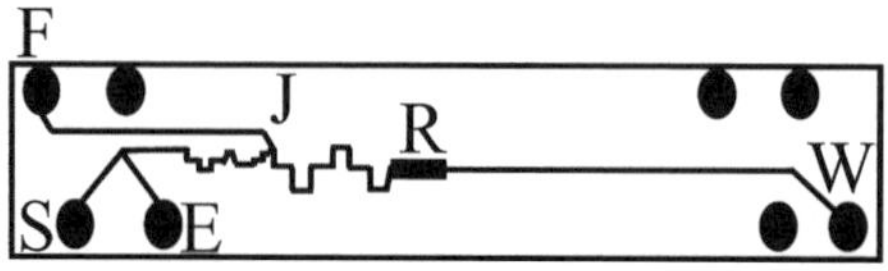

FIGURE 7.19 Microfabricated device for AcChE inhibition assays. S, substrate reservoir; E, enzyme reservoir; F, derivatization reagent reservoir; J, junction of microchannels from F, S, and E; R, reaction segment; W, waste. Reproduced with permission from Reference 49.

presents electropherograms showing the separation of the derivatization reagent (Eo-MI), the derivatized product of the enzyme-catalyzed reaction (Eo-MC), and eosin (Eo), which was used as an internal standard.

Fig. 7.19 is a schematic of the glass microchip device used in this work. For all experiments presented in this paper, the substrate was located in reservoir S, and the enzyme was located in reservoir E. Reservoir F contained the internal standard Eo and the derivatization reagent Eo-MI. Enzyme and substrate mixed and reacted in the section of the channel before the intersection labeled J. At that point the reaction product TC will react with the derivatization reagent. It was necessary to create a short zone of labeled thiocholine so that it could be separated from the derivatization reagent and internal standard as shown in Fig. 7.18; therefore, a plug of Eo-MI and Eo was injected (0.5 s, electrokinetic injection) periodically to determine the concentration of derivatized thiocholine (Eo-TC). The concentration of Eo-TC detected was proportional to the reaction rate for the enzyme-catalyzed reaction. The enzyme-catalyzed reaction was carried out by two methods. In the first method, the entire channel between the mixing point for S and E and the junction J was filled with both enzyme and substrate. In the second approach, only a 3.0 s plug of enzyme was injected from E into a continuous stream of substrate between the S and E mixing points and the junction J. This second method is essentially a form of EMMA using continuous engagement EMMA for the enzyme-catalyzed reaction before the intersection at J. Reaction of the long zone of TC produced with the 0.5 s zone of Eo-MI is probably best thought of as a form of continuous engagement EMMA, though one could also categorize it as plug–plug EMMA with a long plug of TC. Both methods gave similar results, and the authors elected to use the second method for inhibition studies since it consumed less enzyme.

Two inhibitors VX and paraoxon were studied. The inhibitors were added to the enzyme off-chip and introduced from reservoir E. The reaction rate and resulting amount of Eo-TC produced decreased in the presence of the inhibitors. Fig. 7.20 shows a plot of Eo-TC peak height versus [VX]. These data along with data from experiments at higher VX concentrations were used to determine a k_i value of $3.5(\pm 0.2) \times 10^7\ M^{-1}\ min^{-1}$, which is in close agreement with literature values. The detection limit for VX was 4×10^{-10} M. The studies of acetylcholinesterase inhibition by paraxon also produced a k_i value consistent with literature values. This report illustrates the flexibility and versatility of microfabricated microfluidic devices for carrying out enzyme inhibition studies.

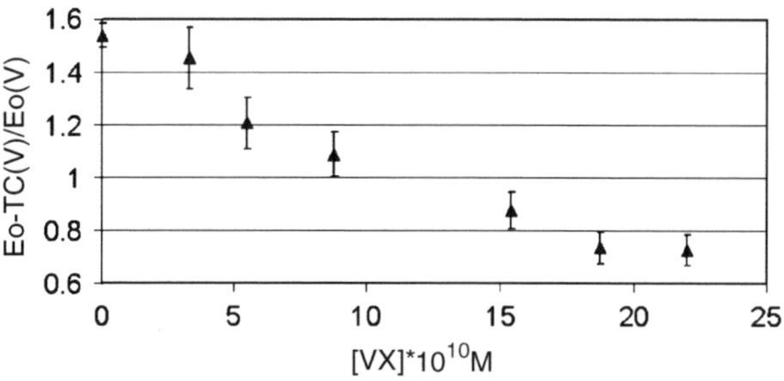

FIGURE 7.20 Product formation (ratio of product peak to internal standard peak) versus VX concentration. Error bars represent standard deviations for three measurements. Reproduced with permission from Reference 49.

REFERENCES

1. Berg JM, Tymoczko JL, Stryer L. Biochemistry. 5th ed. New York: Freeman Press; 2002.
2. Copeland RA. Evaluation of Enzyme Inhibitors in Drug Discovery: A Guide for Medicinal Chemists and Pharmacologists; Hoboken: Wiley & Sons; 2005.
3. Leskovac V. Comprehensive Enzyme Kinetics. New York: Kluwer Academic/Plenum Publishers; 2003.
4. Marangoni AG. Enzyme Kinetics: A Modern Approach. Hoboken: Wiley & Sons; 2003.
5. Mikkelsen SR, Cortón E. Bioanalytical Chemistry. Hoboken: Wiley & Sons; 2004.
6. Copeland RA. Enzymes: A Practical Introduction to Structure, Mechanism, and Data Analysis. 2nd ed.; New York: Wiley & Sons; 2000.
7. Riva S. Trends Biotechnol 2006;24:219–226.
8. Singh R, Sharma R, Tewari N, Geetanjali, Rawat DS. Chem Biodiversity 2006;3: 1279–1287.
9. Wohlgemuth R. Eng Life Sci 2006;6:577–583.
10. Rapley R. The Nucleic Acid Protocols Handbook. Totowa, NJ: Humana Press; 2000.
11. McPherson MJ, Moller SG. PCR. New York: BIOS Scientific Publishers Limited; 2000.
12. Kicman AT, Parkin MC, Iles RK. Mol Cell Endocrinol 2007;260–262:212–227.
13. Liebler DC. Introduction to Proteomics: Tools for the New Biology. Totowa, NJ: Humana Press, 2002.
14. Dzyadevych SV, Soldatkin AP, El'skaya AV, Martelet C, Jaffrezic-Renault N. Anal Chim Acta 2006;568:248–258.
15. Choi MMF. Microchim Acta 2004;148:107–132.
16. Hess FD. Weed Sci 2000;48:160–170.
17. Duff SMG, Chen Y-CS, Fabbri BJ, Yalamanchili G, Hamper BC, Walker DM, Brookfield FA, Boyd EA, Ashton MR, Yarnold CJ, CaJacob CA. Pestic Biochem Physiol 2007;88:1–13.

18. Guerrero A, Rosell G. Curr Med Chem 2005;12:461–469.
19. Pope C, Karanth S, Liu J. Environ Toxicol Pharmacol 2005;19:433–446.
20. Shih T-M, Kan RK, McDonough JH. Chem Biol Interact 2005;157–158:293–303.
21. Bao JJ, Fujima JM, Danielson ND. J Chromatogr B 1997;699:481–497.
22. Van Dyck S, Kaale E, Novakova S, Glatz Z, Hoogmartens J, Van Schepdael A. Electrophoresis 2003;24:3868–3878.
23. Burns KL, May SW. J Chromatogr B 2003;797:175–190.
24. Zhang J, Hoogmartens J, Van Schepdael A. Electrophoresis 2006;27:35–43.
25. Guijt RM, Baltussen E, van Dedem GWK. Electrophoresis 2002;23:823–835.
26. Glatz Z. J Chromatogr B 2006;841:23–37.
27. Wang J. Electrophoresis 2002;23:713–718.
28. Urban PL, Goodall DM, Bruce NC. Biotechnol Adv 2006;24:42–57.
29. Krueger RJ, Hobbs TR, Mihal KA, Tehrani J, Zeece MG. J Chromatogr 1991;543:451–461.
30. Lee K-B, Kim Y-S, Linhardt RJ. Electrophoresis 1991;12:636–640.
31. Okanda FM, Rassi ZE. Electrophoresis 2007;28:89–98.
32. Besanger TR, Hodgson RJ, Green JRA, Brennan JD. Anal Chim Acta 2006;564: 106–115.
33. Hodgson RJ, Chen Y, Zhang Z, Tleugabulova D, Long H, Zhao X, Organ M, Brook MA, Brennan JD. Anal Chem 2004;76:2780–2790.
34. Xue Q, Yeung ES. Anal Chem 1994;66:1175–1178.
35. Xue Q, Yeung ES. Nature 1995;373:681–683.
36. Craig DB, Arriaga EA, Wong JCY, Lu H, Dovichi NJ. JACS 1996;118:5245–5253.
37. Regnier FE, Patterson DH, Harmon BJ. Trends Anal Chem 1995;14:177–181.
38. Bao J, Regnier FE. J Chromatogr 1992;608:217–224.
39. Hadd AG, Raymond DE, Halliwell JW, Jacobson SC, Ramsey JM. Anal Chem 1997;69:3407–3412.
40. Zugel SA, Burke BJ, Regnier FE, Lytle FE. Electrophoresis 2000;72:5731–5735.
41. Xu H, Ewing AG. Anal Bioanal Chem 2004;378:1710–1715.
42. Murakami Y, Morita T, Kanekiyo T, Tamiya E. Biosens Bioelectron 2001;16:1009–1014.
43. Burke BJ, Regnier FE. Electrophoresis 2001;22:3744–3751.
44. Starkey DE, Han A, Bao JJ, Ahn CH, Wehmeyer KR, Prenger MC, Halsall HB, Heineman WR. J Chromatogr B 2001;762:33–41.
45. Xue Q, Wainright A, Gangakhedkar S, Gibbons I. Electrophoresis 2001;22:4000–4007.
46. Cohen CB, Chin-Dixon E, Jeong S, Nikiforov TT. Anal Biochem 1999;273:89–97.
47. Hadd AG, Jacobson SC, Ramsey JM. Anal Chem 1999;71:5206–5212.
48. Duffy DC, Gillis HL, Lin J, Sheppard NF, Kellogg GJ. Anal Chem 1999;71:4669–4678.
49. Heleg-Shabtai V, Gratziany N, Liron Z. Anal Chim Acta 2006;571:228–234.
50. Garcia E, Hasenbank MS, Finlayson B, Yager P. Lab Chip 2007;7:249–255.
51. Schilling EA, Kamholz AE, Yager P. Anal Chem 2002;74:1798–1804.

52. Burke BJ, Regnier FE. Anal Chem 2003;75:1786–1791.
53. Xiong L, Regnier FE. J Chromatogr A 2001;924:165–176.
54. Kheterpal I, Mathies RA. Anal Chem 1999;71:31A–37A.
55. Kan C-W, Fredlake CP, Doherty EAS, Barron AE. Electrophoresis 2004;25:3564–3588.
56. Ma L, Gong X, Yeung ES. Anal Chem 2000;72:3383–3387.
57. He Y, Yeung E. Electrophoresis 2003;24:101–108.
58. Lin S, Fischl AS, Bi X, Parce W. Anal Biochem 2003;314:97–107.
59. Tu J, Anderson LN, Dai J, Peters K, Carr A, Loos P, Buchanan D, Bao JJ, Liu C, Wehmeyer KR. J Chromatogr B 2003;789:323–335.
60. Pang H-M, Kenseth J, Coldiron S. Drug Discov Today 2004;9:1072–1080.
61. Xu HW, Ewing AG. Electrophoresis 2005;26:4711–4717.
62. Dai J, Tu J, Anderson LN, Bao JJ, Liu C, Quay B, Wehmeyer KR. LC-GC 2002;20:606–617.
63. Hoffmann T, Reinhold D, Kaehne T, Faust J, Neubert K, Frank R, Ansorge S. J Chromatogr A 1995;716:355–362.
64. Deloffre L, Sautiere P-E, Huybrechts R, Hens K, Vieau D, Salzet M. Eur J Biochem 2004;271:2101–2106.
65. Iqbal J, Burbiel JC, Mueller CE. Electrophoresis 2006;27:2505–2517.
66. Koval D, Jiraskova J, Strisovsky K, Konvalinka J, Kasicka V. Electrophoresis 2006;27:2558–2566.
67. Sano M, Nishino I, Ueno K, Kamimori H. J Chromatogr B 2004;809:251–256.
68. Yuan J, Wei H, Jin W, Yang X, Wang E. Electrophoresis 2006;27:4047–4051.
69. Krenkova J, Foret F. Electrophoresis 2004;25:3550–3563.
70. Sakai-Kato K, Kato M, Toyo'oka T. J Chromatogr A 2004;1051:261–266.
71. Tang Z-M, Kang J-W. Anal Chem 2006;78:2514–2520.
72. Park S-S, Cho S-I, Kim M-S, Kim Y-K, Kim B-G. Electrophoresis 2003;24:200–206.
73. Rashkovetsky LG, Lyubarskaya YV, Foret F, Hughes DE, Karger BL. J Chromatogr A 1997;781:197–204.
74. Saevels J, Van den Steen K, Van Schepdael A, Hoogmartens J. J Chromatogr A 1996;745:293–298.
75. Tang Z-M, Wang Z-Y, Kang J-W. Electrophoresis 2007;28:360–365.
76. Van Dyck S, Van Schepdael A, Hoogmartens J. Electrophoresis 2001;22:1436–1442.
77. Novakova S, Telnarova M, Glatz Z. J Chromatogr A 2003;990:189–195.
78. Novakova S, Glatz Z. Electrophoresis 2002;23:1063–1069.
79. Papezova K, Nemec T, Chaloupkova R, Glatz Z. J Chromatogr A 2007;1150.
80. Telnarova M, Vytiskova S, Monincova M, Glatz Z. Electrophoresis 2004;25:1028–1033.
81. Belenky A, Hughes D, Korneev A, Dunayevskiy Y. J Chromatogr A 2004;1053: 247–251.
82. Whisnant AR, Johnston SE, Gilman SD. Electrophoresis 2000;21:1341–1348.
83. Whisnant AR, Gilman SD. Anal Biochem 2002;307:226–234.

84. Van Dyck S, Vissers S, Van Schepdael A, Hoogmartens J. J Chromatogr A 2003;986: 303–311.
85. Taga A, Honda S. J Chromatogr A 1996;742:243–250.
86. Van Dyck S, Novakova S, Van Schepdael A, Hoogmartens J. J Chromatogr A 2003;1013:149–156.
87. Iqbal J, Vollmayer P, Braun N, Zimmermann H, Mueller CE. Purinergic Signal 2005;1:349–358.
88. Du W-B, Fang Q, Fang Z-L. Anal Chem 2006;78:6404–6410.
89. Economou A. Trends Anal Chem 2005;24:416–425.
90. Lenehan CE, Barnett NW, Lewis SW. Analyst 2002;127:997–1020.
91. Pu Q, Liu S. Anal Chim Acta 2004;511:105–112.
92. de Boer AR, Bruyneel B, Krabbe JG, Lingeman H, Niessen WMA, Irth H. Lab Chip 2005;5:1286–1292.
93. He B, Burke BJ, Zhang X, Zhang R, Regnier FE. Anal Chem 2001;73:1942–1947.
94. deMello AJ. Nature 2006;442:394–402.
95. Locascio LE. Anal Bioanal Chem 2004;379:325–327.
96. Swinney K, Bornhop D. Crit Rev Anal Chem 2000;30:1–30.
97. Campana AMG, Baeyens WRG, Guzman NA. Biomed Chrom 1998;12:172–176.
98. Mogensen KB, Klank H, Kutter JP. Electrohphoresis 2004;25:3498–3512.
99. Uchiyama K, Nakajima H, Hobo T. Anal Bioanal Chem 2004;379:375–382.
100. Li PCH, In: Cazes J, editor. Microfluidic Lab-on-a-Chip for Chemical and Biological Analysis and Discovery. Vol. 94 Boca Raton, FL, CRC Press; 2005. 187–250.
101. Hodgson RJ, Besanger TR, Brook MA, Brennan JD. Anal Chem 2005;77:7512–7519.
102. Zhang B, Palcic MM, Schriemer DC, Alvarez-Manilla G, Pierce M, Hindsgaul O. Anal Biochem 2001;299:173–182.
103. Ng ESM, Yang F, Kameyama A, Palcic MM, Hindsgaul O, Schriemer DC. Anal Chem 2005;77:6125–6133.
104. Hooper SE, Anderson MR. Electroanalysis 2007;19:652–658.

8

CHEMICAL SYNTHESIS WITHIN CONTINUOUS FLOW MICROREACTORS

Paul Watts

8.1 INTRODUCTION

Until recently, the greatest research effort in the field of miniaturization has been in analytical science with the main aim of this research being to develop miniaturized Total Analytical Systems (μ-TAS) (1). Alongside the continuing development of μ-TAS, a concerted effort has also been underway to establish the benefits that miniaturization can bring to the field of synthetic organic chemistry, the topic of this chapter.

In their simplest form, microreactors consist of a network of micron-sized channels (typically 10–300 μm in diameter) etched into a solid substrate (Fig. 8.1). A number of materials such as silicon, quartz, glass, metals, and polymers have been used to construct microreactors (2), however, important considerations in material choice for this application include chemical compatibility, ease of fabrication, and compatibility with detection methods. Glass (3) is a popular choice for synthetic applications since it is chemically inert, enables the use of a range of visible light detection methods, and fabrication techniques are well established. The other common material for synthetic applications is metal, such as nickel or stainless steel (4).

For solution-based chemistry, the channel networks are connected to a series of reservoirs containing chemical reagents to form the complete device with overall dimensions of a few centimeters. In the microreactor, reagents can be brought

Biological Applications of Microfluidics, Edited by Frank A. Gomez

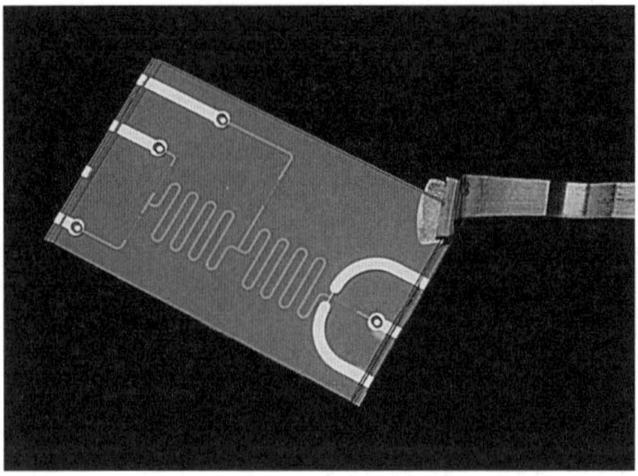

FIGURE 8.1 Glass microreactor with integrated electrodes.

together in a specific sequence, mixed, and allowed to react for a specified time in a controlled region of the channel network. For newcomers to the field of microreaction technology, the easiest way to manipulate fluids through a microchannel network is through the use of pumps, such as syringe pumps. This technique also enables the system to be integrated with online analytics, such as HPLC, providing an automated system suitable for reaction optimization (5). One disadvantage associated with the use of mechanical pumping techniques is that the large external pumps often dwarf the reaction system itself. This can be addressed by the integration of micropumps within the reactor, however, these microfabricated components increase reactor complexity and can be prone to wear and tare. In contrast, the use of nonmechanical pumping mechanisms, which rely on the direct transfer of energy, are advantageous as they are inherently simple, contain no moving parts, enable pulse-free, low flow rates to be obtained and are readily miniaturized. Of these techniques, the most popular nonmechanical pumping mechanism used for fluidic handling within microfabricated devices is electrokinetic flow (6,7). To manipulate fluids using electrokinetic flow, electrodes are inserted into the reagent reservoirs, as illustrated in Fig. 8.1, and an electric field applied. Compared to the use of mechanical micropumps, field-induced flow is advantageous as the electric field acts as both a pump and a valve, providing control over the direction and magnitude of fluid flow. In addition, the technique is readily automated enabling the use of complex flow regimes such as stopped flow. The main disadvantage of this approach however is that the pumps performance is directly linked to the properties of the fluid contained within the microchannel.

8.2 ADVANTAGES OF PERFORMING CHEMICAL SYNTHESIS IN MICROREACTORS

Within most traditional research laboratories, the vast majority of synthetic reactions are still performed using laboratory glassware, employing the same techniques that have been in place for decades, with current production technology

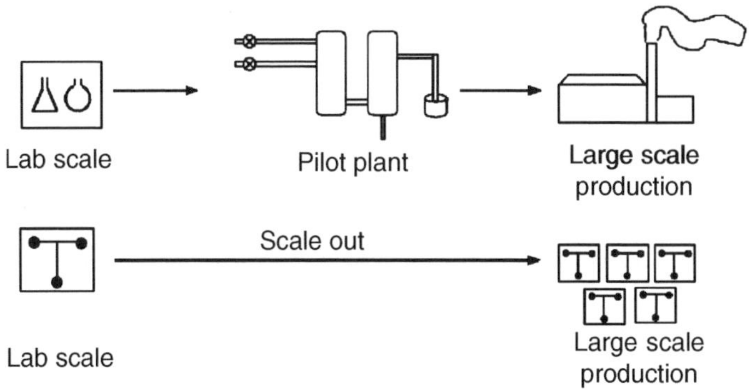

FIGURE 8.2 Schematic comparing the traditional and micro reactor approaches to chemical synthesis.

being based on the scale-up of successful laboratory-scale processes, first to a pilot plant, followed by a final increase in scale to achieve mass production. This approach is however flawed, as at each stage of the scale-up reactor modifications result in changes to the surface to volume ratio, which often have a profound effect on the thermal and mass transportation properties of the reaction. As a result of these variations, it is often necessary to reoptimize the process at each stage of scale-up, leading to severe delays; consequently, the route from laboratory to production is both time-consuming and expensive.

Microreaction technology has been proposed as a means of addressing this problem. Laboratory scale reactions may be performed and optimized using a single microreactor and in order to achieve production, multiple reactors may be simultaneously employed (Fig. 8.2). From a production perspective, this approach to scale-out, rather than scale-up, is advantageous as it enables changes in production volume to be met with ease, that is, by simply increasing or decreasing the number of devices employed in parallel. Furthermore, in comparison to a production plant where reactors are frequently configured for a single function, this flexibility is both advantageous and cost-effective enabling the demands of the customer to be met with ease.

8.3 CHEMICAL SYNTHESIS IN MICROREACTORS

A significant research program is now underway to establish the benefits that microreactors can bring to the field of synthetic organic chemistry. For the purpose of this chapter, a selection of reactions are discussed to highlight the advantages that microreactors provide; however, many reactions have been demonstrated to show enhanced reactivity, product yield, and selectivity when performed in microreactors as compared with conventional benchtop glassware, and readers are directed to more detailed reviews for further information (8–11).

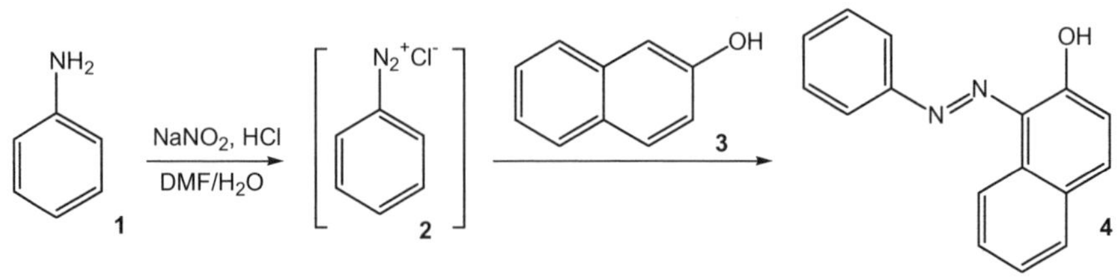

SCHEME 8.1 The *in situ* synthesis and reaction of diazonium salts in a microreactor.

8.3.1 Control of Hazardous Reactions

In traditional chemical reactors fluctuations in temperature are difficult to correct, as any alterations made take time to have an effect on the system as a whole; this consequently leads to the formation of hot spots and can potentially result in thermal runaway and explosion. The diazotization of aromatic amines is an industrially important process that involves the synthesis of highly unstable, explosive intermediates. With this in mind, de Mello and coworkers (12) demonstrated the increased reaction control obtained within a microreactor by converting aniline **1** into the diazonium salt **2** *in situ* prior to its reaction with β-naphthol **3**, affording the product **4** in 52% conversion (Scheme 8.1). The authors used the methodology to produce a range of chemical products with similar efficiency. Other examples of diazotization reactions in microreactors are reported by Harrison et al. (13) and Wille et al. (14).

Similarly, nitration reactions also represent an important but hazardous process, in which excess quantities of concentrated nitric and sulfuric acid are employed. Again, these reactions are extremely exothermic, making it difficult to control the reaction temperature when conducted on a large scale. Performing such reactions in microreactors not only enables excellent temperature control but also enhanced product selectivity (Scheme 8.2). Doku et al. reported the controlled nitration of benzene **5**, in a glass microreactor, obtaining a product distribution of 65% nitrobenzene **6**, 8% 1,3-dinitrobenzene **7** and only 5% of the explosive 1,3,5-trinitrobenzene **8** (15). Pennemann et al. have reviewed other nitration reactions that have been conducted in microreactors (16).

8.3.2 Regioselective Synthesis

Mikami and coworkers (17) reported enhanced regiocontrol when performing a series of Baeyer–Villiger oxidations within a microreactor system. As Scheme 8.3

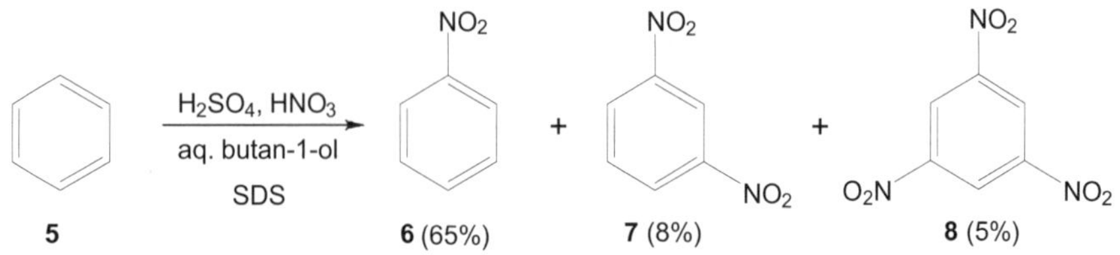

SCHEME 8.2 Nitration of benzene in a microreactor.

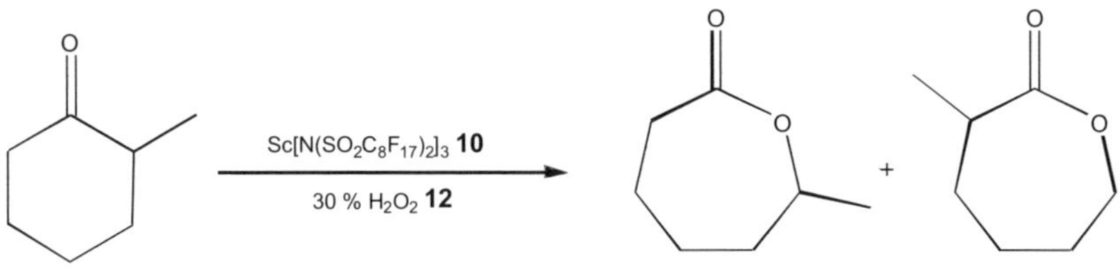

11 9a 9b

SCHEME 8.3 Regioselective Baeyer–Villiger oxidation within a microreactor.

illustrates, a typical reaction involved the insertion of an oxygen atom into a carbon–carbon bond, to afford a lactone (**9a** or **9b**). Under pressure-driven flow a biphasic reagent stream was obtained by introducing a solution of $Sc[N(SO_2C_8F_{17})_2]_3$ catalyst **10** and 2-methylcyclohexanone **11** in benzotrifluoride from one inlet and 30.0% aqueous hydrogen peroxide **12** from a second inlet. The reaction products were subsequently collected in a vial and the organic layer analyzed by GC in order to determine the percentage conversion and reaction regioselectivity. Employing a residence time of 8.1 s the authors report 91.0% conversion of 2-methylcyclohexanone **11** to lactone **9a** with 100.0% regioselectivity. When conducting the comparable reaction in a stirred batch reactor, the authors reported only 28.0% conversion with a product distribution of 69 : 31 (**9a** : **9b**), an observation that the authors attribute to the inhomogeneity obtained in a stirred reactor. Other examples where reactions have been conducted more selectively in microreactors include the Witttig reaction (18), Moffat–Swern oxidations (19), electrochemical alkylation (20), and nitrations (21).

8.3.3 Rapid Reaction Optimization

In addition to increased reaction control, microreaction technology enables the rapid optimization of reactions. Wiles et al. (22) reported the preparation of 1,3-diketone enolates, using an organic base **13**, and their subsequent reaction with a variety of Michael acceptors, such as **14**, to afford 1,4-addition products within a microreactor operated by EOF (Scheme 8.4).

When using a continuous flow of reagents **14** and **15**, only 15% conversion to the product **16** was observed, compared with 56% adduct **17** when diketone **18** was employed. The authors however demonstrated enhancements in conversion through the implementation of a stopped flow technique; this procedure involved the mobilization of reagents through the device whilst employing a periodic stopping of the reagent streams (via removal of the applied field), prior to mobilization of the reagents upon reapplication of a voltage. Using a regime of 2.5 s on and 5 s off, the authors report a 19% increase in conversion to adduct **16**; while lengthening the stopped flow period further, to 10 s, resulted in quantitative conversion to (*E*)-4-benzoyl-5-oxohex-2-enoic acid ethyl ester **16**. The procedure was subsequently repeated for the synthesis of adduct **17**, in which a regime of 2.5 s on and 5 s off resulted in an increase in conversion to 95% (*E*)-4-acetyl-5-oxohex-2-enoic acid

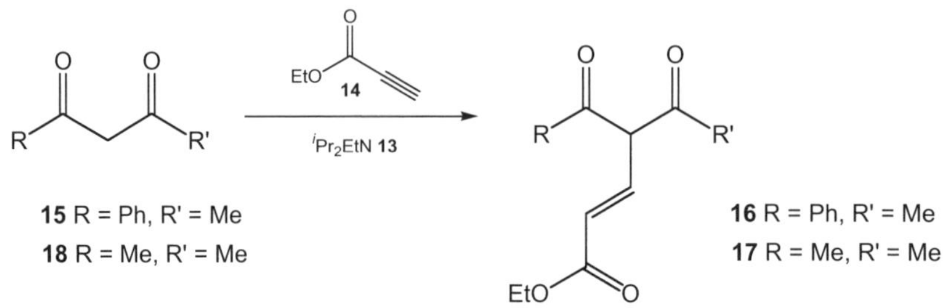

SCHEME 8.4 Michael addition reactions in a microreactor demonstrating reaction optimization.

ethyl ester **17**. The authors proposed that the observed increase in conversion, when employing stopped flow, was due to an effective increase in residence time within the device.

8.3.4 Photochemical Reactions

Photochemical reactions provide an attractive, environmentally friendly technique for the synthesis of complex molecules; however, the ability to transfer reactions from the bench to large scale is met with problems largely associated with the scale-up of light sources. With this in mind, Jensen et al. (23) demonstrated the synthesis of benzopinacol **19** within a silicon/quartz microreactor (Scheme 8.5); whereby UV light passed through the quartz layer to irradiate reagents contained within the micro channel. At flow rates of greater than 10 μL/min, no benzopinacol **19** formation was observed, only recovery of unreacted benzophenone **20** and propan-2-ol **21**. By reducing the flow rate to 4 μL/min, and hence increasing the amount of light absorbed, conversions of up to 60% were reported.

Using a similar approach, Wootton and coworkers (24) investigated the photochemical generation of (explosive) singlet oxygen, within a glass microreactor, demonstrating the synthesis of ascaridole **22** (Scheme 8.6). To perform the reaction, a methanolic solution of α-terpinene **23** and sensitizer (Rose Bengal **24**) was introduced into the reactor (1.0 μL/min) where it was mixed with a stream of oxygen (15 μL/min) prior to irradiation with a 20-W, 6-V tungsten lamp; as a safety precaution, the reaction products can be degassed with nitrogen in order to avoid the accumulation of oxygenated solvents. Using the aforementioned reaction

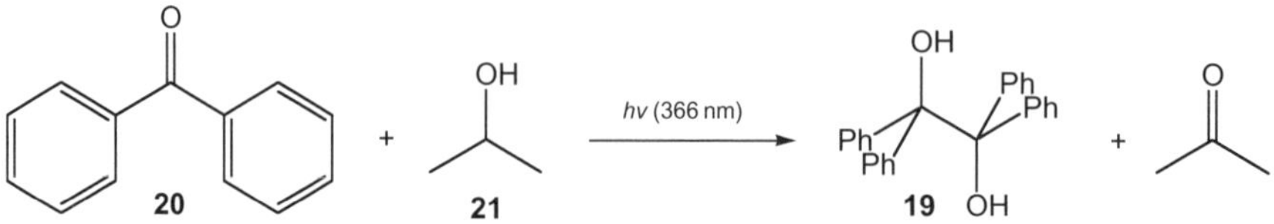

SCHEME 8.5 Photochemical formation of benzopinacol in a microreactor.

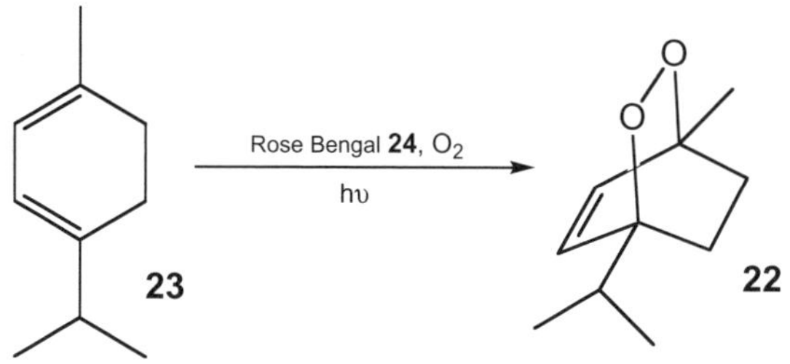

SCHEME 8.6 Photochemical generation and reaction of singlet oxygen in a microreactor.

conditions, the authors report 85% conversion of α-terpinene **22** to ascaridole **23**. These examples clearly demonstrate the efficient photon transfer obtained within continuous flow systems whereby reactor scale-out could be employed in order to increase the quantity of material prepared, whilst still employing commercially available light sources. Other examples where reactions have been conducted photochemically in microreactors include photocycloaddition reactions (25), photocyanation reactions (26) and chlorination reactions (27).

8.3.5 Multistep Reactions

In addition to an array of single step reactions, Watts et al. (28) reported the synthesis of a tripeptide **25**, demonstrating the first solution phase multistep synthesis within an EOF-based microreactor. As Scheme 8.7 illustrates, the synthesis involved coupling a pentafluorophenyl ester **26** and an amine **27** to prepare a dipeptide **28**, which was further reacted with DBU **29** to effect Fmoc deprotection. The resulting amine **30** was subsequently reacted, *in situ*, with a second equivalent of pentafluorophenyl ester **26** to afford the desired tripeptide **25**, obtaining 30% conversion (over three steps). The approach clearly demonstrates the ability to control both the spatial and temporal evolution of reaction intermediates, enabling efficient, multistep processes to be performed in continuous flow reactors. Although the example presented demonstrates the generation of nontoxic intermediates, the technique could be readily applied to the synthesis of toxic intermediates, providing a safe, scalable route to reactions that cannot currently

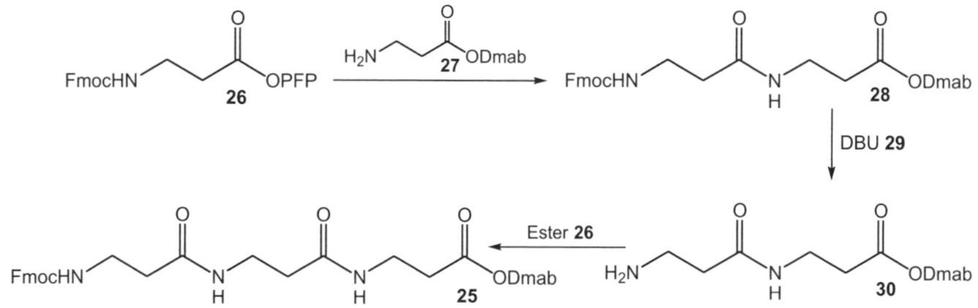

SCHEME 8.7 Multistep peptide synthesis in a microreactor.

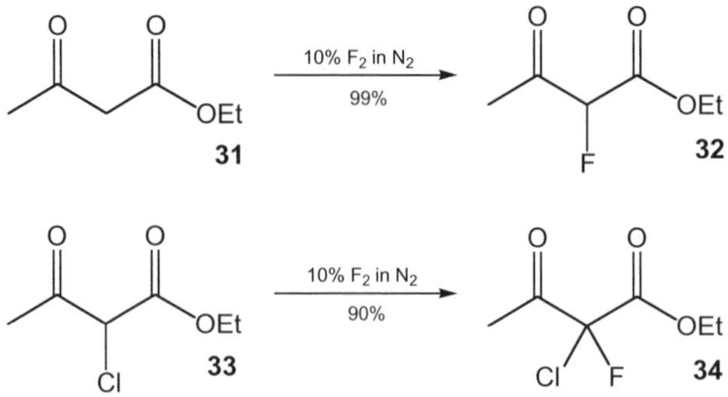

SCHEME 8.8 Fluorination reactions in a microreactor.

be performed on a large scale. Other examples of multistep processes in microreactors include the synthesis of ciprofloxacin (29) and the preparation of oxomaritidine (30).

8.3.6 Biphasic Reactions

In addition to solution phase reactions, biphasic reactions have also benefited from reaction miniaturization. Chambers and coworkers (31,32) demonstrated an early example of a biphasic microreaction, investigating the selective fluorination of 1,3-dicarbonyl compounds. Employing a thin film nickel reactor and 10% fluorine in nitrogen, the authors presented a facile route to an array of monofluorinated compounds. As Scheme 8.8 illustrates, 99% ethyl acetoacetate **31** was converted to ethyl 2-fluoroacetoacetate **32** and ethyl 2-chloroacetoacetate **33** was successfully fluorinated to afford ethyl 2-chloro-2-fluoroacetoacetate **34** in 90% conversion; importantly, no perfluorination products were detected.

Jensen et al. (33) subsequently demonstrated the direct fluorination of aromatic compounds in a silicon/Pyrex microreactor, with nickel-coated microchannels. As Scheme 8.9 illustrates, the fluorination of toluene **35** was selected as a model reaction, whereby monofluorinated toluene was obtained in 96% conversion. Analysis of the reaction mixture, by GC, confirmed the substitution pattern to be 3:1:2 *ortho-* **36**, *meta-* **37**, and *para-* **38** fluorinated toluene.

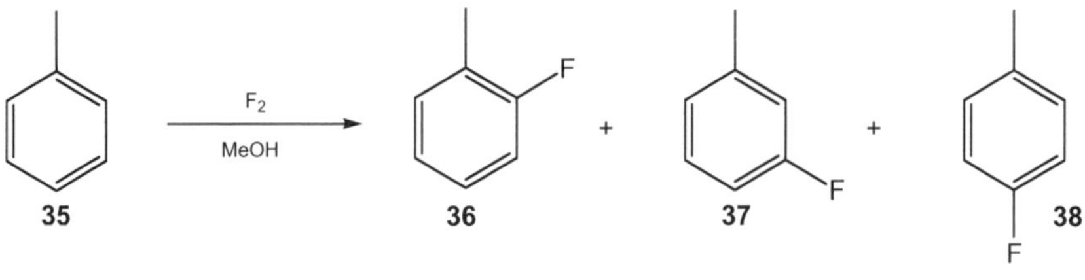

SCHEME 8.9 Aromatic fluorination reactions in a microreactor.

In addition to the gas/liquid systems discussed, Kitamori et al. (34) conducted a phase transfer reaction, demonstrating the diazo coupling of 5-methylresorcinol **39** to afford the azo dye **40** in 100% conversion (Scheme 8.10). The group subsequently employed a similar reactor to perform a 2 × 2 parallel synthesis of amides where the reaction products were collected in the organic layer and any organic salts formed were removed in the aqueous layer; using this approach, >80% yields were reported.

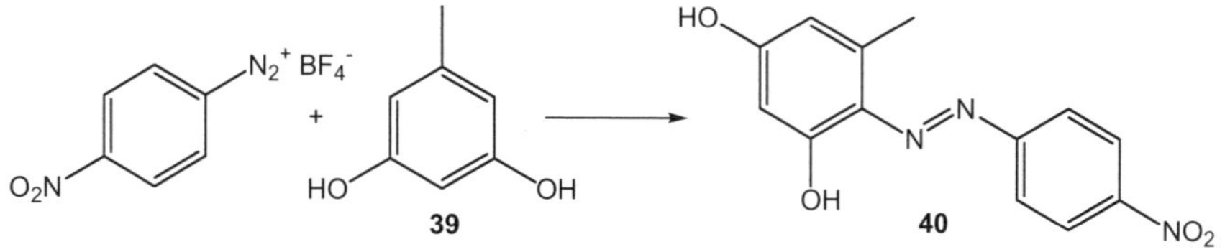

39 **40**

SCHEME 8.10 Diazo coupling reactions in a microreactor.

8.3.7 Synthesis of Analytically Pure Compounds

Although different approaches to compound purification have been evaluated within microreactors, including electrophoretic separation (35) and the aforementioned liquid–liquid extraction (34); the vast majority of reaction mixtures still need purification by traditional techniques, which clearly limits the advantages of microreactors. As an example, Wiles and coworkers (36) recently reported the Knoevenagel condensation within an EOF-based microreactor, demonstrating the quantitative conversion of diethyl malonate **41** and benzaldehyde **42** to 2-benzylidenemalonic acid diethyl ester **43**; the product however remained contaminated with dimethylamine **44** and its organic salt. As with the vast majority of microreactions, a batch-type purification was subsequently required in order to isolate the product **43**, clearly detracting from the advantages associated with reaction miniaturisation (Scheme 8.11).

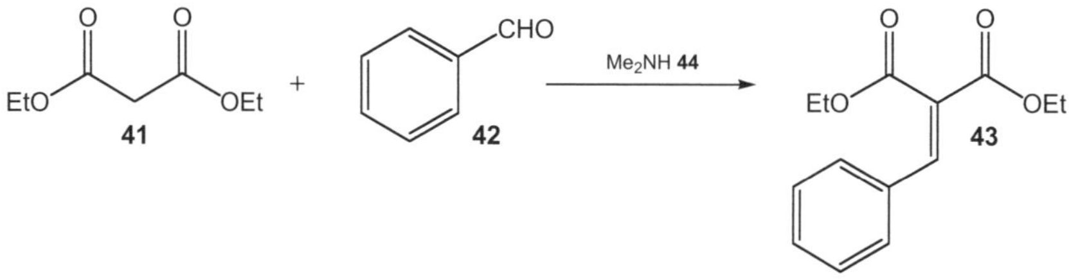

41 **42** **43**

SCHEME 8.11 Solution phase Knoevenagel reaction in a microreactor.

To circumvent this problem, the authors investigated the incorporation of a solid-supported base **45** into the microreactor, enabling the synthesis of analytically pure compounds without the need for additional product purification. EOF was used to mobilize ethyl cyanoacetate **46** and benzaldehyde **48** through the packed-bed where they condensed to afford 2-cyano-3-phenyl-acrylic acid ethyl ester **47** in 99% conversion (Scheme 8.12) (36).

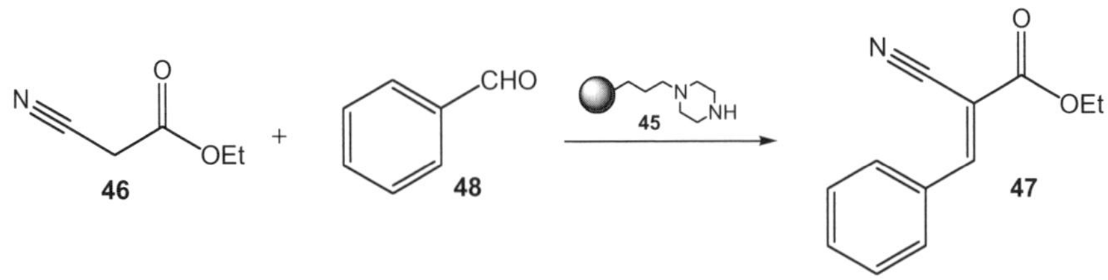

SCHEME 8.12 Knoevenagel synthesis in a microreactor incorporating immobilized reagents.

Using the optimized reaction conditions, a range of aldehydes and activated methylenes were investigated; in all cases excellent product yields and purities were obtained. In addition, the synthesis of 2-cyano-3-phenyl-acrylic acid ethyl ester **47** was subsequently investigated using 3-(dimethylamino)propyl-functionalized silica gel, 3-aminopropyl-functionalized silica gel, 3-(1,3,4,6,7,8-hexahydro-2*H*-pyrimido[1,2-1]-pyrimidino)propyl-functionalised silica gel, and polymer-supported diazabicyclo[2.2.2]octane, demonstrating excellent conversions (>99%). The authors concluded that the supported reagents suffered less physical damage when used within the flow reactor, compared to their use in conventional stirred/shaken reactors, resulting in extended reagent lifetimes. More recently the authors have shown that other catalysts, such as acid catalysts (37) and oxidants (38) may also be included in microreactors.

8.4 LARGE-SCALE MANUFACTURE USING MICROREACTORS

Although the use of microreaction technology is well documented within the academic environment, few papers are in the open literature describing their industrial use. A recent paper by Zhang et al. (39), from Johnson and Johnson Pharmaceuticals, described the use of a stainless steel, CYTOS microreactor system (Cellular Process Chemistry Systems GmbH) to perform a series of reactions employing unstable intermediates and high reaction temperatures for the synthesis of kilogram quantities of product.

In particular, the authors demonstrated the ability to perform a series of exothermic reactions, reporting enhanced reaction control compared to standard batch techniques. One such example was the synthesis of *N*-methoxycarbonyl-L-*tert*-leucine **49** (Scheme 8.13) via the addition of methyl chloroformate **50** to L-*tert*-leucine **51** in the presence of aqueous NaOH. Conducting the reaction in a flow reactor, at −40°C, resulted in the synthesis of *N*-methoxycarbonyl-L-*tert*-leucine **49** in 91.0% yield affording a throughput of 83.0 g/h.

8.5 CONCLUSIONS

Microreactor chemistry is currently showing great promise as a novel method on which to build new chemical technology and processes. Reactions performed in a

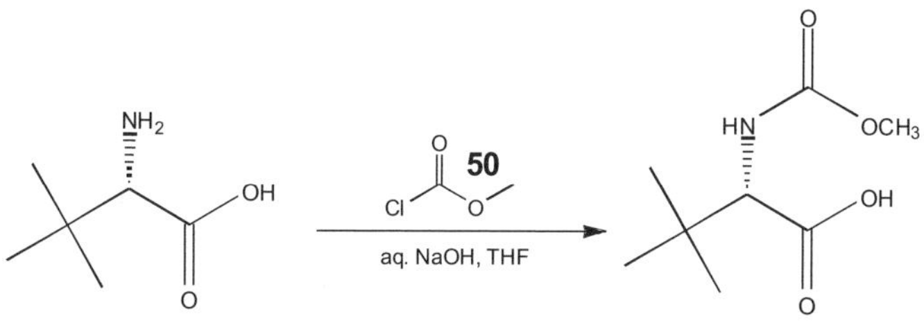

51 49

SCHEME 8.13 Continuous flow synthesis of carbamates.

microreactor invariably generate relatively pure products in high yield, in comparison to the equivalent bulk reactions, in much shorter times. One of the immediate and obvious applications is therefore in combinatorial chemistry and drug discovery, where the generation of compounds either with different reagents or under variable conditions is an essential factor. As an extension of the technology, it has been demonstrated that kilogram quantities of products can be produced without have to "scale up" the synthesis.

REFERENCES

1. Oosterbroek RE, van den Berg A. Lab-on-a-Chip: Miniaturised Systems for (Bio) Chemical Analysis and Synthesis, Elsevier; 2003.
2. Madou M. Fundamentals of Microfabrication Boca Raton: CRC Press; 1997.
3. McCreedy T. TrAC 2000;19:396–401.
4. Ehrfeld W, Hessel V, Lwe H. Microreactors: New Technology for Modern Chemistry, Wiley-VCH; 2000.
5. Zhang X, Wiles C, Painter SL, Watts P, Haswell SJ. Chim Oggi Chem Today 2005;24:43.
6. Wiles C, Watts P, Haswell SJ, Pombo-Villar E. Tetrahedron 2005;61:10757.
7. Rice CL, Whitehead R. J Phys Chem 1965;69:4017.
8. Fletcher PDI, Haswell SJ, Pombo-Villar E, Warrington BH, Watts P, Wong SYF, Zhang X. Tetrahedron 2002;58:4735–4757.
9. Jahnisch K, Hessel V, Lowe H, Baerns M. Angew Chem Int Ed 2004;43:406–446.
10. Schwalbe T, Autze V, Wille G. Chimia 2002;56:636–646.
11. Watts P, Wiles C. Chem Commun 2007;443.
12. Wootton RCR, Fortt R, de Mello AJ. Lab Chip 2002;2:5.
13. Salimi-Moosavi H, Tang T, Harrison DJ. J Am Chem Soc 1997;119:8716.
14. Wille CH, Gabrski HP, Haller TH, Kim H, Unverdorben L, Winter R. Chem Eng J 2004;101:179.
15. Doku GN, Haswell SJ, McCreedy T, Greenway GM. Analyst 2001;126:14.
16. Pennemann H, Watts P, Haswell SJ, Hessel V, Lowe H. OPRD 2004;8:422–439.

17. Mikami K, Islam MN, Yamanaka M, Itoh Y, Shinoda M, Kudo K. Tetrahedron Lett 2004;45:3681.
18. Skelton V, Greenway GM, Haswell SJ, Styring PD, Morgan DO, Warrington B, Wong SYF. Analyst 2001;126:7.
19. Kawaguchi T, Miyata H, Ataka K, Mae K, Yoshida J. Angew Chem Int Ed 2005;44:2413.
20. Suga S, Nagaki A, Yoshida J. Chem Commun 2003:354.
21. Panke G, Schwalbe T, Stirner W, Taghavi-Moghadam S, Wille G. Synthesis 2003;2827.
22. Wiles C, Watts P, Haswell SJ, Pombo-Villar E. Lab Chip 2002;2:62.
23. Lu H, Schmidt MA, Jenson KF. Lab Chip 2001;1:22.
24. Wootton RCR, Fortt R, de Mello AJ. Org Process Res Dev 2002;6:187.
25. Maeda H, Mukae H, Mizuno K. Chemistry Letters 2005;34:66–67.
26. Ueno K, Kitagawa F, Kitamura N. Lab Chip 2002;2:231.
27. Ehrich H, Linke D, Morgenschweis K, Baerns M, Jahnisch K. Chimia 2002;56:647.
28. Watts P, Wiles C, Haswell SJ, Pombo-Villar E. Tetrahedron 2002;58:5427.
29. Schwalbe T, Autze V, Wille G. Chimia 2002;56:636.
30. Baxendale IR, Deeley J, Griffith-Jones CM, Ley SV, Saaby S, Tranmer GK. Chem Commun 2006;2566.
31. Chambers RD, Spink RCH. J Chem Soc, Chem Commun 1999;883.
32. Chambers RD, Holling D, Spink RCH, Sandford G. Lab Chip 2001;1:132.
33. de Mas N, Jackman RJ, Schmidt MA, Jenson KF. IMRET5: Proceedings of the Fifth International Conference on Microreaction Technology, Springer, Berlin;2002, 60.
34. Hisamoto H, Saito T, Tokeshi M, Hibara A, Kitamori T. Chem Commun 2001;2662.
35. George V, Watts P, Haswell SJ, Pombo-Villar E. Chem Commun 2003;2886.
36. Wiles C, Watts P, Haswell SJ. Tetrahedron 2004;60:8421.
37. Wiles C, Watts P, Haswell SJ. Tetrahedron 2005;61:5209.
38. Wiles C, Watts P, Haswell SJ. Tetrahedron Lett 2006;47:5261–5264.
39. Zhang X, Stefanick S, Villani FJ. Org Process Res Dev 2004;8:455.

9

MICROFLUDIC REACTORS FOR SEQUENTIAL AND PARALLEL REACTIONS

YANJU WANG, WEI-YU LIN, ZETA TAK FOR YU AND HSIAN-RONG TSENG

9.1 INTRODUCTION

Chemists carry out reactions and study reaction mechanism using round-bottomed flasks, test tubes, or other specialized apparatus. Although there were significant progresses in discovering new reactions and understanding reaction mechanisms, these conventional macroscopic reaction settings and their associated operation techniques have barely changed during the last century. In the past two decades, significant efforts have been devoted to develop microfluidic reactors (1–9) because they offer a variety of advantages over the macroscopic setting, such as reduced chemical consumption, safety, high surface-area-to-volume ratios, and improved control over mass and heat transfer. In addition, microfluidic devices are amenable to interfacing with a range of analytical systems, for example, microscopy (10,11), X-ray diffraction (12,13), UV–vis spectroscopy (14), mass spectrometry (MS) (15), gas chromatography (GC) (16), liquid chromatography (LC) and LC–MS, for characterization of the reaction products in sequence.

The most cost-efficient way to build a microfluidic reactor is to assemble micrometer (μm)-level tubings to form microchannel networks, where different reagents are mixed and reacted. Fused glass (17,18) and steel (19,20) capillary tubes are often utilized in such a system. The microfabrication technology cultivated by the contemporary semiconductor industry has set a solid foundation, allowing computer-aided design (CAD) and cost-efficient fabrication of a

Biological Applications of Microfluidics, Edited by Frank A. Gomez

microfluidic reactor on a monolithic chip with a size no larger than a business card. Silicon materials, for example, silicon and glass wafers are compatible with the microfabrication processes and have been extensively used for the production of microfluidic reactors. Plastic microfluidic devices have also been developed (21–23). It is noteworthy that after poly(dimethylsiloxane) (PDMS)-based softlithography (24) became available, the fabrication time and cost of microfluidic devices have been drastically reduced. In general, from concept development to a working device, a PDMS-based microfluidic device requires less than three working days when softlithography approach is applied; that is, when a different reaction protocol is selected, a chip can be rapidly designed and produced using a standardized production scheme to meet the new conditions.

Apparently, microfluidic reactors handle reaction solutions at microscale, where flow phenomena are dramatically different from those observed in the macroscopic setting (25–28). For example, inertial effect in microfluidic environment is negligible, thus fluid transport is dictated by fluid viscosity, leading to laminar flow in microchannels. Reynolds number, a ratio between inertial (f_i) and viscous (f_v) forces ($Re = f_i/f_v = Ul/\nu$, where U is the average flow velocity, l is the channel characteristic length, and ν is the kinematic viscosity), is used to characterize this property. A microfluidic platform is, therefore, known to be a turbulence-free system, where mixing is only driven by diffusion. Given the microfluidic nature described above, one of the most challenging issues in the field of microreactor is to achieve controllable mixing in a stand-alone device. There have been some efforts committed to develop micromixers that can be categorized into either active or passive ones (29) in order to facilitate mixing and reaction in microfluidic reactors. Active micromixers require external energy inputs, for example, pressure (30–32), temperature (33), and electrohydrodynamics (34–36), to generate disturbance for facilitating mixing. Alternatively, passive micromixers do not require external power and are often composed of robust channel geometry or microstructures, and are developed upon straightforward concepts of increasing contact surfaces and/or shortening diffusive distance among different fluidic components. Different strategies based on serial (37) and parallel (38–41) multilamination (29), droplet approach (10), and chaotic advection (42,43) have been developed to achieve efficient mixing.

On the basis of a simplified definition to classify (Scheme 9.1) complicated chemical reactions, most chemical reactions fall into a category of either sequential or parallel reactions. Sequential reactions are multiple chemical processes executed in sequence and are often utilized for the production of specific target compounds. On the contrary, parallel reactions consist of a group of similar chemical processes that occur simultaneously and are commonly employed for forming combinatorial library (44–46) or performing chemical screening. In this chapter, we review nine relatively complicated reactions that made good use of the advantageous characteristics of microfluidic reactors to achieve significantly improved outcomes (e.g., reduced sample consumptions, efficiency, improved reaction yields, and conditions) than those obtained in the macroscopic setting. With the reaction classification in mind, we describe these reactions in an order—first the sequential

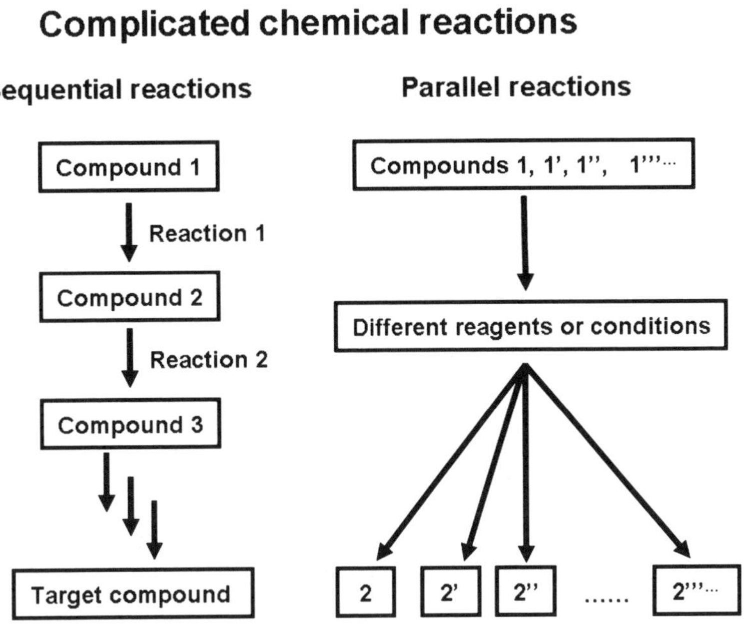

SCHEME 9.1 Side-by-side comparison of sequential reactions and parallel reactions. In sequential reactions, multiple chemical processes are executed in sequence for producing a specific target compound. In contrast, parallel reactions are a group of similar chemical processes that occur simultaneously for forming combinatorial chemical libraries or performing chemical screening.

reactions, followed by the parallel ones. Under each reaction category, we further divide the microfluidic reactors into two different constitutions, namely, continuous flow and integrated microfluidic reactors. Continuous flow microreactors (30,41,47–50) are a class of microreactors with a relatively simple device configuration, for example, branched microchannel networks. In these cases, passive micromixers are often incorporated into the microchannel networks to facilitate mixing and reaction. The devices are operating by introducing reagents and solvents into the microchannels using syringes/syringe pumps or back pressures through tubing connections. In contrast, the integrated microfluidic devices are composed of microchannels and much sophisticated functional microfluidic modules, including pressure-driven isolation valves (31) and peristaltic pumps (30,31). The isolation valves can be utilized to isolate distinct regions and prevent cross contamination from different reactions in a stand-alone device. A peristaltic pump composed of three parallel-oriented isolation valves is capable of delivery, metering, and active mixing of liquid components in devices. Moreover, the integrated microfluidic system can be controlled by a computer via a digital interface for performing sequential and parallel chemical processes on individual chips in an automated fashion.

9.2 SEQUENTIAL REACTIONS IN MICROFLUIDIC DEVICES

9.2.1 Continuous Flow Microfluidic Systems for Sequential Chemical Reactions

9.2.1.1 Room Temperature Swern Oxidation by Using a Continuous Flow Microfluidic Reactor Swern oxidation (51) is one of the most versatile and reliable reaction for converting alcohol compounds to their corresponding aldehyde or ketone products using dimethyl sulfoxide (DMSO) in the presence of an activation reagent (52,53) (e.g., trifluoroacetic anhydride (TFAA)). Usually, the reaction is performed at relatively low temperatures (−50°C or lower) in order to avoid the formation of undesirable side products resulted from the Pummerer rearrangement. Consequently, Swern oxidation is unlikely to be scaled up for industrial application. To overcome this problem, a continuous flow microfluidic reactor, composed of three multilamination-type micromixers and steel tube-based microfluidic network (54) (Fig. 9.1) was developed by Yoshida et al. for performing Swern oxidation at a much higher temperature between −20 and 20°C.

As shown in Fig. 9.1, through connection tubes, a solution of DMSO in CH_2Cl_2 and a solution of TFAA in CH_2Cl_2 were introduced into the first multilamination-type (29)

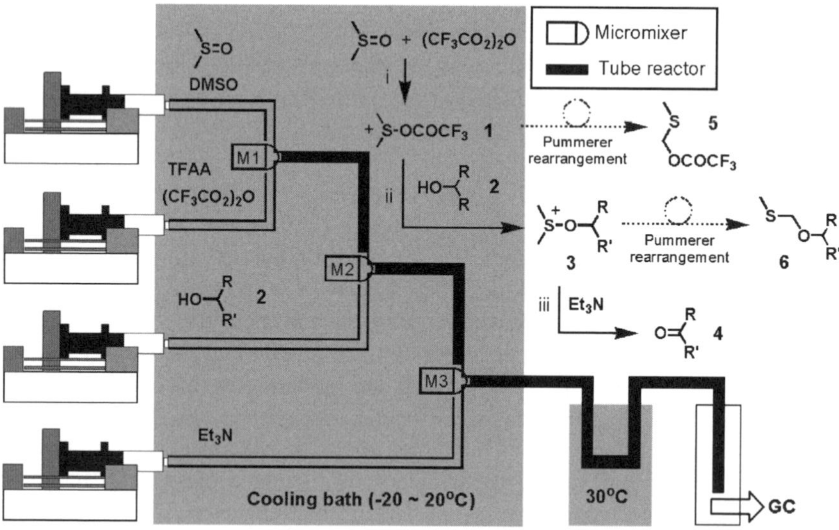

FIGURE 9.1 Schematic representation of a continuous flow microfluidic system utilized for performing Swern oxidation reactions to convert alcohol compounds **2** to the corresponding carbonyl compounds **4**. Three multilamination-type micromixers (M1–3) were connected by steel tubes to give a branched microfluidic network, in which Swern oxidation reaction was accomplished in good yield at a much higher temperature. The undesirable side products **5** and **6** obtained from the respective unstable intermediates **1** and **3** via the Pummerer rearrangement were successfully avoided.

micromixer M1 to give the first reaction mixture, in which DMSO was activated (step i) with TFAA to generate an unstable cationic intermediate **1** in the tube reactor located between the micromixers M1 and M2. The resulting intermediate **1** was sequentially reacted (step ii) with a solution of alcohol **2** to give the second cationic intermediate **3** after mixing through the micromixer M2. The micromixer M3 was then utilized to mix intermediate **3** and triethylamine, to carry out a base-induced elimination reaction (step iii), affording the desired carbonyl compounds **4**. To control reaction temperatures in different sections of the microreactor, the three micromixers and their connecting steel tubes were dipped in a cooling bath to maintain the temperature of the first two steps (steps i and ii) between −20 and 20°C, and the final section of the tube (where the step iii reaction was performed) was maintained at 30°C in a separated bath. Finally, the reaction products were collected and analyzed by GC. The results indicate that the Swern oxidations of primary, secondary, cyclic, and benzyl alcohol were achieved with yields ranging from 70% to 95% in this steel tube-based continuous microfluidic system. For comparison, the same reactions carried out in a conventional macroscopic batch reactor led to poor yields (ca. 11−49%) of desired products. A significant amount of side products, that is, trifluoroacetates **5** and methylthiomethyl ether **6** were obtained instead, due to the Pummerer rearrangement of reaction intermediates. To check the device scalability and durability, this microchip-based Swern oxidation was performed continuously for 3 h at 20°C. The results show that the reaction conversion and product selectivity did not change.

This experiment demonstrated that a continuous microfluidic device can be utilized to facilitate sequential chemical reactions that involve highly unstable intermediates at a much higher temperature compared to the conventional macroscopic setting. Since the device is made of stainless steel that is compatible to the uses of harsh reagents and organic solvents, we envision that a broad variety of sequential reactions requiring different reagents, solvents, and reaction conditions will fit into such a miniaturized reaction environment to achieve improved outcomes.

9.2.1.2 A Microreactor for Sequential Synthesis of β-Peptides There have been extensive studies (7,55,56) on developing microreactors for synthesis of peptide molecules with significant reduction in reaction time. β-Amino acid oligomers (β-peptides) are a different category of peptide molecules that have attracted considerable attentions because of their unique secondary structures, as well as their *in vivo* metabolic stability (57–61). The secondary structures, however, severely hamper the synthesis of these peptide molecules (62). It is well-known that the solid-phase assembly of β-peptides often fails (63) due to the formation of hairpin turn structures. To tackle this problem, the solution-phase fragment coupling approach has been developed, while there is still space to improve the reaction efficiency.

In conjunction with such a fragment coupling approach, a silicon-based continuous flow microfluidic reactor (64) was developed to assemble short peptide fragments, affording desired β-peptide oligomers. This continuous flow

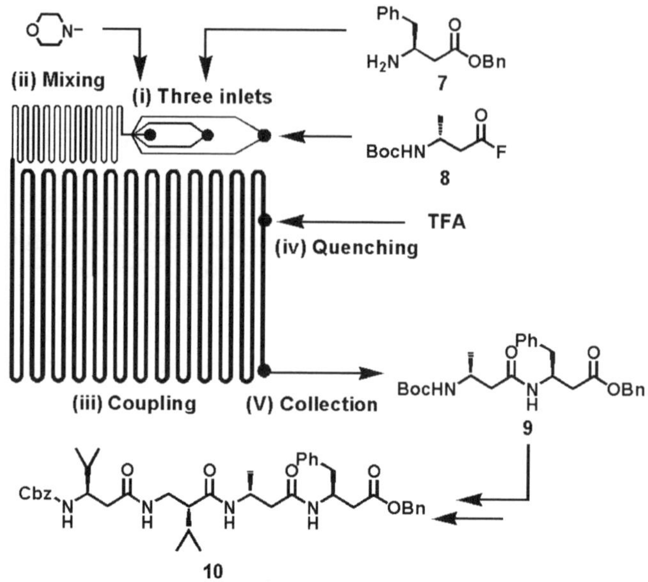

FIGURE 9.2 A graphic representation of a silicon-based continuous flow microfluidic device developed for synthesis of β-oligopeptides. The continuous microchannel can be divided into five functional sections for (i) introducing reagents, (ii) mixing, (iii) coupling reactions, (iv) reaction quenching and (v) product collection, enabling solution-phase fragment coupling of peptides (**7** and **8**) to give dipeptide product **9** in a good yield. Tetrapeptide **10** was obtained by performing stepwise coupling reactions in the same device.

microfluidic reactor (Fig. 9.2) can be divided into five major sections, including (i) three inlets for reagent inputs, (ii) a short mixing channel, (iii) a long reaction loop for peptide coupling reactions, (iv) an inlet for introducing quenching reagents, and (v) an outlet for collecting products. Due to the inert chemical property of silicon oxide layers that cover the microchannel surfaces, this continuous flow microreactor is compatible with a wide range of organic solvents and a broad temperature range (−80 to 150°C). In this study, this microreactor was first employed for quick optimization of reaction conditions in a time-efficient and sample economy fashion. To optimize reaction conditions, synthesis of the dipeptide **9** from benzyl 3-amino-4-phenylbutanoate (**7**) and Boc-β^3-homoalanine fluoride (**8**) was tested repeatedly in the microreactor at controllable temperatures and sample retention times. The resulting reaction mixtures were characterized by LC–MS to identify an optimized condition that gave excellent reaction yield (ca. 91%) of the dipeptide **9**. By performing two additional peptide coupling reactions in the same continuous flow microreactor, the dipeptide **9** was transformed to the tetrapeptide **10**. As control experiments, the sequential reactions for preparing the tetrapeptide **10** from monopeptide **7** were carried out, in conventional solution-phase and solid-phase settings, with the operation times of

3 days and 3.5 h, respectively. Due to the low reaction yield of each reaction in these conventional settings, the final product can only be obtained in a low quantity after the multistep synthesis.

This silicon-based continuous flow microfluidic system demonstrated an effective synthesis of peptides. Sequential synthesis of tetrapeptides can be achieved by performing solution-phase fragment coupling reactions in the same device after three reaction cycles with significantly improved yields. The small reaction volume allowed systematic optimization of reaction parameters for each peptide coupling process by using minimal amounts of reagents. In addition, the inert chemical property of silicon-based microchannel enables the use of a wide range of organic solvents and reagents, as well as a broad range of operation temperature. Therefore, the microchip-based peptide coupling reactions can be performed at unconventionally high temperatures, leading to dramatically reduced reaction times and improved reaction yields. Besides its application in peptide synthesis, this silicon-based continuous flow microfluidic system holds promise to be applied to the assembly of other biopolymers, including oligosaccharides and DNA oligonucleotides. The same device was adopted for the production of several grams of α-disaccharide molecules in the same device (65).

9.2.2 Integrated Microfluidic Systems for Sequential Reactions

9.2.2.1 Multistep Synthesis of Positron Emission Tomography (PET) Imaging Probes Using Integrated Microfluidics Positron emission tomography (66,67) is a noninvasive imaging technology for detailed mapping of biological processes in human and animal bodies. The development of sensitive and specific positron emitting radionuclide-labeled molecular probes is crucial for expanding the capability of target-specific *in vivo* imaging for biological research and drug discovery. The commonly used positron emitting radionuclides have limited lifetimes, for example, ^{18}F, $t_{1/2} = 109.7$ min, ^{11}C, $t_{1/2} = 20.4$ min, ^{13}N, $t_{1/2} = 9.96$ min and ^{15}O, $t_{1/2} = 2.07$ min. The short half-lives of these radionuclides make rapid synthesis of doses essential. Additionally, the preparation of radionuclide-labeled molecular probes provides a conceptual model for the preparation of other molecules (including pharmaceuticals) since it includes common steps required in many chemical synthesis. Finally, the nanogram mass of PET molecular imaging probes administered to subjects is ideal for miniaturized architecture of integrated microfluidics. Thus, the multistep synthesis of PET probes represents an interesting opportunity and a compelling demonstration for microfluidic reactors (68).

An integrated microfluidic device (69) (Fig. 9.3) capable of performing multistep chemical synthesis has been developed for the synthesis of an [^{18}F] radiolabeled molecular imaging probe, 2-deoxy-2-[^{18}F]fluoro-D-glucose ([^{18}F] FDG, **13**). Five sequential processes—(i) concentration of dilute fluoride ion using a miniaturized anion-exchange column located in the square-shaped fluoride concentration loop, (ii) solvent exchange from water to dry MeCN, (iii) fluorination of the D-mannose triflate precursor **11**, (iv) solvent exchange back to water, and (v) acidic hydrolysis of the fluorinated intermediate **12** in the ring-shaped reaction

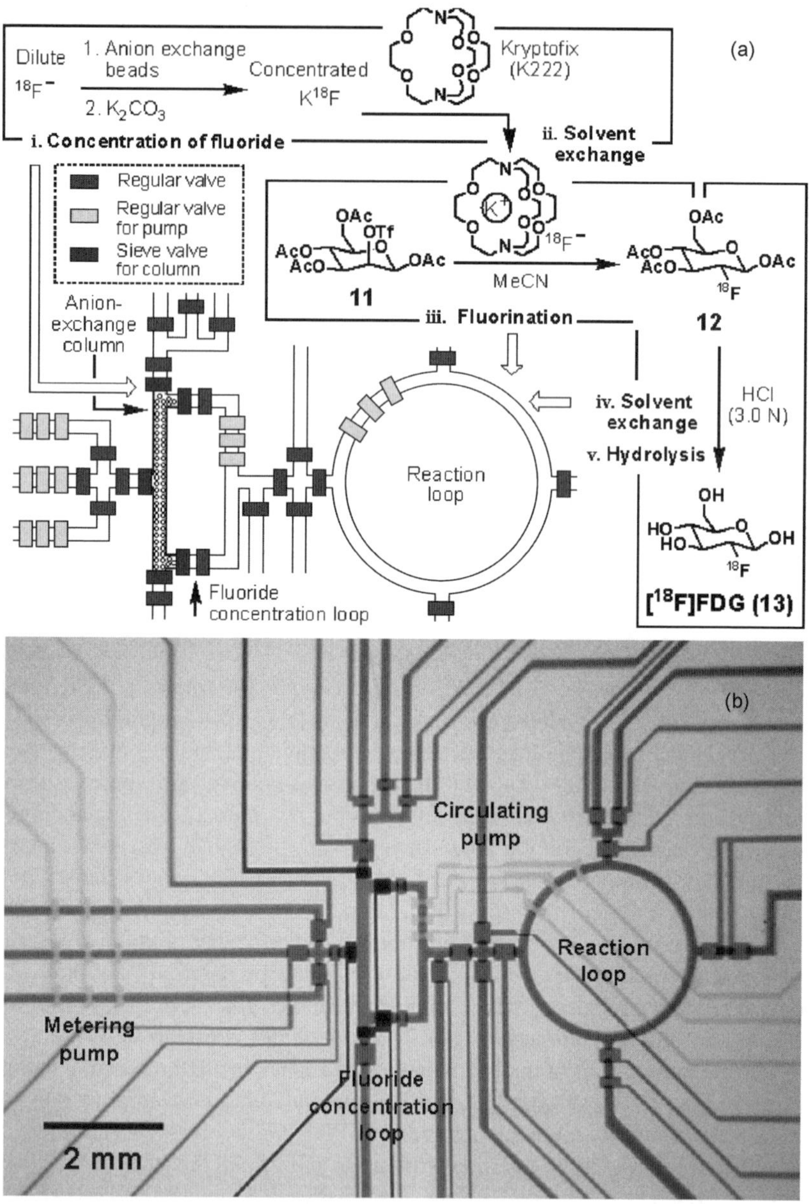

FIGURE 9.3 (**a**) Schematic representation of a PDMS-based microfluidic reactor used in the production of 2-deoxy-2-fluoro-D-glucose ([^{18}F]FDG) (13). Five sequential chemical processes produced nanogram (ng) levels of [^{18}F]FDG (13). The operation of the device is controlled by pressure-driven valves, with their delegated responsibilities illustrated by their colors: red for regular valves (for isolation), gray for pump valves (for fluidic metering circulation), and blue for sieve valves (for trapping anion exchange beads in the column module). (**b**) Optical micrograph of the central area of the microreactor. (See color insert)

loop—produce nanogram (ng) levels of [^{18}F]FDG (**13**). The operation of the microreactor is controlled by pressure-driven valves, with their delegated responsibilities, that is, regular valves for isolation, pump valves for fluidic metering/circulation and sieve valves for trapping anion-exchange beads in the column module. It is worthwhile to note that the gas permeable PDMS matrix allows solvent exchange to occur within the microfluidic channel through direct evaporation, thereby allowing for the sequential execution of anhydrous organic-solvent and aqueous-based chemical reactions. In addition, the device operation was automated under computer control, and yielded the additional rewards of improved and accelerated chemical synthesis and with high radiochemical purity. A simplified and enlarged device was designed and fabricated for production of 2 mCi [^{18}F]FDG (**13**). After neutralizing the resulting [^{18}F]FDG (**13**) solution with phosphate-buffered saline, the reaction mixture was passed through a cartridge containing C18 and Al_2O_3 beads to remove the Kryptofix and [^{18}F]fluoride. One hundred and ninty microcuries of chip-based [^{18}F]FDG (**13**) was sterilized and injected into a tumor-bearing mouse for a micro-PET scan.

The results constitute a proof of principle for performing sequential chemical processes in this PDMS-based microfluidic reactor to produce nanogram-level [^{18}F] FDG with enhanced operational efficiency and chemical economy. The microenvironment created in the integrated device provides many fundamental advantages such as low chemical consumption, precise control over reaction conditions, and faster reaction kinetics. The elastomer-based integrated microfluidic system allows incorporation of many technologically innovative modules, such as ion-exchange columns, active rotary mixers, and high pressure reactors, to achieve favorable reaction conditions (i.e., high temperature, high concentration) that are not possible on a macroscale or even in a continuous flow microfluidic system. The combination of these advantages with those of microfluidic environment provide a powerful tool to generalize, accelerate, and diversify the preparation of radiolabeled imaging probes or any other compound of limited supply or time sensitivity. If the limitations associated with PDMS solvent/chemical resistance can be completely overcome (70,71), it is conceivable that the advantages of such automated chemical reaction circuits can help accelerate the drug discovery process by producing the small mass levels of drugs.

9.2.2.2 Synthesis of DNA Molecules in an Integrated Microfluidic Reactor Made of Solvent Resistant Elastomer To overcome the poor solvent/chemical resistant performance of PDMS materials, a photocurable perfluoropolyether (PFPE) elastomer was developed (71). The PFPE elastomer was produced by performing UV curing of the precursor (i.e., PFPE dimethyacrylate prepared (72) from commercially available PFPE diol and isocyanatoethyl methyacrylate). The PFPE and PDMS precursor and their fully cured elastomers exhibit similar processing and mechanical properties. The softlithography fabrication approach can, therefore, be directly adopted for fabricating integrated microfluidic devices made of PFPE materials. Using PFPE-based microfluidic reactors, broader synthetic applications can be executed in the presence of a variety of organic solvents and reagents.

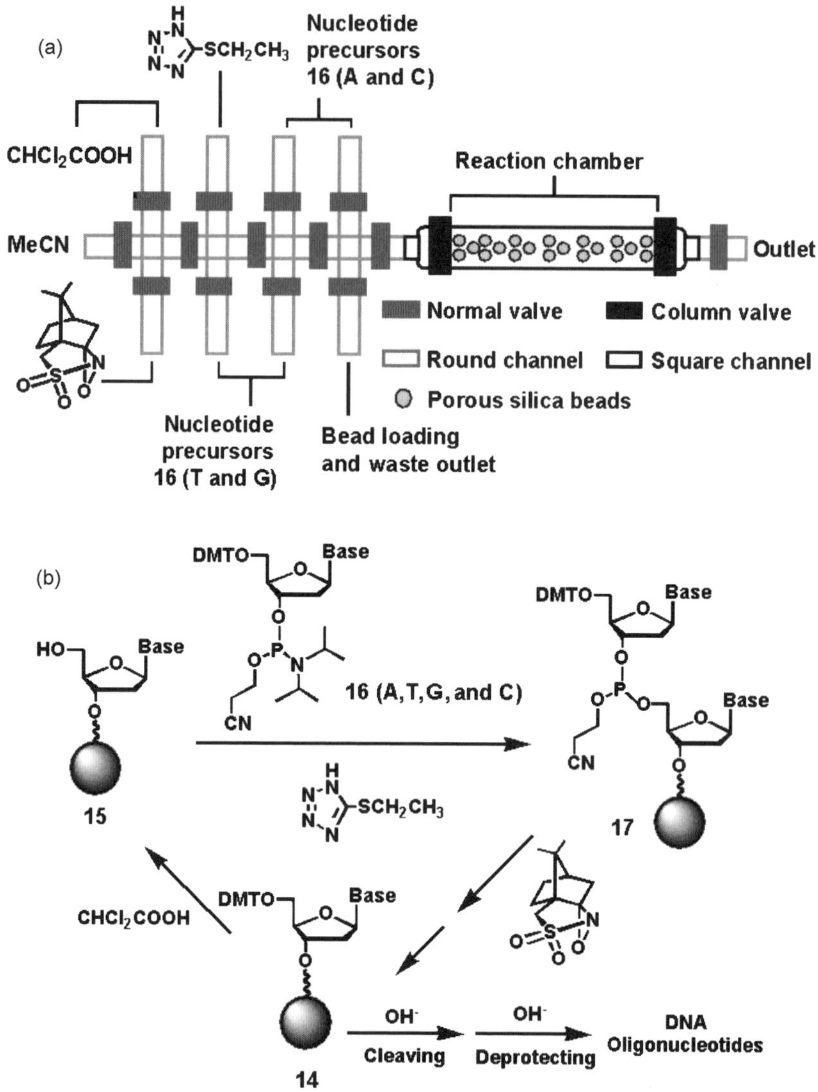

FIGURE 9.4 (**a**) Schematic representation of PFPE-based microfluidic reactors designed for solid-phase synthesis of 20-mer DNA oligonucleotides. The device is composed of nine individually addressable inlet channels for selective accesses of solvents and reagents and a reaction chamber where porous silica beads were confined for performing solid-phase DNA synthesis. (**b**) A reaction cycle for microchip-based DNA synthesis. The reactions started from bead-attached deoxythymidine precursor **14** with 5′-position protected by DMT. A single reaction cycle (including sequential deprotection, coupling and oxidation reactions) for adding a specific nucleotide took about 9 min. The 20-mer DNA oligonucleotides were obtained in 4 h by performing multiple reaction cycles in the device.

On the basis of a solid-phase DNA synthesis approach (73), an integrated microfluidic reactor (74) made of PFPE polymer has been developed for sequential synthesis of DNA oligonucleotides. The device (Fig. 9.4a) is composed of nine individually addressable inlet channels for selective accesses of solvents and reagents (including MeCN, deprotection/oxidation/activation reagents, and phosphoramidite nucleotide precursors), and a reaction chamber where porous silica beads are confined for performing solid-phase synthesis of DNA oligonucleotides. As shown in Fig. 9.4b, the synthesis of DNA oligonucleotides started from the deoxythymidine precursor **14** that was preattached to the porous silica beads with a base-cleavable succinyl linker at the 3′ end. The 5′-position of this surface-attached precursor **14** was protected with dimethoxytrityl (DMT), which can be removed by the deprotection reagent, that is, dichloroacetic acid to give free 5′-position in compound **15** for further coupling with phosphoramidite nucleotide precursor **16**. A specific nucleotide precursor (**16**, for A, T, G, and C nucleotides) can be introduced into the reaction chamber to couple with the surface-attached compound **15** in the presence of 5-ethylthio-1*H*-tetrazole, resulting the product **17**. (1*S*)-(+)-(10-Camphorsulfonyl)oxaziridine (CSO) was utilized as the oxidation reagent (75) to phosphite triesters (**17**) to phosphate triesters, completing one reaction cycle. A single reaction cycle of coupling one DNA nucleotide took about 9 min. The 20-mer DNA oligonucleotides with the sequence of 5′-CCG ACC TGG ATA CTG GCA TT-3′ were synthesized by performing multiple cycles of reaction. The resulting DNA oligonucleotides were released from the beads by base treatments for sequential HPLC/MS characterization. This device is capable of producing 60 pmol DNA oligonucleotides by performing sequential reactions, while only consuming less than 500 nL of 0.1 mol/L nucleotide precursor in each reaction cycle.

This work demonstrated that PFPE materials can be utilized for the fabrication of the integrated microfluidic reactor that is capable of performing sequential synthesis of DNA oligonucleotides in the presence of chemical reagents (e.g., phosphoramidite, dichloroacetic acid and CSO) and organic solvents (i.e., MeCN and CH_2Cl_2). It is also a good example of integrating solid-phase synthetic protocol with an integrated microfluidic setting, which simplified the device operation processes, reduced chemical/solvent consumption, and enabled an automated control with a PC interface. It is conceivable that a variety of integrated microfluidic reactors with similar design will be developed for the synthesis of other biopolymers, for example, oligopeptides and DNA oligonucleotides.

9.3 PARALLEL REACTIONS IN MICROFLUIDIC DEVICES

9.3.1 Continuous Flow Microfluidic Systems for Parallel Reactions

9.3.1.1 2 × 2 Parallel Reactions in a Glass-Based Microchip with Three-Dimensional Microchannel Networks Parallel reactions have been extensively used to create combinatorial libraries (44), which are fundamental elements in combinatorial chemistry for lead compound discovery and optimizations of new drugs. Using a conventional macroscopic reaction setting, it is laborious and sample/time consuming to generate a reasonable size of combinatorial library. A microfluidic system provides a promising platform to carry out combinatorial synthesis from different combinations of reagent pairs. However, for any $m \times n$ (m and $n > 2$) combinatorial library created by $m \times n$ parallel reactions, it is topologically challenging to disentangle all of the two-level crossings to realize these parallel reactions when two-dimensional (2D) microchannel network in a monolithic chip was employed. One of the potential solutions for disentangling the 2D microchannel crossing is to create three-dimensional (3D) microchannel network.

Recently, a continuous microfluidic reactor (76) (Fig. 9.5) with an embedded 3D microchannel network was developed for performing 2×2 parallel reactions. The microreactor was fabricated by sandwiching a glass plate with 10 connecting holes

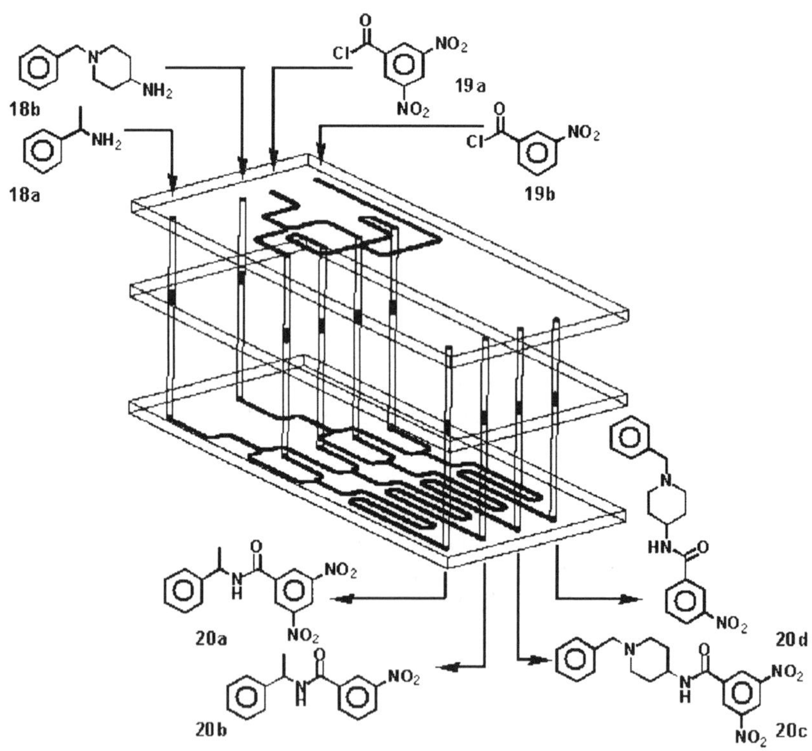

FIGURE 9.5 Schematic illustration of a 3D microchannel network in a glass microchip for performing 2×2 parallel reactions. Three glass plates fabricated by photolithography and wet etching were laminated and bonded thermally to form the device. There are four inlets and four outlets on both sides of the devices. Reagent pairs (amines **18a** and **18b** in aqueous solutions) and (acid chlorides **19a** and **19b** in organic solutions) were introduced through the four inlets to carry out two-phase coupling reactions, producing the respective amides **20a–d** with improved yields.

between two microchannel-etched glass substrates. Conventional photolithography, wet etching, and thermal bonding techniques were employed during the fabrication processes. Fig. 9.5 depicts the configuration and orientation of the 3D microchannel network as well as connecting/inlet/outlet holes in the device. To generate the 2 × 2 combinatorial library, two-phase amide synthesis was conducted by introducing a pair of amines (DL-1-phenylethylamine (**18a**) and 4-amino-1-benzylpiperidine (**18b**) in basic aqueous solutions) and a pair of acid chlorides (3,5-dinitrobenzoyl chloride (**19a**) and 3-nitrobenzoyl chloride (**19b**) in organic solutions) into the microchip through the inlet holes located on the top side of the device. Using the standard T branches in the 3D microchannel network, the flow of each reagent was divided into two streams, and four pairs of divided streams were mixed to form the four combinatorial mixtures (**18a/19a**, **18a/19b**, **18b/19a**, and **18b/19b**), resulting in the respective amides **20a**, **20b**, **20c**, and **20d**. After the chip-based synthesis, the resulting four amide compounds **20a–d** were collected from the organic phases of product mixtures for chromatography analysis. To get an equal distribution of the reagents at the T branches, the channel diameters were designed to be equal for both sides. The chip was connected to capillary tubings through Teflon screws and the solutions were introduced into the device by syringe pumps at variable flow rates. Since the organic and aqueous phases are unmixable, a phase-transfer mechanism suggests that the amine **18** in the aqueous phase diffuses into the organic phase to react with the acid chloride **19**. It is important to note that the microfluidic environment creates higher surface-area-to-volume ratios of the organic and aqueous interfaces. As a result, the high interface areas facilitated phase transfer of the amine **18**, leading to considerably improved yields of the amide product **20**. The control experiments performed in test tubes resulted in products **20** with much lower yields due to the low interface areas in the macroscopic setting.

The 3D microchannel network in a glass microchip demonstrated a practical operation platform for executing parallel reactions to form a 2 × 2 combinatorial library. The two-phase amide synthesis took the advantage of microfluidic characteristic (i.e., high surface-area-to-volume ratios of the inhomogeneous interfaces) to achieve parallel reactions in good yields. In principle, topological equivalence for any 3D microchannel networks, for example, combinatorial systems with much larger sizes of libraries can be built using a similar approach.

9.3.1.2 Synthesis of Three-Member Array of Cycloadducts in a Microchip Under Pressure-Driven Flow As we described in the previous section, it is convincing that 3D microchannel networks are a prime platform for conducting parallel reactions, while the 3D microchannel-based device required complicated fabrication processes, and their fabrication costs are higher than those with 2D microchannel networks. An interesting question to ask is—can a 2D microchannel network be employed for parallel reactions? If the answer is yes, how is the microchannel crossing issue addressed? A continuous flow microfluidic reactor (77) (Fig. 9.6) with a rather unique design of a 2D microchannel network was adopted for conducting parallel reactions to generate a small array of compounds. Different from most of the continuous microfluidic systems in which solutions are delivered

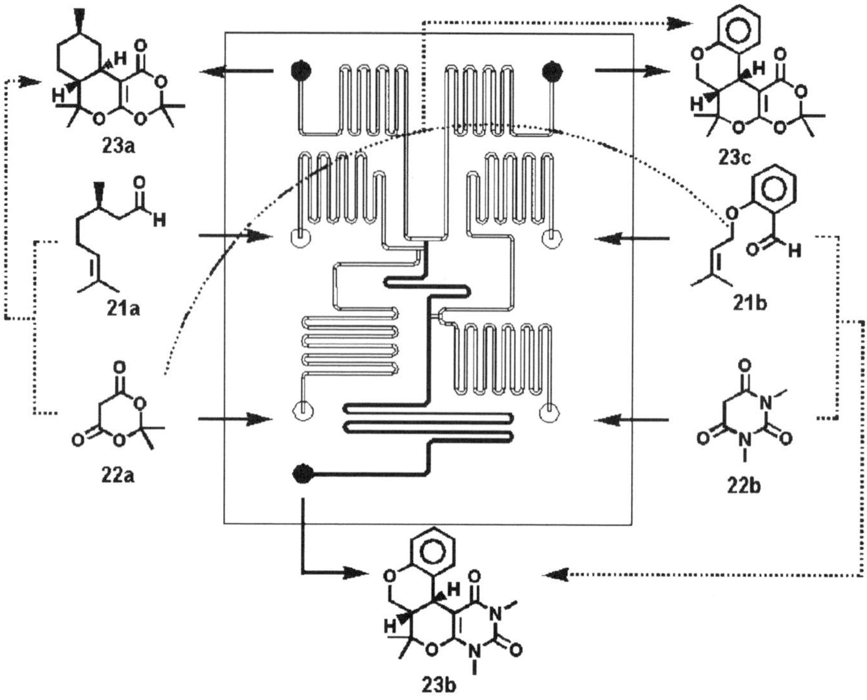

FIGURE 9.6 Schematic representation of a continuous flow microfluidic device for multireaction experiment in a single run. Two different aldehydes **21a/b** and two different 1,3-diketones **22a/b** were introduced into the inlets to obtain three cycloadduct **23a–c** from the three respective outlets.

by syringe pumps, back pressures were utilized to drive the solutions in this 2D microchannel network. An interesting fact is that this device was originally designed for hydrodynamic flow control.

A domino reaction composed of two cascade chemical processes, namely—a Knoevenagel condensation and a subsequent intermolecular hetero-Diels–Alder reaction—was chosen for demonstrating the concept of this pressure-driven device. In this case, two different aldehydes (citronellal (**21a**) and aromatic aldehyde (**21b**)) and two different 1,3-diketones (1,3-dimethylbarbituric acid (**22a**) and Meldrum's acid (**22b**)) were introduced into the device by back pressures through the four respective inlets indicated in Fig. 9.6. Three reactant mixtures **21a/22a**, **21b/22b**, and **21b/22a** were generated specifically to afford the three respective cycloadduct products **23a**, **23b**, and **23c** in the presence of catalyst ethylenediamine acetate. The three products were isolated from the three outlets and analyzed by LC–MS.

With the fact of a 2 × 2 parallel synthesis in mind, it is expected to produce four different products in this study, while only three cycloadduct products (**23a–c**) were obtained due to the intrinsic topological limitation of a 2D microchannel network. Since the chip-design has not been further optimized for other array synthesis, the results show a great potential by using this very unique methodology.

9.3.1.3 7 × 3 Parallel Reactions in an Automated Continuous Flow Microfluidic Reactor Effective miniaturization and automation of an operation platform can dramatically expand the capability and enhance the throughput of drug discovery processes. To test this idea, automated 7 × 3 parallel reactions were performed in a continuous flow microfluidic device (49) (Fig. 9.7) to obtain a combinatorial library. With the support of a number of peripheral components, the analysis of resulting products was conducted in sequence. Knorr synthesis (78) that generates pyrazoles **26** by reacting 1,3-dicarbonyl compounds **24** with hydrazines **25** was utilized for the preparation of library compounds.

The microreactor was fabricated by mounting a borosilicate top plate onto a borosilicate bottom plate with a serpentine microchannel using thermal bonding. There are three inlets and one outlet connected to glass capillary tubings for access to the peripheral components of the system through commercially available capillary connectors. The peripheral components that sustain the automated operation and online analysis are (i) an autosampler for introducing the reactants (**24** and **25**), (ii) a pumping system to drive flow, (iii) a dilution system for diluting

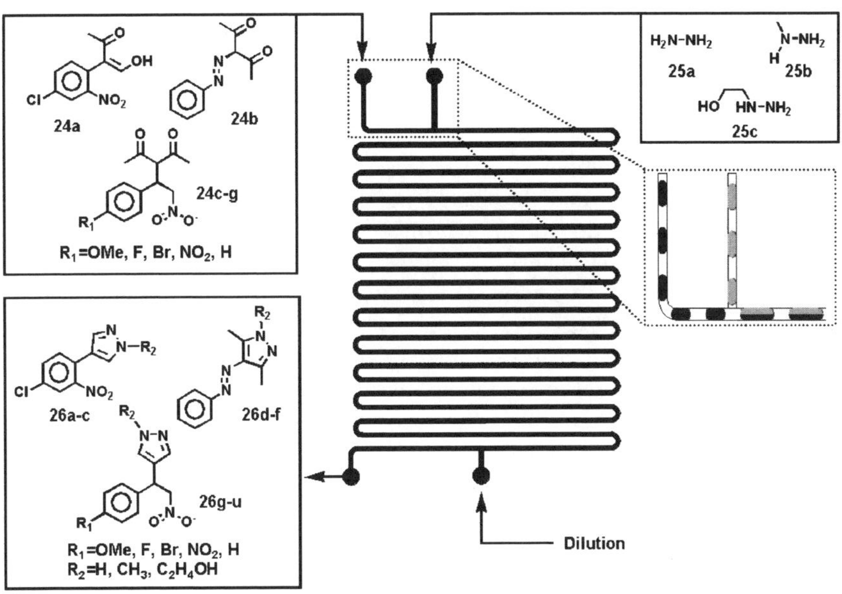

FIGURE 9.7 In conjunction with off-chip peripheral components, including an autosampler, a pumping system, a dilution system, a UV detector and LC–MS analysis systems, a continuous flow microfluidic reactor was developed for synthesis and analysis of a 7 × 3 combinatorial library in an automated fashion. Knorr synthesis of pyrazoles **26** from 1,3-dicarbonyl compounds **24** and hydrazines **25** were utilized for the preparation of the library.

reaction mixture on-chip, (iv) a UV detection system, and (v) a LC–MS for characterizing the reaction mixtures. As show in Fig. 9.7, droplets of the reactants (dicarbonyl compounds **24a–g** and hydrazines **25a–c**) were introduced into two inlets and sequentially mixed and reacted in the microreactor. Quantitative conversions of **26a–u** were detected in this microscale setting by the online analysis of UV and LC–MS. In this proof-of-concept study, a 7 × 3 combinatorial library consisting of 21 pyrazoles **26** has been achieved.

This work demonstrates the capabilities of an automated microfluidic reactor to synthesize and analyze online multiple analog reactions.

9.3.1.4 A Droplet-Based Continuous Flow System for Microgram-Scale Screening of Reaction Conditions Combinatorial methods and high throughput screening can also be utilized for efficient optimization of reaction conditions. Microtiter plates are the commonly used platform for conducting parallel reactions on the milliliter (mL) scale. It is, however, unlikely to carry out large-scale screening in these mL-scale screening platform when the employed compounds and reagents have limited resources. In conjunction with automated samplers, arrayed microwells have been developed to further reduce the reaction volume to submicrolitter (μL) level for each individual screening (79). The inevitable challenge of this miniaturized platform is to deal with the fast evaporation of solvent/compound/reagent in the microwells. Enclosed microchannels in microfluidic devices (49,80,81) prevent sample evaporation and has become an ideal platform for performing parallel reactions in a scale ranging from μL to nL.

One of the good examples is a reaction screening protocol (82) established by Ismagilov et al. This droplet-based (83) continuous flow microreactor (Fig. 9.8) consists of three main components (i) a cartridge that is prefilled with reagent plugs (reagents A, B, C, ...) and fluorinated isolation/carrier fluid, (ii) a PEEK Tee junction for the connection of the cartridge and the inlet tubing containing the substrate, ouabain hexaacetate (**27**), and (iii) a receiving tubing for storing the reaction mixture plugs. Two syringes were used to deliver the reagent plugs and substrate **27** into the PEEK Tee junction where the plugs of screening library (**A** + **27**, **B** + **27**, **C** + **27**, ...) were mixed. After all resulting plugs of screening library flowed into the receiving tubing, the flow was stopped and the receiving tubing was sealed subsequently. To reduce the cross-contamination between adjacent plugs during merging processes to less than 1%, one "blank" solvent plug was introduced between every two reaction mixtures. After the parallel reactions are completed, all the reaction plugs can be deposited onto MALDI (matrix-assisted laser desorption/ionization) plate for characterization by MALDI–MS (the inset in Fig. 9.8). The deacetylation of substrate (ouabain hexaacetate (**27**)) was chosen as a model compound to test this screening system because the ouabain is a poisonous cardiac glycoside, which is commonly used to block the sodium pump (Na–K-ATPase) (84) in the studies of neurons. In this case, various reagents (i.e., amines, inorganic bases, Lewis acids, and lipases) and reaction conditions (e.g., reaction time, solvents, and concentrations) were screened in this miniaturized reaction setting. The formation of deacetylated product, ouabain triacetate (**28**) was

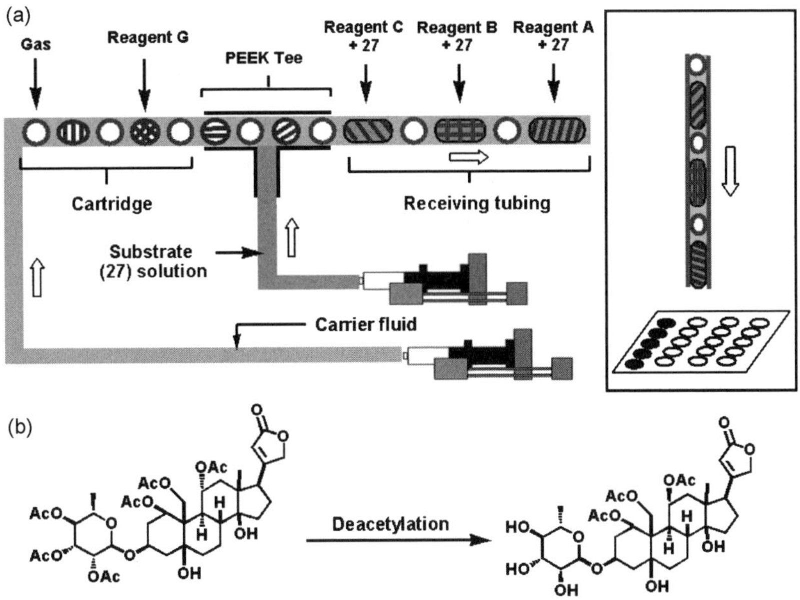

FIGURE 9.8 (**a**) Schematic diagram illustrates a droplet-based continuous flow screening system which was utilized to screen deacetylation conditions of ouabain hexaacetate, (27). This reactor is composed of three major components, including a reagent plug cartridge, a PEEK Tee junction and a receiving tubing. (**b**) Deacetylation reaction of ouabain hexaacetate (**27**) gives ouabain triacetate (**28**).

characterized by MALDI–MS. The results suggested that inorganic bases exhibited the highest reactivity for hydrolysis of the substrate **27**. The deacetylated product **28** was obtained from one of reagents/conditions (0.06 M Na_2CO_3 in H_2O: MeOH = 1:2) at a very high yield of 85%. Using optimized conditions obtained from the screening, unstable ouabain regioisomers (which were difficult to observe in the other systems) can be obtained and detected by quenching reactions at different times in different solvents with substantial isolation yields.

Compared with the commonly used microtiter plates, microfluidic chips or even more advanced robotic arrayed microwells, one of the most remarkable advantages of this continuous flow microfluidic screening system is that it avoids the waste of the screening substrate since the substrate solution were directly mixed with screening reagents through the PEEK Tee junction without indirect transferring processes. In addition, this screening system can be simply constructed by connecting each tubing component together without performing any microfabrication. It is worthy to note that chemical/solvent resistant tubings such as Teflon or glass can be utilized in this system to meet the needs of different reaction conditions, preventing the solvent/ chemical resistant issue (70) of the plastic-based microfluidic reactors. The noticeable shortcoming of this screening system is the dissolution/mixing of reagents in carrier fluid when high reaction temperature is applied.

9.3.2 Integrated Microfluidic Systems for Parallel Reactions

9.3.2.1 Parallel Screening of an* In Situ *Click Chemistry Library in Integrated Microfluidic Chips *In situ* click chemistry is a target-guided synthesis (TGS) (85–100) method for discovering high affinity protein ligands by assembling complementary azide and acetylene building blocks inside the target's binding pockets through 1,3-dipolar cycloaddition (93,94,101–105). As a consequence, the resulting ligands display much higher binding affinities to the protein that induced their formation than the individual fragments. *In situ* click chemistry has been employed recently to synthesize potent inhibitors for acetylcholine esterase (AchE) (93,94,102,104,105), bovine carbonic anhydrase II (bCAII) (103), and HIV protease (103). Similarly, *in situ* click chemistry experiments are also conducted in parallel using 96-well microtiter plates. Inevitably, the experimental conditions result in the consumption of significant amounts of the target biomolecules and reagents, because considerable volumes of the solutions are needed to fill each screening well. Although some target biomolecules are commercially available, the large protein consumption hampers the broad application of *in situ* click chemistry screening, especially when target biomolecules are difficult to obtain. Moreover, the existing approach relies heavily upon manual operation, which limits screening throughput and fidelity.

An integrated microfluidic reactor (106) (Fig. 9.9) was developed to serve as a screening platform for identifying high affinity ligands from an *in situ* click chemistry library because it enables parallelization and automation, and most importantly, the miniaturization allows economical use of target biomolecules and click chemistry fragments. In a stand-alone device, parallel screening of 32 *in situ* click chemistry reactions was tested. Unlike continuous microfluidic system that is topologically challenging to overcome the microchannel crossing issue using 2D design, integrated microfluidic can be utilized to form any $m \times n$ combinatorial library with the cooperation of integrated valves. The known bCAII click chemistry system (103) was selected for this proof-of-concept study. In this case, acetylenic benzenesulfonamide (**29**) ($K_d = 37 \pm 6$ nM) was used as the reactive scaffold ("anchor molecule") for screening a library of 20 complementary azides **30a–t** in the presence of the template protein bCAII. In control experiments, the well-known active site inhibitor, ethoxazolamide (**32**) ($K_d = 0.15 \pm 0.03$ nM), was utilized to suppress the *in situ* click chemistry reactions as evidence for protein-templated product formation.

This microfluidic reactor (Fig. 9.9) comprises four major functional components: (i) a nL-level rotary mixer responsible for selective sampling, precise metering (31), and rotary mixing (30) of nL-amounts of reagents (acetylene **29** and azides **30a–j** (or **30k–t**) with/without inhibitors **32**), (ii) a μL-level chaotic mixer (43) for combining the reagent solutions generated in the active mixer with bCAII solution, (iii) a microfluidic multiplexer (107) for directing each reaction mixture into one of the 32 individually addressable microvessels (component iv), which are pin holes for storing and incubating the reaction mixtures at 37°C. Such microfluidic-based *in situ* click chemistry conditions utilize a reaction volume of

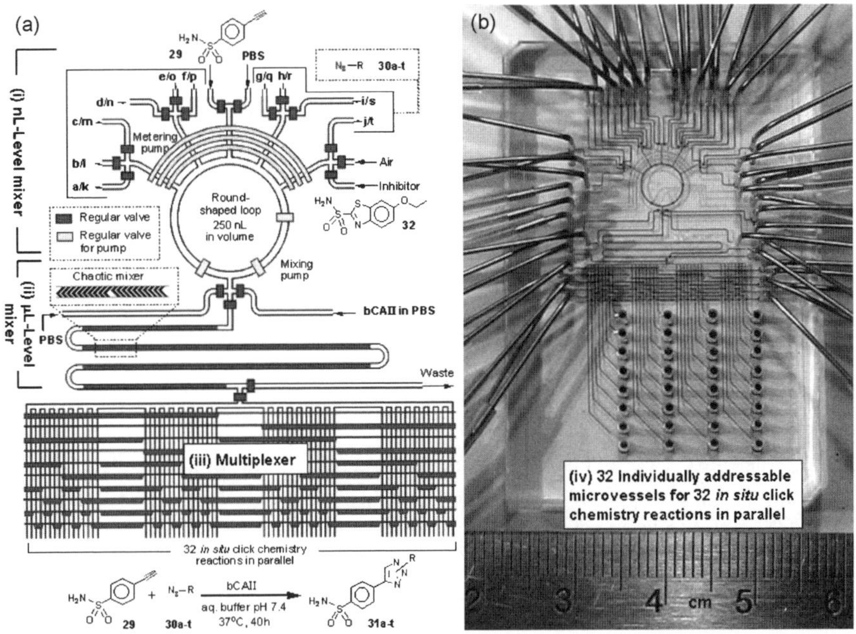

FIGURE 9.9 (**a**) Schematic representation of an integrated microfluidic reactor used for the parallel screening of an *in situ* click chemistry library prepared from **32** independent screening reactions. A series of 10 *in situ* click chemistry reactions between acetylene **29** and azides **30a**–**j** (or **30k**–**t**), 10 control reactions performed in the presence of inhibitor **32**, 10 thermal click chemistry reactions, and two reference solutions were set up in an automated fashion. The operation of the circuit was controlled using pressure-driven valves, whose delegated responsibilities are illustrated by their colors: black for regular valves (for isolation) and white for pump valves (for fluidic metering and circulation). (**b**) Optical image of the actual device; a ruler is present to calibrate the scale. (See color insert)

about 4 μL and only 19 μg of enzyme, 2.4 nmol of the acetylene **29**, and 3.6 nmol of the azide **30** for each reaction. For comparison, the conventional *in situ* click chemistry screening of this bCAII click chemistry system was performed (103) in 96-well microtiter plates, where each well was filled with 100 μL of reaction mixture, containing 94 μg of the enzyme, 6 nmol of the acetylene **29**, and 40 nmol of the azide **30**. Thus, this microfluidic system enables a 2–12-fold sample economy over the conventional approach. After on-chip incubation at 37°C, the resulting reaction mixtures were collected from the respective microvessels and injected into LC/MS for characterization. For reference purposes, the expected 1,4-disubstituted (*anti*) triazoles **30a**–**t** were prepared separately from the corresponding Cu^{I}-catalyzed reactions (103). The LC–MS analyses indicated that 10 out of the 20 reaction combinations had led to the formation of triazole products **31** in the presence of bCAII.

This integrated microfluidic reactor demonstrated an interesting example for performing parallel reactions and promises to make lead discovery through *in situ* click chemistry more convenient and reliable, less expensive, and more diverse, through miniaturization of the reaction system and automation of mixing processes. If a more sensitive analysis technique for identifying the formations of click chemistry products can be developed, further improvement on screening throughput and additional reduction of enzyme and reagent consumption will be achieved. Moreover, it is conceivable that integrated microfluidics not only can be used for *in situ* click chemistry screening, but also extended to other processes that require parallel screening.

9.4 CONCLUSIONS

During the past two decades, the development of microfluidic reactors and exploration of their applications for performing a broad spectrum of fundamental chemical reactions have set a great foundation, on which new-generation devices capable of performing much complicated chemical reactions, that is, sequential synthesis and parallel screenings have emerged and evolved gradually with improvements in sample economy, speed, throughput, complexity, automation, user interface, and integratability with other functional microfluidic modules. In this chapter, we have attempted to describe several newly developed microfluidic reactors, as well as their applications on a number of different aspects. Although a restricted definition was applied to select and classify these microfluidic reactors, the two constitutions of microfluidic systems (i.e., continuous flow and integrated microfluidic reactors) have demonstrated their strengths and uniqueness for conducting sequential synthesis and parallel screenings with significantly improved efficiency and performances compared to those observed in the conventional macroscopic settings. The needs of chemical and pharmaceutical industries are indeed the driving forces behind the fast development of the field of microfluidics. There are many good reasons to believe these new-generation microfluidic platforms will be extended to incorporate with other functioning platforms, for example, biological assays, cell arrays, high throughput analytical devices for a broader exploration beyond a single functioning chemical platform on a chip. We envision new microfluidic platforms for a number of game systems: "Let's Play___," where the blank is "Drug Discovery," "Biomarker Screening," "Biochemical Reaction Kinetics," "Disease Diagnosis," and others where execution is rapid and learning curves are steep.

REFERENCES

1. Jahnisch K, Hessel V, Lowe H, Baerns M. Angew Chem Int Ed 2004;43:406–446.
2. Mcquade T. Abstracts of Papers of the American Chemical Society 2005;229: U932–U932.

3. Watts P, Haswell SJ. Chem Soc Rev 2005;34:235–246.
4. deMello AJ. Nature 2006;442:394–402.
5. Geyer K, Codee JDC, Seeberger PH. Chem—Eur J 2006;12:8434–8442.
6. Juta Kobayashi YMS. umacr Kobayashi, Chem—Asian J 2006;1:22–35.
7. Watts P, Wiles C. Chem Commun 2007;443–467.
8. Watts P, Wiles C. Org Biomol Chem 2007;5:727–732.
9. Jas G, Kirschning A. Chem—Eur J 2003;9:5708–5723.
10. Song H, Tice JD, Ismagilov RF. Angew Chem Int Ed 2003;42:768–772.
11. Sia SK, Whitesides GM. Electrophoresis 2003;24:3563–3576.
12. Hansen CL, Skordalakes E, Berger JM, Quake SR. Proc Natl Acad Sci U S A 2002;99:16531–16536.
13. Zheng B, Tice JD, Roach LS, Ismagilov RF. Angew Chem Int Ed 2004;43:2508–2511.
14. Chan EM, Alivisatos AP, Mathies RA. J Am Chem Soc 2005;127:13854–13861.
15. Su J, Bringer MR, Ismagilov RF, Mrksich M. J Amer Chem Soc 2005;127:7280–7281.
16. Kobayashi J, Mori Y, Okamoto K, Akiyama R, Ueno M, Kitamori T, Kobayashi S. Science 2004;304:1305–1308.
17. Basheer C, Swaminathan S, Lee HK, Valiyaveettil S. Chem Commun 2005;409–410.
18. Comer E, Organ MG. Chem—Eur J 2005;11:7223–7227.
19. Nielsen CA, Chrisman RW, LaPointe RE, Miller TE. Anal Chem 2002;74:3112–3117.
20. Yamamoto S, Hanaoka T, Hamakawa S, Sato K, Mizukami F. Catal Today 2006;118:2–6.
21. Eijkel JCT, Bomer J, Tas NR, van den Berg A. Lab Chip 2004;4:161–163.
22. Ryu KS, Shaikh K, Goluch E, Fan ZF, Liu C. Lab Chip 2004;4:608–613.
23. Fortier MH, Bonneil E, Goodley P, Thibault P. Anal Chem 2005;77:1631–1640.
24. Xia YN, Whitesides GM. Annu Rev Mater Sci 1998;28:153–184.
25. Squires TM, Quake SR. Rev Mod Phys 2005;77:977–1026.
26. Nguyen NT, Wereley ST. Fundamental and Applications of Microfluidic. Boston: Artech House: 2002.
27. Purcell EM. Am J Phys 1977;45:3–11.
28. Brody JP, Yager P, Goldstein RE, Austin RH. Biophys J 1996;71:3430–3441.
29. Nguyen NT, Wu ZG. J Micromech Microeng 2005;15:R1–R16.
30. Chou HP, Unger MA, Quake SR. Biomed Microdev 2001;3:323–330.
31. Unger MA, Chou HP, Thorsen T, Scherer A, Quake SR. Science 2000;288:113–116.
32. Deshmukh A, Liepmann D, Pisano A. In: Solid-State Sensor and Actuator Workshop: Hilton Head, SC;2000.
33. Tsai JH, Lin LW. Sens Actuuators A-Phys 2002;97–98:665–671.
34. Storey BD. Phys D—Nonlinear Phenom 2005;211:151–167.
35. Hoburg JF, Melcher JR. Phys Fluids 1977;20:903–911.
36. El Moctar AO, Aubry N, Batton J. Lab Chip 2003;3:273–280.
37. Branebjerg J, Gravesen G, Krog JP, Nielsen CR. In: Proceedings of IEEE Micro Electro Mechanical Systems, San Diego, California 1996; p 441.

38. Bessoth FG, deMello AJ, Manz A. Anal Commun 1999;36:213–215.
39. Xu Y., Bessoth FG., Eijkel JCT., Manz A. Analyst 2000;125:677–683.
40. Mitchell MC, Spikmans V, de Mello AJ. Analyst 2001;126:24–27.
41. Mitchell MC, Spikmans V, Manz A, de Mello AJ. J Chem Soc-Perkin Transl 2001;514–518.
42. Liu RH, Stremler MA, Sharp KV, Olsen MG, Santiago JG, Adrian RJ, Aref H, Beebe DJ. J Microelectromech Syst 2000;9:190–197.
43. Stroock AD, Dertinger SKW, Ajdari A, Mezic I, Stone HA, Whitesides GM. Science 2002;295:647–651.
44. An HY, Cook PD. Chem Rev 2000;100:3311–3340.
45. Terrett NK. Combinatorial Chemistry:Oxford; New York; Oxford University Press:1998.
46. Wilson SR, Czarnik AW. Combinatorial Chemistry: Synthesis and Application; New York; Wiley:1997.
47. Kikutani Y, Ueno M, Hisamoto H, Tokeshi M, Kitamori T. Qsar Comb Sci 2005; 24:742–757.
48. Jas G, Kirschnimg A. Chem—Eur J 2003;9:5708–5723.
49. Garcia-Egido E, Spikmans V, Wong SYF, Warrington BH. Lab Chip 2003;3:73–76.
50. Brivio M, Verboom W, Reinhoudt DN. Lab Chip 2006;6:329–344.
51. Tidwell TT. Synthesis-Stuttgart 1990;857–870.
52. Omura K, Swern D. Tetrahedron 1978;34:1651–1660.
53. Omura K, Sharma AK, Swern D. J Org Chem 1976;41:957–962.
54. Kawaguchi T, Miyata H, Ataka K, Mae K, Yoshida J. Angew Chem Int Ed 2005;44:2413–2416.
55. Watts P, Haswell SJ, Pombo-Villar E. Chem Eng J 2004;101:237–240.
56. Watts P, Wiles C, Haswell SJ, Pombo-Villar E. Lab Chip 2002;2:141–144.
57. Seebach D, Overhand M, Kuhnle FNM, Martinoni B, Oberer L, Hommel U, Widmer H. Helv Chim Acta 1996;79:913–941.
58. Seebach D, Beck AK, Bierbaum DJ. Chem Biodiv 2004;1:1111–1239.
59. Kimmerlin T, Seebach D. J Peptide Res 2005;65:229–260.
60. Cheng RP, Gellman SH, DeGrado WF. Chem Rev 2001;101:3219–3232.
61. Arvidsson PI, Frackenpohl J, Seebach D. Helv Chim Acta 2003;86:1522–1553.
62. Murray JK, Sadowsky JD, Scalf M, Smith LM, Chen JD, Gellman SH. Biopolymers 2005;80:533–533.
63. Lelais G, Seebach D, Jaun B, Mathad RI, Flogel O, Rossi F, Campo M, Wortmann A. Helv Chim Acta 2006;89:361–403.
64. Flogel O, Codee JDC, Seebach D, Seeberger PH. Angew Chem Int Ed 2006;45: 7000–7003.
65. Karolin Geyer PHS. Helv Chim Acta 2007;90:395–403.
66. Phelps ME. Proc Nat Acad Sci U S A 2000;97:9226–9233.
67. Phelps ME. PET: Molecular Imaging and Its Biological Applications; New York; Springer:2004.
68. HÉlËne A. Angew Chem Int Ed 2007;46:1772–1775.

69. Lee C-C, Sui G, Elizarov A, Shu CJ, Shin Y-S, Dooley AN, Huang J, Daridon A, Wyatt P, Stout D, Kolb HC, Witte ON, Satyamurthy N, Heath JR, Phelps ME, Quake SR, Tseng H-R. Science 2005;310:1793–1796.
70. Lee JN, Park C, Whitesides GM. Anal Chem 2003;75:6544–6554.
71. Rolland JP, Van Dam RM, Schorzman DA, Quake SR, DeSimone JM. J Am Chem Soc 2004;126:2322–2323.
72. Priola A, Bongiovanni R, Malucelli G, Pollicino A, Tonelli C, Simeone G. Macromol Chem Phys 1997;198:1893–1907.
73. Matteucci MD, Caruthers MH. J Am Chem Soc 1981;103:3185–3191.
74. Huang Y, Castrataro P, Lee CC, Quake SR. Lab Chip 2007;7:24–26.
75. Ugi I, Jacob P, Landgraf B, Rupp C, Lemmen P, Verfurth U. Nucleosides Nucleotides 1988;7:605–608.
76. Kikutani Y, Horiuchi T, Uchiyama K, Hisamoto H, Tokeshi M, Kitamori T. Lab Chip 2002;2:188–192.
77. Fernandez-Suarez M, Wong SYF, Warrington BH. Lab Chip 2002;2:170–174.
78. Elderfield RC. Heterocyclic Compounds. New York: Wiley; 1950.
79. Schober A, Schlingloff G, Gross A, Henkel T, Albert J, Mayer G, Wurziger H, Doring M, Tietz H. Microsyst Technol 2004;10:281–292.
80. de Bellefon C, Tanchoux N, Caravieilhes S, Grenouillet P, Hessel V. Angew Chem Int Ed 2000;39:3442–3445.
81. Ratner DM, Murphy ER, Jhunjhunwala M, Snyder DA, Jensen KF, Seeberger PH. Chem Comm 2005;578–580.
82. Hatakeyama T, Chen DLL, Ismagilov RF. J Am Chem Soc 2006;128:2518–2519.
83. Song H, Chen DL, Ismagilov RF. Angew Chem Int Ed 2006;45:7336–7356.
84. Johnson SW, Seutin V, North RA. Science 1992;258:665–667.
85. Huc I, Lehn JM. Proc Nat Acad Sci U S A 1997;94:2106–2110.
86. Ramstrom O, Lehn JM. Chembiochem 2000;1:41–48.
87. Lehn JM, Eliseev AV. Science 2001;291:2331–2332.
88. Bunyapaiboonsri T, Ramstrom O, Lohmann S, Lehn JM, Peng L, Goeldner M. Chembiochem 2001;2:438–444.
89. Ramstrom O, Lehn JM. Nat Rev Drug Discovery 2002;1:26–36.
90. Otto S. Curr Opin Drug Discov Deve 2003;6:509–520.
91. Erlanson DA, Braisted AC, Raphael DR, Randal M, Stroud RM, Gordon EM, Wells JA. Proc Natl. Acad Sci U S A 2000;97:9367–9372.
92. Kehoe JW, Maly DJ, Verdugo DE, Armstrong JI, Cook BN, Ouyang YB, Moore KL, Ellman JA, Bertozzi CR. Bioorg Med Chem Lett 2002;12:329–332.
93. Lewis WG, Green LG, Grynszpan F, Radic Z, Carlier PR, Taylor P, Finn MG, Sharpless KB. Angew Chem Int Ed 2002;41:1053–1057.
94. Bourne Y, Kolb HC, Radic Z, Sharpless KB, Taylor P, Marchot P. Proc Natl Acad Sci U S A 2004;101:1449–1454.
95. Boger DL, Loiseleur O, Castle SL, Beresis RT, Wu JH. Bioorg Med Chem Lett 1997;7:3199–3202.
96. Greasley SE, Marsilje TH, Cai H, Baker S, Benkovic SJ, Boger DL, Wilson IA. Biochemistry 2001;40:13538–13547.

97. Nguyen R, Huc I ,Angew Chem Int Ed 2001;40:1774–1776.
98. Nicolaou KC, Hughes R, Cho SY, Winssinger N, Smethurst C, Labischinski H, Endermann R. Angew Chem Int Ed 2000;39:3823–3828.
99. Nicolaou KC, Hughes R, Cho SY, Winssinger N, Labischinski H, Endermann R. Chem—Eur J 2001;7:3824–3843.
100. Poulin-Kerstien AT, Dervan PB. J Am Chem Soc 2003;125:15811–15821.
101. Kolb HC, Sharpless KB ,Drug Discov Today 2003;8:1128–1137.
102. Manetsch R, Krasinski A, Radic Z, Raushel J, Taylor P, Sharpless KB, Kolb HC. J Am Chem Soc 2004;126:12809–12818.
103. Mocharla VP, Colasson B, Lee LV, Roper S, Sharpless KB, Wong CH, Kolb HC. Angew Chem Int Ed 2005;44:116–120.
104. Krasinski A, Radic Z, Manetsch R, Raushel J, Taylor P, Sharpless KB, Kolb HC. J Am Chem Soc 2005;127:6686–6692.
105. Radic Z, Manetsch R, Krasinski A, Raushel J, Yamauchi J, Garcia C, Kolb H, Sharpless KB, Taylor P. Chem Biol Interact 2005;157:133–141.
106. Wang J, Sui G, Mocharla VP, Lin RJ, Phelps ME, Kolb H, Tseng HR. Angew Chem Int Ed 2006;45:5276–5281.
107. Thorsen T, Maerkl SJ, Quake SR. Science 2002;298:580–584.

10

GENE ISOLATION, GENE TRANSFORMATION, AND ENZYME REACTION ON A CHIP

JONG WOOK HONG

10.1 INTRODUCTION

Recombinant DNA technology is an attractive and functional technology in the fields of genetic engineering, gene therapy, and genetically modified plants and animals. The first step of the technology is the isolation of a gene of interest. The gene is then ligased with a vector and cloned into a cell of interest. On the other hand, it is widely acknowledged that since the development of microfluidics in late 1980s and early 1990s, there has been tremendous advancement in the new research areas. Many of the components of a microfluidic system such as microvalves, micropumps, and various chip materials have been studied extensively. Now the trend of microfluidic research is toward realization of systems for specific applications for biology/medicine, chemistry, space applications, and others.

In this chapter, several microfluidic systems utilized for recombinant DNA technology steps such as gene isolation, gene ligation, and gene transformation along with enzymatic reactions have been described. In a few of these systems, realization of a complete unit operation of some bioprocesses at the level of single cells or single biomolecules has been achieved successfully, which was rarely possible with conventional methods and protocols.

Biological Applications of Microfluidics, Edited by Frank A. Gomez

10.2 DNA/RNA ISOLATION ON A MICROFLUIDIC CHIP

The first step of processing genetic information including recombinant DNA technology is to isolate and purify genes from the target cells, tissues, or organisms. The target cells could be microbial cells including bacteria and yeast, plant cells, and/or mammalian cells. A conventional tube method of nucleic acid isolation requires at least 10^3-10^4 or more identical cells due to its technical limitation. Hence, very limited cells including bacterial cells that are capable of laboratory pure cultivation could be chosen for gene isolation and further downstream processing are possible. Establishing pure culture steps, however, for certain bacterial cells, mammalian or plant cell lines are extremely difficult and are available only for limited cells. Consequently, the studies of unculturable cells including most of the new bacterial cells are hampered by this technology limitation. Breakthrough research on DNA and mRNA isolation from single mammalian cells and multiple bacterial cells has been reported (1,2). Integrated microfluidic formats are used to handle single mammalian cells and multiple bacterial cells in picoliter sample volumes (Fig. 10.1). The chips for single mammalian cell RNA isolation and multiple bacterial cell genomic DNA isolation were fabricated by multilayer soft lithography (MSL) (3). A typical structure of MSL microfluidic device is two layers of fluidic lines and perpendicularly crossing over control lines. You can easily imagine stepping on a garden horse with your feet realizing on and off, sometimes regulated flow too, of the flow. The control or actuation channels are connected to an outside pressure source. The working pressure for the actuation channels is typically several pounds per square inch.

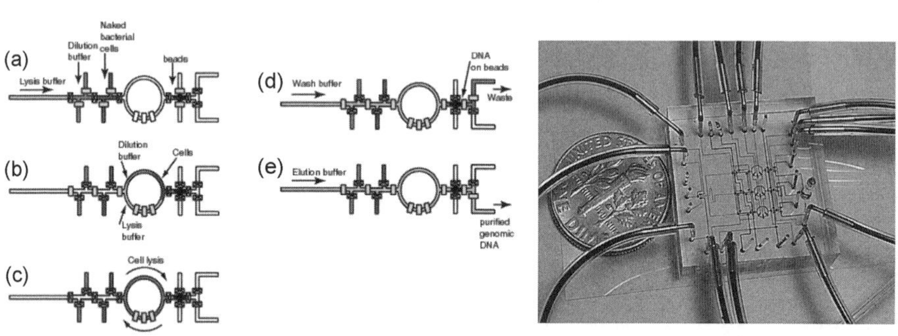

FIGURE 10.1 DNA purification chip and schematic diagram of the complete DNA isolation processes. The chip has multiple layers of fluidic and control or actuation channels. All the processes are operated on a chip by actuating pneumatically controlled microvalves, rectangles. X in rectangle, closed valve. (**a**) Bacterial cells are introduced into the chip. (**b**) The bacterial cells and reagents for cell lysis are introduced into the rotary mixer. (**c**) Cells are lysed with the actuation of the rotary mixer. (**d**) The lysate is introduced over a DNA capturing bead dam, which is constructed prior to the operation of the chip. (**3**) Elution buffer is introduced over the capturing dam. Actual chip for DNA isolation bioprocessing chip is shown in the right side. The actuation channels are filled with a green food coloring and the fluid channels are filled with yellow, blue, and red food based on the functionality differences. The multiple parallel process of DNA recover is possible with small sample volume of 400 pL. (See color insert)

They showed the isolation of β-actin mRNA and abundant zinc finger OZF mRNA from single 3TC mammalian cells on chip without any pre or postsample treatment. All the process of cell loading, cell lysis, purification of target of was completely carried out on a chip. In addition, the isolation and purification of genomic DNA from several bacterial cell on a microfluidic chip has confirmed.

10.3 GENE LIGATION ON A MICROFLUIDIC CHIP

Linking a linear target DNA and circular plasmid DNA, often called a vector, together is an important step in the process of recombinant DNA technology or genetic engineering. During this process unzipped double-stranded DNA fragments from the target and the vector are linked together by the help of specific enzyme called "ligase." Conventional method of the ligation process is to mix DNA, vector, and ligase and then wait for a certain period of time with microliters of samples. An interesting trial of gene ligation has been carried out with complicated microfluidic format. The microfluidic chip was fabricated by a multilayer soft lithography and part of the chip is shown in Fig. 10.2.

The target DNA of green fluorescence protein (GFP) encoding gene was first loaded on the chip through the "DNA" port by sequential operation of valves. By using similar microvalve operation, the vector (linearized vector pNEB205A containing ampicillin resistant gene) was loaded through the "vector" port, and the enzyme (USER™ utrcil-specific excision reagent) was loaded through the "enzyme" port. After the chambers were filled with reagents, nuclease-free water was loaded through the "water" port to push the reagents into the mixing rings from left to right direction in parallel passion. Then three solutions were mixed by opening and closing the three valves consecutively resulting in peristalsis in a flexible microchannel. The incubation process was carried out at 37°C for 15 min. Finally, through sequential valve operation and application of SOC medium through the channels, the ligation products were collected at the "collection" port. The selection process was carried out with a conventional plating method with the blue and white screening method of transformed cells on a Luria–Bertani (LB) agar plate with ampicillin, isopropyl β-D-1-thiogalactopyranoside IPTG, and X-gal. After autoclaving and cooling down

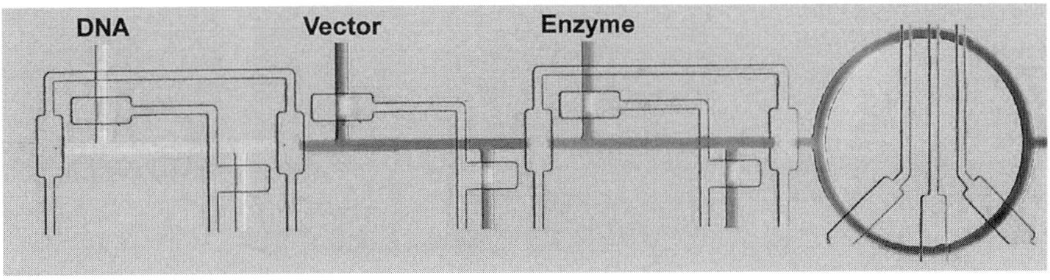

FIGURE 10.2 A part of a microfluidic chip for the single ligation process. Different food coloring represent different potential substrates, such as DNA (yellow), vector (blue), and enzyme (red), and the mixture of yellow, blue, and red coloring in the 6 nL rotary mixer. (See color insert)

the LB agar solution to 55°C, 100 mg of ampicillin (final concentration 100 μg/mL) was added with 1 mL of IPTG (100 mM) and 2 mL of X-gal (20 mg/mL^{-1} in dimethylfomamide) per liter and poured to agar plates. IPTG and X-gal are used to indicate the transformation. The described attempt of gene ligation on a microfluidic chip with picoliter sample volumes, as far as the author knows, was the first trial that demonstrates great possibilities of realization of the whole recombinant DNA processes on a chip.

10.4 GENE TRANSFORMATION ON A CHIP

Inserting linked bundle of target DNA and plasmid DNA into a target bacterial cell is called transformation. The insertion of target foreign materials into mammalian cells is called transfection. A microarray-driven gene transfection system was demonstrated by Zlauddin and Sabatini (4). They described a strategy for the high throughput analysis of gene function in mammalian cells. Using slides printed with sets of cDNAs in expression vectors, they create living microarrays of cell clusters expressing the gene products (Fig. 10.3). Nanolitre volumes of plasmid DNA containing the green fluorescent protein in an aqueous gelatin solution were printed on a glass slide using a robotic arrayer. The spots were first dried, then DNA spots are briefly exposed to a lipid transfection reagent. The lipid transfection reagent is a liposome-based carrier solution that can deliver target DNA into the mammalian cell permeating lipid bilayer cell membranes. The slide was then placed in a culture dish and covered with adherent mammalian cells (HEK 293 cells) in the medium. With time, the cells were growing on the DNA and gelatin spots express the DNA and were divided two to three times in the process of creating a "transfected cell microarray." They called this method a "reverse transfection," as compared with conventional transfection; they have reversed the order of addition of DNA and adherent cells.

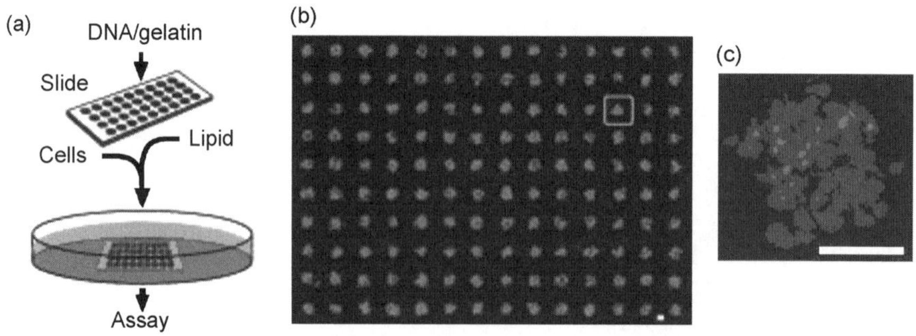

FIGURE 10.3 (**a**) Protocol for making microarrays of transfected cells. (**b**) Scanned image of a GFP-expressing cell microarray made from a slide printed in a 14 × 10 pattern. (**c**) Higher magnification image obtained with fluorescence microscopy of the cell cluster boxed in (**b**). Scale bar, 100 μm (after Reference 4). (See color insert)

The fluorescence of the cells on the transfected cell microarray was detected with a laser fluorescence scanner.

To carry out transformation, bacterial cells having no outer cell walls hence susceptible to accept foreign materials such as DNA called competent cells are used. Since competent cells only have cell membranes, they are very fragile, and special care is required to handle the cells. A similar idea of target gene array was adapted to microfluidic sample (cell) guards for gene tranformation. Nagamine et al. demonstrated continuous chip transformation of bacteria (Fig. 10.4) (5). They combined poly(dimethylsilonane) (PDMS) microfluidic channels for a competent cell supply and a patterned silicon substrate on which different plasmid DNAs were predeposited. The first step of the substrate preparation is to introduce a plasmid DNA solution with small pipettes onto the etched silicon substrates. Then, the plasmid DNA in the solution is dried to form immobilized target DNA arrays. To introduce competent cell, PDMS microfluidic channels were aliened over the dried target plasmid patterns on the microfabricated silicon substrate, resulting in the exposure of the competent *Escherichia coli* cells to the plasmid DNA deposited in the wells. One clever idea is using diluted collagen (type I, Nitta Gelatin Inc) solution to introduce competent *E. coli* JM 109 cells. After 1 h of incubation at 4°C to allowing dissolving of the dried plasmid DNA and contacting of the plasmid DNA with the competent cell. Then, heat shock of 42°C for 90 s was carried out. This heat shock facilitates the plasmid DNA introduction into the competent cells and the gelation of collagen solution in the mixture. Detection of GFP expressed from plasmid pUC-GFP was carried out by using fluorescence microscopy. *E. coli* cells immobilized on the silicon wafer were excited by irradiation with light from a Xe

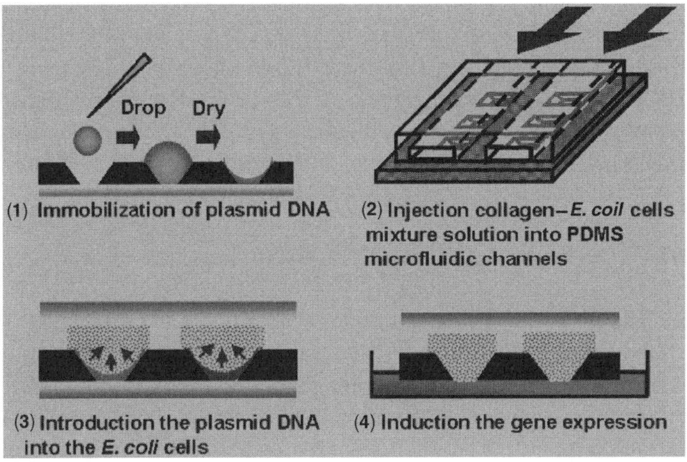

FIGURE 10.4 Protocol of on-chip transformation of *E. coli*. (1) Target DNA of interest is dried onto microfabricated silicon wells. (2) Collagen– *E. coli* mixture solution is introduced into microfluidic channels. (3) Transformation is carried out through heat shock. (4) Gene is expressed (after Reference 5). (See color insert)

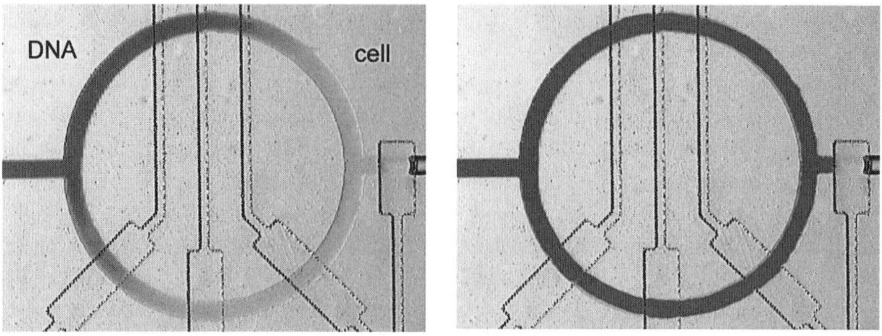

FIGURE 10.5 A part of the gene transformation chip. five nanoliters of a micromixer for the gene transformation before (left) and after (right) the mixing. The blue stands for DNA and the yellow for cells. After on chip transformation the competent cells were recovered from the chip and cultured on an ampicillin containing agar medium. (See color insert)

lamp, and fluorescence (excitation 475 nm/emission 509 nm) was detected using a digital camera.

Gene transformation bioprocess on a complicated microfluidic chip with subnanoliter sample volumes were demonstrated on a PDMS-based chip (6). The design of the chip is similar to DNA isolation chip described in the early part of this chapter except for the rotary mixer sizes. The competent cells (NovaBlue Singles Competent Cells) were loaded on the chip through the sequential operation of the pneumatic valves. After the cell chambers were filled with cells, plasmid DNA of pGFP vector (BD Biosciences) went through the DNA port to push the reagents into the mixing rings. Then two solutions of competent cells and DNA are mixed by opening and closing the pumping valves sequentially (Fig. 10.5). After the mixing the chip was kept at 0°C for 5 min, increased the temperature to 42°C for 30 s, and then decrease the temperature to 0°C for 2 min to allow the transformation of the plasmid DNA and the competent cells to be carried on. Finally, the transformation products were collected through the "collection" port. The transformation on the chip was confirmed by the blue and white screening method.

10.5 ENZYMATIC REACTION ON A CHIP

Conventional ways of enzymatic reactions especially for the evaluation of enzyme kinetic parameters such as K_m and V_{max} are using several millimeters of test tubes. Labor-intensive sample preparations with cumbersome pipetting procedures are required. Therefore, compared to the microfluidic formats, huge amount of samples and reagents are required. In addition, to carry out parallel experiments and achieve intensive data acquisition of enzymatic kinetics, several trained technicians are required. One of the earliest microfluidic system for analyzing the kinetics of an enzymatic reaction was published demonstrated by Hadd et al. (7). Electrokinetic flow control in the microfabricated glass channels was carried out controlling applying voltages in the five different ports for metering substrate, enzyme, and inhibitor in 9 μm in width and 35 μm in depth channels resulting subnanoliter reaction volume control. In particular, a enzyme, β-glactosidase (β-GAL) is metered and mixed with a substrate resorufin β-D-galactopyranoside (RBG), fluorogenic substrate. The kinetics of the reaction was obtained by observing the fluorescence of the hydrolysis product with laser-induced fluorescence (LIF). An attempt to determining enzyme kinetics using continuous-flow microfluidic system was demonstrated by Seong (8). Unlike electrokinetic control of enzyme and reagents, they used pressure-driven continuous flow control on glass–PDMS microchip. They introduced microbeads having immobilized enzymes on them into the chip. The enzymes, horseradish peroxidase (HRP), and β-GAL were coimmobilized onto the microbeads to biotin via amide coupling of an aminocaproyl space and the substrate solutions of H_2O_2 and RBG were passed through microbeads into the microreactor using a syringe pump. The optical and fluorescence images were taken with a fluorescence microscope and fluorescence images were analyzed using precision digital imaging software.

Microplate reader compatible microfluidic device for conducting enzymatic assays was reported by Kang and Park (9). They incorporated a simple sample dilution method that originally developed by the Whiteside group and sequentially mix and make gradient of two solutions, in this case two substrates as shown in Fig. 10.6. The substrates are introduced from right to left direction and the enzyme is

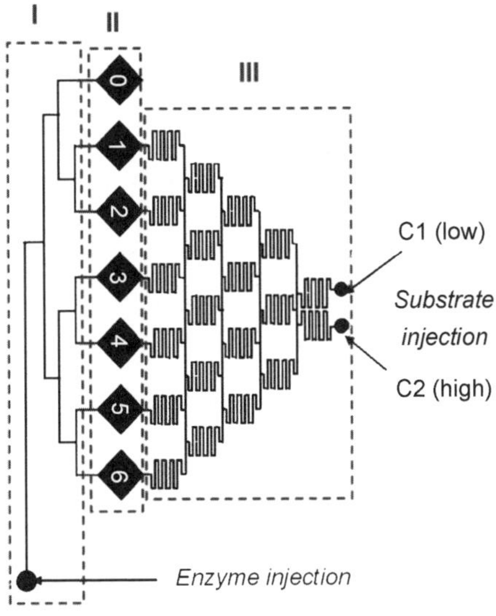

FIGURE 10.6 Schematic diagram of single enzyme assay module in one microfluidic device. One essay module has three parts. Part I has microchannel pattern for injection of enzyme solution. In part II reaction chambers for the enzyme and substrates. Part III contains the channel network for dilution of the substrate solution (figure courtesy of Dr. J.K. Park).

introduced from left to right constructing 60 μL of individual reaction volume. They examined alkaline phosphatase (ALP) and *p*-nitrophenyl phosphate (*p*-NPP) as an enzyme and a substrate, respectively. Although relatively large amounts of enzymes and substrates, hundreds of microliters, are required the device can be used as a simple enzymatic assay for laboratory use with a micoplate reader. Part I has microchannel pattern for injection of the enzyme solution. The reaction chambers for enzyme reaction are positioned in part II (1–6) and the reaction chamber 0 is the control chamber for ensuring the equal distribution of the enzyme sample solution. The channel network for dilution of the substrate solution is placed in part III.

Recently, Jambovane et al (10) have developed a highly integrated and parallel microfluidic chip for enzyme kinetics study. In this chip enzyme/substrate introduction, metering, mixing, and detection were automated in eleven parallel processors. Hence, batch of eleven enzyme–substrate reactions were carried out simultaneously by using less than 5 nL reagents in each processor. The complete design of protein kinetics chip is depicted in Fig. 10.7. Each processor was different in terms of concentration of substrate due to variable mixing ratio of substrate and buffer. The chip was fabricated by using multi layer soft lithography. Two different molds were first fabricated by photolithographic processes to create the fluidic channels and the control channels for actuating the values positioned in the respective layers of the PDMS-based enzyme kinetics chip. The valves and the peristaltic pump in the chip were operated by pneumatic control. The optical detection of enzyme product in each parallel processor was simultaneously detected inside a biochip

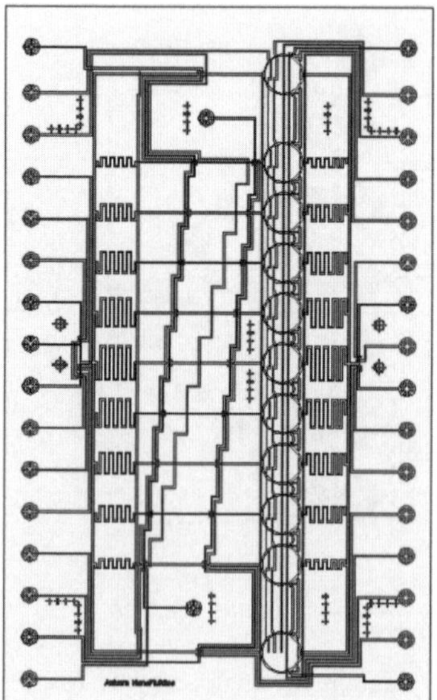

FIGURE 10.7 Design architecture of enzyme kinetics chip. The chip has capability of 11 parallel enzyme processes including positive and negative controls. Hence, kinetic parameters are obtained from a single experiment.

reader (arrayWoRx®, Applied Precision). The data obtained from the reader were transformed to estimate K_m and V_{max} by the Double-reciprocal or Lineweaver–Burk plot. This device could be used in many kinetic study applications such protein–protein binding.

10.6 SUMMARY AND PERSPECTIVE

In this chapter, biological automation and integration in the areas of gene isolation, gene ligation, gene transformation, and enzymatic reactions were described. Today's biology is witnessing unprecedented growth the way computers witnessed in the last century. Hence, automation in biology is the modern and state-of-the-art revolution in engineering. In this revolution, microfluidic chips are playing key roles on account of their distinct advantages like less sample volumes, fast assay time and great opportunity for superior integration and automation. Moreover, many applications that were not possible beforehand would be possible with the new tools.

ACKNOWLEDGMENT

The author would like to thank Sachin Jambovane for his help during the preparation of this manuscript.

REFERENCES

1. Hong JW, Hang G, Anderson WF, Quake SR. Nat Biotechnol 2004;22:435–439.
2. Marcus JS, Anderson WF, Quake SR. Anal Chem 2006;78(9):3084–3089.
3. Unger MA, Thorsen T, Scherer A, Quake SR. Science 2000;288(5463):113–116.
4. Ziauddin J, Sabatini DM. Nature 2001;411:107–110.
5. Nagamine K, Onodera S, Torisawa Y-S, Yasukawa T, Shiku H, Matsue T. Anal Chem 2005;77:4278–4281.
6. Hong JW, Anderson WF, Quake SR. J Phys: Condensed Matter 2006;18(18): S691–S701.
7. Hadd AG, Raymond DE, Halliwell JW, Jacobson SC, Ramsey JM. Anal Chem 1997;69:3407–3412.
8. Seong GH, Heo J, Crooks RM. Anal Chem 2003;75(13):3161–3167.
9. Kang JH. Sen Actuators B: Chemical 2005;107(2):980–985.
10. Jambovane SR, Duin E, Hong JW. A novel nanoliter scale fluidic system for protein kinetics. In preparation.

PART III

SEPARATIONS ON MICROCHIPS

11

CHEMICAL MONITORING IN COMPLEX BIOLOGICAL ENVIRONMENTS USING SEPARATION-BASED SENSORS IN CHIPS

ZECHARIAH SANDLIN AND ROBERT KENNEDY

11.1 SEPARATION-BASED SENSORS

The ability to monitor concentration changes of chemical components in a complex mixture over time is of interest in a diverse array of fields including biomedical research, environmental analysis, industrial manufacturing, biotechnology, and homeland security. Many of these applications require high selectivity and sensitivity to distinguish a desired chemical component from its environment. Chemical sensors have been the classical technology pursued in such applications because of their selectivity, sensitivity, portability and real-time monitoring capability (1–3). By definition, a chemical sensor continuously monitors a specific chemical component over time by reversibly transducing a chemical interaction into an electronic signal for qualitative or quantitative measurement (1). Such sensors contain a detection element for a specific chemical interaction, a transducer to convert either a physical or a chemical change into a measurable signal, and an actuator for display. Typically, chemical sensors are classified by their transducer element into four main categories: electrochemical, optical (optodes), mass sensitive, or thermal sensitive. Biosensors are characterized by the use of a biological interaction (e.g., enzyme/substrate, antibody/antigen, and receptor/hormone) for the detection element. Ideally, chemical

Biological Applications of Microfluidics, Edited by Frank A. Gomez

sensors are selective enough to transduce only information from the chemical component of interest at the exclusion of all others; however, in some cases the removal of interference molecules is necessary. Although well-developed sensors exist for many analytes, it can be a slow and difficult process to develop a sensor with the requisite selectivity, sensitivity, and stability, especially for sensing in complex biological environments (4). Furthermore, most sensors are only capable of monitoring one chemical component at a time, but frequently multiple components are of interest.

Because of these limitations, it is appealing to develop alternative sensing strategies. One such strategy is the "separation-based sensor" (5,6). A separation-based sensor is an analytical system that integrates sampling, sample preparation, sample injection, serial separation, and detection for chemical monitoring. While not a true sensor according to IUPAC definition (7), such devices can potentially offer the same functionality. Furthermore, they can provide significant performance enhancement over chemical sensors when monitoring chemical components in complex environments because of their high selectivity, capacity for multi-component monitoring, and their ability to tune selectivity for monitoring new analytes.

Separation-based sensors can utilize a variety of well-developed chemical separation techniques based on chromatography and electrophoresis for separating chemical components of complex chemical environments. The need to develop a highly selective detector element, as is the case for chemical sensors, is not necessary. Additionally, the selection of appropriate separation conditions can mitigate the need for removing interference molecules from the sample prior to monitoring. A disadvantage of separation-based sensors can be the size and complexity of instrumentation. Another drawback is that the time required to both collect and separate the desired chemical component may result in low temporal resolution and obviate real-time monitoring. Additionally, unlike chemical sensors, which are typically immersed directly into a sample solution or stream, a separation-based sensor has to transport material to the separation component without contributing significantly to band broadening. As reviewed in this chapter, microfluidic technologies ameliorate these problems and can make separation-based sensors a practical alternative to other sensors.

Many separation methods have the potential for use in separation-based sensors. Although chromatographic separation methods, such as high performance liquid chromatography (HPLC) and gas chromatography (GC), have both increased in speed and capacity for automation to the point of being attractive possibilities for separation-based sensors, capillary electrophoresis (CE) has received the most attention. CE is especially well suited for chemical monitoring because of its extremely fast analysis times, highly efficient separations, nanoliter or smaller sample volumes, and ease of automation (5,6). Indeed, separation-based sensors employing CE have been implemented for several applications, such as monitoring cellular secretions (8), enzymatic reactions (9), and most notably *in vivo* monitoring of neurotransmitters (10–16). These instruments integrate several analytical operations including sampling, derivatization, injection, and separation. Although these instruments prove the potential of CE for separation-based sensing, they are based on manually assembled fused silica capillary networks. As such, they require care in assembly and operation making them nonportable and difficult to disseminate.

The inception of microfabricated electrophoretic systems (17) has revolutionized the possibility of integrating the functions of a separation-based sensor into a compact, easy to use, and fully automated device. Indeed, the components of separation-based sensors such as sampling, sample preparation (e.g., filtering and derivatization), and sample injection have already been shown to be amenable to the microfluidic format. Such micro total analysis systems (μ-TAS) or lab-on-a-chip devices have demonstrated fast response times, low sample consumption, and on-site operation on a highly integrated format (18–20). In this chapter we review the instrumentation and theory of fast separations on microchip devices and then discuss examples of systems that integrate such separations into microfluidic separation-based sensors.

11.2 FAST SEPARATIONS WITH SEPARATION-BASED SENSORS

In considering the development of a sensor, the temporal resolution is important. In some cases, concentration dynamics require the ability to monitor changes on the subsecond time scale while in others the capacity to monitor temporal changes over minutes or hours is all that is required. A related issue to temporal resolution is feedback time. True chemical sensors provide rapid conversion of concentration to an observable readout, that is, real-time functionality. To provide high temporal resolution and real-time readout, a fast separation mode is necessary. As stated above, CE is well suited for such an application because of its potential for fast and efficient separations. Fast CE has already been shown to be beneficial in the development of microfluidic separation-based sensors for a diverse range of biological systems such as the monitoring of hormone release from cells (21–23), monitoring of biochemical reaction kinetics (24), and monitoring of neurotransmitter levels in mammalian brains (25–27). In all of these applications, fast separations were important in obtaining sufficient sampling frequencies for good temporal resolution and near real-time output.

11.2.1 Theory of Fast Separations with Capillary Electrophoresis

Before discussing the use of chip-based systems, it is instructive to consider the theoretical underpinnings of achieving fast separations by capillary or channel electrophoresis. In the simple case where diffusion is the only source of band broadening, CE theory dictates that both separation speed and efficiency can be simultaneously increased by increasing the voltage dropped across the separation channel. This principle is demonstrated by the equations for theoretical plates (N)

and migration time (t_{mig}) (28):

$$N = \frac{\mu_{app} V}{2D} \tag{11.1}$$

$$t_{mig} = \frac{l}{v} = \frac{L}{V\mu_{app}} \tag{11.2}$$

where μ_{app} is the apparent electrophoretic mobility, D is the diffusion coefficient of the analyte, l is the separation length, L is the total channel length, v is the analyte velocity, and V is the voltage dropped across L. From these equations, it is apparent that N is proportional to voltage and t_{mig} is inversely proportional to voltage. Peak efficiency is enhanced if diffusional dispersion is small relative to the migration velocity. More importantly, from a chip design perspective, these equations also show that efficiency is independent of channel length; therefore, as long as the voltage dropped across the separation channel is unchanged, shortening of this channel will produce faster separations without decreasing efficiency.

In actual applications, however, other sources of band broadening limit the benefit of higher voltages on efficiency. These sources of band broadening have to be suppressed to obtain diffusion-limited efficiency; their effect can be evaluated by considering the definition of efficiency:

$$N = \frac{t_{mig}^2}{\sigma_{tot}^2} \tag{11.3}$$

where σ_{tot}^2 is the zone variance from all sources of band broadening. In CE, σ_{tot}^2 can be expressed as

$$\sigma_{tot}^2 = \sigma_{diff}^2 + \sigma_{inj}^2 + \sigma_{det}^2 + \sigma_{heat}^2 + \sigma_{ads}^2 + \sigma_{geo}^2 \tag{11.4}$$

where $\sigma_{diff}^2, \sigma_{inj}^2, \sigma_{det}^2, \sigma_{heat}^2, \sigma_{ads}^2, \sigma_{geo}^2$ are the variances caused by longitudinal diffusion, sample injection, detector path length, Joule heating, adsorption, and geometric bends, turns, or changes in separation channel dimensions, respectively. Under diffusion-limited conditions, the expression for σ_{tot}^2 simplifies to

$$\sigma_{tot}^2 = \sigma_{diff}^2 = 2Dt_{mig} \tag{11.5}$$

However, when t_{mig} is small, which reduces the magnitude of σ_{diff}^2, great care must be taken to reduce the contribution from other band broadening sources to retain diffusion-limited efficiency. This problem becomes acute at higher voltages where the resultant increase in velocity will decrease diffusional band broadening but either unchanged or increase band broadening from other sources.

11.2.2 Additional Concerns for Fast Separations on Chips

For a well-developed system with fast separations, efficiencies should be diffusion limited. A review has detailed the methodology behind creating such systems for capillary-based instruments (29). The underlying principles for fast separations in capillaries are relevant for both glass- and polymer-based chips with additional considerations. As stated in the previous section, to achieve fast separations with diffusion-limited efficiencies on chips will require high voltages and a suppression of other sources of band broadening.

11.2.2.1 Broadening from Joule Heating Joule heating is a result of ion resistance to transport through a solvent. Such heating results in a temperature gradient from the center of the channel to the edges where heat is dissipated. This temperature gradient causes band broadening as a result of uneven rates in analyte migration because of viscosity changes and convectional mixing. An absolute rise in temperature is not necessarily problematic as long as the radial temperature gradient is small.

Theoretical models (30–32) and experimental techniques (33–37) have been developed for the evaluation of Joule heating in microchips. In most cases, Joule heating only becomes a concern for inducing band broadening at higher voltages. The magnitude of the temperature increase inside a channel network depends on the power (product of voltage and current) generated during chip operation. The current passed for a given voltage is proportional to the cross-sectional area of a channel and inversely proportional to channel length. Therefore, a reduction in channel cross-sectional area or an increase in channel length will reduce current and reduce the heating effect. The reduction in cross-sectional area also increases the surface-area-to-volume ratio, which improves heat dissipation through the channel surface. A study comparing capillary-based systems and microchip-based systems largely confirmed that glass planar microdevices are less susceptible to Joule heating because of their larger thermal mass and, often, higher surface-area-to-volume ratio (38). Contrarily, polymer chips exhibit lower heat conductivity than silica or glass (35,39) thereby requiring additional attention to heat dissipation (40). Thus, for design of fast separations in chip-based devices the thermal conductivity of the material used and the thermal mass are important considerations.

Although reduced channel dimensions should allow for faster separations at higher voltages before the onset of heat-induced band broadening, this approach places great demands on the detection system because of lower injection and detection volumes. Another approach to combat Joule heating is the reduction of power generation with low conductivity buffers (41). The relationship between buffer conductivity and the dissipated power per volume of buffer is given by (42):

$$W = E^2 \cdot \kappa_b \tag{11.6}$$

where W is the dissipated power per volume of buffer (W/m^3), E is the applied electric field (V/m), and κ_b (W/(m·K)) is the conductivity of the buffer. A decrease in conductivity will only allow for an increase in field strength by the square root of the factor of the conductivity decrease at a constant W.

Another possibility for the reduction of Joule heating is active temperature control or thermostating (43). This technique is commonly employed in commercial capillary-based instruments to increase their effective range of operational voltages.

However, in capillaries it has been shown that significant band broadening can occur from axial temperature gradients caused from variations in the change of heat transfer condition along a capillary's length (44). The electric field in the capillary becomes nonuniform due to temperature-induced changes in fluid conductivity, which causes a pressure gradient to form to meet mass continuity through the capillary. This problem can be avoided by ensuring universal thermostating across a capillary or a microfluidic device. To date, however, work on chips has been focus on designing heaters and coolers for controlling chemical reactions instead of controlling Joule heating. Despite this fact, integrated cooling systems proposed for chips such as endothermic reactions in microchannels (45), thermoacoustic refrigeration (46), and thermoelectric cooling (47,48) based on the Peltier effect might be exploitable for controlling Joule heating.

When Joule heating and other sources of band broadening are minimized, chip-based systems are capable of extremely fast separations using CE. For instance, in the first paper to demonstrate fast separations on chip, fluorescein was separated from fluorescein sulfonate in 4 s with a separation length of 0.75 cm and an applied field of 1875 V/cm (49). In another device, rhodamine B was separated from dichlorofluorescein in 0.8 ms with a separation length of 200 μm and an applied field of 53 kV/cm (Fig. 11.1) (50). To achieve such high potential fields without Joule heating, 83% of the buffer composition contained the zwitterionic organic buffer, HEPES.

11.2.2.2 Broadening from Adsorption Adsorbed molecules at the surface of channels increase band broadening by contributing chromatographic retention to electrophoretic separations. These nonequilibrium chromatographic effects increase

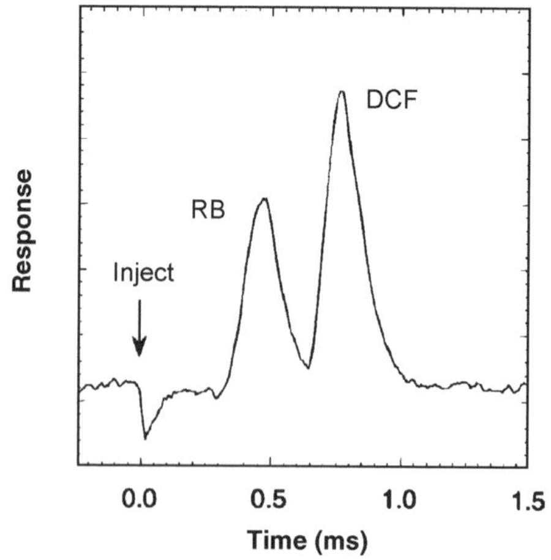

FIGURE 11.1 Electropherogram of a high speed separation of rhodamine B (RB) and dichlorofluorescein (DCF) in 0.8 ms. The separation field strength was 53 kV/cm over a separation length of 200 μm. An arrow demarks the start time at 0 ms. (Reprinted with permission from reference 50. Copyright 1998.)

with faster migration times; therefore, higher voltages might decrease instead of increase resolution. Microfluidic devices have been fabricated from a variety of glasses (51,52) and polymers (53–55), which have a potential of adsorbing an array of compounds. Much work on adsorption modeling has been done for glass surfaces (56–67). This work has shown that analysis speed is largely dependent upon the kinetics of the solute interaction with a surface. Several methods have been employed to reduce these interactions. Polymer-based chips have employed covalent (68–80) and dynamic (81–84) surface coatings and adsorption resistant polymers (85) in an effort to reduce negative effects of adsorption. Unfortunately, attempting to reduce Joule heating induced band broadening by decreasing channel dimensions potentially could increase broadening because a greater surface-area-to-volume ratio will increase the chance of adsorption.

11.2.2.3 Broadening from Geometric Bends It is possible to improve the absolute separation efficiency by increasing the separation channel length while retaining high field strengths in the separation channel. In some cases, microchip devices have turns or bends in a long separation channel to maintain a compact footprint. However, these turns and bends can cause additional zone dispersion by a "race track" effect as first described in coiled capillaries (86). In CE, excess dispersion from geometric bends results from lateral variations in both the migration distance and field strength from the outer radius to inner radius of a turn; electrokinetic transport only exacerbates this problem. For instance, a molecule in the inner radius not only travels a shorter distance but also travels faster than a molecule in the outer radius because of a higher potential field. The amount of dispersion introduced by a turn can be quantified using a one-dimensional model (87):

$$\sigma^2_{\text{geo}} = \frac{\left[2\theta\omega\left(1-\exp\left(-\frac{\omega^2 v_c r_c}{2D\theta(r_c+\omega/2)^2}\right)\right)\right]^2}{24} \tag{11.7}$$

where θ is the angle of the turn in radians, ω is the width of the channel, v_c is the velocity of the analyte, r_c is the radius of curvature along the centerline of the turn, and D is the analyte diffusion coefficient. This equation reveals that the variance can be diminished by reducing the channel width, reducing the flow velocity, increasing the radius of a turn, or increasing the diffusion coefficient (by reducing solution viscosity). Of these methods, only manipulation of the channel width and radius of turn are considered practical. For instance, to decrease the flow velocity in the separation channel the potential field has to be lowered, as shown in Eq. 11.1, a lower applied voltage will cause a concomitant drop in efficiency. In addition, any attempt to increase the diffusion coefficient by decreasing solution viscosity will only cause an increase in flow velocity. Ramsey's group first demonstrated the benefit of using a larger radius of turn for decreasing band broadening in a high efficiency application (Fig. 11.2a) (88). Their design demonstrated no additional band broadening as evidenced by the fact that the effective diffusion coefficient of

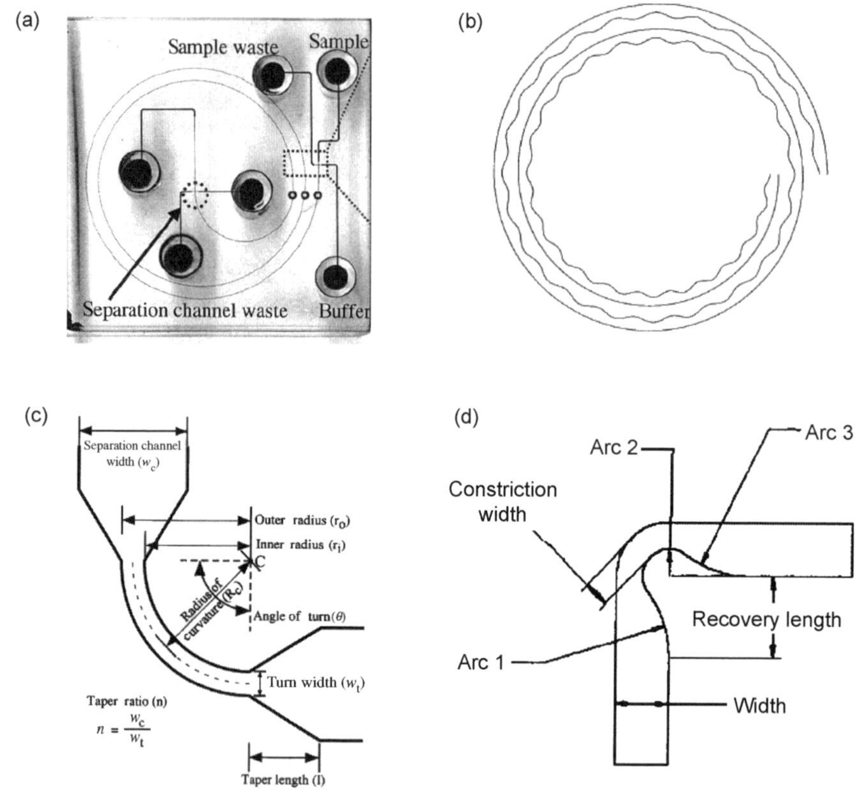

FIGURE 11.2 Various geometries for reducing turn associated band broadening. (**a**) Separation channel with a large radius of curvature. (**b**) Spiral separation geometry with a wavy wall along the inner track of the channel. (**c**) Tapered turn with a reduced channel radius in the turn. (**d**) Parameters for optimized geometries for a reduced channel radius turn. (**a**, reprinted with permission from reference 88. Copyright 2000. **b**, reprinted with permission from reference 91. Copyright 2002. **c**, reprinted with permission from reference 89. Copyright 2000. **d**, reprinted with permission from reference 93. Copyright 2001.)

dichlorofluoroscein was within a few percent of that measured on a microchip with a straight separation channel over a wide range of electric fields. Alternatively, Mathies group first showed that tapered turns (Fig. 11.2c), which reduce the width of a turn, also do not sacrifice electrophoretic performance; however, they stated that geometric bends have no effect on the separation quality of smaller molecules in less viscous media (89). Alternative geometries and strategies have also been explored for both glass (90–94) and polymer chips (95). Fig. 11.2 demonstrates these further attempts to optimize the turn geometries for both large radius turns and reduced channel width turns.

11.2.2.4 Broadening from Finite Injection Volume In the diffusional-limited case, it is assumed that an injected plug is an infinitely narrow zone of unit area. This zone evolves into a Gaussian shape by diffusion over the course of a separation. In actuality, however, introduced sample has a finite volume. The effect of injection volume on variance is described in the following equation (96):

$$\sigma_{\text{inj}}^2 = \frac{l_{\text{inj}}^2}{12} \tag{11.8}$$

where l_{inj} is the length of the sample injection plug. For fast separations, small sample volumes in the hundreds of picoliters or less are required to create narrow sample zones in small cross-sectional area channels. Previous work has shown that for small molecules with weak adsorption to capillary or channel surfaces the variance from injection is often the largest contributor to band broadening (97,98).

Chip-based formats are ideally suited for the manipulation of small volumes for injection purposes. Using electrokinetic pumping, these devices can move charged ions by electrophoresis or neutral species by electroosmotic flow in a precise and highly automated fashion using computer control. Several injection schemes have been developed for chip-based analysis (99) with the two most common being pinched injections (100) and gated injections (101). Depending on the application, one of these injection methods might be more suitable. For instance, pinched injections have temporally stable plugs during sample loading with a size defined largely by the physical dimensions of two intersecting channels. The amount injected can be specifically tailored by adjusting the geometry of this channel intersection. There is some electrophoretic injection bias observed in both the sample loading and dispensing steps with neutral analytes being preferential injected over anionic analytes (102). A drawback for fast chemical sensing, where serial injections are required, however, is the inability to repetitively inject unless an additional control channel is added (103). Contrarily, gated injections have plug dimensions that are time dependent during sample loading and suffer from electrokinetic bias. However, if analysis time is important, gated injections are ideally suited for repetitive injections.

At times, insufficient detector sensitivity might necessitate over injection of sample. As a result, additional band broadening occurs from electromigration dispersion in addition to the broadening from the initial sample plug length (28). This form of broadening can be minimized by using a low concentration of analyte dissolved in the separation buffer. Alternatively, sampling stacking techniques such as field amplified stacking have been developed for microchips to preconcentrate analyte prior to detection (104,105). Such stacking methods, however, typically add time to the overall injection and therefore sacrifice temporal resolution in a sensor application.

11.3 MICRO TOTAL ANALYSIS SYSTEMS WITH ELECTROPHORETIC SEPARATIONS FOR MONITORING OF BIOLOGICAL SYSTEMS

Although a rapid separation may be at the heart of a separation-based sensor, a complete device must integrate sampling, sample preparation (typically derivatization), detection and data output. μ-TAS technologies are well suited for the incorporation of several analytical operations important for separation-based sensors. Examples of such devices have been reported for environmental monitoring (106), enzymatic kinetics (24), and detection of warfare agents (107). In this chapter, we will restrict our discussion to systems for monitoring chemicals in living systems. Two examples will be discussed: (a) monitoring cells in culture and (b) monitoring chemicals *in vivo*.

11.3.1 Separation-Based Sensors on Chips for *In Vitro* Cell Monitoring

A common physiological experiment is to maintain cells or tissues in culture and monitor the release of chemicals from the cells as a way of gauging health or function. Examples include monitoring hormone release from endocrine cells, neurotransmitter release from neurons, growth factors from embryos, and metabolites from cells. Such experiments have application in basic biomedical research and for biotechnology, such as monitoring secreted cell products. Integration of cell culture and separation-based sensing offers an exciting approach for greatly simplifying such experiments and utilizing the advantages of microfluidics.

Many experiments have demonstrated examples of culturing cells on chips. Cell culture systems have been developed that involve recirculation of fluids or static incubation (108–111). Sophisticated designs have allowed integration of multiple cell types on chips as in the cell culture analog (CCA) system (112) used for evaluating pharmacokinetics and chemical toxicity. In this system, up to three different cell types are incubated in separate chambers on a Si-microfluidic chip (113). Drugs can be pumped over the cells while effluent is recirculated through the fluidic system using an off-board peristaltic pump allowing mimicking of cellular interactions through a circulatory system. This system has the potential to test for toxicology and metabolism of drugs in an *in vitro* system that closely mimics the *in vivo* environment.

A less well-developed area has been coupling cell culture with chemical analysis to monitor the physiology of living cells on chips. Examples of coupling chemical analysis and some level of cell culture include use of patch clamp, sensors, immunoassay, and Ca^{2+} imaging (114–118).

The rapid separation capability of chips suggests that it may be possible to monitor cellular secretions in a sensor-like fashion by incorporating cultured cells directly in a chip. An example of such an application is a system developed for monitoring insulin secretion from single islets of Langerhans (21–23). Islets are microorgans found throughout the pancreas and contain clusters of 2000–4000 endocrine cells consisting mostly of β cells (119). β cells facilitate glucose homeostasis by secreting insulin in response to elevated blood glucose concentration. A hallmark of diabetes is a derangement in insulin release from β cells. For different experimental models, the continuous monitoring of insulin is important in elucidating the development of this disease (120).

A separation-based sensor on a glass microfluidic chip has been developed to continuously monitor insulin release from a single islet (Fig. 11.3a) (21,22). In this device, a 300 μm diameter fluidic access hole serves as a culture chamber for an islet

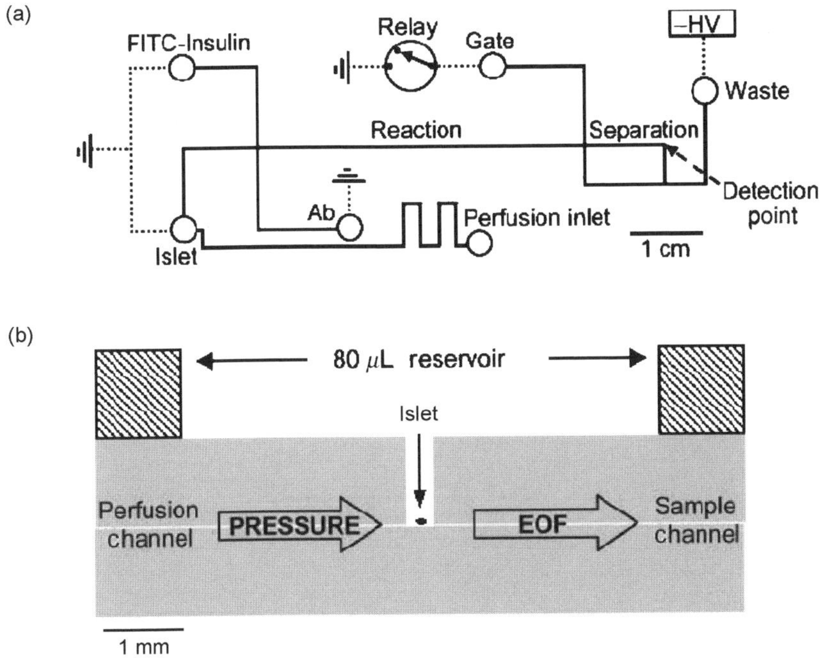

FIGURE 11.3 Microfluidic device with integrated perfusion for the continuous monitoring of insulin release from islets. (**a**) Device layout shows fluidic channels (lines) and access holes (circles) drawn to scale. Dashed lines represent electrical connections. (**b**) A cutaway view of the islet chamber on the chip showing how it is perfused with media and sample via EOF. A single islet is contained in the 300 μm diameter chamber with an 80 μL fluid reservoir above. (Reprinted with permission from reference 22. Copyright 2005.)

(Fig. 11.3b). This chamber is uncovered, which allows for visual inspection of the islet, exchange of gases, and access to cell culture media. Two microfluidic channels intersect the chamber with one serving as a sampling conduit for the analysis portion of the device and the other for continuous perfusion of solutions by gas pressure. A perfusion system allows the contents of the islet chamber to be interchange with cell culture media or various balanced salt solutions for interrogation of the islet. Because of the low fluidic resistance through the islet chamber, pressure-driven flow does not enter the analysis portion of the device. In all other channels, electroosmotic flow (EOF) transports fluids and continuously samples solution from the islet chamber. An online insulin CE-immunoassay (121) is performed in a reaction channel by mixing streams of fluorescein isothiocyanate-labeled insulin (FITC-insulin) and anti-insulin with sample from the islet chamber via EOF. At the end of the reaction channel, its content is diverted to a waste channel; however, upon actuation of a high voltage relay (configured for gated injections) sample enters the separation channel where FITC-insulin and FITC-insulin bound to antibody are separated. Laser-induced fluorescence is used to detect the FITC-labeled compounds 1–2 cm from the injection point using a fluorescent microscope. A resistive heater strip, attached to the underside of the device, heats the chip to 37 °C. Islets require this temperature for insulin secretion; it also serves to facilitate mixing and kinetics of the immunoreaction.

Using the CE-immunoassay, the amount of insulin in the islet chamber is quantitated by determining its effect on the ratio of FITC-insulin bound to antibody to free FITC-insulin. An increase in insulin concentration sampled from the islet chamber competes with FITC-insulin for antibody in the reaction channel resulting in a decrease in bound and an increase in free FITC-insulin. Because the CE-immunoassay uses an antibody specific for insulin, this technique is highly selective. By calculating the change in insulin concentration required to obtain a bound to free ratio change three times the standard deviation of the bound to free ratio at 0 nM insulin in the islet chamber, the limit of detection of this method is 0.8 nM. Furthermore, the relative standard deviation of this assay is <3% over short term and ~6% over 2 h of continuous operation (Fig. 11.4). The temporal resolution of the

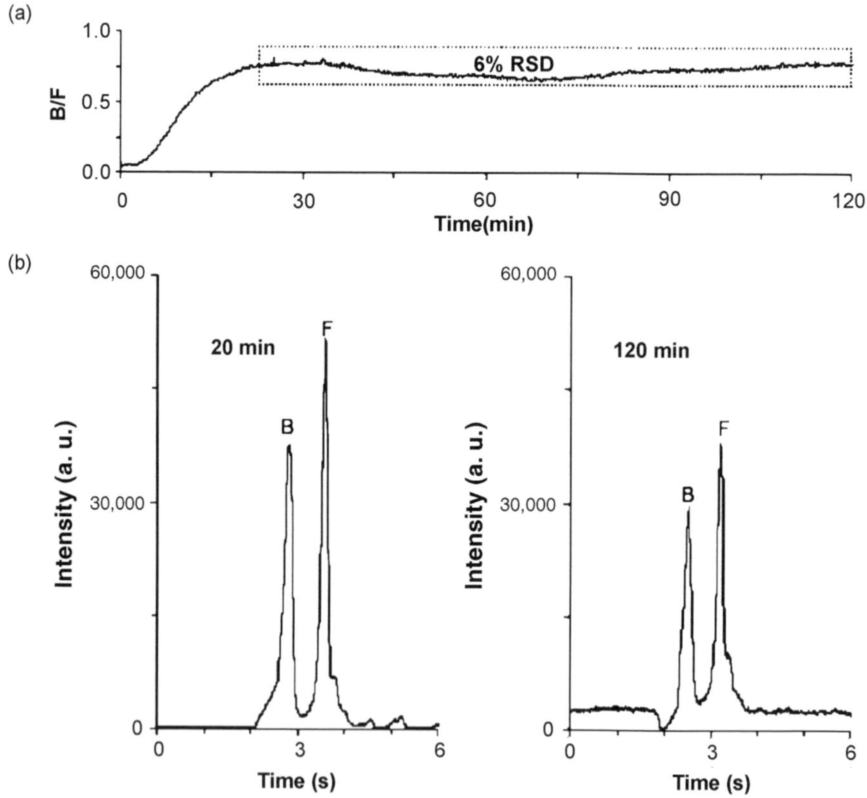

FIGURE 11.4 Demonstration of performance stability of device in Fig. 11.3 during continuous operation. (**a**) Monitoring of bound to free ratios during a continuous 2 h perfusion of 50 nM insulin. Relative standard deviation calculated from region contained in dashed box. (**b**) Electropherograms collected during the experiments depicted in (**a**) at 20 min and 2 h with FITC-insulin (*F*) and FITC-insulin bound to antibody (*B*) peaks marked. (Reprinted with permission from reference 22. Copyright 2005.)

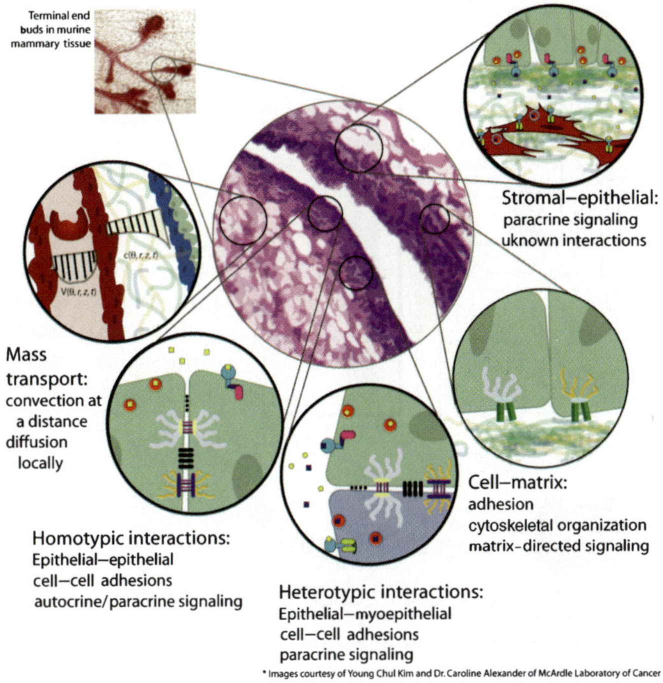

FIGURE 2.1 The *in vivo* microenvironment of mammary epithelial cells. (*See text for full caption.*)

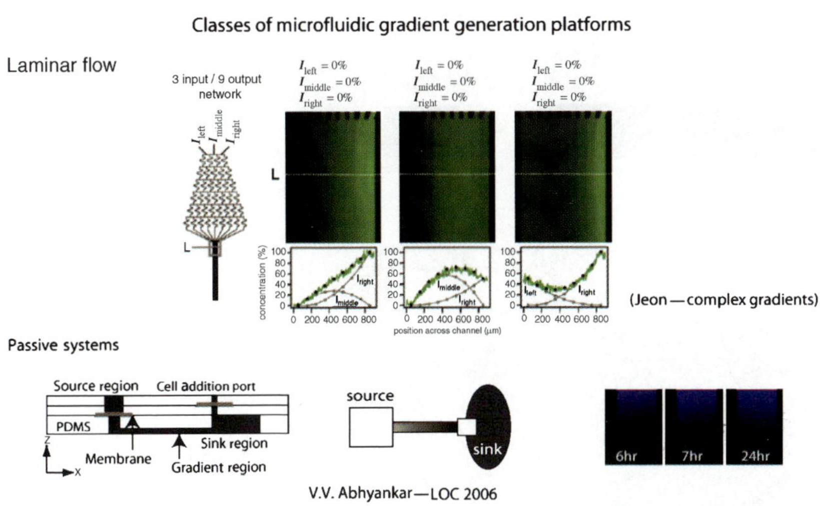

FIGURE 2.2 Schematic representation of microfluidic gradient generation techniques. (*See text for full caption.*)

(a)

	Surface area per culture area (cm²)	Media volume per culture area (mL)	Surface density (cells/cm²)	volume density (cells/cm³)
6-well plate	9.5	8	40,000	47,500
Microchannel	0.08	0.004	40,000	800,000
Fold difference (of 6 well over microchannel)	120x	2,000x	1x	1/17x

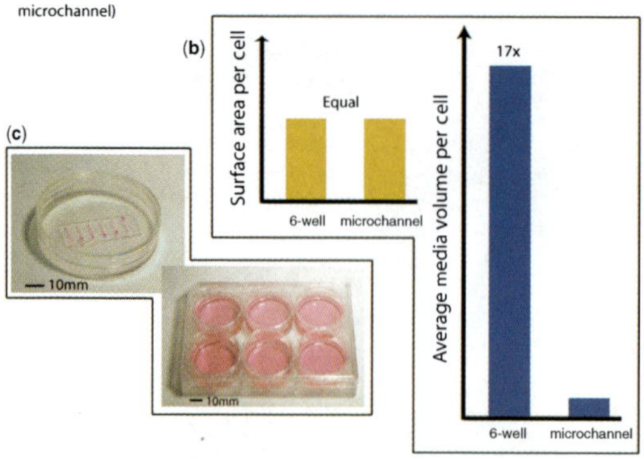

FIGURE 2.3 A comparison between volume densities of culture conditions in traditional, macroscale culture in 6-well plates and in microscale, microchannel culture. (*See text for full caption.*)

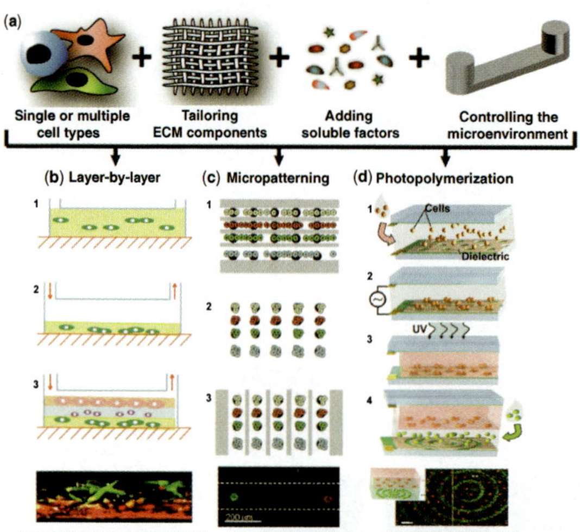

FIGURE 2.4 Mimicking the three-dimensional *in vivo* microenvironment involves the combination of four essential components that have been modeled *in vitro* using various methods by many research groups. (*See text for full caption.*)

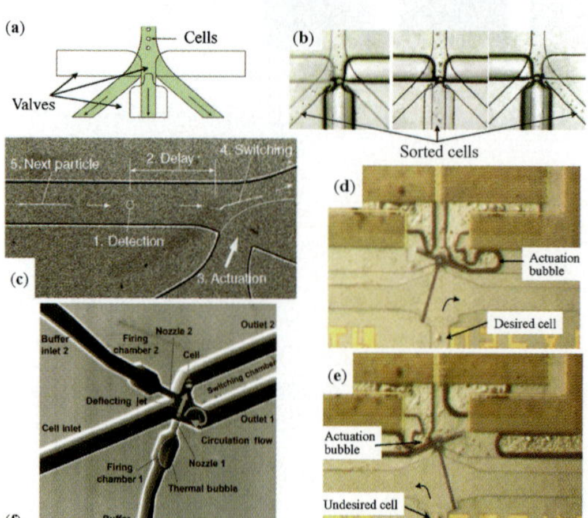

FIGURE 3.1 Sorting by hydrodynamic flow switching. (*See text for full caption.*)

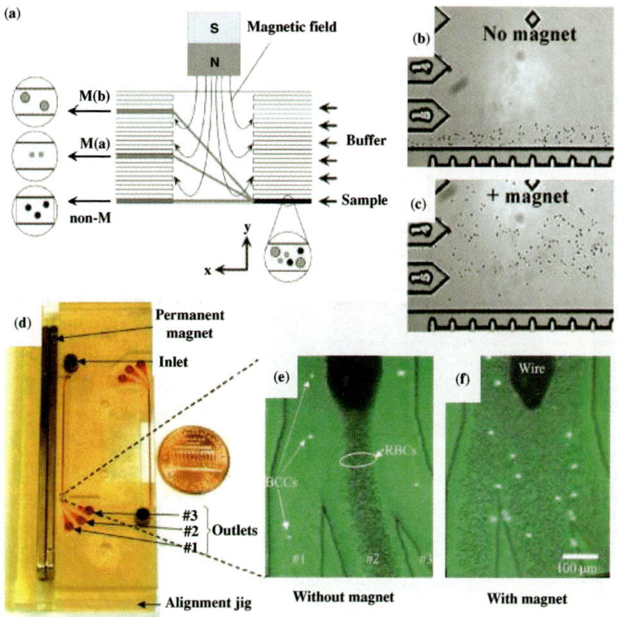

FIGURE 3.3 Flow sorting in a magnetic field. (*See text for full caption.*)

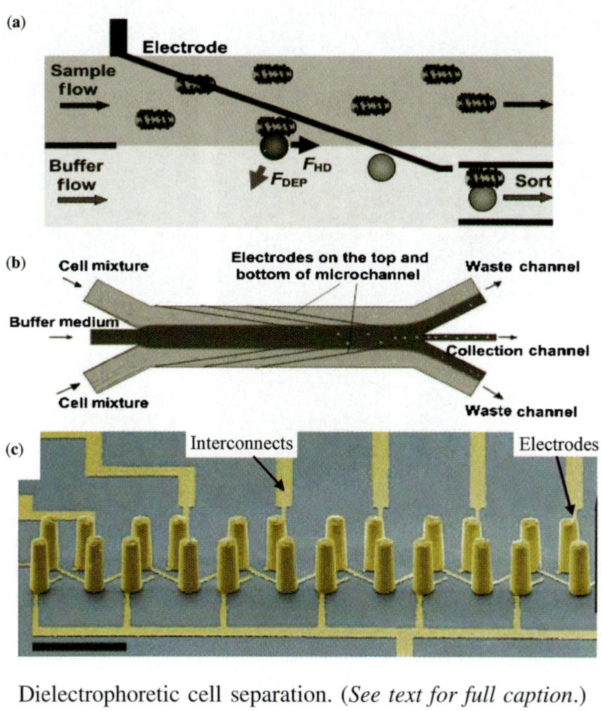

FIGURE 3.5 Dielectrophoretic cell separation. (*See text for full caption.*)

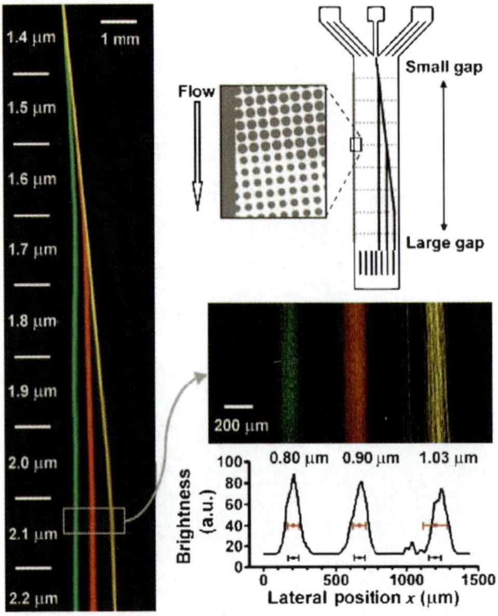

FIGURE 3.7 Bead separation by deterministic lateral displacement as demonstrated by Huang et al. (2004). A matrix composed of offset posts of progressively smaller diameter and an increasing distance between the posts was used to separate fluorescent particles (green, red, and yellow beads with diameters of 0.8, 0.9, and 1.03 μm, respectively). The distance between the centers of the posts is shown on the far left side of the figure. Beads of a given size follow equivalent paths through the array and are segregated into populations based on size. Reprinted with permission from Reference [129].

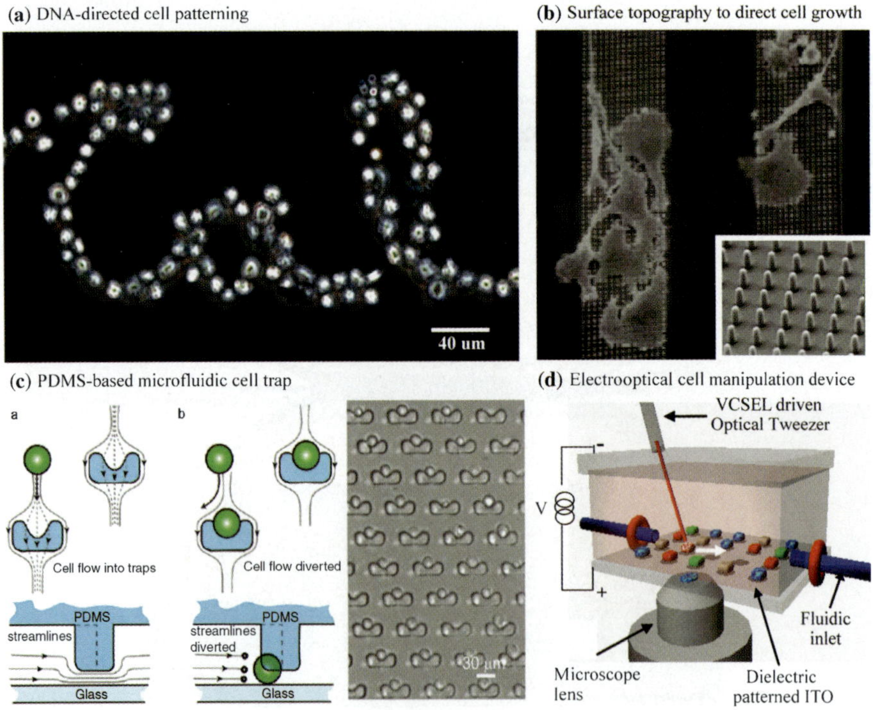

FIGURE 4.5 Cell capture and patterning. (*See text for full caption.*)

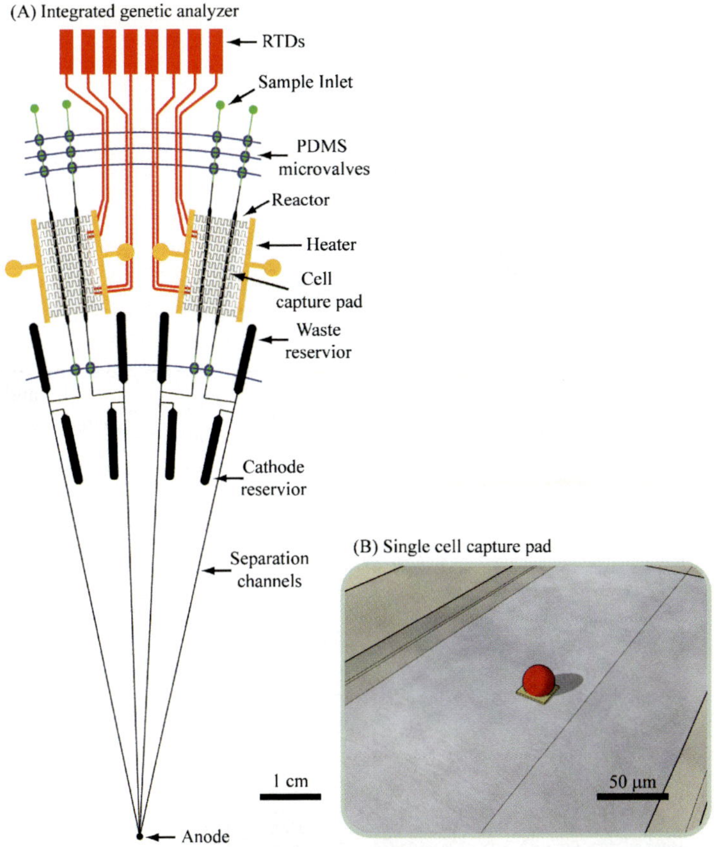

FIGURE 4.8 Fully integrated single cell genetic analyzer. (*See text for full caption.*)

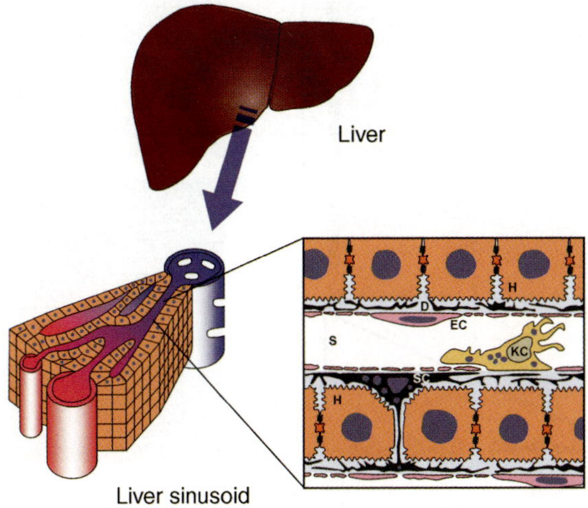

FIGURE 5.1 Physiology is microfluidic. An example of a microfluidic environment presented in human liver. (*See text for full caption.*)

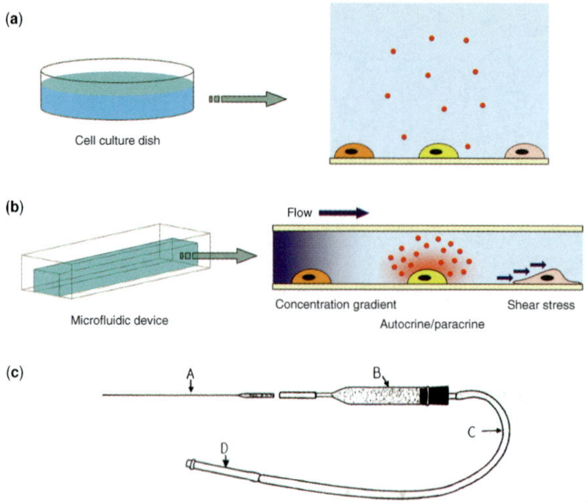

FIGURE 5.2 Cellular microenvironment in a microfluidic culture compared to a conventional 2D culture. (*See text for full caption.*)

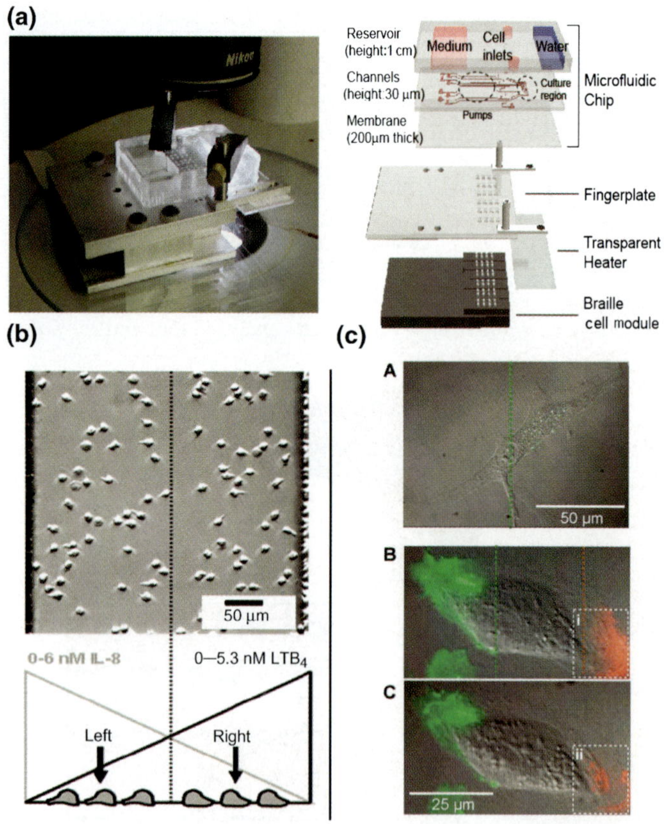

FIGURE 5.3 (**a**) A Braille display-based microfluidic cell culture system [166]. (**b**) Chemotaxis of neutrophils in opposing gradients of IL-8 (0–6 nM) and LTB4 (0–5.3 nM). (**c**) PARTCELL on bovine endothelial cells. (*See text for full caption.*)

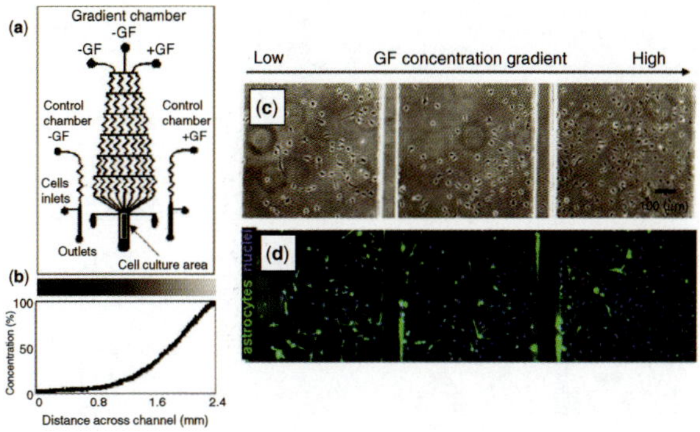

FIGURE 6.1 A gradient-generating microfluidic device for proliferation and differentiation of human neural stem cells. (*See text for full caption.*)

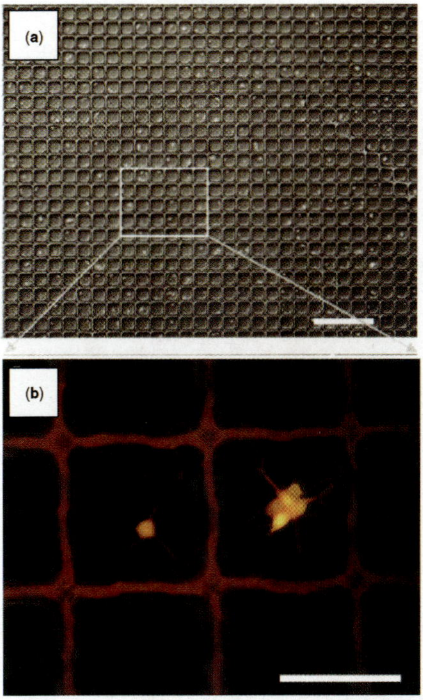

FIGURE 6.2 Differentiation of adult NSCs in the microfabricated array. (*See text for full caption.*)

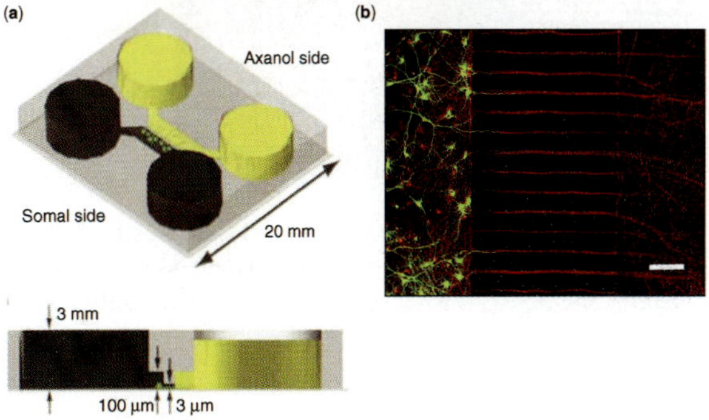

FIGURE 6.3 A microfluidic culture platform for axonal regeneration and transport. (*See text for full caption.*)

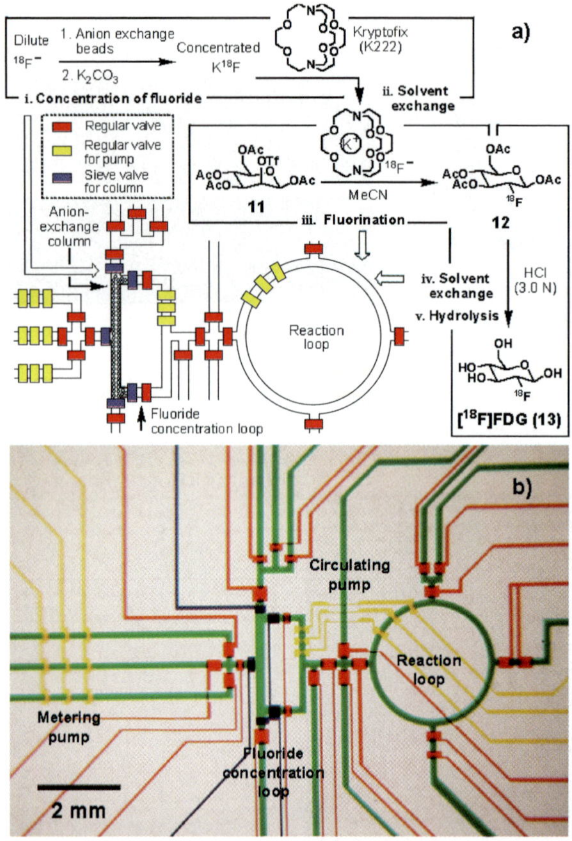

FIGURE 9.3 (*See text for caption.*)

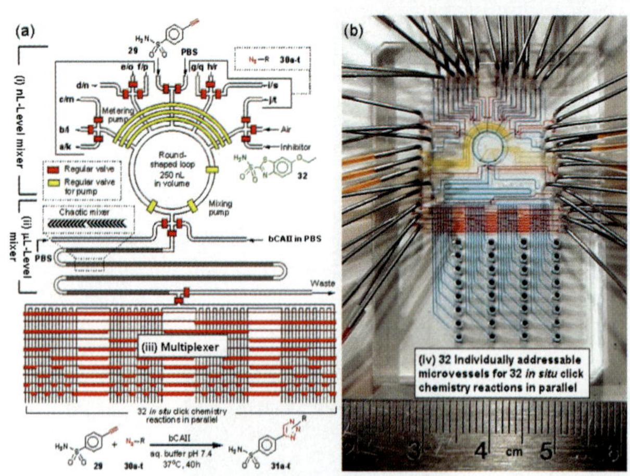

FIGURE 9.9 (**a**) Schematic representation of an integrated microfluidic reactor used for the parallel screening of an *in situ* click chemistry library prepared from **32** independent screening reactions. (**b**) Optical image of the actual device; a ruler is present to calibrate the scale. (*See text for full caption.*)

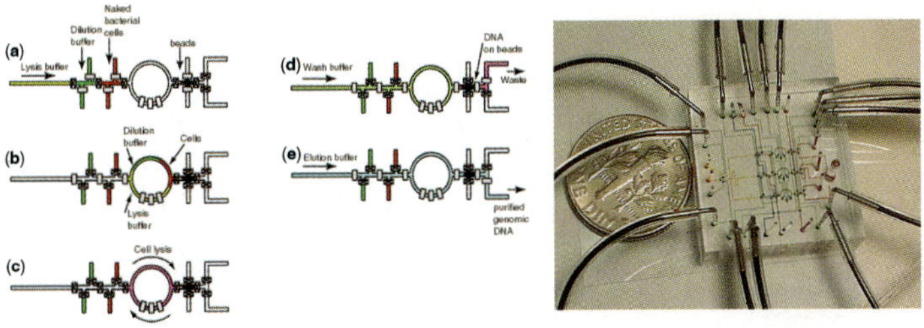

FIGURE 10.1 DNA purification chip and schematic diagram of the complete DNA isolation processes. (*See text for full caption.*)

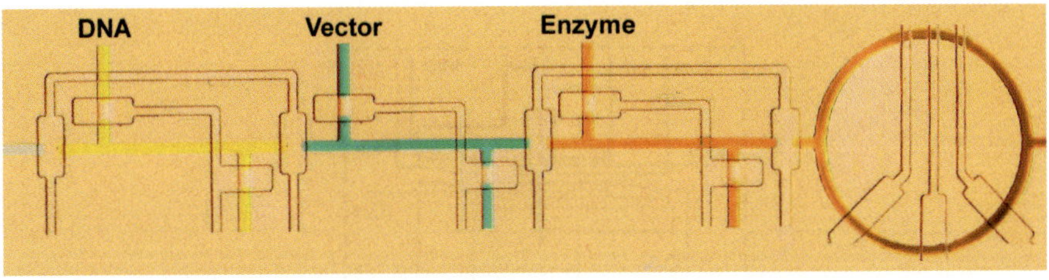

FIGURE 10.2 A part of a microfluidic chip for the single ligation process. Different food coloring represent different potential substrates, such as DNA (yellow), vector (blue), and enzyme (red), and the mixture of yellow, blue, and red coloring in the 6 nL rotary mixer.

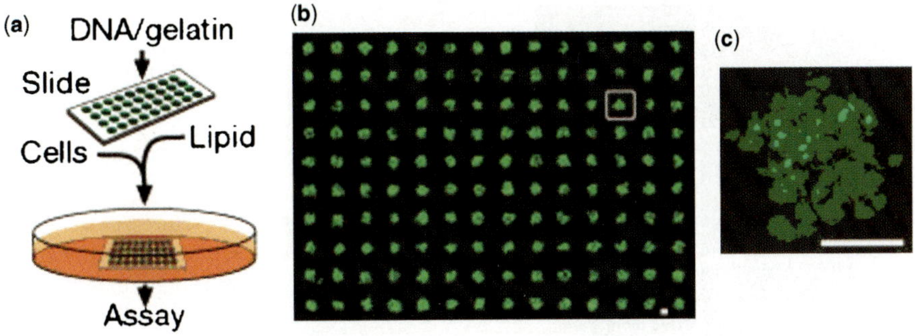

FIGURE 10.3 (**a**) Protocol for making microarrays of transfected cells. (**b**) Scanned image of a GFP-expressing cell microarray made from a slide printed in a 14 10 pattern. (**c**) Higher magnification image obtained with fluorescence microscopy of the cell cluster boxed in (**b**). Scale bar, 100 μm (after Reference 4).

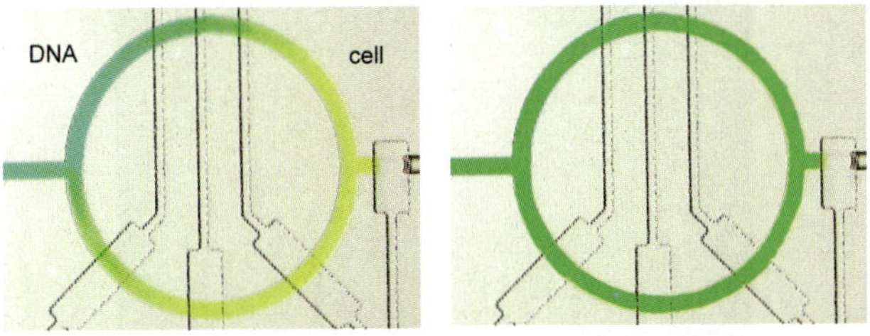

FIGURE 10.5 A part of the gene transformation chip. (*See text for full caption.*)

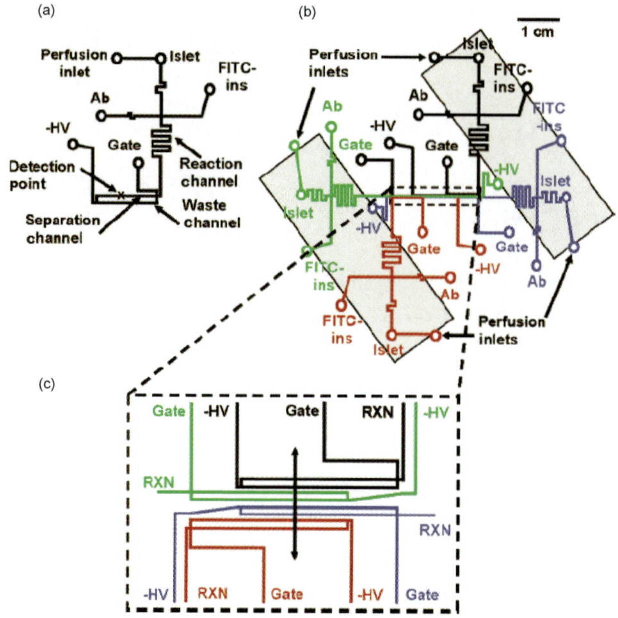

FIGURE 11.5 Microfluidic device with integrated perfusion for the continuous monitoring of insulin secretion from four islets simultaneously. (*See text for full caption.*)

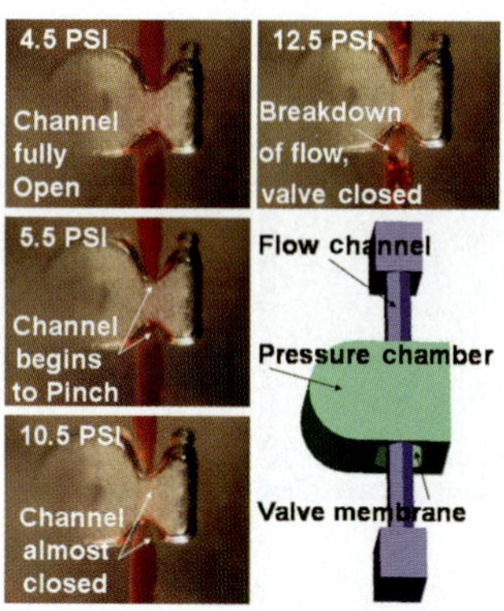

FIGURE 13.3 Three-dimensional valve constructed by wax printing and replication. The actuation pressure is only 62 kPa (9 psi) larger than the channel pressure. The disruption of flow caused by closing the valve is visible as air bubbles in this picture.

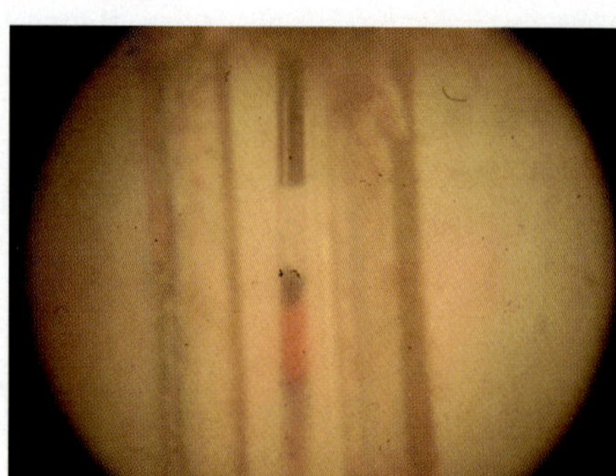

FIGURE 13.8 A microcolumn used for separating two dyes.

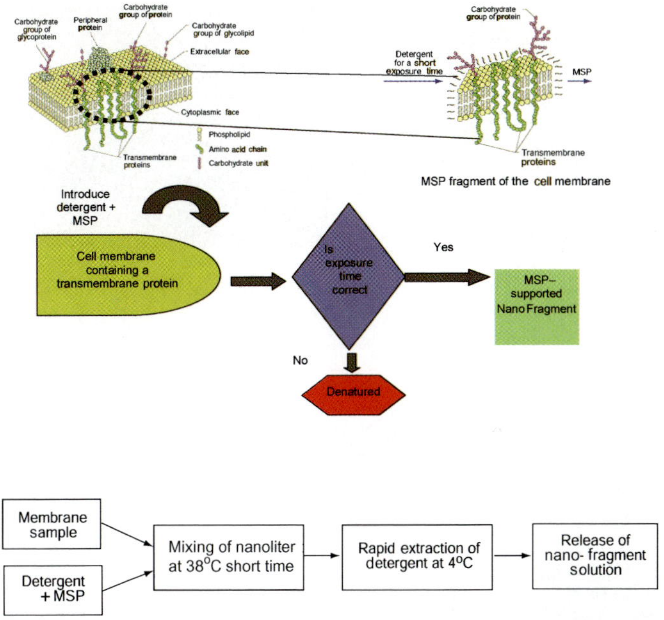

FIGURE 14.1 The functional steps in the demonstration of the microfluidic method for extracting transmembrane proteins from their native lipid bilayers. (*See text for full caption.*)

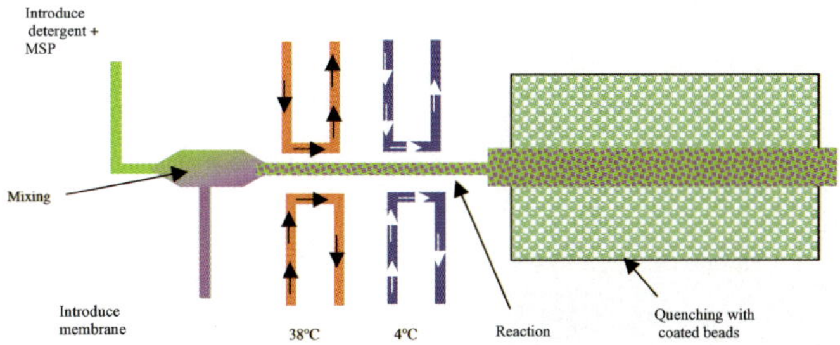

FIGURE 14.2 Schematic design of the microfluidic system for membrane protein extraction.

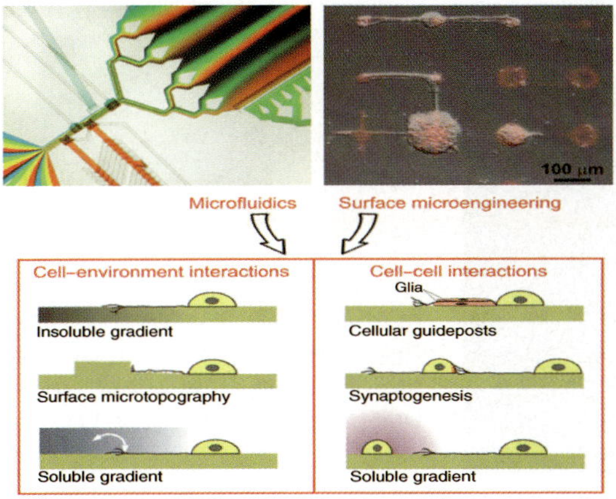

FIGURE 15.1 Applications of microfluidics in developmental neuroscience. Microfluidic devices can be used to influence and mimic cell–environment and cell–cell interactions. Microfluidic technologies allow to control such cellular events at cellular and subcellular levels. Photograph of a microfluidic mixer courtesy of G. Cooksey.

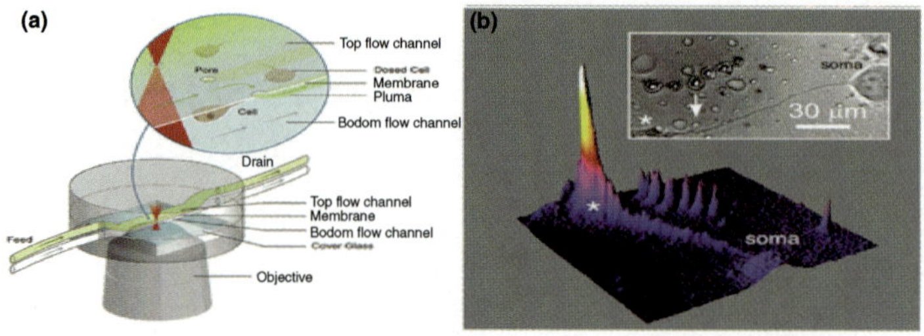

FIGURE 15.5 Cells in dynamic chemical environments. (*See text for full caption.*)

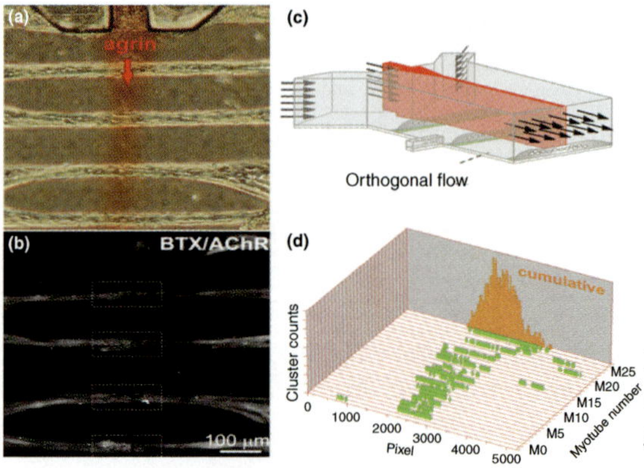

FIGURE 15.6 Synaptogenesis on a microfluidic chip. (*See text for full caption.*)

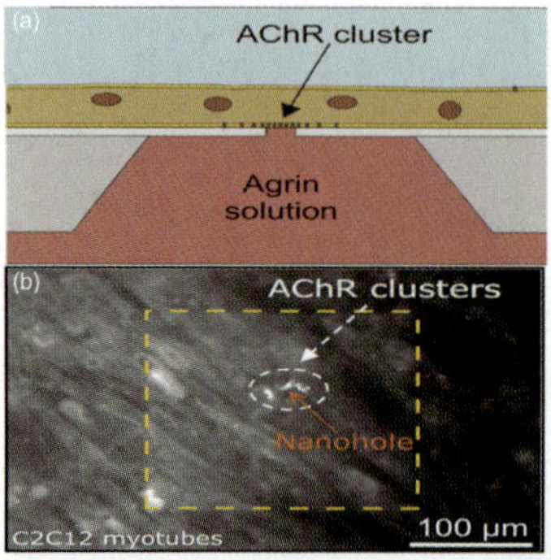

FIGURE 15.7 Focal delivery of agrin to myotubes through nanoholes. (*See text for full caption.*)

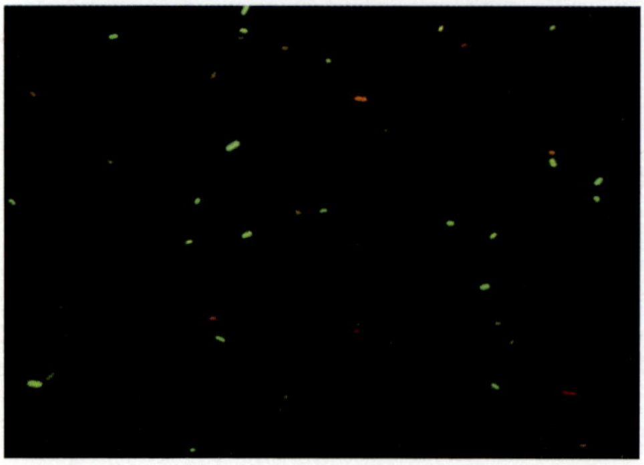

FIGURE 16.8 Fluorescent microscopy over-laid images of red and green stained E. coli on CD from LIVE/DEADBacLight™ Bacterial Viability Assay.

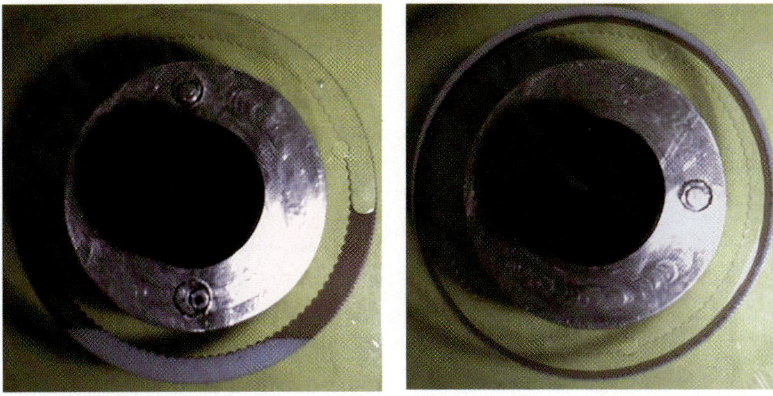

(a) (b)

FIGURE 16.13 Flow patterns for two rotational states of the CD (*See text for full caption.*)

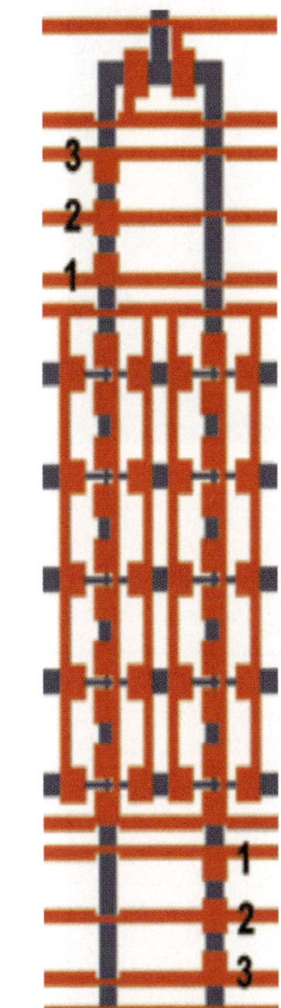

FIGURE 17.1 Multiplexed microfluidic fluorescence immunoassays. (*See text for full caption.*)

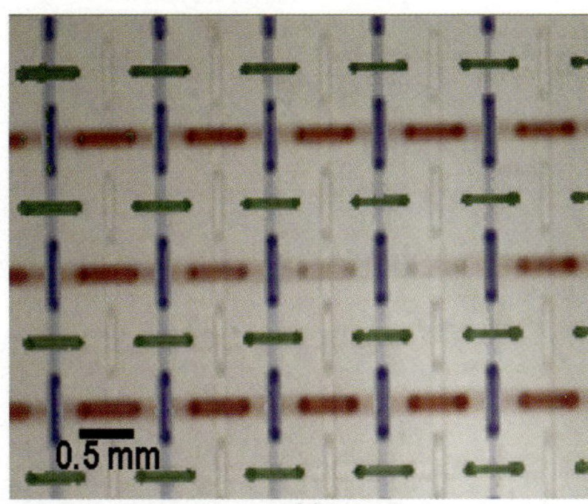

(c)

FIGURE 17.5 Nonrounded vias form successfully over a wider spectrum of dimensions than the rounded vias. The smallest functional nonrounded vias made to date are 25-μm^2: (**c**) A 4-plex design was executed with nonrounded vias at 2500 rpm PDMS spin speed. The four channel sets were filled with solutions containing plain buffer, fluorescein, Cy3, and Cy5 dyes; a colorless image and three false-color fluorescence images in blue, green, and red, respectively, were combined to produce the shown result. Of the 288 sites in the chip, 286 formed vias and 2 formed septa (seen here at the ends of what would have been a red segment, mid row, second from the right).

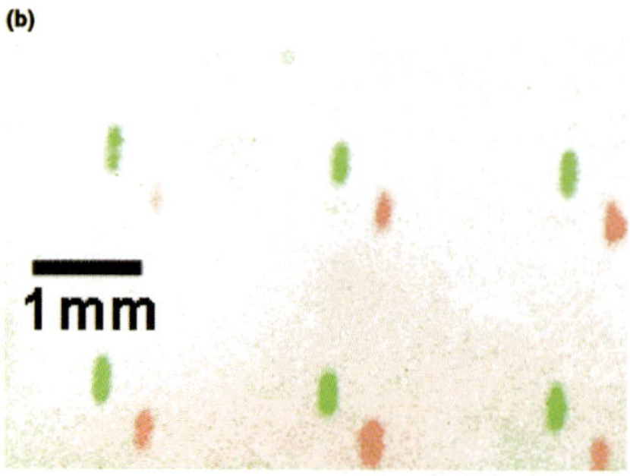

FIGURE 17.6 Printing of nested bioarrays. (**b**) demonstrating the correct assembly of the sandwich immunoassays of the nested bioarrays. If the initial substrate is diced after printing, this technique is a simple, fast, and inexpensive way to mass produce identical nanomedicine chips, each conducting multiple tests on a separate sample. In principle, nested bioarrays of arbitrary size can be printed with three-layer devices (see text).

(a)

(b)

FIGURE 18.4 (**a**) Photo of brass mold created via micromilling. Three twin-T designs are shown, along with block-like "filler" structures. (**b**) Enlarged photo of the offset injection zone and fillers of a twin-T separation channel.

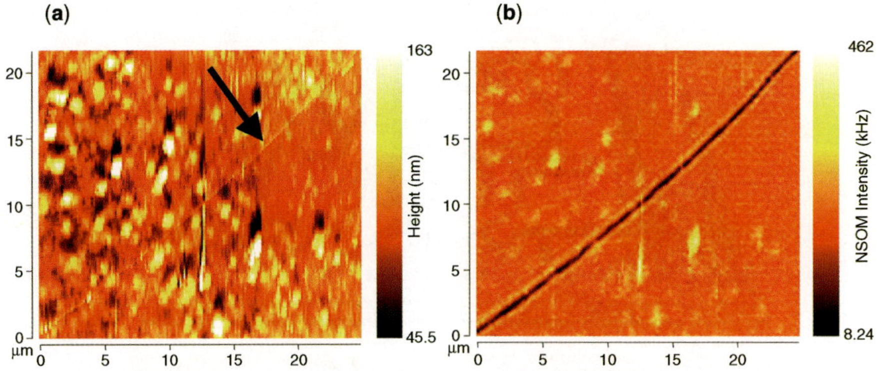

FIGURE 19.3 NSOM pair of images, acquired simultaneously. (*See text for full caption.*)

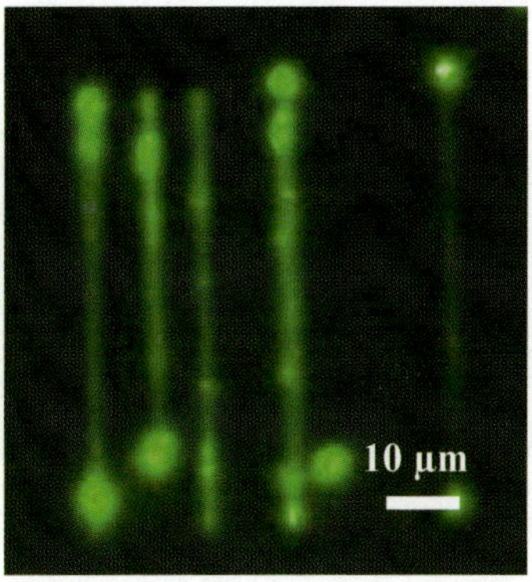

FIGURE 19.4 Lines of fluorescently labeled avidin printed with NFP on aldehyde slides and imaged in far field, with conventional fluorescence microscopy.

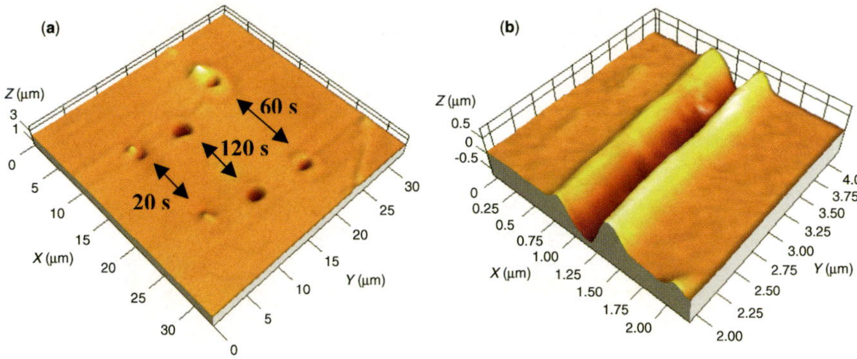

FIGURE 19.5 Depressions created in a BSA hardened substrate as a result of deposition of trypsin. (**a**) Depression volume increases with deposition time. (**b**) Trench created by moving the sample while the pipette is in contact.

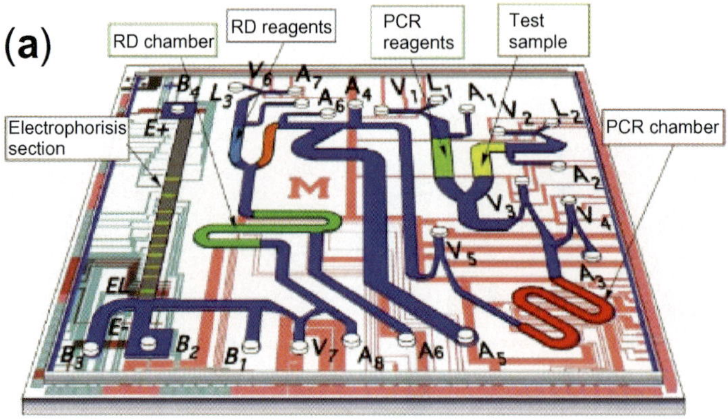

FIGURE 21.4 (**a**) Schematic of integrated chip for influenza strain genotyping showing photopatterned gel electrophoresis zone integrated with on-chip PCR and restriction digest (RD) along with integrated valves, electrodes, and heaters. (*See text for full caption.*)

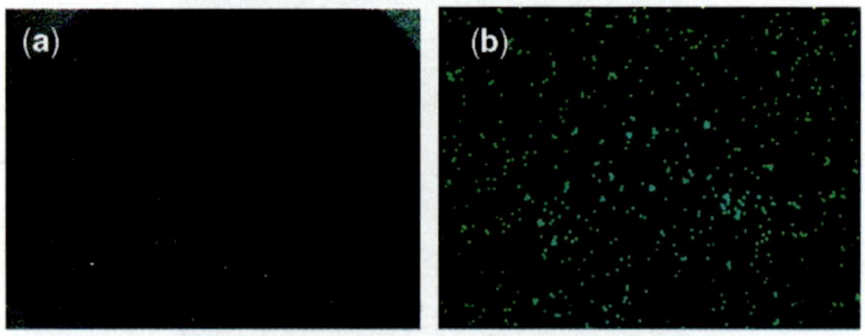

FIGURE 24.8 Fluorescence micrographs of magnetic beads after reacting with 5-FAM $(DA)_3$. (**a**) Non-derivatized and (**b**) derivatized with teicoplanin.

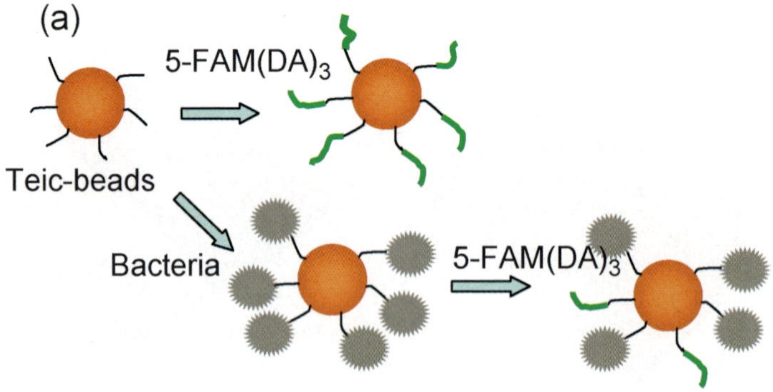

FIGURE 24.9 Gram-positive bacteria binding to teicoplanin bearing magnetic beads. (**a**) Schematic diagram showing the basis of bound bacteria determination. Teic-beads will be incubated with 5-FAM$(DA)_3$ in control. Teic-beads will be incubated with bacteria first and then with 5-FAM$(DA)_3$ for test samples. (*See text for full caption.*)

method is limit by the 5 s required to collect an electropherogram; however, this interval is short enough to resolve dynamic changes in insulin levels upon glucose stimulation of islets.

Another advantage of chip-based systems is their scalability to a multiplex format with little additional cost. The insulin separation-based sensor has recently been converted into a parallel system capable of continuously monitoring four islets simultaneously (Fig. 11.5) (23). This device contains four individual channel

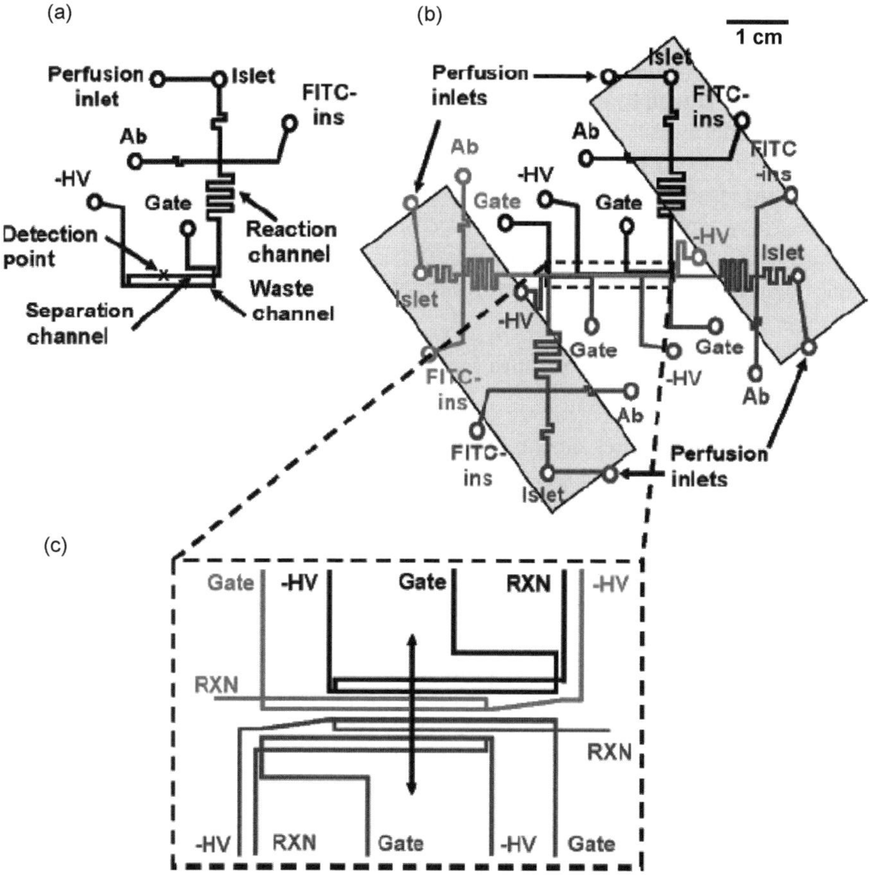

FIGURE 11.5 Microfluidic device with integrated perfusion for the continuous monitoring of insulin secretion from four islets simultaneously. (**a**) Representation of a single islet channel network. Solid lines indicate microfluidic channels whereas circles represent fluidic reservoirs and access holes to the microfluidic channel networks. All channels are 3 μm deep. (**b**) A complete chip showing all four islet networks (drawn to scale). Shaded regions represent heating strips attached to the underside of the device. (**c**) Close-up of the gating and detection region (not to scale). All of the separation channels run parallel to each other, which allow a scanning laser spot to be used for LIF detection. The laser scan path is indicated by an arrow. (Reprinted with permission from reference 23. Copyright 2007.) (See color insert)

networks that sample from their own islet chamber. With this configuration, the device is capable of completing over 1450 immunoassays in less than 40 min allowing parallel monitoring of insulin release from four islets every 6.25 s. Such a device not only allows more islets to be evaluated simultaneously but also is useful in comparing islets harvested from diabetic mouse models (122,123) and wild type controls simultaneously. Because of the limited viability of cultured islets and the expense in breeding genetically modified mice for diabetic models, parallel operation greatly reduces the cost of performing these experiments.

Previous methods used for studying insulin secretion from islets used a perfusion system, fraction collection, and ELISA or RIA of the individual fractions (124,125). Most of these assays lacked the sensitivity to measure insulin from individual islets unless they used nonphysiological conditions, such as highly elevated extracellular Ca^{2+} (124). The difficulty and cost of performing such experiments (i.e., time consuming fraction collection and analysis as well as working at the detection limit for the assay) previously precluded many studies on insulin secretion dynamics. The chip-based system offers to make such measurements routine. Indeed, this chip has already been used in two studies on islet secretion (122,123). Alteration of the experiment may enable it to be used for other cell types as well.

11.3.2 Separation-Based Sensors for Off-Chip Sampling of Cellular Tissues

In addition to cell cultures on chip, microfluidic devices can be coupled to external sampling probes to monitor concentrations from other environments. This configuration allows for studies of more complicated systems that would be infeasible to house on a microfluidic device. For instance, *in vivo* monitoring of changes in neurotransmitter concentration levels in the brain is important in studying the role of neurotransmitters in pharmacological, physiological, and behavioral effects. To accomplish this monitoring on a microfluidic format, two distinct approaches have been used. In one, microdialysis sampling is coupled to a glass-microchip and in the other a miniaturized push-pull sampling probe is coupled to a poly(dimethylsiloxane) (PDMS) device. Both systems have been used to monitor amino acids in the brain of living animals.

11.3.2.1 Microdialysis Sampling Coupled to Glass Microchips Microdialysis has been a popular sampling method for *in vivo* monitoring of neurotransmitters because of its good temporal resolution and its compatibility with sensitive analytical separation techniques such as CE. In microdialysis, a semipermeable membrane probe is inserted into a brain region (or other tissue) for sample collection. The probe is perfused with a solution of similar ionic composition of the brain. Analytes from the extracellular space, which are below the molecular weight cutoff of the semipermeable membrane, diffuse across the membrane into the probe according to their concentration gradient and are collected in the perfusion stream for analysis. The amount of analyte collected in the perfusion stream is dependent on its concentration gradient and the perfusion flow rate. At slow flow rates, the concentration of analyte in the probe divided by the concentration outside the probe approaches 100% allowing for quantitative monitoring; although most experiments are performed at higher flow rates where the recovery is well under 50%. The sampling area of a microdialysis probe is typically cylindrical with a 200–500 μm outer diameter and 1–4 mm in length.

As discussed above, microdialysis has been coupled to capillary electrophoresis for continuous *in vivo* monitoring. In these approaches, the sample stream is pumped into a tee where it is mixed with fluorogenic reagent and reacted while flowing through a reaction capillary. A sample of this stream is periodically injected onto a separation capillary where the analytes of interest are rapidly separated by electrophoresis and detected by LIF. The first basic coupling of a microdialysis probe to a chip has been demonstrated for monitoring an enzymatic assay (24). This system, however, did not replicate all the functions necessary for *in vivo* monitoring.

More recently, microdialysis has been coupled to a microfluidic chip that performs all of the functions necessary for *in vivo* monitoring of primary amine neurotransmitters in a living brain (26). The device (Fig. 11.6) contains a sample introduction channel (SIC) for interfacing to a microdialysis probe, a precolumn reactor for derivatization, a flow-gated injector, and a separation channel. After implantation of a microdialysis probe into a rat brain, artificial cerebral spinal fluid is perfused through the probe with a syringe pump. Primary amino neurotransmitters collected in the perfusate is transported to the SIC on the microfluidic device. The SIC contains a 360 μm diameter hole at its junction with the electrokinetic network.

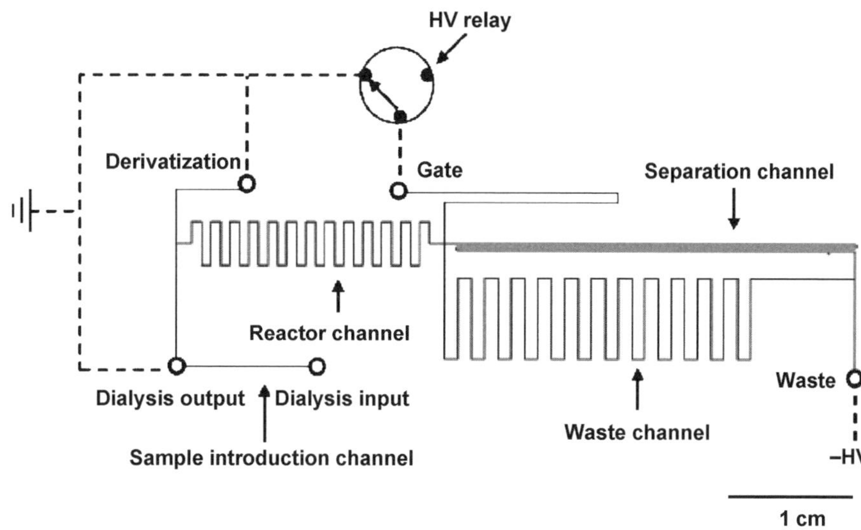

FIGURE 11.6 Glass microfluidic device coupled to microdialysis for online monitoring of primary amine neurotransmitters. Channels (lines) are drawn to scale with electrical connections represented by dashed lines. Fluidic access holes were 360 μm in diameter. (Reprinted with permission from reference 26. Copyright 2005.)

This hole serves as a low resistant outlet for the hydrodynamically flowing perfusate resulting in negligent flow splitting into the electrokinetic network. EOF transports sample from the SIC to a reactor channel where primary amines react with *o*-phthaldialdehyde (OPA) and β-mercaptoethanol (BME) to form fluorescent isoindols (126). Fluorescent product is diverted to a waste channel; however, upon actuation of a high voltage relay a sample plug is injected onto an electrophoresis channel where the OPA-derivatized primary amine neurotransmitters are separated. An epi-fluorescent confocal microscope is used to detect the separated analyte zones with LIF about 9 cm down the separation channel. Depending on the applied potential field, electropherograms can be collected every 95 or 20 s allowing for the monitoring of dynamic changes in neurotransmitter levels.

This separation-based sensor has an *in vitro* limit of detection of ~200 nM for primary amine neurotransmitters. Relative standard deviation in peak heights is 2% with efficiencies of 156,000 theoretical plates at 1320 V/cm (Fig. 11.7). With *in vivo* studies, a slower probe perfusion rate is used to maximize analyte recovery at the probe. The slower flow rate limits temporal resolution of the system to 130 s. It is important to note that the temporal resolution is limited not by the separation speed but by band broadening of the concentration zones due to Taylor dispersion as they travel from the sampling probe to the chip. Nonetheless, the sensor is able to detect changes in glutamate basal levels upon pharmacological manipulation of glutamate levels.

The performance of this device could be improved in a few areas. For instance, the temporal resolution could be enhanced by reducing system volume associated with capillary tubing that connects the microdialysis probe to the device and by reducing the volume of the fluidic access hole on the chip that interfaces with the capillary tubing. Smaller system volumes will allow material to be transported from the microdialysis probe to the chip faster at a given microdialysis perfusion flow rate. Additionally, increasing this flow rate will also improve temporal resolution by reducing the time available for Taylor dispersion. An improvement in detection limits, which might be possible with the use of an alternative fast-reaction

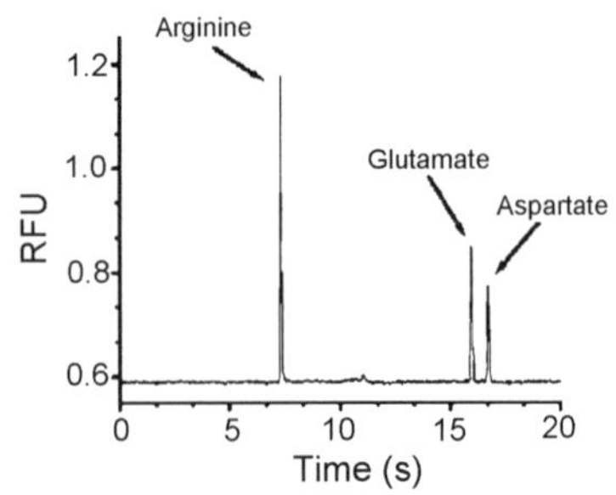

FIGURE 11.7 *in vitro* electropherogram of high speed separation of three amino acids at 50 μM in aCSF (online precolumn derivatization with *o*-phthaldialdehyde and β-mercaptoethanol) in 20 s with device in Fig. 11.6. (Reprinted with permission from reference 26. Copyright 2005.)

fluorogenic reagent such as 2,3-dicarboxaldehyde (NDA) with BME (127), will allow for less sample injection thereby reducing the contribution of injection volume to band broadening. A chip-based device coupled to microdialysis already has demonstrated the potential advantage of using NDA/BME for separation-based sensing (128). Also, adding a separation buffer additive, such as cyclodextrins, potentially could enhance the selectivity of this sensor (16,129). Taking advantage of the microfluidic format, this system potentially could be converted to a multiplex format allowing for the monitoring of different classes of neurotransmitters simultaneously or allowing for higher throughput of animals. Furthermore, the incorporation of optics and miniaturization of controlling electronic would make a portable system more feasible.

11.3.2.2 Low Flow Push-Pull Sampling Coupled to Polymer Microchips Although microdialysis sampling is extremely popular, the relatively large probe size results in low spatial resolution of the measurement, especially compared to miniaturized electrochemical sensors. Spatial resolution is important in neuroscience applications because many brain regions with distinct functionality are quite small. This problem is exacerbated in small experimental species such as the mouse. For this reason, interest in miniaturized sampling probes has emerged. One such technique is low flow push-pull perfusion (LF-PPP) (130). A LF-PPP probe consists of two capillary tubes (30–60 μm i.d. and 40–90 μm o.d. each) placed side by side. Sample is pulled in through one tube while aCSF is pushed back out through the other tube to replenish the sample region. Flow rates through the tubes are 10–50 nL/min. Sampling occurs at the tip of the probe resulting in a nearly 100-fold greater spatial resolution over microdialysis. Furthermore, the small size of the probe results in much less tissue damage in the sampled region. Although LF-PPP has important advantages over microdialysis, it places great demands on the analytical system. Fluid must be withdrawn at low flow rates and only tiny quantities are available for analysis. Microfluidics can meet these demands because of its capability of moving fluids at low flow rates and its incorporation of highly miniaturized analytical systems. In terms of separation-based sensors, the coupling of LF-PPP and online microfluidic analysis yields a device with extraordinary spatial resolution, temporal resolution, and multianalyte capability.

The device shown in Fig. 11.8 couples LF-PPP to online electrophoretic analysis for *in vivo* monitoring (27). Unlike microdialysis sampling, where a pump on the inlet side of the probe (and therefore not in-line with the analytical system) controls the flow, with LF-PPP a pump must be placed between the sampling probe and the analytical system. This device is made from PDMS using multilayer soft lithography to take advantage of the simple pumps that can be made. In these systems, application of air pressure to a control channel results in compression and closure of the fluidic channel. Actuation of three or more of these control channels in appropriate sequences generates pumping. The flow rate is dependent on the channel dimensions and actuation frequency. The pumps shown in Figure 11.8 produce flow rates in the range of 40–130 nL/min through a single channel. With the arrangement of two channels as shown, the pump can control both the "push" and "pull" flow.

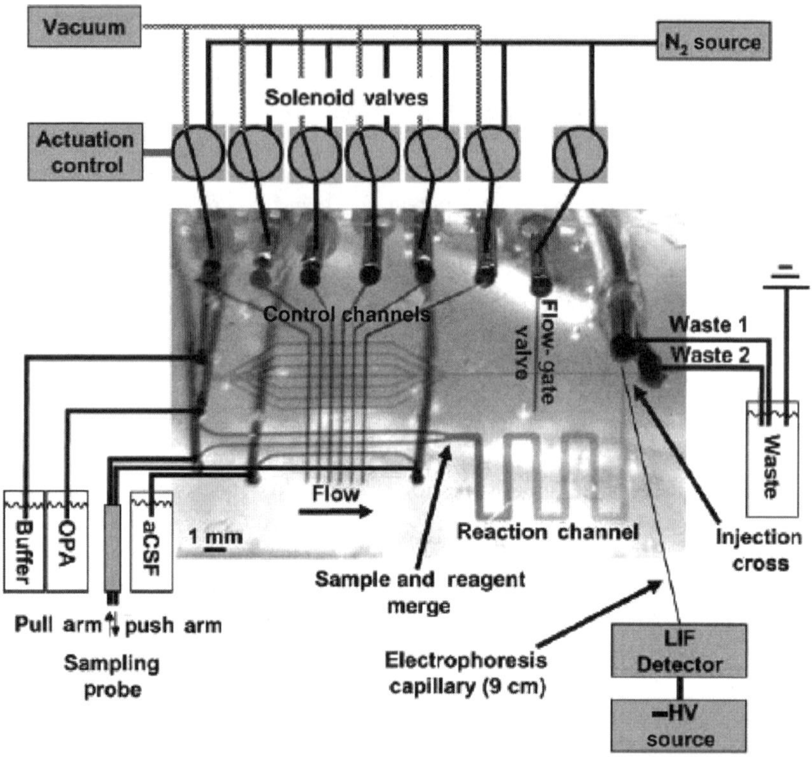

FIGURE 11.8 PDMS microfluidic device coupled to push-pull perfusion for online monitoring of primary amine neurotransmitters. The chip handles sampling, derivatization, flow gating, and injection onto an embedded fused-silica capillary. Sample, buffer, and reagent are pumped by peristalsis from left to right by six control channels. An additional control channel serves as a flow-gate valve to stop flow in the separation channel to create injections. All control valves are actuated pneumatically with N_2 delivered by solenoid valves. (Reprinted with permission from reference 27. Copyright 2006.)

The peristaltic pumps also control the flow of solutions throughout the PDMS device. Sample from the push-pull perfusion tip is mixed with OPA and BME to form fluorescent isoindols (126). These fluorescent products are pumped to a flow-gated injector that is control with pneumatic valves. The flow-gated injector is supplied with electrophoresis buffer through a multichannel pump. This multichannel pump provides flow rates sufficiently high to gate the flow for injection control. Separation occurs in a piece of fused-silica capillary inserted into the PDMS device. LIF is used for detection with detection limits of ~40 nM. Efficiencies of 250,000 plates and relative standard deviations of ~4% for glutamate and aspartate are achievable *in vitro*. *In vivo* electropherograms have been collected that resolved eight primary amines every 30 s for over 4 h of continuous operation (Fig. 11.9). As demonstrated in Fig. 11.10, this device is capable of detecting a change in peak intensities of three

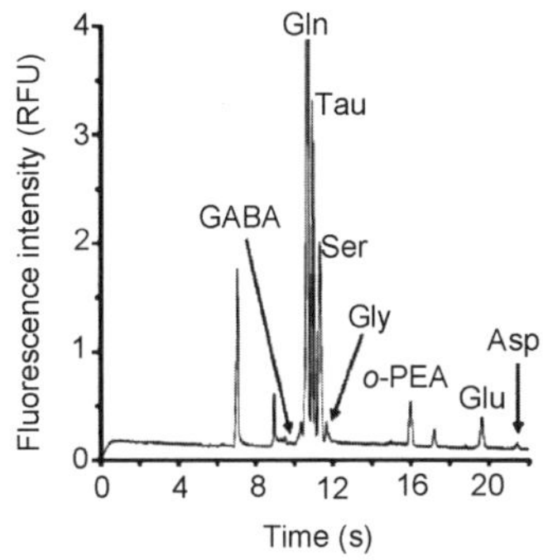

FIGURE 11.9 *in vivo* electropherogram of high speed separation of GABA, glutamine (Gln), taurine (Tau), serine (Ser), glycine (Gly), *o*-PEA, glutamate (Glu), and aspartate (Asp) with device in Fig. 11.8. (Reprinted with permission from reference 27. Copyright 2006.)

amino acid neurotransmitters, *in vivo*, after a series of K^+ stimulations, which depolarize neurons causing a local change in extracellular neurotransmitter levels.

This device provides a good combination of high temporal and spatial resolution for monitoring. Further improvements are necessary to make the device as easy-to-use as a conventional sensor. An important goal is to incorporate a separation channel and detector components into the chip to make it more compact.

Previously, separation-based sensors for the monitoring of neurotransmitters have been capillary-based systems. Although these systems have great performance with

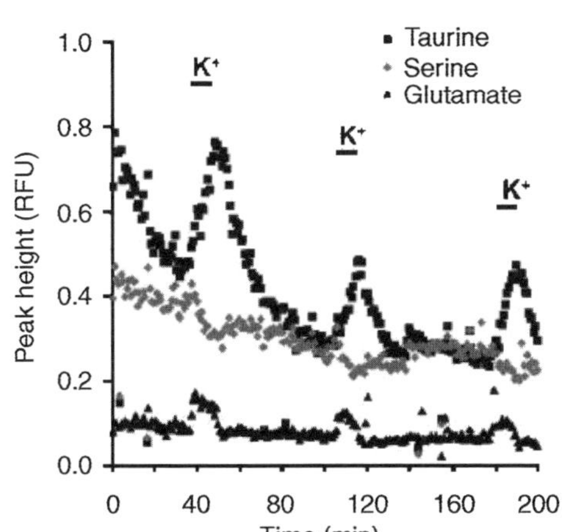

FIGURE 11.10 *in vivo* monitoring of neurotransmitter levels with device in Fig. 11.8. Peak intensities for glutamate, serine, and taurine are plotted as a function of time during a four hour experiment demonstrating the sensor-like capabilities of this device. Three, 10 min K^+ stimulations were performed to show that multiple stimulations can be performed and detected during an experiment. High K^+ concentrations depolarize neurons causing local changes in extracellular neurotransmitter levels. (Reprinted with permission from reference 27. Copyright 2006.)

sampling frequencies of ~10 s and efficiencies of 500,000 plates, they require complicated fluidic connections and require significant space. The chip-based systems may eventually replace these devices. Successful development of chip-based sensors may also lead to other applications besides *in vivo* monitoring.

11.4 MINIATURIZATION AND INTEGRATION OF SEPARATION-BASED SENSOR COMPONENTS

In many ways, chemical sensors are still more amenable to portability than separation-based sensors. For the most part, the portability claimed by proponents of microfluidics has yet to be realized because the components used to operate devices are either sizable, require significant power, or are prohibitively expensive. If separation-based sensors are to become easily disseminated or become ubiquitous in on-site monitoring applications, all of its components (e.g., separation component, detector, and controlling electronics) require miniaturization for incorporation into a handheld device or other portable format. The difficulty, however, in developing a portable device is highly dependent on the application. For instance, a simple device such as one developed for cell sorting only requires gravity to pump cells passively while exploiting their physiology to create a separation mechanism (131). Although simple and easily amenable to portability, this separation mechanism is impractical for many other applications. In recent years, there has been a trend in developing the tools necessary to create truly portable devices.

11.4.1 Miniaturization of Controlling Electronics

As discussed previously, CE is used as the separation mechanism in most separation-based sensors because of its favorable properties. CE separations, however, require a high voltage power source, relay, and electrical connections. Recently, work has been devoted on miniaturizing these components for incorporation into a microfluidic device. For instance, Jackson et al. have developed a battery-powered, dual source high voltage power supply and interface circuits for CE capable of producing ±1 kV at 380 μA for 15 h of continuous operation. It is interfaced to a chip with a circuit containing two miniature relays, diodes, and resistors configured to operate a pinched injection valve. This system had a combined size of 10 cm × 15 cm × 2.5 cm and weight of 0.35 kg. Garcia et al. have developed a battery operated three-channel high voltage power supply capable of 0 to ±4 kV (132). A transformer is included to recharge the batteries and feed the high voltage relays. This system is configurable for both pinched and gated injections. Collins et al. have developed a high voltage power supply that included a voltage-divider network and two high voltage relays capable of producing 15 kV and operating a gated injection valve (133). Finally, a reusable integrated high voltage power supply has been developed that is powered by an onboard MN21 cell battery (134). All of these methods have demonstrated the feasibility of miniaturizing the controlling electronics necessary for CE separations.

11.4.2 Miniaturization of Detection Components

The detection method used in CE-based microchips depends on the analyte being monitored and the sensitivity required. For the most part, LIF is used for detection in microchips because of its great sensitivity and compatibility with devices. However, to date, most microfluidic devices employ the use of LIF detectors that take up considerable space and are unsuitable for transportation. To alleviate this problem, work has focused on miniaturizing and incorporating the components necessary for LIF detection into microfluidic chips. Although the miniaturization of LIF components is important for portable devices, the degree of integration of these LIF components can vary to (a) an external light source and external optical transducers, (b) an external light source and integrated optical transducers, and (c) an integrated light source and integrated optical transducers. An important review has been written that discusses the integration of these higher degrees of optical functions for microchips in great detail (135). Already, completely integrated portable systems have been developed based on LIF detection (106,136–139).

In addition to LIF, other detection systems have a potential for use in portable devices. For instance, electrochemical detection is well suited for miniaturization and integration into microchips (140). Indeed, a portable device has been fabricated capable of amperometric, potentiometric, and conductometric detection with CE separation (141). Furthermore, a review by Viskari and Landers details other potential detection methods for microchips such as infrared detection, Raman detection, nuclear magnetic resonance (NMR) detection, surface plasmon resonance (SPR) detection, absorbance detection, thermal lens detection, chemiluminescence and electrochemiluminescence detection, RI detection, and microplasma-optical emission detection (OED) (142). Of these methods, thermal lens detection and SPR have already been developed and miniaturized for portability.

The integration of detectors and controlling electronics with microfluidic devices has allowed for the construction of portable separation-based sensors. Portability coupled with further development in separation performance and temporal resolution will make separation-based sensors even more popular for real-time monitoring.

REFERENCES

1. Cattrall RW. Chemical Sensors. Oxford, UK; Oxford University Press;1997.
2. Eggins BR. Chemical Sensors and Biosensors. John Wiley & Sons;2002.
3. Fogt EJ. Clin Chem 1990;36:1573–1580.
4. Frost M, Meyerhoff ME. Anal Chem 2006;78:7370–7377.
5. Davies MI, Cooper JD, Desmond SS, Lunte CE, Lunte SM. Adv Drug Deliv Rev 2000;45:169–188.
6. Xu HW, Ewing AG. Anal Bioanal Chem 2004;378:1710–1715.
7. Manz A, Eijkel JCT. Pure Appl Chem 2001;73:1555–1561.
8. Tao L, Aspinwall CA, Kennedy RT. Electrophoresis 1998;19:403–408.

9. Jameson EE, Cunliffe JM, Neubig RR, Sunahara RK, Kennedy RT. Anal Chem 2003;75:4297–4304.
10. Hogan BL, Lunte SM, Stobaugh JF, Lunte CE. Anal Chem 1994;66:596–602.
11. Zhou SY, Zuo H, Stobaugh JF, Lunte CE, Lunte SM. Anal Chem 1995;67:594–599.
12. Lada MW, Schaller G, Carriger MH, Vickroy TW, Kennedy RT. Anal Chim Acta 1995;307:217–225.
13. Lada MW, Kennedy RT. J Neurosci Methods 1995;63:147–152.
14. Lada MW, Vickroy TW, Kennedy RT. Anal Chem 1997;69:4560–4565.
15. Lada MW, Kennedy RT. J Neurosci Methods 1997;72:153–159.
16. Bowser MT, Kennedy RT. Electrophoresis 2001;22:3668–3676.
17. Manz A, Graber N, Widmer HM. Sens Actuators B–Chemical 1990;1:244–248.
18. Reyes DR, Iossifidis D, Auroux PA, Manz A. Anal Chem 2002;74:2623–2636.
19. Auroux PA, Iossifidis D, Reyes DR, Manz A. Anal Chem 2002;74:2637–2652.
20. Vilkner T, Janasek D, Manz A. Anal Chem 2004;76:3373–3385.
21. Roper MG, Shackman JG, Dahlgren GM, Kennedy RT. Anal Chem 2003;75:4711–4717.
22. Shackman JG, Dahlgren GM, Peters JL, Kennedy RT. Lab Chip 2005;5:56–63.
23. Dishinger JF, Kennedy RT. Anal Chem 2007;79:947–954.
24. Huynh BH, Fogarty BA, Martin RS, Lunte SM. Anal Chem 2004;76:6440–6447.
25. Cellar NA, Burns ST, Meiners JC, Chen H, Kennedy RT. Anal Chem 2005;77:7067–7073.
26. Sandlin ZD, Shou MS, Shackman JG, Kennedy RT. Anal Chem 2005;77:7702–7708.
27. Cellar NA, Kennedy RT. Lab Chip 2006;6:1205–1212.
28. Jorgenson JW, Lukacs KD. Anal Chem 1981;53:1298–1302.
29. Kennedy RT, German I, Thompson JE, Witowski SR. Chem Rev 1999;99:3081–3131.
30. Cifuentes A, Poppe H. Chromatographia 1994;39:391–404.
31. Ermakov SV, Jacobson SC, Ramsey JM. Anal Chem 1998;70:4494–4504.
32. Xuan XC, Li DQ. J Chromatogr A 2005;1064:227–237.
33. Ross D, Gaitan M, Locascio LE. Anal Chem 2001;73:4117–4123.
34. Swinney K, Bornhop DJ. Electrophoresis 2002;23:613–620.
35. Erickson D, Sinton D, Li D.Q. Lab Chip 2003;3:141–149.
36. Chein RY, Yang YC, Lin YS. Electrophoresis 2006;27:640–649.
37. Evenhuis CJ, Guijt RM, Macka M, Marriott PJ, Haddad PR. Electrophoresis 2005;26:4333–4344.
38. Petersen NJ, Nikolajsen RPH, Mogensen KB, Kutter JP. Electrophoresis 2004;25:253–269.
39. Tang GY, Yang C, Chai CK, Gong HQ. Anal Chim Acta 2004;507:27–37.
40. Zhang Y, Bao N, Yu XD, Xu JJ, Chen HY. J Chromatogr A 2004;1057:247–251.
41. Hjerten S, Valtcheva L, Elenbring K, Liao JL. Electrophoresis 1995;16:584–594.
42. Grushka E, McCormick RM, Kirkland JJ. Anal Chem 1989;61:241–246.
43. Veraart JR, Gooijer C, Lingeman H. Chromatographia 1997;44:129–134.
44. Xuan XC, Li DQ. Electrophoresis 2005;26:166–175.

45. Guijt RM, Dodge A, van Dedem GWK, de Rooij NF, Verpoorte E. Lab Chip 2003; 3:1–4.
46. Nika P, Bailly Y, Guermeur F. Int J Heat Mass Transfer 2005;48:3773–3792.
47. Yang RG, Chen G, Snyder GJ, Fleurial JP. J Appl Phys 2004;95:8226–8232.
48. Rosengarten G, Mutzenich S, Kalantar-zadeh K. Exp Thermal Fluid Sci 2006;30:821–828.
49. Harrison DJ, Fan ZH, Seiler K, Manz A, Widmer HM. Anal Chim Acta 1993;283: 361–366.
50. Jacobson SC, Culbertson CT, Daler JE, Ramsey JM. Anal Chem 1998;70:3476–3480.
51. Manz A, Harrison DJ, Verpoorte EMJ, Fettinger JC, Paulus A, Ludi H, Widmer HM. J Chromatogr 1992;593:253–258.
52. Harrison DJ, Fluri K, Seiler K, Fan ZH, Effenhauser CS, Manz A. Science 1993;261: 895–897.
53. Becker H, Locascio LE. Talanta 2002;56:267–287.
54. Sia SK, Whitesides GM. Electrophoresis 2003;24:3563–3576.
55. McDonald JC, Whitesides GM. Acc Chem Res 2002;35:491–499.
56. Schure MR, Lenhoff AM. Anal Chem 1993;65:3024–3037.
57. Belder D, Ludwig M. Electrophoresis 2003;24:3595–3606.
58. Ghosal S. J Fluid Mechanics 2003;491:285–300.
59. Towns JK, Regnier FE. Anal Chem 1992;64:2473–2478.
60. Bello MS, Zhukov MY, Righetti PG. J Chromatogr A 1995;693:113–130.
61. Minarik M, Gas B, Rizzi A, Kenndler E. J Capillary Electrophoresis 1995;2:89–96.
62. Stedry M, Gas B, Kenndler E. Electrophoresis 1995;16:2027–2033.
63. Gas B, Stedry M, Rizzi A, Kenndler E. Electrophoresis 1995;16:958–967.
64. Ghosal S. Anal Chem 2002;74:4198–4203.
65. Ghosal S. Anal Chem 2002;74:771–775.
66. Castelletti L, Verzola B, Gelfi C, Stoyanov A, Righetti PG. J Chromatogr A 2000;894: 281–289.
67. Zhukov MY, Ermakov SV, Righetti PG. J Chromatogr A 1997;766:171–185.
68. Wang B, Abdulali-Kanji Z, Dodwell E, Horton JH, Oleschuk RD. Electrophoresis 2003;24:1442–1450.
69. Wang B, Chen L, Abdulali-Kanji Z, Horton JH, Oleschuk RD. Langmuir 2003;19: 9792–9798.
70. Wang B, Oleschuk RD, Horton JH. Langmuir 2005;21:1290–1298.
71. Xiao DQ, Zhang H, Wirth M. Langmuir 2002;18:9971–9976.
72. Hu SW, Ren XQ, Bachman M, Sims CE, Li GP, Allbritton NL. Langmuir 2004;20:5569–5574.
73. Lahann J, Balcells M, Rodon T, Lee J, Choi IS, Jensen KF, Langer R. Langmuir 2002;18:3632–3638.
74. Xiao DQ, Van Le T, Wirth MJ. Anal Chem 2004;76:2055–2061.
75. Hu SW, Ren XQ, Bachman M, Sims CE, Li GP, Allbritton NL. Anal Chem 2004;76:1865–1870.

76. Hu SW, Ren XQ, Bachman M, Sims CE, Li GP, Allbritton N. Anal Chem 2002;74: 4117–4123.
77. Roman GT, Hlaus T, Bass KJ, Seelhammer TG, Culbertson CT. Anal Chem 2005;77: 1414–1422.
78. Lahann J, Balcells M, Lu H, Rodon T, Jensen KF, Langer R. Anal Chem 2003;75: 2117–2122.
79. Hu SW, Ren XQ, Bachman M, Sims CE, Li GP, Allbritton N. Electrophoresis 2003;24: 3679–3688.
80. Slentz BE, Penner NA, Regnier FE. J Chromatogr A 2002;948:225–233.
81. Liu Y, Fanguy JC, Bledsoe JM, Henry CS. Anal Chem 2000;72:5939–5944.
82. Yang TL, Jung SY, Mao HB, Cremer PS. Anal Chem 2001;73:165–169.
83. Badal MY, Wong M, Chiem N, Salimi-Moosavi H, Harrison DJ. J Chromatogr A 2002;947:277–286.
84. Roman GT, McDaniel K, Culbertson CT. Analyst 2006;131:194–201.
85. Liu J, Sun X, Lee ML. Anal Chem 2007;79:1926–1931.
86. Kasicka V, Prusik Z, Gas B, Stedry M. Electrophoresis 1995;16:2034–2038.
87. Culbertson CT, Jacobson SC, Ramsey JM. Anal Chem 1998;70:3781–3789.
88. Culbertson CT, Jacobson SC, Ramsey JM. Anal Chem 2000;72:5814–5819.
89. Paegel BM, Hutt LD, Simpson PC, Mathies RA. Anal Chem 2000;72:3030–3037.
90. Griffiths SK, Nilson RH. Anal Chem 2002;74:2960–2967.
91. Dutta D, Leighton DT. Anal Chem 2002;74:1007–1016.
92. Fu LM, Yang RJ, Lee GB. Electrophoresis 2002;23:602–612.
93. Molho JI, Herr AE, Mosier BP, Santiago JG, Kenny TW, Brennen RA, Gordon GB, Mohammadi B. Anal Chem 2001;73:1350–1360.
94. Griffiths SK, Nilson RH. Anal Chem 2001;73:272–278.
95. Johnson TJ, Ross D, Gaitan M, Locascio LE. Anal Chem 2001;73:3656–3661.
96. Liu JP, Dolnik V, Hsieh YZ, Novotny M. Anal Chem 1992;64:1328–1336.
97. Moore AW, Jorgenson JW. Anal Chem 1993;65:3550–3560.
98. Tao L, Thompson JE, Kennedy RT. Anal Chem 1998;70:4015–4022.
99. Wenclawiak BW, Puschl R. Anal Lett 2006;39:3–16.
100. Jacobson SC, Hergenroder R, Koutny LB, Warmack RJ, Ramsey JM. Anal Chem 1994;66:1107–1113.
101. Jacobson SC, Koutny LB, Hergenroder R, Moore AW, Ramsey JM. Anal Chem 1994; 66:3472–3476.
102. Alarie JP, Jacobson SC, Ramsey JM. Electrophoresis 2001;22:312–317.
103. Thomas CD, Jacobson SC, Ramsey JM. Anal Chem 2004;76:6053–6057.
104. Osbourn DM, Weiss DJ, Lunte CE. Electrophoresis 2000;21:2768–2779.
105. Chien RL. Electrophoresis 2003;24:486–497.
106. Fruetel JA, Renzi RF, VanderNoot VA, Stamps J, Horn BA, West JAA, Ferko S, Crocker R, Bailey CG, Arnold D, Wiedenman B, Choi WY, Yee D, Shokair I, Hasselbrink E, Paul P, Rakestraw D, Padgen D. Electrophoresis 2005;26:1144–1154.
107. Pumera MJ Chromatogr A 2006;1113:5–13.
108. Park TH, Shuler ML. Biotechnol Prog 2003;19:243–253.

109. Zhu XY, Chu LY, Chueh BH, Shen MW, Hazarika B, Phadke N, Takayama S. Analyst 2004;129:1026–1031.

110. Gu W, Zhu XY, Futai N, Cho BS, Takayama S. Proc Natl Acad Sci. U S A 2004;101: 15861–15866.

111. Prokop A, Prokop Z, Schaffer D, Kozlov E, Wikswo J, Cliffel D, Baudenbacher F. Biomed Microdev 2004;6:325–339.

112. Sweeney LM, Shuler ML, Babish JG, Ghanem A. Toxicol *In Vitro* 1995;9:307–316.

113. Sin A, Chin KC, Jamil MF, Kostov Y, Rao G, Shuler ML. Biotechnol Prog 2004;20:338–345.

114. Cai XX, Klauke N, Glidle A, Cobbold P, Smith GL, Cooper JM. Anal Chem 2002;74: 908–914.

115. Sinclair J, Olofsson J, Pihl J, Orwar O. Anal Chem 2003;75:6718–6722.

116. Pantoja R, Nagarah JM, Starace DM, Melosh NA, Blunck R, Bezanilla F, Heath JR. Biosen Bioelectron 2004;20:509–517.

117. Werdich AA, Lima EA, Ivanov B, Ges I, Anderson ME, Wikswo JP, Baudenbacher FJ. Lab on a Chip 2004;4:357–362.

118. Rocheleau JV, Walker GM, Head WS, McGuinness OP, Piston DW. Proc Natl Acad Sci U S A 2004;101:12899–12903.

119. Kulkarni RN. Int J Biochem Cell Biol 2004;36:365–371.

120. Orahilly S, Turner RC, Matthews DR. N Eng J Med 1988;318:1225–1230.

121. Schultz NM, Kennedy RT. Anal Chem 1993;65:3161–3165.

122. Gerin I, Dolinsky VW, Shackman JG, Kennedy RT, Chiang SH, Burant CF, Steffensen KR, Gustafsson JA, MacDougald OA. J Biol Chem 2005;280: 23024–23031.

123. Ueki K, Okada T, Hu J, Liew CW, Assmann A, Dahlgren GM, Peters JL, Shackman JG, Zhang M, Artner I, Satin LS, Stein R, Holzenberger M, Kennedy RT, Kahn CR, Kulkarni RN. Nat Genet 2006;38:583–588.

124. Gilon P, Shepherd RM, Henquin JC. J Biol Chem 1993;268:22265–22268.

125. Westerlund J, Gylfe E, Bergsten P. J Clin Invest 1997;100:2547–2551.

126. Roth M. Anal Chem 1971;43:880–882.

127. Gilman SD, Ewing AG. Anal Methods Instrum 1995;2:133–141.

128. Huynh BH, Fogarty BA, Nandi P, Lunte SA. J Pharm Biomed Anal 2006;42:529–534.

129. Shou MS, Ferrario CR, Schultz KN, Robinson TE, Kennedy RT. Anal Chem 2006;78:6717–6725.

130. Kottegoda S, Shaik I, Shippy SA. J Neurosci Methods 2002;121:93–101.

131. Cho BS, Schuster TG, Zhu XY, Chang D, Smith GD, Takayama S. Anal Chem 2003;75:1671–1675.

132. Garcia CD, Liu Y, Anderson P, Henry CS. Lab on a Chip 2003;3:324–328.

133. Collins GE, Wu P, Lu Q, Ramsey JD, Bromund RH. Lab on a Chip 2004;4:408–411.

134. Erickson D, Sinton D, Li DQ. Lab on a Chip 2004;4:87–90.

135. Verpoorte E. Lab on a Chip 2003;3:42N–52N.

136. Culbertson CT, Tugnawat Y, Meyer AR, Roman GT, Ramsey JM, Gonda SR. Anal Chem 2005;77:7933–7940.

137. Renzi RF, Stamps J, Horn BA, Ferko S, VanderNoot VA, West JAA, Crocker R, Wiedenman B, Yee D, Fruetel JA. Anal Chem 2005;77:435–441.
138. Skelley AM, Scherer JR, Aubrey AD, Grover WH, Ivester RHC, Ehrenfreund P, Grunthaner FJ, Bada JL, Mathies RA. Proc Natl Acad Sci U S A 2005;102: 1041–1046.
139. Lagally ET, Scherer JR, Blazej RG. Toriello NM, Diep BA, Ramchandani M, Sensabaugh GF, Riley LW, Mathies RA. Anal Chem 2004;76:3162–3170.
140. Martin RS, Ratzlaff KL, Huynh BH, Lunte SM. Anal Chem 2002;74:1136–1143.
141. Kappes T, Galliker B, Schwarz MA, Hauser PC. Trends Anal Chem 2001;20:133–139.
142. Viskari PJ, Landers JP. Electrophoresis 2006;27:1797–1810.

12

ANALYTICAL STRATEGIES TOWARD THE ANALYSIS OF PHENOLIC COMPOUNDS BY CAPILLARY ELECTROPHORESIS AND MICROCHIP CAPILLARY ELECTROPHORESIS

MARIA F. MORA AND CARLOS D. GARCIA

12.1 INTRODUCTION

Ranging from simple molecules to compounds with molecular weights greater than 30 kDa, the occurrence of phenolic compounds is extremely variable in nature (1–7). Depending upon their basic structure, phenolic compounds can be generally divided into 10 main types: simple phenols, phenolic acids, coumarins and isocoumarins, naphthoquinones, xanthones, stilbenes, anthraquinones, flavonoids, and lignins (8). Because the biological effects and possible applications of these compounds depend directly on the structure and the respective concentration, the analysis of phenols is critical in biological (9), pharmaceutical (6), environmental (3,5), and industrial fields (10). Among others, phenol and some phenolic compounds are found in industrial and natural waste and are able to confer unpleasant flavors to drinking water, affect the bioavailability of metals, and produce a wide variety of negative effects on biological processes (11,12). Other phenols, however, are reported to exhibit beneficial health effects (13) such as anticarcinogenic, anti-inflammatory, anti-atherogenic, antithrombotic, immune modulating, and analgesic

Biological Applications of Microfluidics, Edited by Frank A. Gomez

activities. Because most phenols can be oxidized by oxygen, particular phenolic compounds can be also added to cosmetics, food, and plastics as antioxidants (4).

Regardless of their toxicity and persistency in the environment, the determination of the concentration of phenols is a key parameter to adopt corrective actions and to enforce regulations (14–18). Various methods have been reported for the determination of phenolic compounds in numerous samples including spectrophotometric (14), chromatographic (8,19–21), electrochemical (22–24), flow-through (16,25,26), and electrophoretic (27,28). Numerous biosensors have been also presented (28–32). Additionally, several preconcentration (9,16,33–36) and sample handling strategies (37,38) have been recently considered in order to improve the selectivity and/or the sensitivity of the analysis. Although a great amount of information can be obtained by most of the aforementioned methodologies, they are typically time consuming, need large sample volumes, generate significant amounts of waste, or require bulky and expensive instrumentation.

Capillary electrophoresis (CE) is an alternative method for the separation of phenols (39–52) that can provide inherent advantages. In CE, the separation takes place inside a small capillary filled with an electrolyte solution. When a potential difference is applied across the capillary, bulk flow of the solution is generated by a process referred to as electroosmotic flow (EOF) (53). Analytes, introduced at one extreme of the capillary, migrate with a velocity that is proportional to their charge/mass ratio, the magnitude of the applied potential, and the solution conditions. In general terms, the overall analysis time results from a combination of the migration velocity and the EOF. Therefore, CE allows the possibility of using a wide range of solution conditions to control the EOF, providing fast, high throughput, highly efficient, and reliable separations (54–58). Because CE does not require mechanical pumps, it can be easily miniaturized while still offering high performance, versatility, reagent economy, and automation capabilities (59–73).

Phenolic compounds have been analyzed by CE in conjunction with a wide variety of detectors such as UV-vis (42,74), chemiluminescence (54), MS (40,75,76), ICP-MS (45), and electrochemical (44,77,78). Electrochemical detection (ECD) can provide inherent miniaturization, automation, and portability, particularly if microchips are used (55,79). Among others, amperometry is one of the most widely reported ECD methods used in conjunction with CE microchips. However, when a constant potential is applied to the detection electrode, oxidation products can be deposited on the electrode surface producing a significant decrease in signal magnitude (80). To overcome this drawback, many approaches have been presented (15,81,82), including pulsed amperometric detection (PAD) (83,85). PAD utilizes a simple three potential waveform where the electrode is first oxidized at a high positive potential (which cleans the electrode), reactivated at a negative potential, and then used to detect the analyte at a moderate positive potential. Advantages of PAD include a sensitivity comparable to (or better than) DC amperometry, speed, and the possibility of performing direct detection of analytes that are nonelectrochemically active by other techniques (83,84,86,87).

In the following sections, different strategies toward the determination of phenolic compounds using CE and microchip-CE will be discussed. This chapter will be first focused on the effects of the background electrolyte and capillary surface on the electrophoretic separation of phenolic compounds. Then, different aspects related to the electrochemical detection of phenols will be discussed (1,2). Finally, a series of applications using CE and microchip-CE will be presented. For the first application, we will describe the analysis of a series of environmentally important phenols including three EPA priority contaminants (phenol, pentachlorophenol, and 4,6-dinitro-*o*-cresol) (2) and six phenolic acids (ferulic, *p*-coumaric, salicylic, 4-hydroxyphenylacetic, vanillic, and 4-hydroxybenzoic acids) (3). The second application includes the analysis of four synthetic phenolic antioxidants (propyl gallate, octyl gallate, lauryl gallate, and nordihydroguaiaretic acid) (4) used in food industry. The third example of the capabilities of these devices includes the analysis of a set of over-the-counter drugs (salicylic acid, acetaminophen, diflunisal, and diclofenac) (6).

12.2 EXPERIMENTAL SECTION

12.2.1 Chemicals and Apparatus

SU-8 2035 photoresist was purchased from Micro Chem Co. (Newton, MA), Sylgard 184 silicone elastomer and curing agent were obtained from Dow Corning (Midland, MI). Aqueous solutions were prepared using analytical grade reagents and 18 MΩ·cm resistance water (NANOpure Diamond, Barnstead; Dubuque, IA). Pentachlorophenol (PCP), 4,6-dinitro-*o*-cresol (DNOC), 2-nitrophenol (NP), ferulic acid (FA), *p*-coumaric acid (CA), salicylic acid (SA), 4-hydroxyphenylacetic acid (HPA), vanillic acid (VA), 4-hydroxybenzoic acid (HBA), acetaminophen (AP), diflunisal (DFS), diclofenac (DCF), catechol, dopamine, and epinephrine were purchased from Sigma (St Louis, MO, USA). *O*-Aminophenol (AP) was purchased from Eastman Kodak Co. (Rochester, NY). Sample stock solutions of analytes were prepared by dissolving the desired amount in methanol (Fisher; Fairlawn, NJ). Electrolytes were prepared by weighing the desired amount of sodium borate ($Na_2B_4O_7 \cdot 10H_2O$) and adjusting the pH with either 0.1 M NaOH or 0.1 M HCl (Fisher; Fairlawn, NJ). The pH measurements were performed with a combined glass electrode and a digital pH meter (Orion 420A+, Thermo; Waltham, MA). A home-made high voltage power supply (HVPS) (88) was used for separation and injection in all the experiments. The HVPS consists of two positive and one negative DC–DC converters, a microprocessor-controlled timer and a power supply. This arrangement allows the control of the potentials applied in the 0 to ± 4000 V range with noise values in the nanoamperes range.

Electrochemical detection (amperometry and pulsed amperometric detection) was performed using a CHI812 electrochemical detector (CH Instruments; Austin, TX). A 25-μm gold wire was used as the working electrode according to a previously presented arrangement (2–4,6,58,87,89,90), defining a detection area of 0.0039 mm^2 (25 μm electrode diameter, 50 μm channel width). Ag wire and Pt electrodes were used as pseudoreference and auxiliary electrodes, respectively, and placed in the waste reservoir. For PAD, the waveform applied to the electrode consists of a cleaning step (1.2 V—0.05 s), a reconstruction step (−0.5 V—0.025 s), and a detection step (0.6 V—0.15 s).

12.2.2 Fabrication of the PDMS Microchip

A previously described procedure was used to fabricate the molds (91). Briefly, a 100-mm silicon wafer (Silicon Valley Microelectronics Inc.; Santa Clara, CA) was cleaned and oxidized with piranha solution (1:1 H_2SO_4:H_2O_2), rinsed with deionized water, and dried at 200 °C for 15 min. The wafer was then coated with SU-8 2035 negative photoresist using a spin coater (Laurell Technologies; North Wales, PA) followed by two preexposure baking steps at 65 and 95 °C for 5 and 10 min, respectively. A digitally produced mask (The Photoplot Store; Colorado Springs, CO) containing the channel pattern was placed on the coated wafer, exposed to light via a near-UV flood source (365 nm—13 mW/cm^2, Optical Associates Inc.; San Jose, CA), and then baked at 95 °C for 60 min. The positive relief was developed, rinsed with methanol, and dried under a N_2 stream. Two poly(dimethylsiloxane) (PDMS) layers were fabricated by pouring a degassed mixture of Sylgard 184 silicone elastomer and curing agent (10:1) onto either a molding master or a blank wafer, followed by curing for at least 2 h at 65 °C. The cured PDMS layers were separated from the mold and reservoirs were made at the end of each channel using a circular punch. Then, the detection electrode was aligned at the end of the separation channel in a perpendicular channel designed *ad hoc* (2–4,89,90). Next, the two PDMS layers were irreversible sealed and an electrical connection to the working electrode was made using silver paint (Structure Probe Inc; West Chester, PA) and a copper wire.

A double-T injector with a 500 μm gap between side channels was used for all experiments defining a 1.2 nL sample plug. The separation channel was 52 mm long. A schematic drawing of the microchip is included in Fig. 12.1.

The standard amount of buffer dispensed on each reservoir (buffer, sample waste, and waste) was 50 μL, except for the sample reservoir where only 45 μL

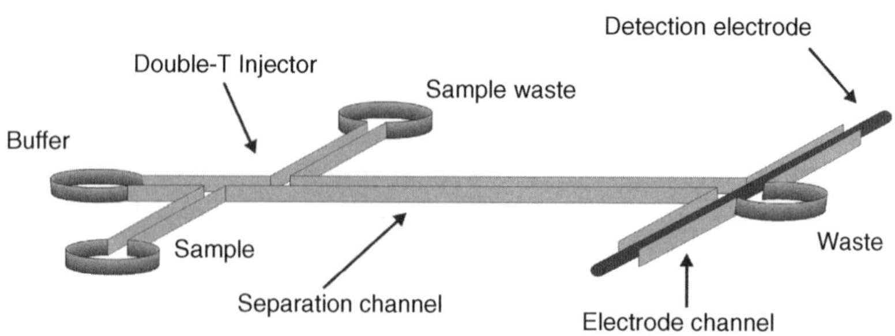

FIGURE 12.1 Schematic drawing of the microchip-CE with PAD. Channels width: 50 μm, channels depth: 50 μm, double-T arms: 4 mm long, double-T volume: 1.25 nL, separation channel length: 70 mm, solution reservoirs: 6 mm diameter, detection electrode: 25 μm diameter. Adapted from Reference 23.

were dispensed in order to avoid leakage into the separation channel. During the sample injection, potentials of +500 V, −50 V and +500 V were applied to the sample, sample waste, and buffer reservoirs, respectively. To avoid Joule heating in the sample-to-buffer channel during the injection, a resistor (1 MΩ) was included in series with the chip (88,90,92). During the separation step, the potential applied to the buffer reservoir was raised to +1200 V (or to the corresponding separation potential) while the potential applied to the sample waste reservoir was changed to +500 V. The waste reservoir was always grounded. Electrical connections to the microchip were made with platinum electrodes placed into the reservoirs at the ends of each channel. To minimize noise, a separate ground electrode, positioned in the waste reservoir, was used for the separation system.

12.3 RESULTS AND DISCUSSION

12.3.1 Separation of Phenolic Compounds by CE and Microchip-CE

The successful separation of phenolic compounds requires an accurate control of the background electrolyte as well as the capillary surface. In this regard, one of the most important variables in capillary electrophoresis is the pH and concentration of the background electrolyte. Although under alkaline conditions most phenolic compounds become anions (pK_a phenol = 9.98) (93) and migrate according to their charge to mass ratio (52), one of the most challenging aspects of microchip-CE is to achieve the separation of analytes in a relatively short channel (~5 cm long). Therefore, the challenge is to achieve the separation of the selected analytes using the simplest solution conditions and requires a precise control of the solution pH. Initially, the effect of solution pH on the separation of pentachlorophenol, 4,6-dinitro-*o*-cresol, and phenol was studied between pH 10 and 12.6 using 5.0 mM phosphate solutions. Although poor resolution was achieved between pH 11.0 and 12.0 ($R = 0.33$, phenol-PCP, pH 11.6), baseline separation of the three compounds was achieved at pH 12.4 ($t_{M\text{-}PCP} = 94 \pm 1$ s, $t_{M\text{-}DNOC} = 101 \pm 1$ s, $t_{M\text{-}Ph} = 108 \pm 1$ s).[2] As Fig. 12.2 shows, the migration order reflects a clear dependence of the electrophoretic mobility with respect to the molecular weight of the selected compounds.

The effect of the pH on the migration time of six organic acids was also investigated using a 5-mM borate solution as background electrolyte (3). The results presented in Fig. 12.3 show that most of the selected organic acids can be separated in the studied pH range (9.6–11.0), preserving the migration order. The most remarkable change observed was the comigration of HPA and SA at pH 11.0 ($t_{M\text{-}HPA} = 144 \pm 2$ s/$t_{M\text{-}SA} = 143.1 \pm 0.7$ s).

As a consequence of the increase in ionic strength of the background electrolyte, lower EOF values and higher migration times were observed at higher pH values. Therefore, in order to maximize the separation of the analytes, pH 10.0 was selected as the optimum value and used for the following experiments, which allows the separation of ferulic acid, *p*-coumaric acid, salicylic acid, 4-hydroxyphenylacetic acid, vanillic acid, and 4-hydroxybenzoic acid in less than 3 min.

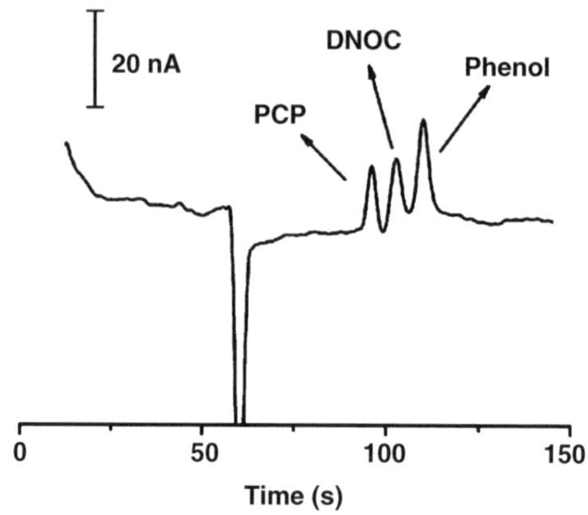

FIGURE 12.2 Separation and PAD of 67.7 μM phenol, 39.4 μM pentachlorophenol (PCP), and 25.0 μM 4,6-dinitro-*o*-cresol (DNOC). Conditions: 5 mM phosphate buffer pH 12.4, $T_{INJECTION} = 10$ s, $E_{DETECTION}$: +0.7 V, $E_{SEPARATION} = +1200$ V. Adapted from Reference 2.

If changes in solution pH do not provide enough resolution, a second possibility to improve separations in CE is to control the properties of the capillary wall (94–97). By coating the capillary surface it is possible to reduce analyte–wall interactions, modify the EOF, improve the reproducibility, and enhance the resolution (97). Taking advantage of the fact that phenols do not adsorb to negatively charged clays, we have developed a simple and inexpensive procedure to coat fused silica capillaries with poly(diallyldimethyl ammonium chloride) and montmorillonite (5). The coated capillaries allowed the separation of nine environmentally

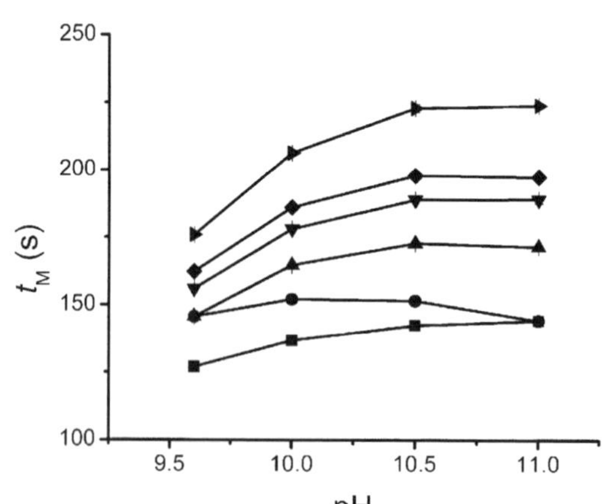

FIGURE 12.3 Effect of background electrolyte pH on the migration time of 3.4 μg/mL HPA (–■–), 5.8 μg/mL SA (–●–), 2.3 μg/mL FA (–▲–), 4.0 μg/mL CA (–▼–), 2.2 μg/mL VA (–◆–), 3.1 μg/mL HBA (–▶–), Other conditions: 5 mM borate (pH 10.0), $E_{SEPARATION} = +1200$ V (B–W), $T_{INJECTION} = 10$ s, $E_{OXIDATION} = 1.6$ V, $E_{REGENERATION} = -1.1$ V, $E_{DET} = +0.8$ V. With permission from Reference 3.

important phenolic compounds while showing a significant improvement in the resolution, when compared to bare fused silica capillaries. The coated capillaries also showed a highly stable EOF (run-to-run RSD less than 1.5%, $n = 20$), allowing continuous use for several days without conditioning. Since poly(diallyldimethyl ammonium chloride) adsorbs to PDMS, the possibility of using the same procedure to coat PDMS microchannels was also demonstrated.

Simple (one component) alkaline aqueous electrolytes are not suitable for the separation of every phenolic compound due to problems associated with solubility and stability (15,22,81). In some of those cases, micellar electrokinetic chromatography (MEKC) has proven to be a very attractive separation method (39,98–100). MEKC is performed by including surfactants in the background electrolyte at concentrations above the critical micelle concentration (CMC). Generally, the partition coefficient of the analyte between the background electrolyte and the micelle is the main factor determining selectivity. Taking advantage of this possibility, sodium dodecyl sulfate (SDS) was used for the separation of a group of alkyl gallates and nordihydroguaiaretic acid. In Fig. 12.4, the dependence of the migration time of these phenolic antioxidants as function of SDS concentration is shown.

As can be observed, a clear decrease in the migration time (t_M) was obtained for all the selected compounds upon the first addition of SDS (10 mM). This decrease in t_M is the result of an increase in electroosmotic flow produced by the adsorption of SDS onto the PDMS surface. As the SDS concentration was progressively increased, longer migration times but constant migration order were obtained. At 30 mM SDS, all of the peaks were baseline separated. It is also worth mentioning

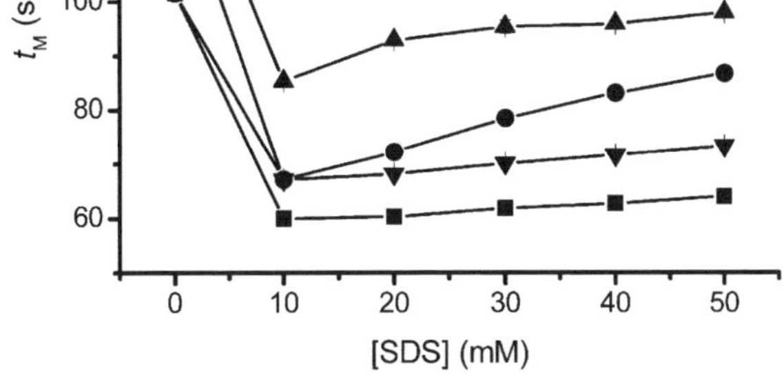

FIGURE 12.4 Effect of SDS concentration on the migration time for 28 µM propyl gallate (PG, –■–), 25 µM octyl gallate (OG, –●–), 81 µM lauryl gallate (LG, –▲–) and, 20 µM nordihydroguaiaretic acid (NDGA, –▼–). Other conditions: 10 mM borate (pH 9.6), $E_{SEPARATION} = +1200$ V (B–D), $T_{INJECTION} = 5$ s, $E_{DETECTION} = +0.5$ V. With permission from Reference 4.

that SDS also significantly improves the electrochemical detection of phenolic compounds. Even though several groups have observed this effect (22,58,87,90,92,101), the precise mechanism is not yet clear.

An additional alternative to improve the separation of phenolic compounds is to add organic solvents to the background electrolyte. Although some additives such as THF, alcohols, or acetonitrile have provided significant improvements in separation efficiency (102–105), the improvement in separation is typically hampered by a decrease in signal/noise ratio (when ECD is used) (89) or changes in the structure of the microchip substrate (swelling or even dissolution) (106).

12.3.2 Electrochemical Detection of Phenolic Compounds

Phenols can be electrochemically oxidized under a wide range of conditions (60,65,67,68,107,108). However, a decrease in signal magnitude is frequently observed if the electrode is not coated or frequently polished (15,22,81,109). This current decrease has been attributed to the polymerization of the oxidation products on the electrode surface (80) and, as expected, depends on the electrode material, the applied potential, the solution pH, and both the concentration and structure of the analyte (110). PAD can minimize the electrode fouling and increase the signal stability. One disadvantage of PAD, however, is the larger number of parameters, which typically requires a more exhaustive optimization procedure than other electrochemical techniques (84,87). Although all the parameters of PAD have a large effect on the magnitude and stability of the signal, probably the most significant is the detection potential (E_{DET}). The optimum E_{DET} for a given set of compounds can be obtained from a hydrodynamic voltammogram (I_p versus E_{DET}), as shown in Fig. 12.5.

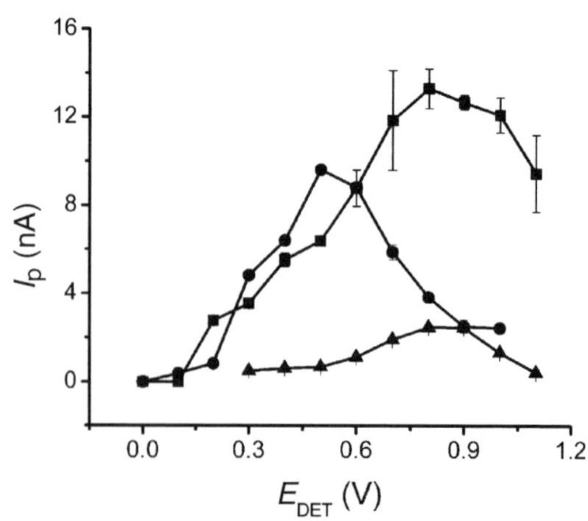

FIGURE 12.5 Hydrodynamic voltammograms corresponding to phenol (–■–), propyl gallate (–●–), and 4-hydroxyphenylacetic acid (–▲–). Adapted from Reference 2–4,23.

As can be observed, the peak current (I_p) for phenol, propyl gallate, and 4-hydroxyphenylacetic acid increases as the detection potential increases, until a maximum is obtained at around +0.6 V. The current decrease observed at higher potentials (>0.8 V) has been attributed to the formation of a passivating oxide layer on the electrode surface (101,111). Depending on the structure and analytical conditions, different phenolic compounds present slight differences in E_{DET} (±0.15 V). This behavior has been attributed to the ability of some compounds to stabilize the oxidation products, reduce the overvoltage, and hence facilitate the electron transfer reaction (112). It is worth mentioning that although these experiments were performed using different background electrolytes, similar behaviors were obtained for all the studied compounds, suggesting a similar oxidation mechanism (80,113). As stated before, the oxidization (cleaning) and reduction (reconstruction) potentials, and the time length for each step also have a significant effect on the peak current. According to our experience, better results can be obtained if the electrode is oxidized at +1.6 V for 50 ms, followed by reconstruction step at −0.7 V for 25 ms. To illustrate the efficiency of PAD to preserve the signal magnitude, a series of more than 100 repetitive injections of 90 μM phenol were performed (Fig. 12.6). As can be observed, a minimum difference (1.6%) was obtained between the initial and final responses.

A disadvantage of PAD, however, is that specific instrumentation is required. To overcome this problem, without compromising signal stability, carbon electrodes have been used. Along with amperometry, carbon electrodes offer low cost, good compatibility with a variety of solvents (109,114,115), low background currents and relative stability and sensitivity toward the oxidation of organic couples (116). Although macroelectrodes can be used, most of the fabricated carbon electrodes have been deposited using photolithography or screen-printing techniques (55,117–119).

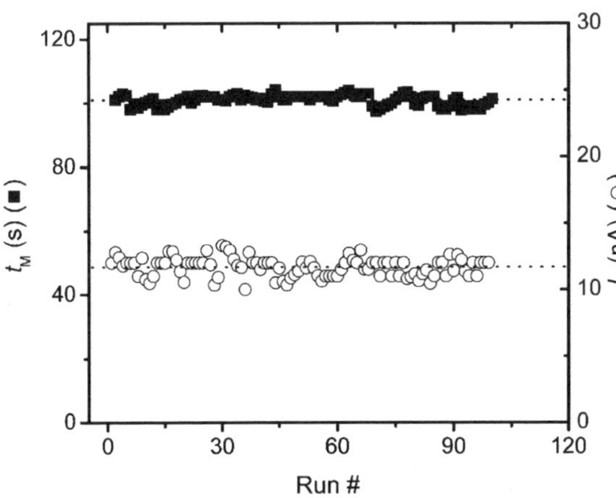

FIGURE 12.6 Electrochemical response and migration time obtained with repetitive injections of 90 μM phenol. Conditions: 5 mM phosphate buffer pH 12.4, $T_{INJECTION}$ = 10 s, $E_{DETECTION}$ = +0.7 V, $E_{SEPARATION}$ = +1200 V. With permission from Reference 2.

Alternatively, carbon fibers have also been used (116,120–123), but single carbon fibers are generally too thin (<7 μm) and brittle to be included as stand-alone detection electrodes in microchips. Our group has recently reported the possibility of using a low-cost conductive carbon ink to coat metallic wires (1). In this way, the electrodes combine the sensitivity provided by the carbon surface with the geometrical advantages of the microwire (23,84,89,124) design. The electrodes were then integrated in a poly(dimethylsiloxane) microchip and used for the analysis of phenolic compounds. Using the optimum conditions, the analysis of a mixture of dopamine, epinephrine, catechol, and 4-aminophenol was achieved in less than 4 min, showing good linear responses ($R^2 = 0.999$) in the 0.1−190 μM range, and limits of detection (without the use of stacking or a decoupler) of 140 and 105 nM for dopamine and epinephrine, respectively.

12.4 APPLICATIONS

The suitability of CE and microchip-CE for separating low levels of phenolic compounds in real applications was also demonstrated. As described before, Fig. 12.2 shows an electropherogram obtained with a standard mixture of phenols (pentachlorophenol, 4,6-dinitro-*o*-cresol, and phenol). The same conditions were used to analyze different real samples. Clear peaks were observed for the selected compounds when spiked city water or sore throat medicine samples were analysed (2). To further demonstrate the aptitude of these devices to deal with environmental samples, the degradation of a set of six compounds (salicylic acid, 4-hydroxyphenylacetic acid, vanillic acid, and 4-hydroxybenzoic acid) by aquatic plants was studied (3). These compounds, which originate from the decomposition of the biopolymers, microbial synthesis, root exudates, and many industrial processes, play important roles in aquatic ecosystems (125,126) and have been identified as indicators of dissolved organic matter from terrestrial sources (127). Using the optimum conditions the mixture was separated in less than 3 min (Fig. 12.7).

Then, a control (no plant) and two batches containing aquatic plants (*Elodea densa* or *Cabomba caroliniana*) and 5 μg/mL of each compound were analyzed at different times (0, 8, 24, 48, and 72 h) using the microchip-CE−ECD. As the result of a combination of processes that may include the uptake, transport, conjugation, and storage (128,129), a significant decrease in the concentration of phenolic acids was observed in the flasks containing plants. We observed that the selected aquatic plants faded all phenolic acids at approximately the same rate, without showing a significant preference for one acid over the others. This device could be used for on-site monitoring of phytoremediation processes, avoiding the mobilization of contaminated samples to the laboratory.

The capabilities of the developed microchip-CE with ECD to analyze clinically important samples was demonstrated by developing a simple and rapid method for the analysis of four of the most commonly used over-the-counter drugs to treat fever, inflammation, and pain (salicylic acid, acetaminophen, diflunisal, and diclofenac) (130). When used above the therapeutic levels, these drugs or their metabolites can cause a wide variety of adverse effects (131–134) and their fast analysis could have a significant impact in treatment and recovery of the patients. Because of these reasons, several conditions including separation potential, pH, and concentration of the electrolyte solution were optimized. In this study, salicylic acid, acetaminophen, diflunisal, and diclofenac were separated in less than 2 min using a 5 mM borate buffer at pH 11.5 and a separation potential of +1200 V. Linear relationships were obtained between the concentration and peak current in the 0.5–15.3 μg/mL range and detection limits around 0.26 μg/mL. After 30 consecutive injections, the stability of both the response and migration time of the analytes showed relative standard deviations of less than 4.6% and 1.0%, respectively. Fig. 12.8 shows the electropherograms obtained with a standard mixture of the four selected drugs and a spiked serum sample.

Good recoveries (calculated by comparing the concentrations in the spiked samples with those in a standard under the same conditions) were obtained for acetaminophen (90 ± 5%), diclofenac (99 ± 6%), and diflunisal (81 ± 6%). Lower yields were obtained for salicylic acid (72 ± 5%), probably due to nonspecific adsorption to proteins in the serum, which were precipitated before the analysis.

Finally, a simple and fast method for the simultaneous determination of the antioxidants propyl gallate (PG), octyl gallate (OG), lauryl gallate (LG), and nordihydroguaiaretic acid (NDGA) has been established by using microchip micellar electrokinetic chromatography with pulsed amperometric detection. These compounds are considered phenolic synthetic antioxidants and are used (alone or as mixtures) in many cosmetic, pharmaceutical, and food-processing industries (135,136). Besides the beneficial effects, some antioxidants have been linked to

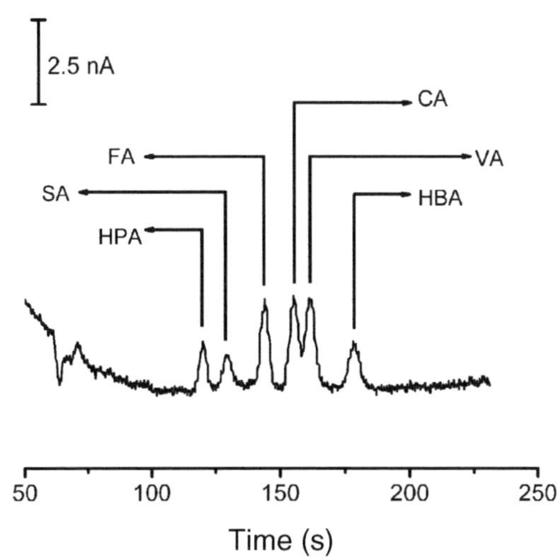

FIGURE 12.7 Electropherograms corresponding to a mixture of 1.4 μg/mL HPA, 3.7 μg/mL SA, 3.1 μg/mL FA, 3.0 μg/mL CA, 4.5 μg/mL VA, and 2.0 μg/mL HBA. Conditions: 5 mM borate (pH 10.0), $E_{SEPARATION} = +1200$ V (B–W), $T_{INJECTION} = 10$ s, $E_{OXIDATION} = 1.6$ V, $E_{REGENERATION} = -1.1$ V, $E_{DETECTION} = +0.8$ V. With permission from Reference 3.

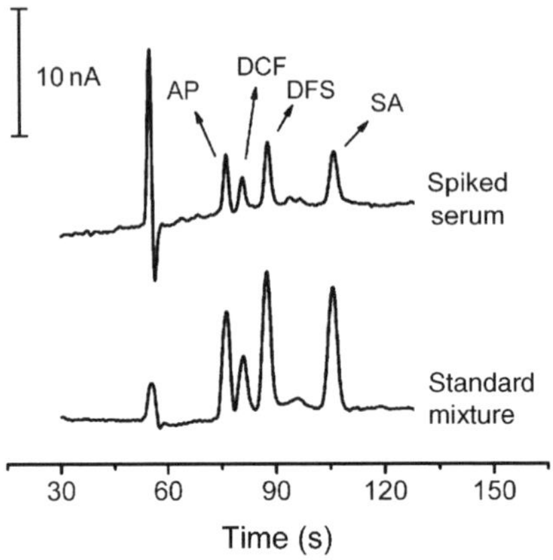

FIGURE 12.8 Electropherograms corresponding to a standard mixture, and the serum sample spiked with four drugs. Conditions: 5 mM borate (pH 11.5), $E_{SEPARATION} = +1200$ V (B–W), $T_{INJECTION} = 5$ s, $E_{OXIDATION} = 1.6$ V, $E_{REGENERATION} = -0.9$ V, $E_{DETECTION} = +0.7$ V. AP, acetaminophen; DCF, diclofenac; DFS, diflunisal; SA, salicylic acid. With permission from Reference 6.

allergies, carcinogenicity, mutagenesis, and cytotoxicity (137–139). For this reason, the acceptable intake of these compounds has been regulated by different organizations. Fig. 12.9 shows the electropherograms corresponding to a standard mixture of antioxidants, a commercial sample (gravy mix) containing propyl gallate, and the same sample spiked with the four antioxidants.

As can be observed, the baseline separation of the four selected antioxidants was achieved in less than 2 min. Using the optimized conditions, a gravy mix sample was also analyzed (B), and the corresponding amount of propyl gallate was calculated to be 45 mg of PG/kg of sample.

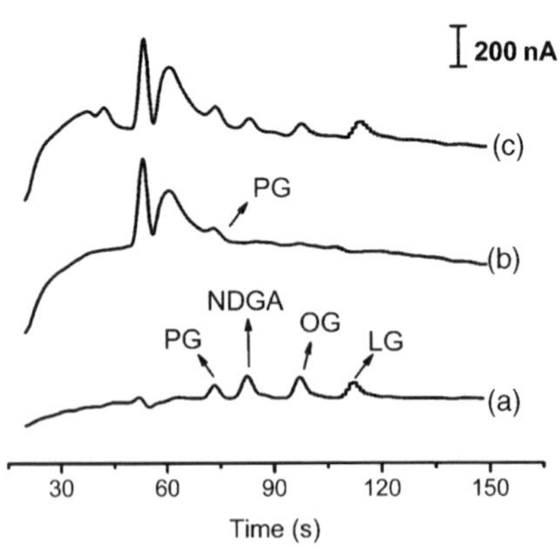

FIGURE 12.9 Electropherograms corresponding to a standard mixture (**a**), a food sample (**b**) and the food sample spiked with propyl gallate, nordihydroguaiaretic acid (NDGA), octyl gallate (OG) and lauryl gallate (LG) (**c**). Conditions: 30 mM Borate, 30 mM SDS, pH 9.6, $E_{DETECTION} = 0.5$ V, $T_{INJECTION} = 5$ s. With permission from Reference 4.

12.5 CONCLUSIONS

This report describes different analytical strategies to improve the separation and detection of phenolic compounds using CE and microchip-CE. The analytical effects of performing rational modifications of the surface channel, background electrolyte, and detection method have been discussed. In all the cases, the described methodologies illustrate practical, simple, rapid, and inexpensive methods for the simultaneous analysis of phenolic compounds with a variety of biological effects. After the optimization of the analytical parameters, mixtures of phenolic compounds can be separated in a few minutes and detected above the low nanomolar range. Although most of the recently presented systems offer inexpensive, portable and customized solutions for the analysis of very small (nL) samples, the detection limits and resolution must be improved in order to allow effective applicability to complex samples. Certainly, microchip-CE–ECD has the potential to become a widely accepted methodology for the analysis of phenolics in the near future.

ACKNOWLEDGMENTS

The authors thank The University of Texas at San Antonio and the San Antonio Area Foundation for financial support.

REFERENCES

1. Ding Y, Ayon A, García CD. Anal Chim Acta 2007;584:244–251.
2. Ding Y, García CD. Analyst 2006;131:208–214.
3. Ding Y, García CD. Electrophoresis 2006;27:5119–5127.
4. Ding Y, Mora MF, García CD. Anal Chim Acta 2006;561:126–132.
5. Mora MF, García CD. Electrophoresis 2007;28:1197–1203.
6. Ding Y, García CD. Electroanalysis 2006;18:2202–2209.
7. Bravo L. Nutr Rev 1998;56:317–333.
8. Gomez-Caravaca AM, Gomez-Romero M, Arraez-Roman D, Segura-Carretero A, Fernandez-Gutierrez A. J Pharm Biomed Anal 2006;41:1220–1234.
9. Naczk M, Shahidi F. J Pharm Biomed Anal 2006;41:1523–1542.
10. Rodriguez Couto S, Toca Herrera JL. Biotechnol Adv 2006;24:500–513.
11. Harrison MAJ, Barra S, Borghesi D, Vione D, Arsene C, Iulian Olariu R. Atmos Environ 2005;39:231–248.
12. Dabrowski A, Podkoscielny P, Hubicki Z, Barczak M. Chemosphere 2005;58: 1049–1070.
13. Shahidi F, Ho C-T. Phenolic Compounds in Foods and Natural Health Products. Washington, DC: American Chemical Society. 2005.
14. AOAC. Official Method of Analysis 983, 15,1994.

15. García CD, Ortiz PI. Electroanalysis 1998;10:832–835.
16. Kuban P, Berg M, García CD, Karlberg B. J Chromatogr A 2001;912:163–170.
17. Ahlstrom L-H, Sparr Eskilsson C, Bjorklund E. TrAC Trends Anal Chem 2005;24: 49–56.
18. US Food and Drug Administration. Code of Federal Regulations. 2003;9:625–651.
19. Willfor SM, Smeds AI, Holmbom BR. J Chromatogr A 2006;1112:64–77.
20. de Rijke E, Out P, Niessen WMA, Ariese F, Gooijer C, Brinkman UAT. J Chromatogr A 2006;1112:31–63.
21. Bravo R, Caltabiano LM, Fernandez C, Smith KD, Gallegos M, Whitehead JRD, Weerasekera G, Restrepo P, Bishop AM. J Chromatogr B 2005;820:229–236.
22. Hompesch RW, García CD, Weiss DJ, Vivanco JM, Henry CS. Analyst 2005;130: 694–700.
23. Mora MF, Ding Y, Mejia E, García CD. J Cap Elec Microchip Tech 2007;10:7–18.
24. Blasco AJ, Rogerio MC, Gonzalez MC, Escarpa A. Anal Chim Acta 2005;539: 237–244.
25. Costin JW, Barnett NW, Lewis SW, McGillivery DJ. Anal Chim Acta 2003;499: 47–56.
26. Behzad H, Reza D. Anal Bioanal Chem 2006;384:1246–1253.
27. Hernandez-Borges J, Gonzalez-Hernandez G, Borges-Miquel T, Rodriguez-Delgado MA. Food Chem 2005;91:105–111.
28. Jac P, Polasek M, Pospisilova M. J Pharm Biomed Anal 2006;40:805–814.
29. Kim MA, Lee W-Y. Anal Chim Acta 2003;479:143–150.
30. Mello LD, Sotomayor MDPT, Kubota LT. Sens Actuators B 2003;96:636–645.
31. Li Y-F, Liu Z-M, Liu Y-L, Yang Y-H, Shen G-L, Yu R-Q. Anal Biochem 2006;349: 33–40.
32. Shin HJ, Park HH, Lim WK. J Biotechnol 2005;119:36–43.
33. Shiddiky MJA, Park H, Shim YB. Anal Chem 2006;78:6809–6817.
34. Kawaguchi M, Sakui N, Okanouchi N, Ito R, Saito K, Izumi S, Makino T, Nakazawa H. J Chromatogr B 2005;820:49–57.
35. Mahugo Santana C, Sosa Ferrera Z, Santana Rodriguez JJ. J Chromatogr Sci 2005; 43:282–288.
36. Chen J-L, Liu C-Y. Anal Chim Acta 2005;528:83–88.
37. Robards K. J Chromatogr A 2003;1000:657–691.
38. Tura D, Robards K. J Chromatogr A 2002;975:71–93.
39. Guan Y, Chu Q, Fu L, Wu T, Ye J. Food Chem 2006;94:157–162.
40. Huck CW, Stecher G, Scherz H, Bonn G. Electrophoresis 2005;26:1319–1333.
41. Vaher M, Ehala S, Kaljurand M. Electrophoresis 2005;26:990–1000.
42. Priego Capote F, Ruiz Jimenez J, de Castro MD. J Chromatogr A 2004;1045:239–246.
43. Carrasco Pancorbo A, Cruces-Blanco C, Segura Carretero A, Fernandez Gutierrez A. J Agric Food Chem 2004;52:6687–6693.
44. Peng Y, Chu Q, Liu F, Ye J. J Agric Food Chem 2004;52:153–156.
45. Kannamkumarath SS, Wuilloud RG, Jayasinghe S, Caruso JA. Electrophoresis 2004;25: 1843–1851.

46. Volpi N. Electrophoresis 2004;25:1872–1878.
47. Cabovska B, Kreishman GP, Wassell DF, Stalcup AM. J Chromatogr A 2003;1007:179–187.
48. Vaher M, Koel M. J Chromatogr A 2003;990:225–230.
49. Porras SP, Kuldvee R, Palonen S, Riekkola ML. J Chromatogr A 2003;990:35–44.
50. Bonoli M, Montanucci M, Gallina Toschi T, Lercker G. J Chromatogr A 2003;1011: 163–172.
51. Tsukagoshi K, Kameda T, Yamamoto M, Nakajima R. J Chromatogr A 2002;978: 213–220.
52. Canals I, Bosch E, Rosés M. Anal Chim Acta 2002;458:355–366.
53. Rathore AS, Guttman A. Electrokinetic Phenomena. New York, NY: Marcel Dekker, 2003.
54. Yin X-B, Wang E. Anal Chim Acta 2005;533:113–120.
55. Wang J. Electroanalysis 2005;17:1133–1140.
56. Dolnik V. Electrophoresis 2004;25:3589–3601.
57. Trojanowicz M, Szewczynska M, Wcislo M. Electroanalysis 2003;15:347–365.
58. García CD, Dressen BM, Henderson A, Henry CS. Electrophoresis 2005;26: 703–709.
59. Ng JM, Gitlin I, Stroock AD, Whitesides GM. Electrophoresis 2002;23:3461–3473.
60. Scampicchio M, Wang J, Mannino S, Chatrathi MP. J Chromatogr A 2004;1049: 189–194.
61. Wang J, Chen G, Chatrathi MP, Musameh M. Anal Chem 2004;76:298–302.
62. He FY, Liu AL, Xia XH. Anal Bioanal Chem 2004;379:1062–1067.
63. Wang J, Chen G, Chatrathi MP, Fujishima A, Tryk DA, Shin D. Anal Chem 2003;75:935–939.
64. Deng G, Collins GE. J Chromatogr A 2003;989:311–316.
65. Wang J, Pumera M, Chatrathi MP, Escarpa A, Konrad R, Griebel A, Dorner W, Lowe H. Electrophoresis 2002;23:596–601.
66. Tanyanyiwa J, Hauser PC. Anal Chem 2002;74:6378–6382.
67. Klett O, Nischang I, Nyholm L. Electrophoresis 2002;23:3678–3682.
68. Wang J, Ibanez A, Chatrathi MP, Escarpa A. Anal Chem 2001;73:5323–5327.
69. O'Neill AP, O'Brien P, Alderman J, Hoffman D, McEnery M, Murrihy J, Glennon JD. J Chromatogr A 2001;924:259–263.
70. Lacher NA, Garrison KE, Martin RS, Lunte SM. Electrophoresis 2001;22:2526–2536.
71. Hibara A, Tokeshi M, Uchiyama K, Hisamoto H, Kitamori T. Anal Sci 2001;17:89–93.
72. Tokeshi M, Minagawa T, Kitamori T. J Chromatogr A 2000;894:19–23.
73. McEnery M, Tan A, Glennon JD, Alderman J, Patterson J, O'Mathuna SC. Analyst 2000;125:25–27.
74. Wang SP, Lee WT. J Chromatogr A 2003;987:269–275.
75. Klampfl CW. J Chromatogr A 2004;1044:131–144.
76. Peri-Okonny UL, Kenndler E, Stubbs RJ, Guzman NA. Electrophoresis 2003;24: 139–150.
77. Holland LA, Leigh AM. Electrophoresis 2002;23:3649–3658.

78. Guan Y, Wu T, Ye J. J Chromatogr A 2005;821:229–234.
79. Chen G, Lin Y, Wang J. Talanta 2006;68:497–503.
80. Ferreira M, Varela H, Torresi RM, Tremiliosi-Filho G. Electrochim Acta 2006;52: 434–442.
81. García CD, Ortiz PI. Electroanalysis 2000;12:1074–1076.
82. de Carvalho RM, Kubota LT, Rath S. J Electroanal Chem 2003;548:19–26.
83. LaCourse WR, Modi SJ. Electroanalysis 2005;17:1141–1152.
84. García CD, Henry CS. Electroanalysis 2005;17:1125–1131.
85. Kuhnline CD, Gangel MG, Hulvey MK, Martin RS. Analyst 2006;131:202–207.
86. García CD, Henry C. In: Wang W, Soper SA, editors. BioMEMS: Technologies and Applications. Boca Raton, FL: Taylor and Francis Group; 2007. p 265–298.
87. García CD, Henry CS. Electroanalysis 2005;17:223–229.
88. García CD, Liu Y, Anderson P, Henry CS. Lab Chip 2003;3:331–335.
89. García CD, Henry CS. Anal Chem 2003;75:4778–4783.
90. García CD, Henry CS. Anal Chim Acta 2004;508:1–9.
91. García CD, Henry CS, Microchip Capillary Electrophoresis: Methods and Protocols Vol. 339. Totowa, NJ: Humana Press; 2006. p 27–36.
92. García CD, Engling G, Herckes P, Collettr JL, Henry CS. Environ Sci Technol 2005;39: 618–623.
93. Shen S, Chang Z, Liu H. Sep Purif Technol 2006;49:217–222.
94. Belder D, Deege A, Husmann H, Kohler F, Ludwig M. Electrophoresis 2001;22: 3813–3818.
95. Chiari M, Cretich M, Damin F, Ceriotti L, Consonni R. Electrophoresis 2000;21:909–916.
96. Figeys D, Aebersold R. J Chromatogr B Biomed Sci Appl 1997;695:163–168.
97. Horvath J, Dolnik V. Electrophoresis 2001;22:644–655.
98. Boyce MC. J Chromatogr A 1999;847:369–375.
99. Delgado-Zamarreno MM, Sanchez-Perez A, GonzalezMaza I, Hernandez-Mendez J. J Chromatogr A 2000;871:403–414.
100. Guan Y, Chu Q, Fu L, Ye J. J Chromatogr A 2005;1074:201–204.
101. García CD, Henry CS. Analyst 2004;129:579–584.
102. Tonin FG, Jager AV, Micke GA, Farah JP, Tavares MF. Electrophoresis 2005;26: 3387–3396.
103. Jager AV, Tonin FG, Tavares MF. J Sep Sci 2005;28:957–965.
104. Roman GT, Hlaus T, Bass KJ, Seelhammer TG, Culbertson CT. Anal Chem 2005;77: 1414–1422.
105. Roman GT, McDaniel K, Culbertson CT. Analyst 2006;131:194–201.
106. Lee JN, Park C, Whitesides GM. Anal Chem 2003;75:6544–6554.
107. García CD, Ortiz PI. Anal Sci 1999;15:461–465.
108. Schwarz MA, Galliker B, Fluri K, Kappes T, Hauser PC. Analyst 2001;126:147–151.
109. García CD, De Pauli CP, Ortiz PI. J Electroanal Chem 2001;510:115–119.
110. García G, García CD, Ortiz PI, De Pauli CP. J Electroanal Chem 2002;519:53–59.
111. LaCourse WR, Pulsed Electrochemical Detection in High-Performance Liquid Chromatography. New York: Wiley J. & Sons; 1997.

112. Wen X-L, Jia Y-H, Liu Z-L. Talanta 1999;50:1027–1033.
113. Arslan G, Yazici B, Erbil M. J Hazard Mater 2005;124:37–43.
114. Woods LA, Ewing AG. Chemphyschem 2003;4:207–211.
115. Wang Y, Chen H. J Chromatogr A 2005;1080:192–198.
116. Gawron AJ, Martin RS, Lunte SM. Electrophoresis 2001;22:242–248.
117. Wang J. Talanta 2002;56:223–231.
118. Kovarik ML, Li MW, Martin RS. Electrophoresis 2005;26:202–210.
119. Wang J. Biosens Bioelectron 2006;21:1887–1892.
120. Weng Q, Xu G, Yuan K, Tang P. J Chromatogr B 2006;835:55–61.
121. Zeng Y, Chen H, Pang DW, Wang ZL, Cheng JK. Anal Chem 2002;74:2441–2445.
122. Lunte SM, O'Shea TJ. Electrophoresis 1994;15:79–86.
123. Xu JJ, Bao N, Xia XH, Peng Y, Chen HY. Anal Chem 2004;76:6902–6907.
124. Liu Y, Vickers JA, Henry CS. Anal Chem 2004;76:1513–1517.
125. García CD, Ortiz PI. Talanta 2003;61:547–556.
126. Ji R, Chen Z, Corvini PFX, Kappler A, Brune A, Haider K, Schaffer A. Chemosphere 2005;60:1169–1181.
127. Deng Y, Fan X, Delgado A, Nolan C, Furton K, Zuo Y, Jones RD. J Chromatogr A 1998;817:145–152.
128. Komoβa D, Langebartels C, Sandermann H. In: Trapp S, Farlane JCM, editors. Plant Contamination. Boca Raton, FL: CRC Press;1995.
129. Zachritz WH, Lundie LL, Wang H. Ecol Eng 1996;7:105–116.
130. Aronoff DM, Oates JA, Boutaud O. Clin Pharmacol Ther 2006;79:9–19.
131. Dargan PI, Wallace CI, Jones AL. Emerg Med J 2002;19:206–209.
132. Tanaka E, Yamazaki K, Misawa S. J Clin Pharm Ther 2000;25:325–332.
133. James LP, Wells E, Beard RH, Farrar HC. J Pediatr 2002;140:522–526.
134. Oaks JL, Gilbert M, Virani MZ, Watson RT, Meteyer CU, Rideout BA, Shivaprasad HL, Ahmed S, Chaudhry MJ, Arshad M, Mahmood S, Ali A, Khan AA. Nature 2004;427:630–633.
135. Manzocco L, Calligaris S, Mastrocola D, Nicoli MC, Lerici CR. Trends Food Sci Technol 2000;11:340–346.
136. Lupo MP. Clin Dermatol 2001;19:467–473.
137. Fries E, Puttmann W. Water Res 2002;36:2319–2327.
138. Botterweck AAM, Verhagen H, Goldbohm RA, Kleinjans J, van den Brandt PA. Food Chem Toxicol 2000;38:599–605.
139. Bauer A, Dwyer-Nield L, Keil K, Koski K, Malkinson A. Exp Lung Res 2001;27: 197–216.

13

CHEMICAL SEPARATIONS IN 3D MICROFLUIDICS

George Manuel Maltezos, Axel Scherer, Alvaro Gomez and Frank A. Gomez

13.1 INTRODUCTION

Over the last several years, lithographic techniques have been successfully applied toward the miniaturization of fluidic elements, such as valves, pumps, and simple three-dimensional structures (1–9). The integration of many devices on a single fluidic chip has enabled the development of powerful and flexible analysis systems with applications ranging from cell sorting to protein synthesis (10–14). Through replication molding and embossing from photolithographically patterned dies, inexpensive fluidic systems with pneumatic actuation have been developed, by several groups (15–18). Hermetically sealed valves, pumps, and flow channels can be formed in poly(dimethylsiloxane) (PDMS) and related compounds (RTV, etc.), and in multilayer soft lithography, two or more replication-molded layers are aligned and subsequently bonded to create systems of pneumatic actuation channels controlling flow within a layer of flow channels (8). As the geometry of the pneumatic valve determines the actuation pressure, it is possible to define pneumatic multiplexing geometries that permit the control of many valves on a microfluidic chip by a much smaller number of control valves off-chip (19). Unfortunately, the two-dimensional nature of the flow channel arrangement limits the interconnection of this kind of two-dimensional fluidic system. Moreover, multilayer soft lithography requires the use of elastomeric materials that can bond well to each other to avoid delamination of the pneumatic film layer from the fluid flow layer. Our approach in this report was to

Biological Applications of Microfluidics, Edited by Frank A. Gomez

create a truly three-dimensional mold that could be embedded in a polymer, and the melted out in order to eliminate the requirement of traditional microfluidics that the chip be multilayer and bonded to a substrate in order to form a complete channel with valves. We also sought to eliminate the two-dimensional restrictions on fluid and control line routing and alignment by creating three-dimensional vias to enable crossed paths. Finally, we sought to make a completely embedded chip with control lines and fluid lines made only of polymer.

13.2 FABRICATION

To avoid the limitations of two-dimensional layered systems and to enable more flexible microfluidic plumbing topologies, we developed a three-dimensional replication molding technique that permits the construction of valves and pumps that are interconnected in all dimensions. To create appropriate three-dimensional replication dies, a commercial wax printing system (Solidscape T66) is used to define features as small as 12.5 μm high by 115 μm wide (Fig. 13.1) when printing directly on a glass or silicon substrate. The apparent surface roughness of the wax mold has little influence on valve performance other than to slightly increasing the closing pressure required for a leak tight valve. The combination of printed wax droplets with precise milling heads and stage positioning enables wax molds to be constructed with feature sizes comparable to those made by photolithography. The wax mold can be computer designed and printed directly onto a flat substrate without the need for any photolithography masks. The designer can fabricate three-dimensional microfluidic components interconnected with great flexibility. Three-dimensional connections between fluidic layers offer very flexible design opportunities that are inaccessible with planar techniques. Printing is carried out with two wax sources—one that forms the structures that are "negatives" of the desired microchannels and the other that

FIGURE 13.1 SEM picture showing a 115 μm wide wax mold for a fluid line printed on a glass slide obtained using three-dimensional replication molding from max molds.

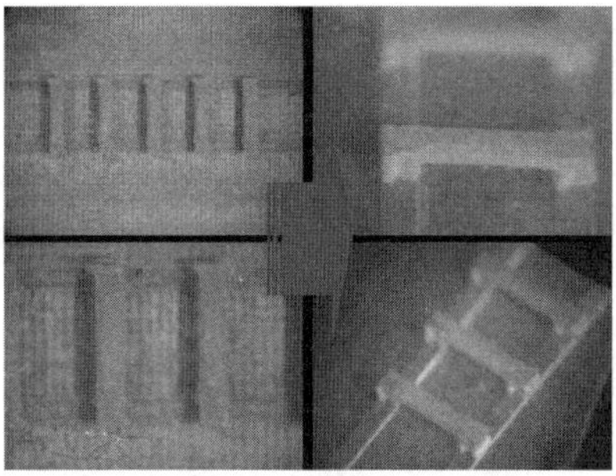

(1) Wax mold is designed and printed (2) Elastomer is filled into mold (3) Wax mold is melted away at 150°C

FIGURE 13.2 Wax printed via holes connecting two flow channels before and after removal of the wax. The basic fabrication steps are described: (1) First the mold is designed and printed, the mold is then placed in solvent (mineral oil) in order to dissolve the sacrificial wax and then dried. (2) PDMS is poured over the wax mold and allowed to cure at 60°C. (3) The wax is then melted out of the PDMS at 150°C, leaving a three-dimensional microfluidic chip.

serves as a sacrificial material and temporarily supports suspended structures during fabrication. After dissolution of the sacrificial wax material, PDMS or another prepolymer can be poured onto the mold and cured. Finally, the sacrificial wax mold is melted away to provide the completed microfluidic device geometry (Figs.13.2 and 13.3). This approach was used to construct fluidic conduits that require structural supports only every few centimeters, as well as robust, tunable, three-dimensional valves (Figs. 13.3 and 13.4) that can control flow pressures of over 220 kPa (33 psi). The 3-D replication-molded microfluidic design is also very resistant to swelling caused by aggressive solvents.

Three-dimensional printing eliminates the need for bonding the pneumatic control layer to the flow layer as both are formed in the same monolithic mold. This enables the use of elastomers that can be bonded only once or do not satisfy the adhesion requirements of multilayer fabrication such as the highly solvent-resistant perfluoropolyether (PFPE). The elimination of multiple bonding steps also avoids the need for aligning multiple elastomeric layers and compensation for polymer shrinkage. Additionally, components can be embedded into the device in a three-dimensional fashion and pin input holes can be formed as part of the mold in situations where punching would crack brittle polymer layers. Solvent-resistant microfluidic components enable the use of organic solvents incompatible with PDMS, thus opening up a vast array of potential microfluidics applications in organic chemistry and combinatorial synthesis.

13.3 RESULTS AND DISCUSSION ON 3D VALVES

We have designed a truly three-dimensional normally open valve geometry (Fig. 13.3). A suspended fluid flow tube is surrounded by a doughnut-shaped

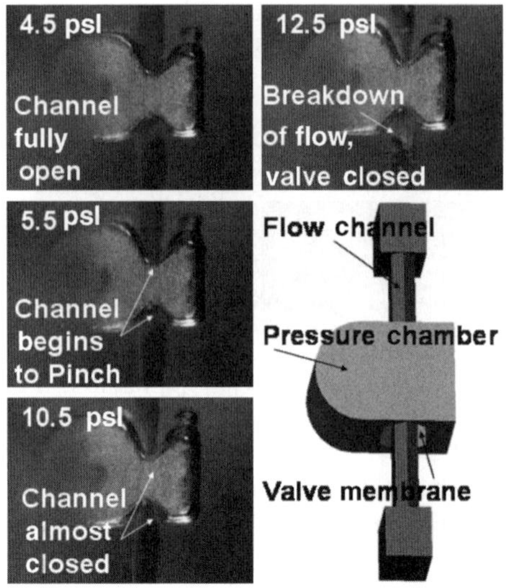

FIGURE 13.3 Three-dimensional valve constructed by wax printing and replication. The actuation pressure is only 62 kPa (9 psi) larger than the channel pressure. The disruption of flow caused by closing the valve is visible as air bubbles in this picture. (See color insert)

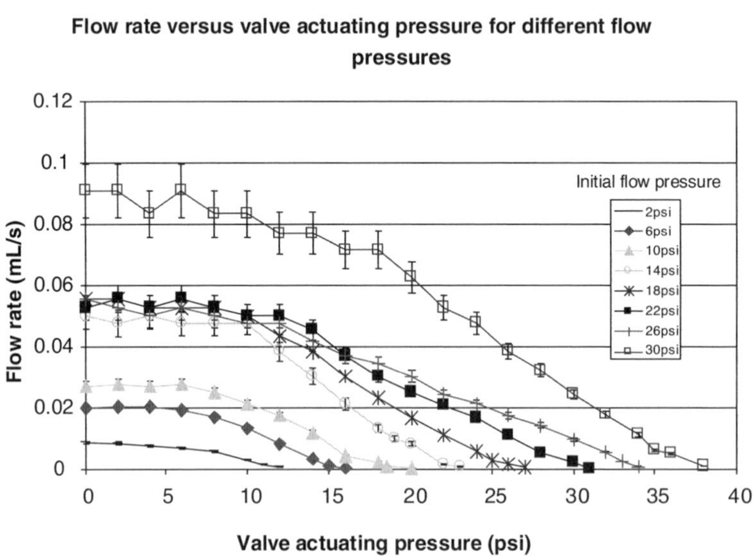

FIGURE 13.4 The valve enters a tunable region in which the flow pressure is strongly affected by the actuating pressure. Toward the right of the graph, the cutoff region with leak-tight flow of less than 0.1 mL over 1 h of testing is observed.

pressure chamber. The valve is actuated by increasing the pressure in the doughnut chamber surrounding the fluid flow tube by a predictable amount dependent on the precise valve geometry. Figure 13.3 has a schematic diagram of the wax mold, onto which PDMS is poured and cured, and the wax is then melted out. The closing pressure in the "pressure chamber" communicates with the "flow channel" via a thin "valve membrane" formed by a gap in the wax mold. Typical flow curves of a three-dimensional pneumatic valve constructed in PDMS can be observed. In Fig. 13.4, the flow rate is shown as a function of the actuating pressure of the pneumatic ring for various flow pressures applied to the fluidic channel. From this data, it is evident that the 3D valve is able to perform even at relatively very high pressures of ~250 kPa (35 psi). At all tested pressures, the valve can be closed by applying a pneumatic pressure 62 kPa (9 psi) above the flow pressure applied to push fluids through the flow channel. The maximum pressure range as well as the control over the valve actuating pressure compares very favorably with traditional planar valves constructed through multilayer soft lithography. In comparison, our multilayer soft lithography layers delaminate at approximately 82 kPa (12 psi). Figure 13.4 also shows that the 3D valve can be predictably tuned over a large range of flow rates by controlling the actuating pressure and initial flow pressure. The graph depicts a family of curves that represent a variety of different initial flow pressures. In general, we observe three important regions: (a) toward the left part of the valve response plot, at low actuation pressures, we note a region within that the valve is unaffected by the actuating pressure. In this case, the flow pressure is significantly larger than the actuating pressure. As the actuating pressure becomes comparable to the flow pressure, (b) the valve enters a tunable region where the flow is linearly sensitive to actuation pressure. Finally, (c) the valve is pinched off when the difference between the actuating pressure and flow pressure reaches 62 kPa (9 psi). Flow rates were experimentally measured with a 10-mL graduated cylinder and a stopwatch. After the pneumatic valve actuator pressure was established, the fluid flow valve was opened and simultaneously a timer was used to measure flow rates. When 1.0 mL of fluid flowed through the valve and was accumulated in a collection reservoir, the time was measured and a flow rate was calculated. Measurements were conducted for several devices to confirm good reproducibility. Flow hysterisis was found to be negligible and did not influence the measurements as the valve was always open at the beginning of each experiment.

Our pneumatic 3D valve was also tested in solvents that are known to deteriorate PDMS channels. For example, the valve performance was evaluated when metering toluene, a material known to result in swelling of PDMS and deterioration and distortion of conventional PDMS fluidic systems. As our 3D valve definition procedure does not rely on multilayer PDMS films that could delaminate, no leakage or deterioration could be observed in our 3D valve after exposure to toluene. Although the tenability suffered due to swelling over time, the valve performance was not influenced. Figure 13.5 shows an optical micrograph of a three dimensional multiplexer mold consisting of an integrated bilevel array of 18 3D pneumatic valves. From this image, it is clear that large plumbing systems consisting of integrated arrays of microfluidic valves can be constructed by 3D microvalve definition. In such

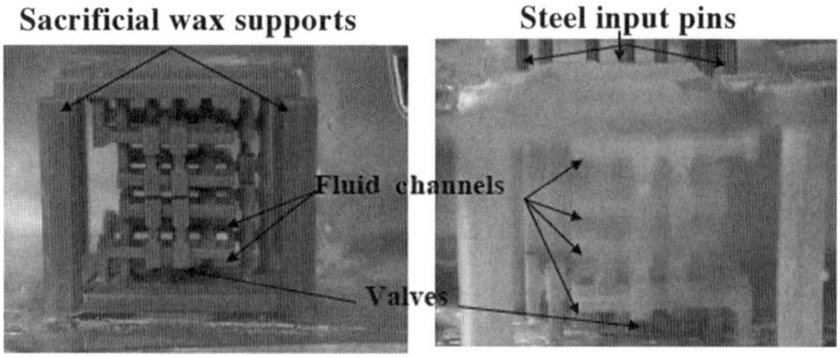

FIGURE 13.5 Micrograph of a wax mold before and after PDMS replication molding showing the geometry of the flow channels and the pneumatic actuation valves for a 36 valve, 16 to 1 fluidic multiplexer. The chip is made entirely of PDMS without the need for bonding to glass, and pressure inputs are made via steal pins on both sides (only the topside show).

a valve network, the density of fluidic elements can be significantly increased beyond what is available for more traditional 2D microfluidic networks constructed from PDMS. In such 3D fluidic chips, the smallest flow pressure line that can be defined by the lateral and vertical resolution of the wax printer is 115 μm wide by 12.5 μm high (although some difficult geometries require more material strength and must be made larger). These dimensions match well with geometries suitable for the definition of useful microfluidic "laboratory-on-a-chip" applications. It is well known that in geometries much below 10 μm the Reynold's number is very low and the differential fluid flow velocity in the center of the channel versus at the channel wall can be significant. These effects limit the usefulness of fluidic systems with dimensions much below the 100×10 channels and valves that we describe here.

13.4 MICROFLUIDIC THREE-DIMENSIONAL SEPARATION COLUMNS

Another application of three-dimensional fluidics is the definition of microfluidic separation columns. The separation of mixtures using column chromatography is a widely used method in analytical chemistry. In column chromatography, molecules travel through a gel (stationary phase) placed in a column and the mobile phase, a solvent in which the materials to be separated are miscible in, flows down the column. Different substances will flow at different speeds, allowing for the mixture to be separated. Here, a method known as flash chromatography is demonstrated in a microfluidic system. It differs from common gravity chromatography in that is uses pressure to flow molecules through the packed column. In this case, the three-dimensional "lost wax" approach for microfluidics described above provides a simple method for miniaturizing a chromatographic column.

First we created a 3D computer model (Fig. 13.6) and then fabricated it using our rapid prototype machine (Solidscape T66). The microcolumn has dimensions of

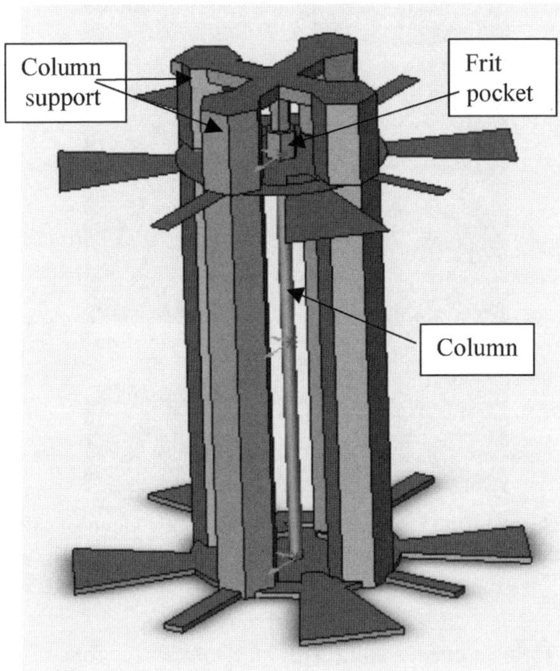

FIGURE 13.6 A computer model of the wax structure used to make the microfluidic chromatography columns.

600 μm in diameter by 15,000 μm in length. The ratio of these dimensions (1/25) is comparable to a very long benchtop separation column and used to ensure good separation. To make such a thin vertical column with good reliability, it was necessary to position larger supports around the column and to define a crossbar over the top of the entire structure to secure it during the lost wax processing steps. At the end of the process, this connection was cut, and just the column remains. The wax model also contains a "frit pocket" to hold a sodium silicate frit filter in place. The wax model was subsequently embedded into a PDMS elastomer again. Once the wax was melted away, a microcolumn was created inside the PDMS elastomer with the dimensions corresponding to the wax model. A glass frit-like material was embedded into the column to serve as a base for the silica that supports the stationary phase. It was very difficult to embed a traditional glass frit into the monolithic structure of the PDMS with the necessary leak-tight fit, and therefore an alternative method introducing sodium silicate was developed. Sodium silicate is liquid at room temperature, and this allows us to construct the frit by placing a wire of the appropriate diameter into the column, up to the point where a space has been made for the frit (the frit pocket in Fig. 13.6). The wire serves as a seal to hold the liquid sodium silicate in the frit pocket. Sodium silicate is then injected into the frit pocket and is then heated to 170°C for 10 min. Under these conditions, sodium silicate

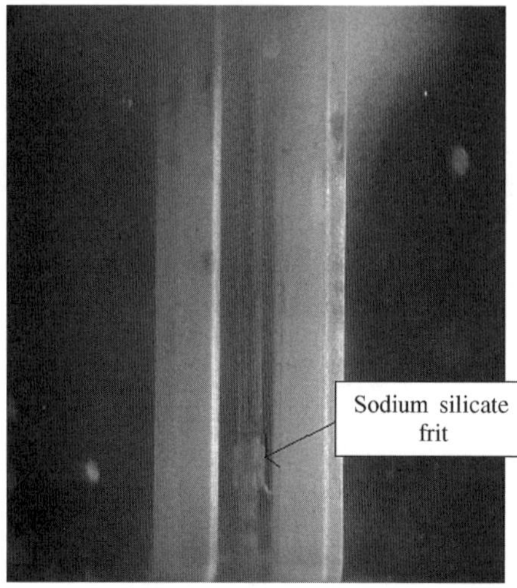

FIGURE 13.7 The completed column cast in PDMS. The sodium silicate frit is visible and the column is packed with silica.

crystallizes and forms a porous solid with ideal properties for use as a column frit. Silica is then carefully poured into the column, either in dry form or by means of a liquid slurry (Fig. 13.7).

13.5 RESULTS ON LIQUID CHROMATOGRAPHY

Our initial experiment performed on the chromatography column consisted of separating two food dyes. They were mixed and introduced into the column. Two pounds per square inch of pressure was then connected via a 23 gage steel pin and used to move the dyes through the column. A typically separation took less than 20 min from start to finish.

A 0.2-μL plug of a mixture of dyes was introduced to the top of the tube via pipette. The dyes were observed to separate in the column and were collected at the bottom. The pattern was similar to thin layer chromatography done on the same dyes. We have not yet found ideal conditions and solvents optimized for separating these dyes, and therefore our initial separations have been slightly "smeared." Nevertheless, the columns do show promise. Collection is also a concern due to the fact that the column lacks a mouth from where the purified material can drip easily from. This was solved by using microslides (concave slides commonly used in microbiology) attached to the bottom of the column allowing it to drip and be collected for further TLC plate studies.

Future research will see better separation using a larger number of solvents. These columns can be made in fluorinated polymers such as SIFEL and PFPE and should allow the majority of organic solvents to be used. Other types of columns can also be

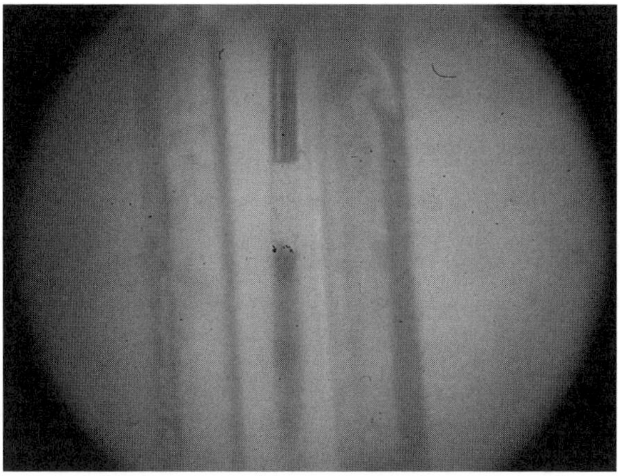

FIGURE 13.8 A microcolumn used for separating two dyes. (See color insert)

made by the technique described here, and these include elution and affinity columns. Moreover, a multilayer stack of two- and three-dimensional microfluidics can be aligned to the columns in order to introduce fluids into the columns and extract these afterward. Finally, it is quite easy to make large arrays of these columns, and the possibilities for combinatorial chemistry and other applications in which multiple reactions need to be carried out and purified are apparent (Fig. 13.8).

13.6 CONCLUSIONS

Three-dimensional soft lithography offers many advantages over the more conventional multilayer soft lithography, which is based on 2D valve and pump definition. One key advantage of developing devices from 3D replication molding is that it enables the use of a wide variety of elastomers and plastics that are more corrosion resistant than PDMS. Moreover, the pressure in the flow channels can be increased and the actuation pressure of the pneumatic lines can be decreased by implementing designs that do not involve layers that may delaminate and can close the valve by applying pneumatic pressure from all sides. The most important opportunity obtained from 3D definition is the increase in interconnectivity of the fluidic components and improvement in the flow channel integration in all three dimensions through the use of via holes that can jump over a fluidic layer with a commercially available wax molding system. Our first results on this new kind of microfluidics indicate that denser integration with larger numbers of components and more complex fluidic multiplexing systems can be implemented through 3D replication molding. In addition to creating valves and pumps, it is also possible to miniaturize chromatography columns and to easily introduce filtration systems into fluidic systems.

REFERENCES

1. Brody JP, Yager P. Hilton Head 1996;8:105–108.
2. Quake SR, Scherer A. Science 2000;290:1536–1540.
3. Chou HP, Unger MA, Scherer A, Quake SR. Biomed Microdev 2001;3:323–330.
4. Liu J, Enzelberger M, Quake SR. Electrophoresis 2007. Forthcoming.
5. Jeon NL, Chiu DT, Wargo CJ, Wu HK, Choi IS, Anderson JR, Whitesides GM. Biomed Devi 2002;4:117–121.
6. Wu HK, Odom TW, Chiu DT, Whitesides GM. J Am Chem Soc 2003;125:554–559.
7. Unger MA, Chou HP, Thorsen T, Scherer A, Quake SR. Science 2000;288:113.
8. Ikuta K, Maruo S, Fukaya Y, Fujisawa T, MEMS 2000;23–27:131–136.
9. Ikuta K, Hasegawa T, Adachi T, Maruo S, MEMS 1998;25–29:739–744.
10. Morgan H, Green NG, Hughes MP, Monaghan W, Tan TC. J Micromech Microeng 1997;7:65–70.
11. Fiedler S, Shirley SG, Schnelle T, Fuhr G. Anal Chem 1998;70:1909–1915.
12. Kopp MU, deMello AJ, Manz A. Science 1998;280:1046–1048.
13. Waters LC, Jacobson SC, Kroutchinina N, Khandurina J, Foote RS, Ramsey JM. Anal Chem 1998;70:158–162.
14. Li PH, Harrison DJ. Anal Chem 1997;69:1564–1568.
15. Hadd AG, Raymond DE, Halliwell JW, Jacobson SC, Ramsey JM. Anal Chem 1997;69:3407–3412.
16. Hadd AG, Jacobson SC, Ramsey JM. Anal Chem 1999;71:5206–5212.
17. Chiem NH, Harrison DJ. Electrophoresis 1998;19:3040–3044.
18. Fu AY, Spence C, Scherer A, Arnold FH, Quake SR. Nat Biotechnol 2000;18:309–312.
19. Thorsen T, Maerkl SJ, Quake SR. Science 2002;298:580–584.

14

ENABLING FUNDAMENTAL RESEARCH IN PROTEOMICS

Naresh Menon

14.1 INTRODUCTION

Membrane proteins play a crucial role in many cellular and physiological processes. They are essential mediators of the transfer of material and information between cells and their environment, between compartments within cells, and between compartments comprising the organ systems. Functionally, normal membrane proteins are vital to health, and specific defects are associated with many known disease states. Membrane proteins are the targets of a large number of pharmacologically and toxicologically active substances and are responsible, in part, for their uptake, metabolism, and clearance. In contrast to the remarkable advances that have been made in the understanding of soluble proteins, understanding of the structure, function, and mechanisms of integral membrane proteins has been limited by the paucity of available high resolution structures.

While this compelling problem in modern biology persists, over the past decade, microfluidic systems have been shown to have a great potential in a diverse array of biological applications including biomolecular separations (5–8), enzymatic assays (6–12), immuonhybridization reactions (13–15), and polymerase chain reaction (16–18). The use of nanoliter reaction volumes and parallel sample processing are potential advantages of microfluidic devices, making them ideally suited to chemical analysis, high throughput screening applications, and other cases where reagents are limited. A major challenge associated with realizing the desired scales in microfluidic devices is to reduce simultaneously the number of pipetting steps needed

Biological Applications of Microfluidics, Edited by Frank A. Gomez

to load the devices while amortizing the sample volume over several reaction or separation steps.

In this chapter, we discuss the technical challenges associated with extracting membrane proteins and how the potential of microfluidic LOC devices to revolutionize proteomics by extracting membrane proteins in a solubilized, monodispersed form that maintains their phospholipid–protein native structure and functionality by a rapid, reliable, and efficient method. The method described below is applicable to a wide range of rapidly occurring catalytic reactions that, if precisely controlled, could dramatically increase the process yield.

14.2 MEMBRANE PROTEIN EXTRACTION

It has been proven extremely difficult to separate functional proteins into the physiologically effective three-dimensional conformations needed for understanding "molecular machines" of prokaryotic and eukaryotic cells and their possible modifications. This is especially true of membrane proteins. A major hurdle to understanding the structure and function of membrane proteins has been the unavailability of a simple and reproducible procedure for isolating them in a monodispersed form that maintains the phospholipid–protein native structure and functionality (1). As in other organisms, proteins in the cell membranes are critical to the organism's functioning but are impossible to isolate for study without fully or partially denaturizing them. Techniques are available to study the molecular structure and function of soluble proteins, but no technique is available to study membrane proteins.

The only new method of isolating membrane proteins, by means of nanodisks, has recently emerged (2). In this method, a cell membrane preparation containing the target membrane protein is totally solubilized with detergent in the presence of membrane scaffold protein (MSP). After detergent removal by dialysis or Bio-Beads, a soluble MSP-supported nanodisk is formed with the target incorporated into the resulting phospholipid bilayer (3). Unfortunately, in the nanodisk-based techniques, the amphiphilic detergent (such as 0.5 M cholic acid) that solubilizes the membrane proteins simultaneously separates the proteins from their native lipid bilayer, affecting their structural and functional features. In this chapter, we will discuss a new method performed in a lab-on-chip (LOC) format that enables the controlled solubilization of membrane proteins while preserving its supporting innate lipid bilayers. A microfludic system is at the heart of our new method as it allows us to control the solubilization process with high accuracy and prevent denaturation of the protein, permitting a more accurate and complete analysis of the protein functions by giving access to both extracellular and intracellular segments of the protein. This process makes it possible to research and develop biochemical methodologies previously restricted to soluble proteins. Furthermore, this generic method for using LOCs for controlling a catalysis reaction can be used for other similar processes where rapid reaction quenching is critical.

The fundamental reason why our method is applicable is that a transmembrane protein locally compresses the membrane bilayer, increasing the local surface tension energy and decreasing the stretching energy (4). We discuss this in greater detail in Section 14.2.2.1. The resulting thermodynamic condition encourages the penetration of amphiphilic detergent molecules between membrane lipid molecules in regions free of transmembrane proteins and discourages their penetration in the vicinity of these proteins. At the early stage of interaction, detergent molecules dissolve the cell lipid bilayer preferentially at a distance from membrane proteins while leaving the membrane intact in the vicinity of transmembrane proteins. The dissolution speed depends on the nature of the detergent and on the type of membrane, temperature, and detergent concentration. By using microfluidic technology that precisely controls molecular interactions between membrane structures and detergent—by controlling the time, temperature, and concentration of the detergent interacting with a membrane sample, as well as the speed and the method of detergent evacuation—MPS process carves out nanofragments of innate membrane supporting each membrane protein without disturbing the lipid bilayer. The system of microfluidic channels on the MPS chip ensures rapid mixing of membrane sample with detergent and MSP at 38°C followed by a steep cooling to 4°C, and then rapid removal of the detergent by absorbance beads placed into the flow.

The precise control of the reaction between membrane proteins (sample) and a mixture of detergent and MSP (reagent) to dissolve the lipid bilayers into soluble nanofragments of transmembrane proteins surrounded by the native lipid bilayers while retaining their functionality—all on an inexpensive lab-on-chip format—was a key novelty of our research. This system for dissolution of lipid bilayers was precisely timed by the flow rate, temperature controlling elements, and detergent quenching beads – all integrated into the lab-on-chip system for the first time. The methodology as investigated is illustrated in a functional diagram in Fig. 14.1.

14.2.1 Design and Fabrication of the Lab-on-Chip System

14.2.1.1 Design Consideration The following theoretical considerations of membrane protein–lipid bilayer interactions are at the foundation of the new method and illustrate the need for precise reaction control at the protein interaction length scale, only possible with micro/nanofluidics. The equilibrium bilayer of a cell membrane is locally flat and consists of two identical monolayers characterized by a thickness L_m, which is related to surface density Σ_0 (area per molecule) through an equation of state. In this analysis, we take this equation of state to be the condition of incompressibility of the hydrophobic layer so that $L_m\Sigma_0 = N\nu_0$, where ν_0 is a transmembrane protein segment volume and N is the number of segments. The energy per protein (per unit width) is then given by

$$F_d = -\gamma L_p + F_M, \tag{14.1}$$

where γ is the surface tension of the protein hydrophobic region, describing the energy difference (per unit area) between the proteins in solution and embedded in the bilayer. L_p is the protein height, and F_M is the protein-induced membrane

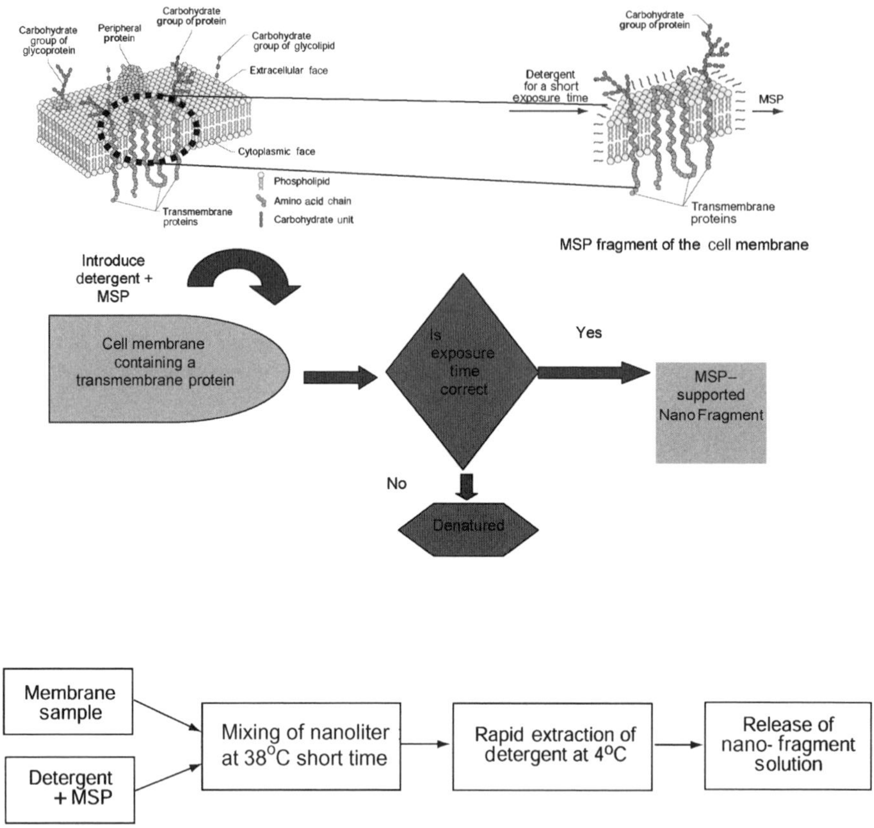

FIGURE 14.1 The functional steps in the demonstration of the microfluidic method for extracting transmembrane proteins from their native lipid bilayers. Each step in the catalysis-quencher reaction is controlled precisely by auxiliary microfluidic channels for temperature control and controlled introduction of catalysts and quenchers. Timing is controlled by adjusting flow rates. (See color insert)

perturbation energy, given by (19)

$$F_{\mathrm{M}} = \int_{0}^{\infty} \frac{L_{\mathrm{m}}}{v}\left(B\Delta^2 + \left(\kappa - \kappa'\sum\nolimits_{0}\right)L_{\mathrm{m}}\frac{d^2\Delta}{dz^2} + K\left[\frac{d^2\Delta}{dz^2}\right]^2\right)dz \qquad (14.2)$$

This energy is a function of the local deformation of the membrane, which is characterized by a dimensionless change in its thickness, relative to the equilibrium thickness. The deformation is defined by $\Delta(z) = (L_{\mathrm{m}}\text{-}\ L(z)) \approx L_{\mathrm{m}}$; where $L(z)$ is the local thickness of the perturbed monolayer and z is the distance from the protein boundary. Δ_0 defines the perturbation at the protein boundary, equal to the normalized difference between the protein and the bilayer thickness. The volume of the hydrophobic block v is given by Nv_0. B is the monolayer compressibility, the

energetic cost associated with perturbation of the area per chain from Σ_0 (each monolayer is assumed to bend with the same sign of curvature in contrast to the usual bending modes of a bilayer, in which the two monolayers have nearly equal and opposite curvatures). The monolayer spontaneous curvature κ determines the sign and the magnitude of the free interface curvature of the monolayer at a water–oil interface. The change in spontaneous curvature as a function of density is given by $\kappa' = \partial\kappa/\partial\Sigma$ evaluated at the equilibrium bilayer surface density Σ_0. The bending modulus K is the energetic penalty for bending the monolayer. All energies are given in units of kT, where k is the Boltzmann coefficient and T the temperature, and all length scales are dimensionless, normalized by L_m.

The bilayer perturbation energy defines the membrane resistance to protein incorporation; however, transmembrane proteins are driven into the bilayer by the unfavorable interactions between their hydrophobic regions and the aqueous solution. Bilayer compression by protein incorporation results in an increase in the local surface tension energy and a decrease in the stretching energy (4). This is why incorporation of amphiphilic detergent molecules will be attracted to areas free of transmembrane proteins more than regions in the vicinity of these proteins, to the membrane that is away from proteins will be broken down sooner than the membrane adjacent to a protein. Therefore, by applying the detergent just long enough to dissolve the nonadjacent membrane and then removing it suddenly before it can dissolve the adjacent membrane, we can free transmembrane proteins without causing them to denature. Depending on the membrane protein, the dissolution time may be as small as a few milliseconds to as large as tens of seconds. No reliable and high efficiency laboratory method exists for extracting transmembrane proteins where the reaction times are so fast.

14.2.1.2 Microfabrication of Microfluidic Devices Pneumatically actuated microfluidic devices are fabricated at both the Caltech Nanofabrication lab and in the labs at CSULA from two patterned flexible polymer layers—the "flow layer" (containing the channels that carry reagents) and the "control layer" (containing the channels that cause valve closure when pressurized). Each layer is fabricated from two-part-cure silicone (Dow Corning Sylgard 184) cast on a photoresist mold. The master molds are made by spin-coating photoresist onto a silicon wafer and then patterned with high resolution transparency masks (typically 3386, 8000, or 20,000 dpi). The fluid layer mold uses a photoresist that can be reflowed by baking at high temperatures (e.g., Shipley SJR 5740). Reflowing creates a rounded channel crosssection that allows complete closure during valve actuation. The mold is spin coated with 20:1 part A:B Sylgard at 2500 rpm for 1 min, resulting in a layer thickness of around 30 μm. The channels in the control layer do not require a specific geometry, so any photoresist can be used. Onto this mold, about 0.5 cm 5:1 part A:B Sylgard is cast to form the control layer. Each layer is cured for 30 min in an oven (80°C). Control channel interconnect holes are punched through the thick layer (after release from the mold), and the layer are aligned and sealed (channel-side down) onto the flow layer (still on the mold). Bonding is realized by curing the devices together for an additional 45–60 min (80°C). The resulting two-layer devices are cut from the mold and mounted onto RCA-cleaned glass microscope slides and bonded overnight at 80°C. Layers are exposed to plasma oxidation rendering their surfaces hydrophilic due to the presence of silanol groups. This greatly reduces adsorption of species in the channels allowing wetting of the surfaces and lessening nucleation of bubbles.

14.2.1.3 Fluid Mixing Microfluidic systems for sample mixing have been addressed by numerous researchers over the years. Circular reagent mixers that was prototyped by others included pneumatic valves that were sequentially activated to mix fluids within the ring. These devices could hold a maximum of 11 nL of solution based on channel dimensions of 100 μm width, 25 μm height, and 4300 μm circumference. The size of the mixer was arbitrary and other diameters of mixers can be varied depending on the amount of volume of solution required for the experiments. Preliminary work focusing on determining the cycle time of the chips was performed and showed that mixing of solution takes a few seconds and is reproducible. This was determined to be *too slow* for our reaction process. As a result, we decided to pursue a laminar flow-based mixing system. In a laminar flow, mixing time is determined by the diffusion time between sample and the reagent flows. Given that the sample and the reagents react as long as they are in contact with each other, by controlling the distance (channel length) they travel, the channel cross-sectional area that determines the volume of interaction and the rate at which they travel (flow rate), we can accurately control the diffusion-limited reaction time as well as the amount of material that undergoes a reaction. Figure 14.2 shows the functional architecture of the lab-on-chip device used during this research program. It consists of a mixing chamber where the membrane proteins (sample) are brought in contact with a mixture of detergent and MSP (reagent). The fluids travel from left to right into a trough carved into the PDMS chip that is filled with 600 μm quenching beads. Separate microchannels circulate warm and chilled fluids to

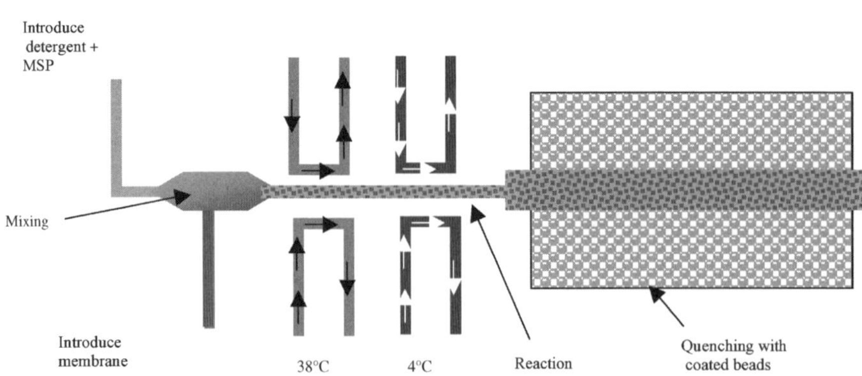

FIGURE 14.2 Schematic design of the microfluidic system for membrane protein extraction. (See color insert)

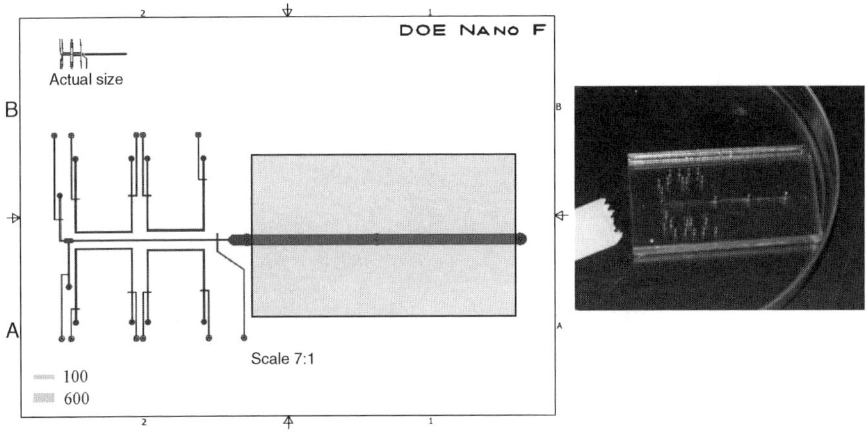

FIGURE 14.3 MPS Microfluidic layout (AutoCad) (**a**) and a fabricated chip (**b**). On the basis of our architecture, the chip allows for controlled reaction of 5 nL of sample + reagent within the chip. The channels in red are for pneumatic control of the valves and in blue are the microfluidic channels. The orange region is indicated as the section that is carved out of PDMS to include the microbeads.

produce a steep temperature gradient that is designed to allow for sample and reagent to react over a short (few milimeter) microfluidic path length. Furthermore, the fluids collected in the quenching region are also chilled to stop the reaction.

Figure 14.3a shows the AutoCAD layout drawing for the MPS lab-on-chip prototype and 14.3b is a photograph of a fabricated chip. In Fig. 14.3a, the fluidic channels are all in blue while the valve control channels are in red. All channels have an aspect ratio of 1:10 height to width—a parameter dictated by the elastomer material composition and has been empirically shown to prevent the channels from collapsing. The fluidic channels come in two widths—100 μm and 600 μm. The wider channels are used to introduce and trap the microbeads used for quenching the detergents. The reaction region for the sample and the reagent is ~10 mm, resulting in a reaction volume of ~10 nL. Figure 14.4 shows the full three-dimensional drawing of the entire reaction chip.

Figure 14.5 shows two photographs with colored dye to study the variations in mixing caused by varying the flow rates relative to each other. Notice that the laminar flow is maintained while maintaining the relative volumes in the flow.

The design was implemented in multiple PDMS chips and extensive testing was conducted to validate our hypothesis.

14.2.1.4 Conventional Testing Protocol We conducted conventional nanodisk protocol to solubilize the plasma membrane of rat liver cells to measure the activity of the transmembrane enzyme alkaline phosphatase. As a control, we used un-solubilized rat liver membranes. Also, we carried modified nanodisk protocol to solubilize membrane of the *Shewanella oneidensis* that was verified by measuring

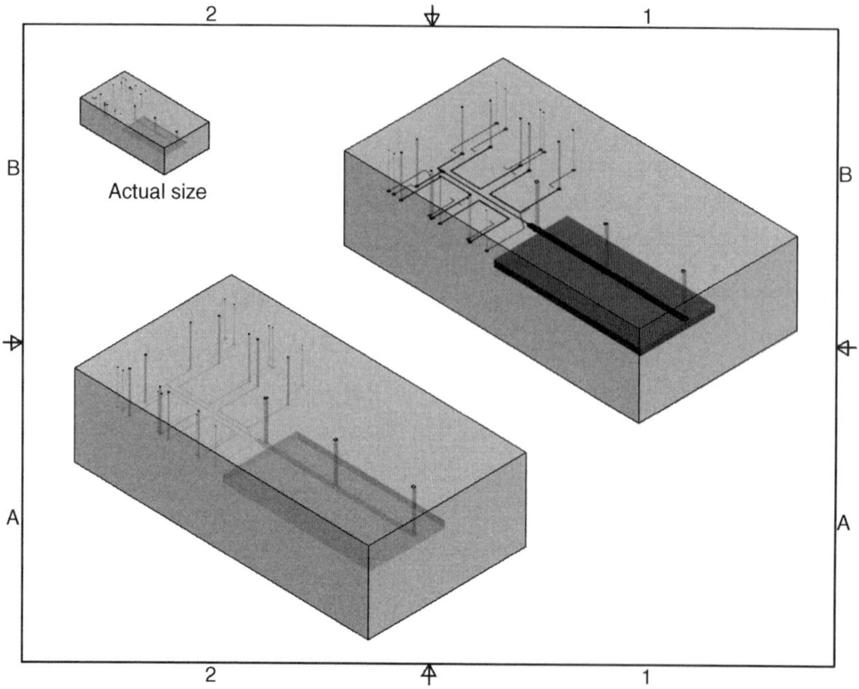

FIGURE 14.4 Three-dimensional image of the MPS microfluidic lab-on-chip system. In purple is the region where the quenching microbeads are placed. In blue are the microfludic channels and in red are the pneumatic valve control channels. Vertically drilled holes are shown where the fluids are accessed.

the increase of the transmembrane cytochromes (heme)-related absorbance at 410 nm and evaluating the heme-linked peroxidase activity of c-type cytochromes.

Rat membranes were custom isolated by The Burnham Institute (La Jolla, CA) according to the protocol developed by Meier et al. in 1984 (10). Rats are killed and their livers are rapidly removed and chilled on ice. Ten-gram portions of liver are cut into small pieces, washed three times in 80 mL cold 1 mM $NaHCO_3$ (pH 7.4), and homogenized in the same volume of 1 mM $NaHCO_3$ with a loose (Type A) dounce homognizer (seven up-and-down strokes). The homogenate is further diluted to 1800–2000 mL with cold $NaHCO_3$ and filtered twice through two layers of 60-grade cheesecloth. Centrifugation at $1500 \times g$ for 15 min (GSA rotor, E.I. du Pont de Nemours & Co., Inc., Sorvall Instruments Div., Newtown, CT) gives a "crude nuclear pellet," which is resuspended in 5.5 pellet volumes of 56% sucrose (wt/wt; density = 1.2623) and stirred for 15 min to disrupt membrane aggregates. The sample is then dynamically loaded onto a 100-mL cushion of 56% (wt/wt) sucrose with a variable speed Sorvall pump into the zonal rotor TZ-28 (Sorvall Instruments) and overlaid by 400 mL of 44% sucrose (wt/wt; density = 1.1972) and 200 ml of 36.5% sucrose (wt/wt; density = 1.1587), respectively. Finally, the rotor

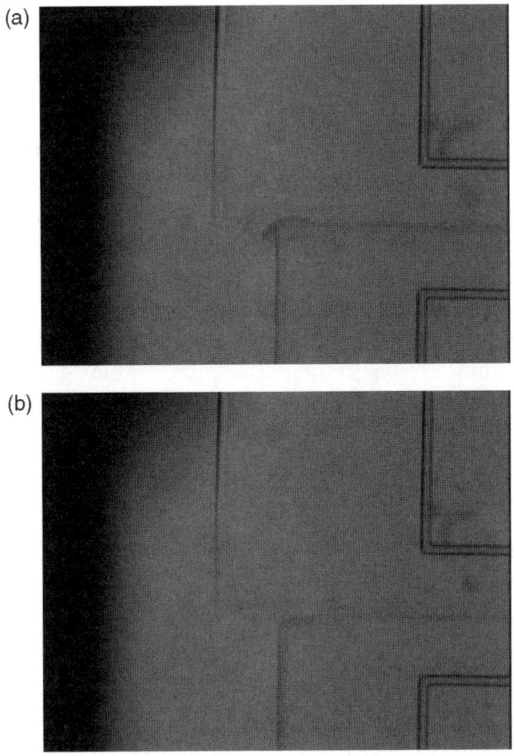

FIGURE 14.5 Studies analyzing the impact of microfluidic flow rates in laminar flow mixing where the relative flow rates are varied. In both figures, the fluid flow is from left to right. In (**a**) the flow rate of the colored dye is greater (2X) than in (**b**). Note that the reaction chamber and the subsequent mixed flow region retain a proportional ratio of colored versus noncolored fluids—indicating that the laminar flow is maintained.

is filled to its total volume capacity (1350 mL) with 0.25 M sucrose. During the entire loading procedure, the rotor is running at 3000 rpm. The completed discontinuous sucrose gradient system is centrifuged at 20,000 rpm for 120 min. After slow deceleration to a complete stop, seventy 15-mL fractions are collected from the bottom of the rotor and analyzed for density, turbidity (absorbance at 700 nm), and marker enzyme (alkaline phosphatase) activities. Fractions containing the bulk of plasma membrane fragments (e.g., fraction 54–64 out of 70) are combined and diluted with $NaHCO_3$ up to 1000 mL. The suspension is centrifuged at $7500 \times g$ for 15 min, the pellet is gently resuspended (by vortex mixing) in 250 mL $NaHCO_3$, and the material representing the liver plasma membranes is resedimented at $2700 \times g$ for 15 min. We received 5 g of the concentrated rat liver membrane sample in $NaHCO_3$ solution.

The Membrane Samples of the S. oneidensis were obtained from the Molecular Biosciences Department of Pacific Northwest National Laboratory. The membranes of *S. oneidensis* were prepared according to the method of Giometti et al. (2003) and separated from the soluble components by high speed centrifugation.

Cell Culture *S. oneidensis* strain MR-1 was grown in Luria-Bertani (LB) medium at 25°C. For aerobic growth (high oxygen), 20 mL of LB broth inoculated with 1 mL of overnight culture was placed in a 250-mL flask and incubated on a rotary shaker at a speed of 200 rpm. The cells were harvested at the exponential phase (OD600 = 1). For low oxygen growth, the speed of the rotary shaker was reduced to minimize oxygen uptake.

Membrane Sample Preparation Frozen cells were thawed and immediately mixed with two volumes of lysis buffer containing 50 mM glucose, 10 mM EDTA, 25 mM Tris–HCl, pH 8.0, and 4 mg/mL lysozyme in the presence of protease inhibitors (complete mini protease inhibitor cocktail; Roche Diagnostics, Indianapolis, IN, USA). Lysozyme was added to the lysis buffer immediately before use. Cells were first sonicated and then shaken at 4°C for 1–2 h. The resulting lysate was centrifuged for 40 min at 12,000 × *g* in a refrigerated Eppendorf centrifuge to remove unbroken cells. The supernatant was collected and centrifuged at 160,000 × *g* for 30 min in a Beckman (Beckman-Coulter, Fullerton, CA, USA) TL100 tabletop ultracentrifuge, followed by an additional 10 min at 435,000 × *g*. The final supernatant was collected as the soluble cytosolic fraction, and the pellet was suspended in lysis buffer containing 0.5% w/v CHAPS (Bio-Rad, Hercules, CA, USA). The resuspended pellet was transferred into a chilled Dounce homogenizer and evenly dispersed with 10 slow strokes of the pestle. The dispersed pellet was collected as the crude membrane preparation. The crude membrane fractions were stored at −80°C until further analysis. *S. oneidensis* MR1 strain samples (~200 milligrams) were received in a concentrated sample from Pacific Northwest National Laboratory.

Preparation of the Nano Disk Samples RAT LIVER MEMBRANE SAMPLE. In the first step, the concentrated rat liver membrane sample was diluted 100 times by 0.05 mM Tris–HCl buffer, pH 8.0. After this, in accordance with the procedure for self-assembly of nanodiscs (11), the rat liver membrane was detergent solubilized with 0.5 M cholic acid and mixed with human apoprotein A1 in the approximate ratio of 1:100:200 for Apo/membrane/detergent. The reconstitution samples, which include ~100 nmol scaffold protein, 10 μmol membrane, and 20 μmol cholate were preincubated for 1.5 h at 4°C. The detergent was removed by incubating with Bio-Beads® SM-2 adsorbent (~0.4 g Bio-Beads/1 mL reconstitution mixture) for 1.5 h at 4°C.

S. ONEIDANSIS MEMBRANE SAMPLE. Prior to carrying out nanodisk protocol, the concentrated *S. oneidansis* membrane sample was diluted 100 times by 0.05 mM Tris–HCl buffer, pH 8.0. Subsequently, we mixed the diluted sample with 0.5 M cholic acid and human apoprotein A1 in the approximate ratio of 1:100:200 for Apo/membrane/detergent and incubated at 38°C for 1, 2, 3, 4, 5, and 6 h. The detergent was removed by incubating with SM-2 adsorbent beads for 2 h at room temperature. All samples were filtered through 30 nm pore filter and the concentration/absorbance of the solubilized cytochromes was measured in ELISA reader at 410 nm. The concentration of the released transmembrane cytochromes gradually increased with the increase of the incubation time of the sample reaching its plateau at 5 h of the incubation.

14.2.1.5 Measurement Protocol

Preparation of the Membrane Fragments

RAT LIVER MEMBRANE SAMPLE. Rat liver membrane sample mixed with ApoA1 and the 0.5 M cholic acid (in the approximate ratio of 1:100:200 for Apo/sample/detergent) and preheated to 38°C were injected into separate ports of the microfluidic LOC chip into the mixing chamber. Diffusion limited laminar flow mixing controlled by fluid flow rates and reaction temperature were allowed to occur within the microfluidic chip. Following the mixing chamber, the mixture was introduced to the macrofluidic compartment filled with SM-2 adsorbent beads (Bio-Rad Laboratories, Hercules, CA, USA) (~0.4 g Bio-Beads/1 mL) and cooled to 4°C into outflow port. The reaction was performed at different incrementally increasing flow rate: 1, 5, 10, 20, and 50 μL/min. See Fig. 14.6 to compare flow rate time and actual reaction time. A desirable temperature gradient could not be achieved with temperature controlling channels embedded within the LOC chip. We substituted the temperature gradient required to start–stop the reaction by preheating the samplereagent to 38°C while keeping the chip chilled at 4°C. The processed samples were collected in centrifuge tubes at the outflow port and analyzed.

S. ONEIDANSIS MEMBRANE SAMPLE. Prior to injecting into the LOC chip, the *S. oneidansis* membrane sample was preincubated in 0.5 M sodium cholate at 38°C

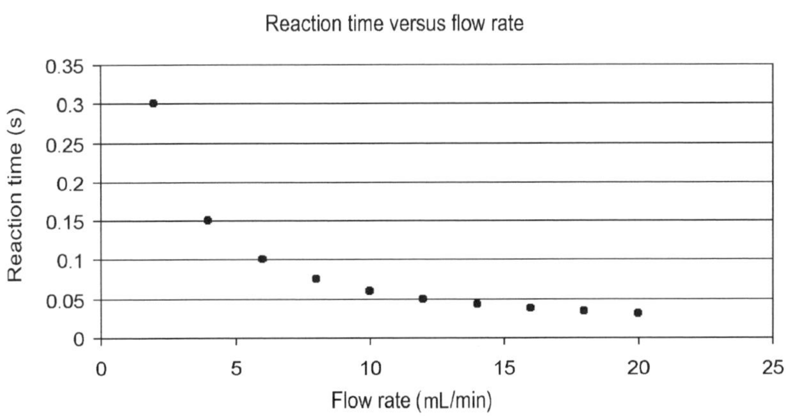

FIGURE 14.6 Estimating the reaction region to have ~10 nL of material, based on the fluid flow rate, we compute the approximate reaction time in seconds. The theoretical estimate for optimal time for extracting rat liver membrane is ~0.1 s.

for 2 h. After that, the treated sample was mixed with Apo A1 and again preheated to 38°C as well as 1.5 M sodium cholate. Both of them were injected into the mixing chamber of the microfluidic chip followed by the macrofluidic compartment. We carried our MPS process at low flow rates: 0.1, 0.5, 0.8, 1.0, and 2.0 μL/min. Again, the processed samples were collected in centrifuge tubes at the outflow port and analyzed.

14.2.1.6 Analysis and Results *Analytical Procedures* Prepared eukaryotic and microbial fragments, and assembled nanodiscs were size separated by filtration through 30 nm polycarbonate filter. Also, we filtered unprocessed membrane samples at its working dilution.

Rat Liver Membrane Sample We measured activity of the alkaline phosphatase in the nanofragments solubilized at differed flow rate and in the assembled nanodiscs, as well as in the filtered and unfiltered membrane sample. POC conducted a comparative assay on an equal molar (per protein) quantity of nanofragments versus nanodiscs, unprocessed filtered and unfiltered sample was conducted. To ensure equal molar quantity of the solubilized membrane proteins, in particular alkaline phosphatase, in all processed and unprocessed samples, we started with the same original diluted working sample in all experiments and adjusted/balanced dilution in the final samples.

Alkaline phosphatase activity was measured in ELISA microplates. 150 μL of each of the eight samples–nanofragments prepared at 1, 5, 10, 20, and 50 μL/min flow rate, nanodisc, unprocessed filtered, and unprocessed unfiltered sample–were added to the six wells of the microplate. The reaction time associated with the flow rates is estimated by Fig. 14.6. After that, 150 μL of the reaction mixture/incubation medium consisting of 0.5% *p*-nitrophenylphosphate and 1 mM $MgCl_2$ in 0.05 mM Tris–HCl buffer pH 8.0 was added to the triplicates (of six wells) of each experimental well.

The left triplicates of each sample served as controls for the enzymatic reaction: only buffer was added to them. Additionally, twice-diluted incubation medium without samples was used as a control of the reaction mixture. The microplates enclosed in a plastic bag with a zip lock to prevent water evaporation were incubated at 37°C. The enzymatic reaction was stopped in 15 min by adding 100 μL of 1 M Potassium phosphate solution. The incubation time of 15 min fell in the middle of the linear part of the calibration curve for this sample. The optical density of the formed p-nitrophenol in all wells of the microplate was measured in ELISA reader at 410 nm (Fig. 14.7).

In summary, we reach the following conclusions:

(1) The methodology works as we had expected. By altering the flow rates, we are able to vary the reaction time (see the correlation between flow rate and reaction time in Fig. 14.6).

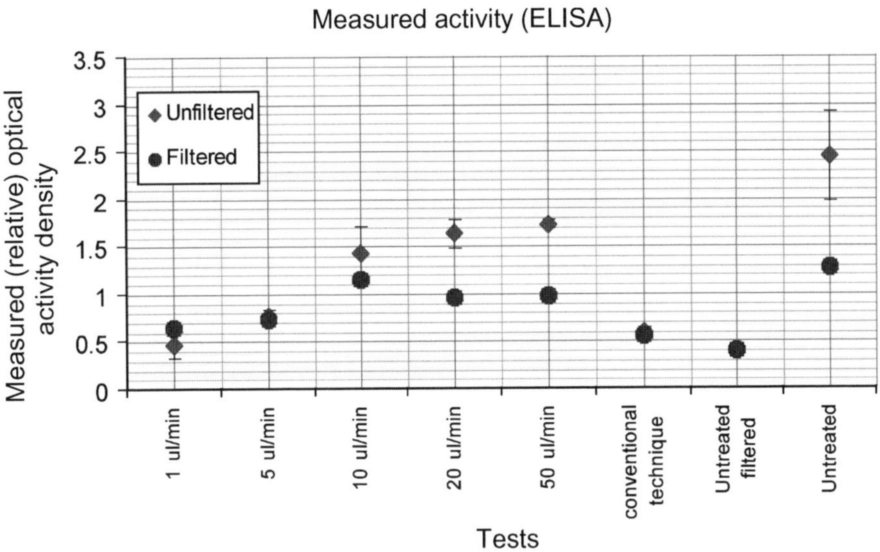

FIGURE 14.7 Activity of the alkaline phosphatase in eight membrane samples of rat liver: MPS sample prepared at 1, 5, 10, 20, and 50 μL/min flow rates in the MPS chip and nanodisc sample filtered through 30 nm pore filter (blue) or unfiltered (red), unprocessed filtered, and unprocessed unfiltered sample measured in ELISA microplates by means of ELISA reader.

(2) Preliminary data on the method is a factor of 2 (100%) better than the conventional method.

(3) There is still room for improvement since, ideally, an optimized process would have alkaline phosphatase activity close to the unprocessed and unfiltered membrane sample.

Analysis of the results showed that alkaline phosphatase activity (Fig. 14.8) was better preserved in the rat liver membrane MPS particles that in nanodisc ones. Also, there is an open opportunity to optimize the productivity of the MPS system by adjustment.

S. oneidensis Membrane Sample As with the rat liver membrane sample, we analyzed activity of a transmembrane enzyme in the nanofragments solubilized at differing flow rates and in assembled nanodiscs, as well as in the filtered and unfiltered membrane sample. We measured peroxidase activity of the c-type cytochromes using commercial SIGMA FAST OPD (*o*-phenylenediamine dihydrochloride) peroxidase substrate (Sigma, P-9187). SIGMA FAST OPD tablets were developed for use as a soluble substrate for the detection of peroxidase activity in enzyme immunoassays (EIA and ELISA). The tablets require no additional buffers or steps to prepare an active substrate solution. One OPD tablet and one

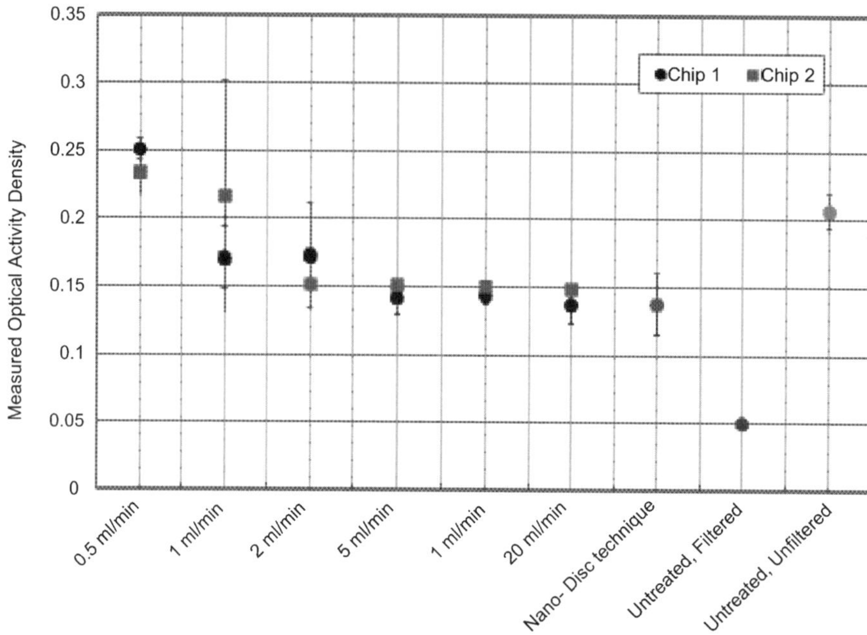

FIGURE 14.8 Activity of the peroxidase (optical density of the products of the reaction) in 15 membrane samples of the *S. oneidensis*: Nanofragment sample at 0. 5, 1.0, 2.0, 5.0, 10.0, and 20.0 μL/min flow rate obtained from two Nanofragment chips, Nanodisc sample (all filtered through 30 nm membrane filter), unprocessed filtered, and unprocessed unfiltered sample measured in ELISA microplates by means of an ELISA reader at 450 nm.

urea hydrogen peroxide/buffer tablet dissolved in 20 mL of water yields 20 mL of ready-to-use substrate. When the tablet is dissolved in this manner, the final concentrations will be 0.4 mg/mL OPD, 0.4 mg/mL urea hydrogen peroxide, and 0.05 M phosphate–citrate, pH 5.0.

The peroxidase activity was measured in ELISA microplates. To six wells of the microplate we added 150 μL of each of the 15 samples that were prepared at 0.5, 1.0, 2.0, 5.0, 10.0, and 20.0 μL/min flow rates from two LOC chips, Nanodisc sample (all these samples were filtered through 30-nm membrane filter), unprocessed filtered, and unprocessed unfiltered sample was added. After this, 150 μL of the incubation medium was added to the triplicates of the six wells. The left triplicates of each sample served as controls for the enzymatic reaction: only buffer was added to them. The microplates enclosed in a plastic bag with a zip lock to prevent water evaporation were incubated at 37°C. The incubation time of 30 min was selected because it fell in the middle of the linear part of the calibration curve for this sample. The optical density of the products of the reaction in all wells of the microplate was measured in an ELISA reader at 450 nm (see Fig. 14.8).

In summary, we observe that because of the very lengthy reaction times needed for extracting *S. oneidensis* membranes, the microfluidic lab-on-chip method proposed here is the best technique available to determine accurately the reaction time and, therefore, devise an optimal protocol to analyze and characterize this critical bacteria that promises to usher in a new era in soil sequestration.

14.3 CONCLUSION

In this chapter, we have shown a unique problem in modern biology that can be addressed, for the first time, by microfluidic lab-on-chip systems. The general technique described in this chapter is applicable to other catalytic processes that require precise reaction control, which may be achieved by enzymatic (quenchers) or conditional (temperature) methods. Microfluidics platform, therefore, not only provides tremendous cost savings from reduced usage of reagents/samples but also has the potential to provide precise reaction control that, in some cases, is essential for sample processing.

In summary, for the sample extraction method to which we applied our LOC method, we reached the following conclusions:

(1) The LOC cell-membrane method works not only for eukaryotic but also for prokaryotic cell membranes. By adjusting the flow rates, we are able to optimize the performance of the LOC system for each specific membrane sample.
(2) The preliminary method is a factor of 1.6 better than the conventional nanodisc technique.
(3) There is still room for improvement, since an ideally optimized process would have peroxidase activity close to or higher than the unprocessed and unfiltered microbial membrane sample.

ACKNOWLEDGMENTS

We would like to thank Drs. Frank A. Gomez and Meneka Piyasena, and Mark Goldberg for their collaboration during this research program. This program was funded by the Department of Energy and the Genomes to Life Program.

REFERENCES

1. Bayburt TH, Sligar SG. Protein Science 2003;12:2476.
2. Shaw AW, McLean MA, Sligar SG. FEBS Lett 2004;556:260.
3. Civjan NR, Bayburt TH, Schuler MA, Sligar SG. Biotechniques 2003;35:556.
4. Pata V, Dan N. Biophys J 2003;85:2111.
5. Lazar IM, Ramsey RS, Ramsey JM. Anal Chem 2001;73:1733.
6. Fintschenko Y, Choi W-Y, Ngola SM, Shepodd TJ. Anal Chem 2001;73:849.
7. MacDonald MP, Spalding GC, Dholakia K. Nature 2003;426:421.

8. Herr AE, Molho JI, Drouvalakis KA, Mikkelsen JC, Utz PJ, Santiago JG, Kenny TW. Anal Chem 2003;75:1180.
9. Seong GH, Heo J, Crooks RM. Anal Chem 2003;75:3161.
10. Mao HB, Yang TL, Cremer PS. J Am Chem Soc 2002;124:4432.
11. Hadd AG, Jacobson SC, Ramsey JM. Anal Chem 1999;71:5206.
12. Leach AM, Wheeler AR, Zare RN. Anal Chem 2003;75:967.
13. Chiem NH, Harrison D. J Clin Chem 1998;44:591.
14. Bromberg A, Mathies R. Anal Chem 2003;75:1188.
15. Wang J, Ibanez A, Chatrathi MP. Electrophoresis 2002;23:3744.
16. Lagally ET, Mednitz I, Mathies RA. Anal Chem 2001;73:565.
17. Khandurina J, McKnight TE, Jacobson SC, Waters LC, Foote RS, Ramsey JM. Anal Chem 2000;72:2995.
18. Kopp. MT, deMello AJ, Manz A. Science 1998;280:1046.
19. Dan N, Pincus P, Safran SA. Langmuir 1993;9:2768.

PART IV

BIOMEDICAL APPLICATIONS OF MICROFLUIDICS

15

MICROENGINEERING NEURAL DEVELOPMENT

Anna Tourovskaia, Nianzhen Li and Albert Folch

15.1 INTRODUCTION

The goal of developmental neuroscience is to understand the dynamic processes that contribute to the wiring of the nervous system, from neurogenesis and neuronal migration to axonal pathfinding and synapse formation. Studying these processes *in vivo* is a daunting task due to the system's inherent biophysical and biochemical complexity (so it is not easily amenable to experimental manipulation), hence many efforts have focused on developing *in vitro* systems. However, as our knowledge of neuronal development gains depth, conventional cell culture technologies become increasingly limiting. A traditional cell culture involves plating dissociated neurons in a homogenous and static environment onto flat surfaces, and while this cell culture paradigm is inexpensive and straightforward, it is so at the expense of a reduced physiological relevance because the dynamic microscale organization of cells and biomolecules is lost. Microfluidic devices, on the contrary, due to their accessibility to small dimensions (comparable to or smaller than a single cell) and unique physicochemical behavior of liquids at the microscale (such as laminar flow, large surface-to-volume ratio, and nonnegligible surface tension) can interact with target cells in controlled physiologically relevant ways not possible with macroscopic techniques and thus can reduce confounding effects inherent to traditional neuronal cultures.

Microfluidic systems are also advantageous because (a) they greatly reduce the consumption of precious/hazardous reagents; (b) with little redesign, they can be integrated with other microfluidic components (e.g., microvalves) to extend

Biological Applications of Microfluidics, Edited by Frank A. Gomez

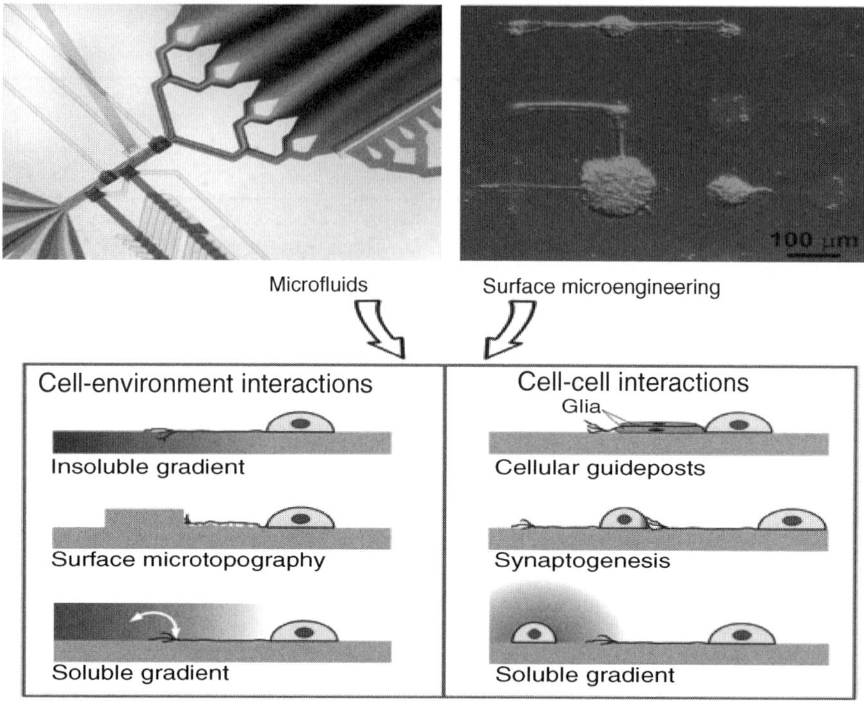

FIGURE 15.1 Applications of microfluidics in developmental neuroscience. Microfluidic devices can be used to influence and mimic cell–environment and cell–cell interactions. Microfluidic technologies allow to control such cellular events at cellular and subcellular levels. Photograph of a microfluidic mixer courtesy of G. Cooksey. (See color insert)

functionalities; (c) they can be mass produced in low cost, portable units; and (d) due to the nonstochastic nature of laminar flow in microchannels, modeling of flow velocity profiles, pressure drops, and diffusion is straightforward: commercially available finite-element fluid dynamics modeling packages are inexpensive and require only a few days to master (by a graduate-level engineering student unfamiliar with the software). Even though microfluidics holds great promise for advancement of cell culture methodology, the integration of microfluidics with neuron cultures is still in its infancy.

This chapter will discuss some of the current applications of microfluidics that illustrate the potential of this technology for neural science, as conceptually depicted in Fig. 15.1.

15.2 MICROENGINEERING GUIDANCE OF AXONS TO THEIR TARGETS

During development, axons navigate their way through a three-dimensional dynamic maze of biomolecular cues to find their correct destinations. At any given time the growth cone is sensing a wide range of cues, either diffusible or immobilized on the substratum or on adjacent cell surfaces. Anatomical structures and extracellular matrix itself also provide navigational information for axonal outgrowth. Biochemical guidance cues can be attractive or repulsive as well as immobilized (contact dependent) or soluble (diffusible) (1,2). Growth cones integrate a complex set of cues to direct the route of growing axons by assembly (or disassembly) of the cytoskeleton at the leading edge of the growth cone. With traditional *in vitro* methods, it is extremely difficult to generate a precise, dynamic extracellular microenvironment; moreover, traditional assays generally study one guidance molecule at a time and yield results at a very low throughput. Microfluidic systems are being developed that allow mimicking of such complex environments. This section discusses several specific microfluidics approaches to create microscale patterns of biochemical guidance cues and gradients of substrate-bound or diffusible factors.

15.2.1 Microfluidic Patterning of Biomolecules for Probing Neuronal Adhesion and Growth

Microfluidic conduits (microchannels) are formed when a mold with recessed features (most often made of PDMS) is sealed against the surface of a substrate (3). This watertight reversible seal is formed due to the hydrophobicity and elastomeric nature of PDMS allowing it to conform to any smooth dry surface. With this method, the material is added at sites where the PDMS does not come into contact with the surface. The microchannels can be filled with fluids using "passive" forces such as capillary action (for small shallow microchannels) (4) or pressure-driven flows (5). Rapid depletion of proteins from solution can occur in very small microchannels (i.e., with high surface-to-volume ratio), which can be an inconvenience but also allows for forming a protein surface gradient (6). Usually, the microfluidic channels are removed to expose the patterned area. Similarly, microfluidic patterning can also be used to selectively remove materials from the surface by filling the microchannels with liquid (3) or gas etchants (7). PDMS can also be permanently bonded to any surface containing hydroxyl groups including PDMS, glass, and silicon after activating/cleaning surfaces with oxygen plasma prior bringing them into contact (8).

Microfluidic networks have been used for patterning of proteins and other molecules that promote neuronal attachment, growth, and structural polarity to mimic some of the *in vivo* architecture and to control size, shape, and position of cells on the surface (9). Protein and cell patterning with microfluidic channels is an attractive alternative to other patterning tools because (a) deposited material binds to the surface from solution and is not exposed to denaturing conditions; (b) proteins or other molecules can be physisorbed with standard well-established protocols or chemically immobilized at the experimenter's choice; and (c) different biomolecules can be simultaneously patterned using different channels (10). However, microfluidic patterning is restricted to interconnected geometries, which may be a limitation for certain applications.

Romanova et al. used microfluidic channels to create chemisorbed patterns of poly-L-lysine (PDL) and collagen IV on glass and achieved the controlled growth of *Aplysia* bag cell neurons (11). The adhesive lines defined the preferred orientation of neurite outgrowth and determined their branching patterns (confirming similar observations with other micropatterning techniques and other types of neurons, for example (12,13). Neurons responded to the spatial constraints by profound changes in morphology: neurite branching was diminished, neurites were shorter and thicker, and the number of neurites per cell was also reduced (i.e., overall, neurite morphology was simpler than in neuronal cultures on homogeneous substrates) (Fig. 15.2). The morphological changes on micropatterned substrates were dictated by the shape of the patterns but were independent of whether the adhesive coating was PDL or collagen. Interestingly, micropatterning of PDL and collagen also altered the electrophysiological behavior of the neurons: a decrease was observed in action potential broadening (on PDL micropatterns) and action potential amplitude (on collagen micropatterns) in comparison with uniform surfaces. These results are a reminder that geometrical confinement and/or, more generally, surface chemistry can influence not only cell morphology (i.e., axonal growth and branching) but also, in more subtle ways, other critical neuronal functions such as electrical excitability.

Our group studied neuronal growth on physisorbed micropatterns of PDL and diluted Matrigel (1:10) on glass substrates (14). Matrigel is a basal lamina extract composed of several ECM proteins, mostly laminin, and is also a permissive substrate for neuronal growth (15). We found that most neurons preferred to adhere and grow on PDL regions as compared to the adjacent Matrigel regions.

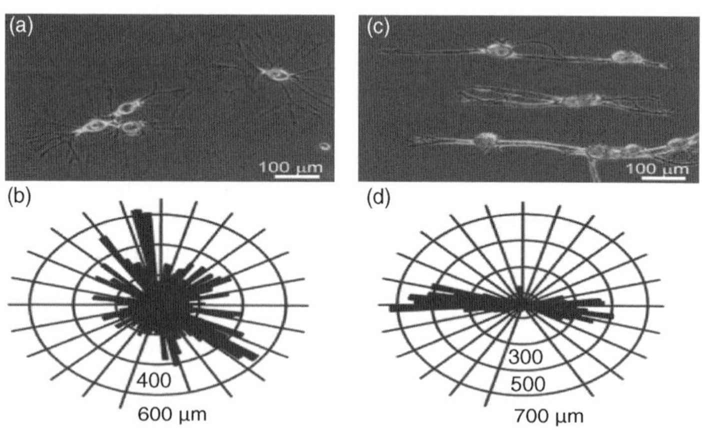

FIGURE 15.2 Engineering the morphology of cultured neurons. (**a**) *Aplysia* bag cell neurons cultured on uniform substrates of PLL form randomly directed neurites with a relatively even amount of outgrowth in all directions as demonstrated by a polar histogram (**b**). When cultured on micropatterned substrates, neurons conform to the growth-guiding patterns of poly-L-lysine and exhibit pattern-defined morphology (**c**). The polar histogram in (**d**) demonstrates that the amount of outgrowth for neurons on patterned surface is greater in the direction aligned with the pattern. Adapted from Reference 11 (courtesy of E.V. Romanova and J.V. Sweedler).

Occasionally, some somas adhered on Matrigel regions and some neurons extended axons from PDL to Matrigel regions suggesting that Matrigel is a growth permissive substrate although less attractive than PDL. This preference for PDL changed, however, when neurons were challenged by an additional topographic cue–substrate "straightness" (flipped PDMS microchannels were used as microgrooves to mimic anatomical guidance cues). Likewise, axonal growth responses to microtopographical cues alone were also different from responses to simultaneous biochemical and microtopographical cues (14).

Grattarola and coworkers micromachined conduits in silicon (rhomboid pits connected by 50 μm wide and 250 μm long channels) and microfluidically deposited PDL in the pits and channels to confine (both physically and chemically) the growth of dissociated neurons (16). This method provides a low cost means to create neuronal networks of small interconnected populations of neurons. However, silicon chips are transparent only to infrared light (wavelength >1100 nm) and thus are not compatible with transmitted-light microscopy, such as brightfield, phase-contrast, darkfield, DIC, which imposes limits on the use of this approach.

As mentioned above, microfluidic patterning is mostly limited to two dimensions due to the geometrical constraints of the microfluidic conduits, which are often closed structures. This limitation can be circumvented through the use of more complex microfluidic devices made by stacking multiple layers or by the use of three-dimensional multilayered molds. Degenaar et al. created three-dimensional microfluidic arrays ("microfluidic stencils") for micropatterning of adhesive chemistries to selectively direct the growth of primary neurons on an open substrate (17). This approach combines topographic microstructures with chemical substrate patterning and allows to achieve guidance of individual neurites.

Glial cells also perform many important functions of the nervous system. Astrocytes, for example, provide neuronal migration cues and modulate extracellular environment, neuronal activity, and synaptic transmission. These functions are mediated, in part, by calcium waves: Neurotransmitters have been found to evoke an increase in intracellular calcium concentration in astrocytes (18). This rise in calcium levels then spreads to other neighboring astrocytes resulting in an intercellular calcium wave. Both gap junctions and extracellular diffusible messengers have been implicated in the propagation of a calcium wave. To distinguish between the two pathways (gap junctions versus diffusible messengers), Takano et al. used an elegant micropatterning approach. Agarose stripes were molded onto glass with a microfluidic device, and then the background (bare glass) was coated with (cell-adhesive) PDL (19). Cortical astrocytes selectively attached and grew on the PDL tracks flanked by agarose. When an astrocyte within a given PDL track was mechanically stimulated with a glass pipette, a calcium wave propagated to the cells within the same track as well as across the agarose barrier to the adjacent pattern, confirming the involvement of an extracellular messenger pathway. Following the addition of an ATP receptor antagonist, however, the calcium wave was reduced in the same lane and completely blocked in adjacent tracks, proving that there is a component of the calcium waves that can be primarily attributed to signaling through gap junctions.

15.2.2 Microfluidic Surface Patterning for Studying Axogenesis

A hallmark of a neuron is its unique morphology featuring the structurally, molecularly, and functionally distinct membrane extensions known as axons and dendrites. Some time after being born, the neuron must decide (or "specify") which extension will become an axon; this "axon-specification" process, or axogenesis, breaks the symmetry of the cell, establishing a "polarity." Axogenesis is still not well understood and very little is known about the signals that instruct initially indistinguishable neurites to acquire axonal characteristics.

Dertinger et al. used a microfluidic gradient generator device to study the influence of substrate-bound gradients of laminin on neuronal guidance (20). The device features a gradient mixer consisting of two inlet channels (one for analyte, one for buffer) that continue into long zig-zag channels (20) or into a chaotic mixer (21) to facilitate mixing by diffusion (Fig. 15.3a). These channels are split several times to produce a series of dilutions that eventually merge into a single channel, combining the fluid streams and forming a gradient. The gradient forms perpendicular to the direction of flow and is deposited onto poly-L-lysine (PLL)-coated PDMS substrates. Rat hippocampal neurons cultured on these substrate-bound gradients of laminin formed axons, the majority of which was oriented toward the direction of increasing surface density of laminin (above a certain concentration threshold) (Fig. 15.3b) suggesting that a laminin gradient can promote axon specification and/or axon guidance.

Banker and coworkers hypothesized that local presentation of extracellular signals can direct axon specification. To test this hypothesis they cultured neurons on surfaces with stripes of alternating chemistries created using microfluidic patterning (Fig. 15.3c), shown is a hippocampal neuron with its soma and dendrites growing on PLL, and axon growing on a stripe of neuron–glia cell adhesion molecule or NgCAM). It was demonstrated that when neurons were growing on surfaces with alternating stripes of PLL and laminin or NgCAM, the vast majority of axons formed on laminin or NgCAM. Time-lapse imaging revealed that when neurites crossed from PLL into an adjacent stripe of laminin or NgCAM, this contact resulted in faster neurite growth rate, which lead to axon formation. When stripes of laminin and NgCAM were alternated, cells with their somas on laminin mostly formed axons on NgCAM, while those with somas on NgCAM generally formed axons on laminin, indicating that the change from one axon-promoting substrate to another during neurite outgrowth also provided sufficient signal to specify axon among undifferentiated neurites. These findings strongly support the hypothesis that local contact of immature neurites with extracellular matrix can stimulate axon specification.

15.2.3 Microfluidic Surface Patterning for Contact Guidance Studies

Using microfluidic patterning, Turney and Bridgman studied intracellular mechanisms underlying the turning of axons at the transition between laminin-1 and polyornithine (PLO) stripes (22). Using knockout-mice neurons and specific

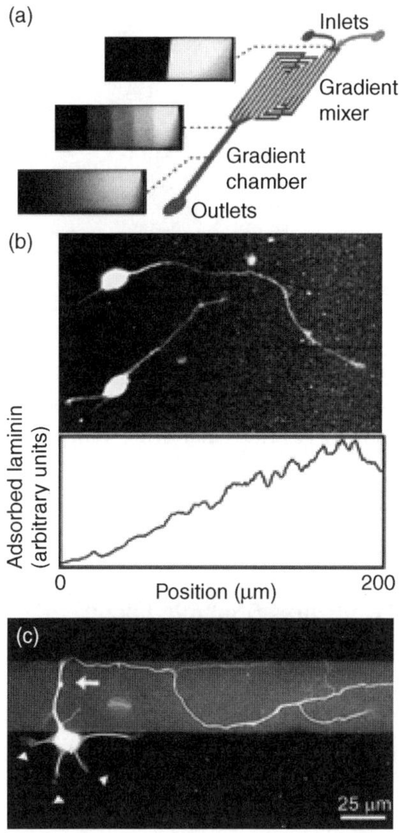

FIGURE 15.3 Microfluidic surface patterning for studying axogenesis. (**a**) Schematic drawing of a microfluidic gradient generator. Adapted from Reference 21 (courtesy of S. Dertinger). (**b**) Hippocampal neuron (immunostained for tubulin) growing on a gradient of laminin (revealed by immunostaining) extending the axon in the direction of increasing laminin density on the surface. The graph below shows the shape of the immobilized laminin gradient. Adapted from Reference 20 (courtesy of S. Dertinger). (**c**) Hippocampal neuron growing on alternating stripes of PLL and NgCAM. While soma and dendrites (arrowheads) are attached to PLL (dark), the neurite that contacted the NgCAM (fluorescent) formed an axon (arrow). The pattern of NgCAM is revealed by immunostaining and the neuron is immunolabeled for tubulin. Reprinted from Reference 42. Copyright ©1999 by the Society for Neuroscience.

inhibitors, it was shown that myosin II activity is essential for the turning response at laminin-1/PLO borders, while the amount and/or isoform of myosin II controls the efficiency and speed of turning. It was also demonstrated that turning depended on substrate-bound laminin and not on the abrupt change in outgrowth rates between different substrates, suggesting the possibility that myosin II-dependent contraction in growth cone was induced by signaling from laminin-activated integrins (23).

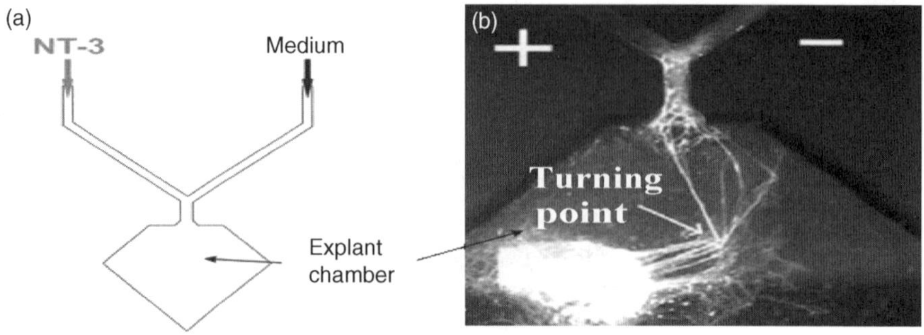

FIGURE 15.4 Guidance by soluble gradient of NT-3. (**a**) Schematic drawing of the microfluidic gradient-generating device. (**b**) Rat spiral ganglion explant (stained with antineurofilament antibody) grown in the device for 72 h and exposed to the gradient of NT-3 (+) in neuronal medium (−). Neurites outgrowing from the explant exhibit sharp turning toward the microchannel. Reprinted with permission from Elsevier from Reference 43. Copyright © 2004 Elsevier B.V.

15.2.4 Guidance by Gradients of Soluble Factors

Ryan et al. used a microfluidic device to study spinal ganglion (SG) neurite guidance by a soluble gradient of neurothrophin-3 (NT-3) (24). This device consists of two microchannels that merge into a single channel, which then opens into a trapezoidal region where a SG explant is placed (Fig. 15.4a). When the two laminar fluid streams (here, buffer and NT-3) join, diffusion orthogonally to the flow between the two streams establishes a concentration gradient (of NT-3) across the main channel. The gradient then becomes progressively shallower when the fluid streams enter the large trapezoidal explant chamber. Therefore, the neurites extending from the explant are first presented a diffuse gradient in the chamber and a steep gradient as they grow into the microchannel. Ryan et al. observed that the majority of SG neurites turned and grew toward higher concentrations of NT-3 (Fig. 15.4b). These experiments suggested that developing SG neurites could be guided by a single gradient of NT-3 in the absence of other guidance cues.

Shear and colleagues demonstrated a novel strategy to produce dynamic chemical environments for the studies of axon guidance and neuronal networks (25). In this approach cells are cultured on top of a thin polymeric membrane separating two stacked microfluidic channels, one of which houses cells and another carries molecules of interest (Fig. 15.5a). A hole in the membrane communicates the two microfluidic channels and acts as a point source for stimulating the cells. With a clever *in situ* (i.e., "cell-friendly") microfabrication approach based on laser ablation and protein photocrosslinking, the holes could be opened and closed (Fig. 15.5b). Holes are created in the presence of cells and virtually any subcellular domain or features can be specifically targeted with a stream of a molecule of interest.

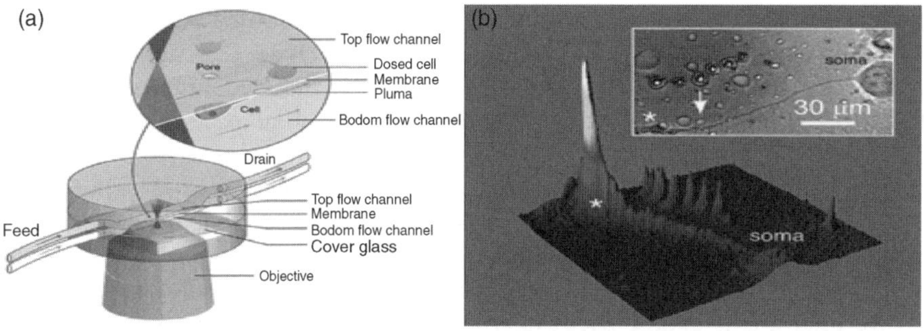

FIGURE 15.5 Cells in dynamic chemical environments. (**a**) Schematic illustration of the microfluidic method to produce dynamic chemical environments. Cells are cultured on a membrane that partitions two microfluidic channels. The membrane is perforated using a femtosecond-pulsed laser beam and reagent from the top channel is forced through the pore, locally stimulating a cell domain of interest. (**b**) Fluorescent intensity surface plot and DIC image (inset) demonstrate labeling of neuroblastoma glioma cell process with laminar streams formed through a set of six pores. Cell was labeled with a membrane-permeant dye. Arrow indicates the direction of fluid flow. Reprinted in part with permission from Reference 23. Copyright © 2006 American Chemical Society. (See color insert)

15.3 SYNAPTOGENESIS ON A MICROFLUIDIC CHIP

Synapses are membrane specializations that allow nerve cells to communicate with each other or to relay signals to other types of cells, such as muscle cells. Hence, the formation of synapses, or synaptogenesis, is a central process in the development of the nervous system. Synaptogenesis is a highly localized and interactive process: during development, synapses are formed by reciprocal signaling between the presynaptic neuron and its specific postsynaptic target resulting in precise apposition of pre- and postsynaptic elements (26,27). Nerve-derived postsynaptic signals direct the target cells to form clusters of neurotransmitter receptors and other postsynaptic elements in precise juxtaposition with the nerve terminal; conversely, target-derived signals promote local differentiation of presynaptic axons into functional nerve terminals (27). While the formation and maintenance of synapses are critical for proper nervous system function and have received a lot of attention, the lack of suitable methods for focal application of molecules to target cells has limited the majority of *in vitro* studies to bathing cells in homogeneous solutions—clearly a poor mimick of the physiology of synaptogenesis.

15.3.1 Microfluidic Focal Delivery of Synaptogenic Factors Using Multiple Laminar Flows

The neuromuscular synapse or neuromuscular junction (NMJ) has provided a convenient model for synaptic development studies (26,28) because of its experimental accessibility, large size, relatively simple anatomy (generally each muscle fiber is innervated by only one motor axon), and its structural and functional similarity to the synapses of the central nervous system (27). The hallmark of NMJ is its high density of acetylcholine receptors (AChRs) clustered postsynaptically on the muscle cell membrane. The clustering of AChRs to the postsynaptic site was previously thought to be induced by the local release of agrin, a heparan sulfate proteoglycan, from the presynaptic nerve terminal (26,29–32). However, the exact role of neural agrin is still not clear because all the parameters that define the role of agrin (onset, concentration, and duration) cannot be easily manipulated *in vivo* and, so far, could not be controlled *in vitro* simultaneously using existing methods.

We used microfluidic devices to recapitulate a physiologically relevant, local presentation of synaptogenic molecules to arrays of microengineered myotubes (Fig. 15.6a and c). Essentially, the presynaptic neuron is replaced with a microfluidic device. The attachment, proliferation, and fusion of myoblasts, which lasts on the order of 1 week, is confined to microfabricated tracks of cell-adherent proteins separated by a thin layer of cell-repellent polymer spaces to produce linear myotubes in the direction perpendicular to the flow. Then, a stimulatory

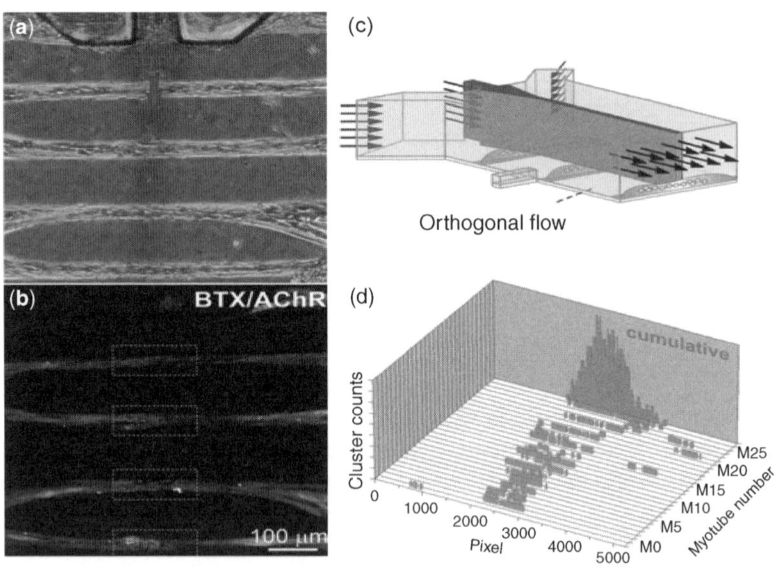

FIGURE 15.6 Synaptogenesis on a microfluidic chip. (**a**) Phase-contrast micrograph taken during perfusion with agrin; also schematically shown in 3D in (**c**); the agrin-containing stream appears red because a food-coloring dye was added to the agrin solution for visualization. (**b**) Corresponding fluorescence micrograph of AChR cluster staining (fluorescently conjugated α-bungarotoxin, BTX). Boxed are areas of agrin-induced AChR clustering (AChR clusters appear as discrete brightly fluorescent specks). (**d**) Mapping of spatial distribution of AChR clusters on myotubes in the device (part of which is shown in (**b**)) shows that the vast majority of AChR clusters localize around the centerline of the microchannel. See Reference 34 for details. (See color insert)

laminar-flow stream is positioned over specific cellular domains of the muscle cells (33,34). This precise alignment of large numbers of myotubes with the focal stimulus and with the imaging field makes the results readily amenable to quantitative, statistically rich analysis. We found that following a focal agrin stimulus, the vast majority of AChR clusters localize to the area where the agrin stream contacted the myotubes, showing a gradual increase of the AChR clustering area toward downstream, most likely due to the broadening of the diffusion profile of the agrin stream (Fig. 15.6b and d) (34).

15.3.2 Planar Nanoholes Devices

Membranes that feature micron- and submicron-scale apertures (nanofabricated holes or "nanoholes") can be used as cell culture substrates that provide access to a small area of the cell(s) just above the nanohole(s). Planar nanoholes have been used for electrophysiological probing of single cells (35–38), for pharmacological testing of large numbers of cells as well as for biochemical stimulation of cells (39–41).

Kosar et al. used cell culture substrates containing nanofabricated apertures (2–8 μm diameter) to focally stimulate myotubes with agrin (Fig. 15.7a) (40). Myoblasts were allowed to proliferate and fuse on thin, high transparency silicon nitride membranes suspended over micromachined silicon cavities. The nanoholes

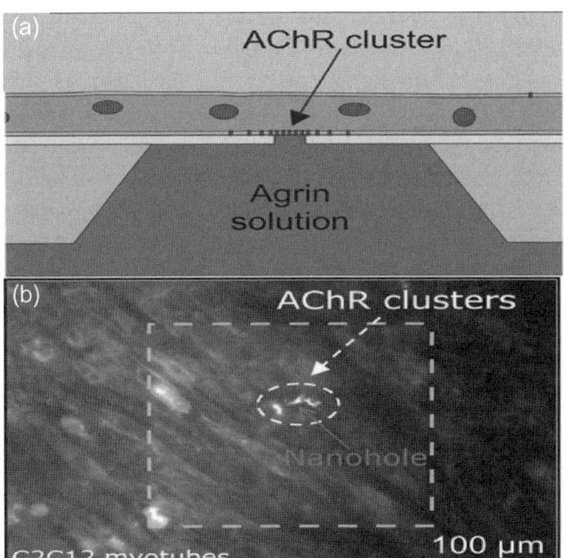

FIGURE 15.7 Focal delivery of agrin to myotubes through nanoholes. (**a**) Schematic illustration of the experimental design. (**b**) Fluorescent micrograph of myotubes showing localized aggregation of AChR in the vicinity of a nanohole. Box outlines the silicon nitride membrane window in the device. Fluorescently conjugated α-bungarotoxin was used to detect AChR. See Reference 40 for details. Figure courtesy of F. Kosar. (See color insert)

were defined using electron beam lithography and chemical etching, allowing for the parallel fabrication of devices containing many different nanohole sizes with submicron resolution. The cavities (thus, the nanoholes too) could be microfluidically addressed (from either side) in an array format, allowing multiple experiments to be run in parallel on one device. The method allows to control the starting time, the duration, and the concentration of the agrin stimulus. AChR clustering in cultured myotubes was observed in the vicinity of the nanoholes (Fig. 15.7b), a process reminiscent of AChR clustering during muscle innervation (40). This technique has wide applicability in cell–cell communication studies and cell-based bioassays.

Similar devices were used for subcellular-scale delivery of a neurotransmitter (glutamate) to a micropatterned network of rat cortical neurons (41). Localized chemical stimulation of neurons near the nanoholes generated action potentials and induced changes in the spontaneous activity patterns in cortical neurons within a network. The authors proposed that such a system could also be used to study the influence of local chemical gradients on neural network properties to understand some of the basic principles of synaptic plasticity.

15.4 CONCLUSIONS

Broadly speaking, microscale technologies and microfluidic systems constitute an approach with immense promise for advancing the many subdisciplines of neuroscience. Microfluidic systems, in particular, offer a unique set of tools for micropatterning neurons and glial cells, for localized or graded chemical stimulation of cells, and for generating linear or complex chemical gradients of soluble or substrate-bound molecular factors. Microfluidic tools are also naturally suited for high throughput assays, fast exchange of fluids, and multiphenotype cell-based screening. Even though microfluidics is only beginning to merge with neurobiology, this fusion has already provided valuable insights into how axons navigate through surrounding microenvironments and establish synaptic connections as well as into how axons interact with surrounding cells. However, the formidable complexity of the nervous system's cellular architecture poses massive obstacles for *in vitro* studies, which inherently rely on a huge simplification of the *in vivo* architecture. Yet microfabricated cell culture systems allow for (a) a more quantitative design of the cellular microenvironment than traditional petri dish studies, and (b) high throughput experimentation and data acquisition at low cost. Thus, we envision that micropatterning and microfluidic technologies will eventually become ubiquitous tools in *in vitro* neuroscience research.

REFERENCES

1. Tessier-Lavigne M, Goodman CS. Science 1996;274:1123.
2. Gilbert SF, Developmental Biology. 7th ed. Sunderland, MA. Sinauer Associates; 2003.

3. Kane RS, Takayama S, Ostuni E, Ingber DE, Whitesides GM. Biomaterials 1999;20:2363.
4. Delamarche E, Bernard A, Schmid H, Michel B, Biebuyck H. Science 1997;276:779–781.
5. Folch A, Toner M. Biotechnol Prog 1998;14:388–392.
6. Delamarche E, Juncker D, Schmid H. Adv Mater 2005;17:2911–2933.
7. Tourovskaia A, Barber T, Wickes B, Hirdes D, Grin B, Castner D, Healy KE, Folch A. Langmuir 2003;19:4754–4764.
8. McDonald JC, Whitesides GM. Acc Chem Res 2002;35:491–499.
9. Li N, Tourovskaia A, Folch A. Crit Rev Biomed Eng 2003;31:423–488.
10. Folch A, Toner M. Annu Rev Biomed Eng 2000;2:227–256.
11. Romanova EV, Fosser KA, Rubakhin SS, Nuzzo RG, Sweedler JV. FASEB J 2004;18:1267–1269.
12. Clark P, Britland S, Connolly P. J Cell Sci 1993;105:203–212.
13. Tai HC, Buettner HM. Biotechnol Prog 1998;14:364–370.
14. Li N, Folch A. Exp Cell Res 2005;311:307–316.
15. Kleinman HK, Martin GR. Semin Cancer Biol 2005;15:378–386.
16. Martinoia S, Bove M, Tedesco M, Margesin B, Grattarola M. J Neurosci Methods 1999;87:35–44.
17. Degenaar P, Le Pioufle B, Griscom L, Tixier A, Akagi Y, Morita Y, Murakami Y, Yokoyama K, Fujita H, Tamiya E. J Biochem 2001;130:367–376.
18. Haydon PG. Nat Rev Neurosci 2001;2:185–193.
19. Takano H, Sul JY, Mazzanti ML, Doyle RT, Haydon PG, Porter MD. Anal Chem 2002;74:4640–4646.
20. Dertinger SK, Jiang X, Li Z, Murthy VN, Whitesides GM. Proc Natl Acad Sci U S A 2002;99:12542–12547.
21. Jiang X, Xu Q, Dertinger SK, Stroock AD, Fu TM, Whitesides GM. Anal Chem 2005;77:2338–2347.
22. Turney SG, Bridgman PC. Nat Neurosci 2005;8:717–719.
23. Turney SG, Culican SM, Lichtman JW. J Neurosci Methods 1996;64:199–208.
24. Ryan AF, Wittig J, Evans A, Dazert S, Mullen L. Audiol Neurootol 2006;11:134–143.
25. Nielson R, Shear JB. Anal Chem 2006;78:5987–5993.
26. Sanes JR, Lichtman JW. Annu Rev Neurosci 1999;22:389–442.
27. Kandel ER, Schwartz JH, Jessell TM. Principles of Neural Science, 4th ed. , New York: McGraw-Hill;2000.
28. Hall ZW, Sanes JR. Cell 1993;72:99–121.
29. McMahan UJ. Cold Spring Harb Symp Quant Biol 1990;55:407–418.
30. McMahan UJ, Horton SE, Werle MJ, Honig LS, Kroger S, Ruegg MA, Escher G. Curr Opin Cell Biol 1992;4:869–874.
31. Bezakova G, Rabben I, Sefland I, Fumagalli G, Lomo T. Proc Natl Acad Sci U S A 2001;98:9924–9929.
32. Kummer TT, Misgeld T, Sanes JR. Curr Opin Neurobiol 2006;16:74–82.
33. Tourovskaia A, Figueroa-Masot X, Folch A. Lab Chip 2005;5:14–19.

34. Tourovskaia A, Kosar TF, Folch A. Biophys J 2006;90:2192–2198.
35. Fertig N, Tilke A, Blick RH, Kotthaus JP, Behrends JC. Appl Phys Lett 2000;77:1218–1220.
36. Lehnert T, Gijs MAM, Netzer R, Bischoff U. App Phys Lett 2002;81:5063–5065.
37. Pantoja R, Nagarah JM, Starace DM, Melosh NA, Blunck R, Bezanilla F, Heath JR. Biosens Bioelectron 2004;20:509–517.
38. Kosar TF, Chen C, Stucky NL, Folch A. J Biomed Nanotechns 2005;1(2):161–167.
39. Peterman MC, Noolandi J, Blumenkranz MS, Fishman HA. Proc Natl Acad Sci U S A 2004;101:9951–9954.
40. Kosar TF, Tourovskaia A, Figueroa-Masot X, Adams ME, Folch A. Lab Chip 2006;6:632–638.
41. Mourzina Y, Steffen A, Kaliaguine D, Wolfrum B, Schulte P, Bocker-Meffert S. J Soc Interface 2006;3:333–343.
42. Esch T, Lemmon V, Banker G. J Neurosci 1999;19:6417–6426.
43. Wittig JH, Ryan AF, Asbeck PM. J Neurosci Methods 2004;144(1):79–89.

16

APPLICATIONS OF CENTRIFUGAL MICROFLUIDICS IN BIOLOGY

HORACIO KIDO, JIM ZOVAL AND MARC MADOU

16.1 INTRODUCTION

With microfluidics, the miniaturization of analytical equipment may potentially alleviate the shortcomings associated with large and expensive instrumentation through the reduction in reagent volumes, favorable scaling properties of several important instrument processes (basic theory of hydrodynamics and diffusion predicts faster heating and cooling and more efficient chromatographic and electrophoretic separations in miniaturized equipment), and batch-fabrication that may enable low cost, disposable instruments to be used once and then thrown away to prevent sample contamination (1). Micromachining (MEMS) might also allow cofabrication of many integrated functional instrument blocks. Tasks that are now performed in a series of conventional benchtop instruments could then be combined into one unit, reducing labor and minimizing the risk of sample contamination.

Today it appears that sensor array development in electronic noses and tongues has slowed down because of the lack of highly stable chemical and biological sensors: Too frequent recalibration of the sensors and relearning of the pattern recognition software is putting a damper on the original enthusiasm for this sensor approach. In the case of miniaturization of instrumentation through the application of microfluidics, progress was made in the development of platforms for high throughput screening (HTS) as evidenced by new products introduced by, for example, Caliper and Tecan Boston (2,3). In contrast, progress with miniaturized analytical equipment remains limited. There have been platforms developed for a limited amount of human and veterinary diagnostic test that do not require complex

Biological Applications of Microfluidics, Edited by Frank A. Gomez

fluidic design, for example, Abaxis (4). In this chapter we are, in a narrow sense, summarizing the state of the art of compact disc (CD)-based microfluidics and in a broader sense we are comparing the technical barriers involved in applying microfluidics to sensing and diagnostic as opposed to applying such techniques to high throughput screening. It will quickly become apparent that the former poses the more severe technical challenges and as a result the promise of lab-on-a-chip has not been fulfilled yet.

16.2 WHY USE CENTRIFUGAL FORCE FOR FLUID MANIPULATION?

There are various technologies for moving small quantities of fluids or suspended particles from reservoirs to mixing and reaction sites, to detectors, and eventually to waste or to a next instrument. Methods to accomplish this include syringe and peristaltic pumps, electrochemical bubble generation, acoustics, magnetics, DC and AC electrokinetics, centrifuge. etc. In Table 16.1 we compare four of the more important and promising fluid propulsion means (5). The pressure that mechanical pumps have to generate to propel fluids through capillaries is higher the narrower the conduit. Pressure and centripetal force are both volume-dependent forces, which scale as L^3 (in this case L is the characteristic length corresponding to the capillary diameter). Piezoelectric, electroosmotic, electrowetting, and electrohydrodynamic (EHD) pumping (the latter two are not shown in Table 16.1) all scale as surface forces (L^2), which represent more favorable scaling behavior in the microdomain (propulsion forces scaling with a lower power of the critical dimension become more attractive in the microdomain) and lend themselves better to pumping in smaller and longer channels. In principle, this should make pressure- and centrifuge-based systems less favorable but other factors turn out to be more decisive; despite better scaling of the nonmechanical pumping approaches in Table 16.1, almost all biotechnology equipment today remain based on traditional external syringe or peristaltic pumps. The advantages of this approach are that it relies on well-developed, commercially available components and that a very wide range of flow rates is attainable. Although integrated micromachined pumps based on two one-way valves may achieve a precise flow control on the order of 1 μL/min with fast response, high sensitivity, and negligible dead volume, these pumps generate only modest flow rates and low pressures, and consume a large amount of chip area and considerable power.

Acoustic streaming is a constant (DC) fluid motion induced by an oscillating sound field at a solid/fluid boundary. A disposable fluidic manifold with capillary flow channels can simply be laid on top of the acoustic pump network in the reader instrument. The method is considerably more complex to implement than electroosmosis (see next) but the insensitivity of acoustic streaming to the chemical nature of the fluids inside the fluidic channels and its ability to mix fluids make it a potentially viable approach. A typical flow rate measured for water in a small metal pipe lying on a piezoelectric plate is 0.02 cm^3/s at 40 V, peak to peak (6). Today acoustic streaming as a propulsion mechanism remains in the research stage.

TABLE 16.1 Comparison of Microfluidics Propulsion Techniques

	Fluid propulsion mechanism			
Comparison	Centrifuge	Pressure	Acoustic	Electrokinetic
Valving solved?	Yes for liquids no for vapor	Yes for liquids and vapor	No solution shown yet for liquid or vapor	Yes for liquids, no for vapor
Maturity	Products available	Products available	Research	Products available
Propulsion force influenced by	Density and viscosity	Generic	Generic	pH, ionic strength
Power source	Rotary motor	Pump, mechanical roller	5–40 V	10 kV
Materials	Plastics	Plastics	Piezoelectrics	Glass, plastics
Scaling	L^3	L^3	L^2	L^2
Flow rate	From less than 1 nL/s to greater than 100 μL/s	Very wide range (less than nL/s to L/s)	20 μL/s	0.001–1 μL/s
General remarks	Inexpensive CD drive, mixing is easy, most samples possible (including cells). Better for diagnostics	Standard technique. Difficult to miniaturize and multiplex	Least mature of the four techniques. Might be too expensive. Better for smallest samples	Mixing difficult. High voltage source is dangerous and many parameters influence propulsion, better for smallest samples (HTS)

Electroosmotic pumping (DC electrokinetics) in a capillary does not involve any moving parts and is easily implemented. All that is needed is a metal electrode in some type of a reservoir at each end of a small flow channel. Typical electroosmotic flow velocities are on the order of 1 mm/s with a 1200 V/cm applied electric field. For example, in free-flow capillary electrophoresis work by Jorgenson, electroosmotic flow of 1.7 mm/s was reported (7). This is fast enough for most analytical purposes. Harrison et al. achieved electroosmotic pumping with flow rates up to 1 cm/s in 20 μm capillaries that were micromachined in glass (8). They also demonstrated the injection, mixing, and reaction of fluids in a manifold of micromachined flow channels without the use of valves. The key aspect for tight valving of liquids at intersecting capillaries in such a manifold is the suppression of convective and diffusion effects. The authors demonstrated that these effects can be controlled by the appropriate application of voltages to the intersecting channels simultaneously. Some disadvantages of electroosmosis are the required high voltage (1–30 kV power supply) and direct electrical-to-fluid contact with resulting sensitivity of flow rate to the charge of the capillary wall and to the ionic strength and pH of the solution. It is consequently more difficult to make it into a generic propulsion method. For example, liquids with high ionic strength cause excessive Joule heating; it is therefore difficult or impossible to pump biological fluids such as blood and urine.

Using a rotating disc, centrifugal pumping provides flow rates ranging from less than 10 nL/s to greater than 100 μL/s depending on disc geometry, rotational rate (RPM), and fluid properties (see Fig. 16.1) (9). Pumping is relatively insensitive to physicochemical properties such as pH, ionic strength, or chemical composition (in contrast to AC and DC electrokinetic means of pumping). Aqueous solutions, solvents (e.g., DMSO), surfactants, and biological fluids (blood, milk, and urine) have all been pumped successfully. Fluid gating, as we will describe in more detail further below, is accomplished using "capillary" valves in which capillary forces pin fluids at an enlargement in a channel until rotationally induced pressure is sufficient to overcome the capillary pressure (at the so-called burst frequency) or by hydrophobic methods. Since the types and the amounts of fluids one can pump on a centrifugal platform spans a greater dynamic range than for electrokinetic and

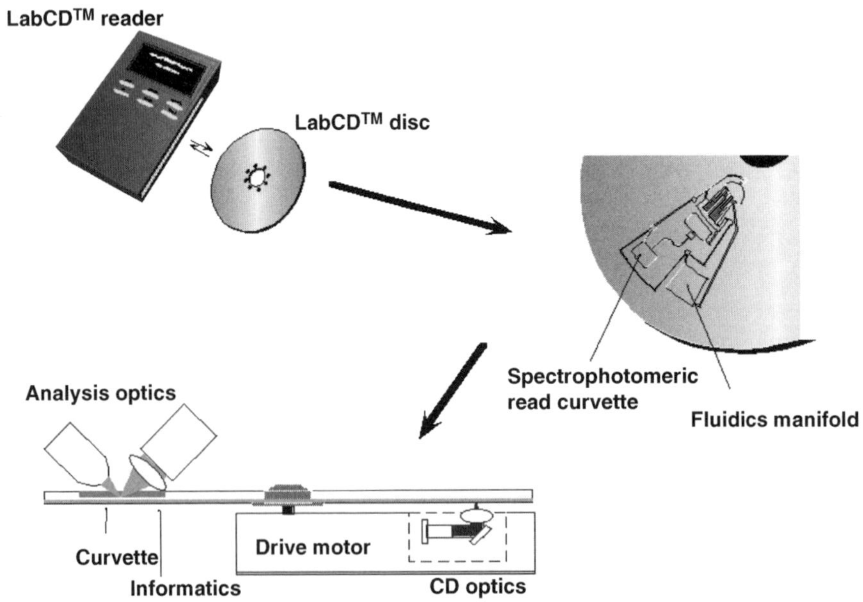

FIGURE 16.1 LabCD™ Instrument and disposable disc. Here, the analytical result is obtained through reflection spectrophotometry.

acoustic pumps, this approach seems more amenable to sample preparation tasks than electrokinetic and acoustic approaches. Moreover, miniaturization and multiplexing are quite easily implemented. A whole range of fluidic functions including valving, decanting, calibration, mixing, volume definition, sample splitting, and separation can be implemented on this platform and analytical measurements may be electrochemical, fluorescent, or absorption based and informatics embedded on the same disc could provide test-specific information.

A most important deciding factor in choosing a fluidic system is the ease of implementing valves; the method that most elegantly solves the valving issue is already commercially accepted, even if the scaling is not the most favourable, namely, in the use of traditional pumps. In traditional pumps two one-way valves form a barrier for both liquids and vapors. In the case of the microcentrifuge, valving is accomplished by varying rotation speed and capillary diameter. Thus, there is no real physical valve required for stopping water flow, but as in the case of acoustic and electrokinetic pumping there is no simple means to stop vapors from spreading over the whole fluidic platform. If liquids need to be stored for a long time on the disposable, as often is the case for use in sensing and diagnostics, valves must be barriers for both liquid and vapor. Some timid attempts at implementing vapor barriers on the CD will be reported in this chapter.

From the preceding comparison of fluidic propulsion methods for sensing and diagnostic applications, centrifugation in fluidic channels and reservoirs crafted in a CD-like plastic substrate as shown in Fig. 16.1 constitutes an attractive fluidic platform.

16.3 HOW CENTRIFUGAL MICROFLUIDIC PLATFORMS WORK

CD fluid propulsion is achieved through centrifugally induced pressure and depends on rotation rate, geometry and location of channels and reservoirs, and fluid properties. Madou et al. (10) and Duffy et al. (11) characterized the flow rate of aqueous solutions in fluidic CD structures and compared the results to simple centrifuge theory. The average velocity of the liquid (U) from centrifugal theory is given as

$$U = D_{\mathrm{h}}^2 \rho \omega^2 \bar{r} \Delta r / 32\, \mu \mathrm{L} \tag{16.1}$$

mathDisplayand the volumetric flow rate (Q) as

$$Q = UA \tag{16.2}$$

mathDisplaywhere D_{h} is the hydraulic diameter of the channel (defined as $4A/P$, where A is the cross-sectional area and P is the wetted perimeter of the channel), ρ is the density of the liquid, *omega* is the angular velocity of the CD, $\bar{r}$ is the average distance of the liquid in the channels to the center of the disc, Δr is the radial extent of the fluid, μ is the viscosity of the solution, and L is the length of the liquid in the

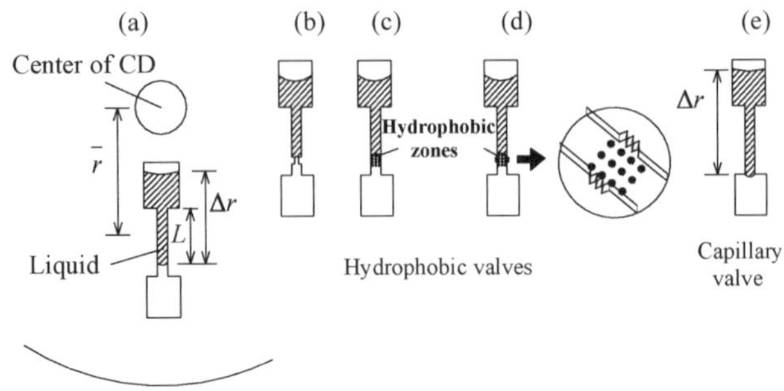

FIGURE 16.2 Schematic illustrations for the description of CD microfluidics. (**a**) Two reservoirs connected by a microfluidic chamber. (**b**) Hydrophobic valve made by a constriction in a chamber made of hydrophobic material. (**c**) Hydrophobic valve made by the application of hydrophobic material to a zone in the channel. (**d**) Hydrophobic channel made by the application of hydrophobic material to a zone in a channel made with structured vertical walls (see inset). (**e**) Capillary valve made by a sudden expansion in channel diameter such as when a channel meets a reservoir.

capillary channel (see also Fig. 16.2a). Flow rates, ranging from 5 nL/s to 0.1 mL/s, have been achieved by various combinations of rotational speeds (from 400 to 1600 rpm), channel widths (20–500 μm), and channel depths (16–340 μm). The experimental flow rates were compared to rates predicted by the theoretical model and exhibited an 18.5% coefficient of variation. The authors note that experimental errors in measuring the highest and lowest flow rates made for the largest contribution to this coefficient of variation. The absence of systematic deviation from theory validates the model for describing flow in microfluidic channels under centripetal force. Duffy et al. (11) measured flow rates of water, plasma, bovine blood, three concentrations of hematocrit, urine, dimethyl sulfoxide (DMSO), and PCR products and reported that centrifugal pumping is relatively insensitive to such physiochemical properties as ionic strength, pH, conductivity, and the presence of various analytes, noting good agreement between experiment and theory for all the liquids.

16.4 CD APPLICATIONS

16.4.1 Two-Point Calibration of an Optode-Based Detection System

A CD-based system with ion-selective optode detection and a two-point-calibration structure for the accurate detection of a wide variety of ions was developed. The microfluidic architecture, depicted in Fig. 16.3, is comprised of channels, five solution reservoirs, a chamber for colorimetric measurement of the optode membrane, and a waste reservoir, all manufactured onto a poly(methyl methacrylate) disc. Ion-selective optode membranes, composed of plasticized poly(vinyl chloride) impregnated with an ionophore, a chromoionophore, and a lipophilic anionic additive, were cast, with a spin-on device, onto a support layer and then

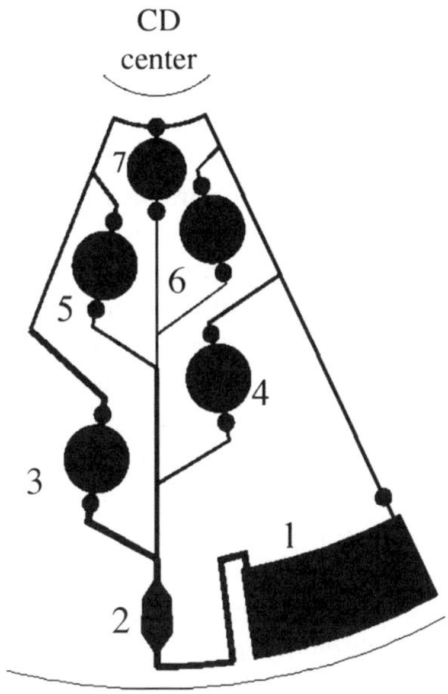

FIGURE 16.3 Schematic illustration of the microfluidic structure employed for the ion selective optode CD platform. The fluidic structure contains five solution reservoirs (numbered 3–7), a detection chamber (2), and a waste reservoir (1). Reservoirs (3) and (5) contain the first and second calibrant, respectively, reservoirs (4) and (6) contain wash solutions, and reservoir (7) contains the sample. Upon increasing rotation rates, calibrant 1, wash 1, calibrant 2, rinse 2, and then sample were serially gated into the optical detection chamber. Absorption of the calibrants and sample was measured.

immobilized on the disc. With this system, it is possible to deliver calibrant solutions, washing buffers, and "unknown" solutions (e.g., saliva, blood, urine) to the measuring chamber where the optode membrane is located. Absorbance measurements on a potassium optode indicate that optodes immobilized on the platform exhibit the theoretical absorbance response. Samples of unknown concentration can be quantified to within 3% error by fitting the response curve for a given optode membrane using an acid (for measuring the signal for a fully protonated chromoionophore), a base (for fully deprotonated chromoionophore), and two standard solutions. Further, the ability to measure ion concentrations employing one standard solution in conjunction with acid and base, and with two standards alone were studied to delineate whether the current architecture could be simplified. Finally, the efficacy of incorporating washing steps into the calibration protocol was investigated.

This work was further extended to include anion-selective optodes and fluorescence rather than absorbance detection. Furthermore, in addition to employing a standard excitation source where a fiber optic probe is coupled to a lamp, LASER diodes were evaluated as excitation sources to enhance the fluorescence signal.

16.4.2 CD Platform For Enzyme-Linked Immunosorbant Assays (ELISA)

The automation of immunoassays on microfluidic platforms presents multiple challenges because of the high number of fluidic processes and the many different liquid reagents involved. Often there is also the need for highly accurate quantitative results at extremely low concentration and care must be taken to prevent nonspecific binding of reporter enzymes and to deliver well defined volumes of reagents consistently. An enzyme-linked immunosorbant assay (i.e., ELISA) is one of the most common immunoassay methods and is often carried out in microtiter plates using labor-intensive manual pippetting techniques. Recently, Lee et al. have implemented an automated enzyme-linked immunosorbant assay on the CD platform. This group used a five-step flow sequence in the same CD design illustrated in Fig. 16.3. A capture antibody (anti-rat IgG) was applied to the detection reservoir (reservoir 2 in Fig. 16.3) by adsorption to the PMMA CD surface, then the surface was blocked to prevent nonspecific binding. Antigen/sample (rat IgG), wash solution, second antibody, and substrate solutions were loaded into reservoirs 3–7 (Fig. 16.3) respectively. Using capillary valving techniques, the sample and reagents were pumped, one at a time, through the detection chamber. First, the sample was introduced for antibody–antigen binding (reservoir 3), then a wash solution (reservoir 4), then an enzyme-labeled secondary antibody (reservoir 5), then another wash solution (reservoir 6), and finally the substrate was added (reservoir 7). The U-shaped bend in the fluidic path allows the solutions to incubate in the capture zone/detection chamber until the next solution is released into the chamber. Detection of the fluorescence is performed after the substrate is introduced into the detection reservoir (reservoir 2). Endpoint measurements (completion of enzyme–substrate reaction) were made and compared to conventional microtiter plate methods using similar protocols. The CD ELISA platform was shown to have advantages such as lower reagent consumption, and shorter assay times, explained in terms of larger surface to volume ratio's that favor diffusion-limited processes. Since the reagents were all loaded into the CD at the same time, there was no need for manual operator interventions in between fluidic assay steps. The consistent control and repeatability of liquid propulsion removes experimental errors associated with inconsistent manual pippetting methods, for example, rinsing/washing can be carried out not only with equal volumes but equal flow conditions.

16.4.3 Multiple Parallel Assays

The ability to obtain simultaneous and identical flow rates, incubation times, mixing dynamics, and detection makes the CD an attractive platform for multiple parallel assays. Kellogg et al. have reported on a CD system that performs multiple (48) enzymatic assays simultaneously by combining centrifugal pumping in microfluidic channels with capillary valving and colorimetric detection. The investigation of multiplexed parallel enzyme inhibitor assays are needed for high throughput screening in diagnostics and in screening of drug libraries. For example, enzymatic dephosphorylation of colorless *p*-nitrophenol phosphate by alkaline phosphatase results in the formation of the yellow colored *p*-nitrophenol and inhibition of this reaction may be quantified by light absorption measurement. Theophylline, a known inhibitor of the reaction, was used as the model inhibitory compound in Kellogg et al.'s feasibility study. A single assay element on the CD contains three reservoirs: one for the enzyme, one for the inhibitor, and one for the substrate. Rotation of the CD allows the enzyme and inhibitor to pass capillary valves, mix in a meandering 100-μm wide channel, and then move to a point where flow is stopped by another capillary valve. A further increase in the rotational speed allows the enzyme/inhibitor mixture and substrate to pass through the next set of capillary valves where they are mixed in a second meandering channel and emptied into an on-disc planar cuvette. The CD is slowed, and absorption through each of the 48 parallel assay cuvettes is measured by reflectance, all in a period of 60 s, the entire fluidic process including measurement took about 3 min. The CDs were fabricated using PDMS replication techniques, with the addition of a white pigment to the poly(dimethylsiloxane) (PDMS) polymerization for enhanced reflectivity in the colorimetric measurements. The flow rates and meandering channel widths were selected such that the diffusion rate would allow 90% mixing of the solutions.

The variation in performance between the individual fluidic CD structures was quantified by carrying out the same assay 45 times simultaneously on a CD. The background corrected absorbance was measured and the coefficient of variation in the assay was ~3.2%. When the experiment was repeated on different discs the coefficient of variation ranged from 3% to 3.5%. Furthermore, variation of absorption across a single cuvette was less than 1%, confirming complete mixing. In experiments to show enzyme inhibition, 45 simultaneous reactions were carried out on the CD using fixed concentrations of enzyme and substrate and 15 concentrations of theophylline in triplicate and a complete isotherm was generated for the inhibition of alkaline phosphatase. The three remaining structures were used for calibration with known concentrations of *p*-nitrophenol. A dose response was seen over 3 logs of theophylline concentration in the range of 0.1–100 mM. The authors conclude that a large number of identical assays, with applications in rapid, high throughput screening, can be carried out on the CD platform simultaneously because of the symmetric force acting on the fluids in high quality identical microfluidic structures and that detection was simplified by rotating all the reaction mixtures under a fixed detector. In later work, the same group has extended the number of assays to 96 per CD and has investigated fluorescent enzymatic assays (Fig. 16.4).

16.4.4 Cellular-Based Assays on CD Platform

Cell-based assays are often used in drug screening (12) and rely on labor-intensive microtiter plate technologies. Microtiter plate methods may be difficult to automate without the use of large and expensive liquid handling systems and they present problems with evaporation when scaled down to small volumes. Thomas et al. have reported on a CD platform-based automated adherent cell system. This adherent cell assays involved introducing the compounds to be screened to a cell culture, then determining if the cells were killed (cell viability assay).

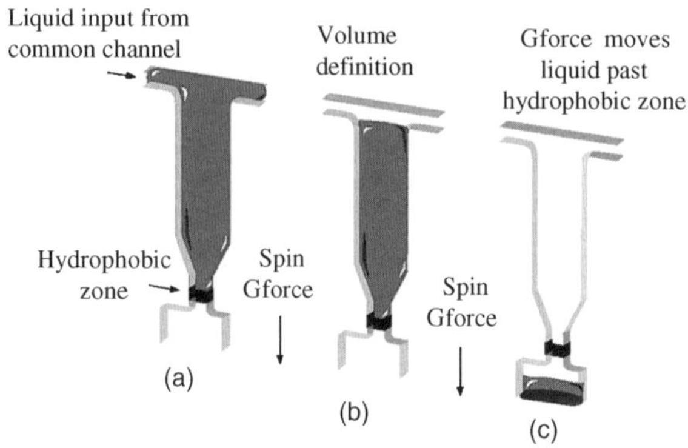

FIGURE 16.4 Schematic illustration of liquid volume definition. (**a**) The common distribution channel and liquid volume definition reservoirs are filled (by capillary forces) with a reagent to be metered. Liquid entering the reservoir does not pass the hydrophobic zone (valve) because of surface tension forces. (**b**) The CD is rotated at a rate that supplies enough centripetal force to empty the common distribution but not enough to force the liquid through the hydrophobic zone. The volume of the fluid metered is determined by the volume of the reservoir. (**c**) A further increase in the rotational speed provides enough force to move the well-defined volume of solution past the hydrophobic valve. (Application Report 101, Gyrolab MALDA SP1, Gyros AB, Uppsala, Sweden.)

Reagents for cell growth, rinsing and viability staining were serially loaded into an annular common distribution chamber and centripetal force was used for reagent loading, exchange, and rinsing of the cell growth chamber (Fig. 16.5). Individual inlets are for the addition of compounds to be screened. The plastic channels, Fig. 16.5b, were capped with a polydimethylsiloxane sheet capable of fast gas transport in and out of the culture reservoirs.

HeLA, L929, CHO-M1, and MRC-5 cell lines were cultivated on the CD device. Cell viability assays were performed, on the CD, by removing the growth medium from the cells, washing the cells with PBS, and introducing a solution of the fluorescence assay reagents into the growth chamber. The LIVE/DEAD® Viability Assay (Molecular Probes, Inc., Eugene, OR) uses a mixture of calcein green-fluorescent nucleic acid stain and the red-fluorescent nucleic acid stain, ethidium. The assay performance is based on the differing abilities of the stains to penetrate healthy bacterial cells. The calcein green-fluorescent dye will label all cells, live or dead. The red red-fluorescent ethidium stain will only label cells with damaged membranes. The red stain causes a reduction in the green stain fluorescence when both dyes are present. When the appropriate mixture of green and red stains is used, cells with intact membranes will have a green fluorescence and cells with damaged membranes will have a red fluorescence. The background remains almost completely nonfluorescent (Fig. 16.6). All liquid transfers were carried out using centripetal force from CD rotation with angular frequencies between 200 and 600 rpm. Quantitative detection of multiple cell viability assays, within 30 s, was carried out by measurement of calcein fluorescence with a CCD-based fluorescence

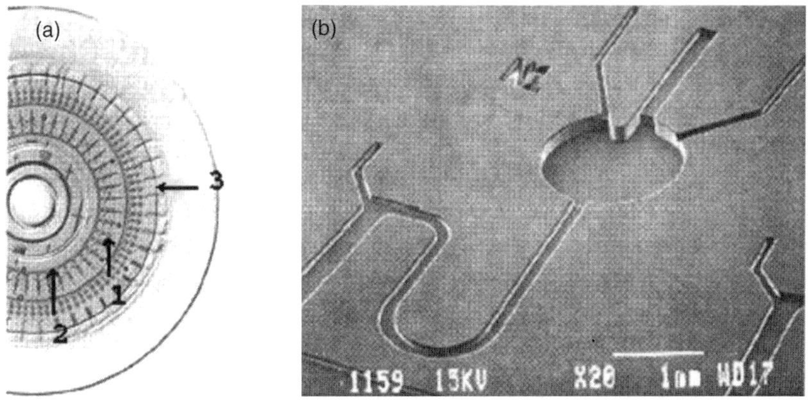

FIGURE 16.5 Microfabricated cell culture CD. (**a**) The CD caries a number of cell growth chambers (1) radially arranged around a common distribution channel (2) and is sealed with a silicone cover (3). (**b**) SEM close-up of an individual cell growth chamber and microfluidic connections. (Micro Total Analysis Systems 2000©, (2000) pp. 249–252, Nick Thomas, Anette Ocklind, Ingrid Blikstad, Suzanne Griffiths, Michael Kenrick, Helene Derand, Gunnar Ekstrand, Christel Ellström, Anders Larsson and Per Anderson "Integrated Cell Based Assays in Microfabricated Disposable CD Devices," Academic Publishers.)

imaging system. These experimental results show linear fluorescence intensity across the range of 200–4000 cells and give an indication of the potential of this platform for miniaturized quantitative cell-based assays.

In the same work, the authors reported the results of experiments designed to investigate the effect on cells of using centripetal force to move liquids. The cells tested were shown to be compatible with centripetal forges of at least $600 \times g$, much larger than the $50–100 \times g$ needed for filling and emptying cell chambers. Furthermore, it was reported that cells grown in such devices appear to show the same cell morphology as cells grown under standard conditions.

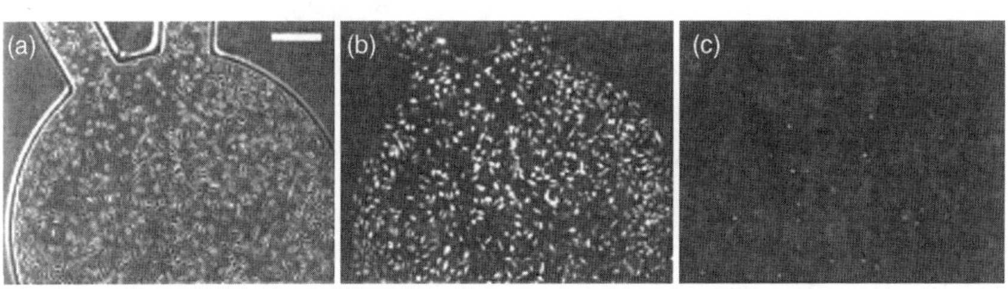

FIGURE 16.6 L929 fibroblasts cultured for 48 h in CD growth chambers. (**a**) Phase contrast (scale bar 100 μm), (**b**) epifluorescence image of calcein stained viable cells, (**b**) epifluorescence image of ethidium stained nonviable cells. (Micro Total Analysis Systems 2000©, (2000) pp. 249–252, Nick Thomas, Anette Ocklind, Ingrid Blikstad, Suzanne Griffiths, Michael Kenrick, Helene Derand, Gunnar Ekstrand, Christel Ellström, Anders Larsson and Per Anderson "Integrated Cell Based Assays in Microfabricated Disposable CD Devices," Academic Publishers.)

In separate work done by our group in collaboration with NASA Ames et al., the LIVE/DEAD® BacLight™ Bacterial Viability Kit (Molecular Probes, Inc., Eugene, OR) has been integrated to a completely automated process on CD. Disposable and reusable CD structures, hardware, and software were developed for the LIVE/DEAD assay.

The CD design for assay automation must have the following functions or properties: contain separate reservoirs for each dye and the sample, retain those solutions in the reservoir until the disc is rotated at a certain velocity, evenly and completely mix the two dyes, evenly and completely mix the dye-mixture with the sample containing the cells, collect this final mixture in a reservoir with good optical properties. Two methods for quick fabrication of prototype CDs were used. One method used molded PDMS structures. In a second method, a dry film photoresist (DF 8130, Think & Tinker, Palmer Lake, CO) was laminated onto a 1 mm thick polycarbonate disc with predrilled holes for sample introduction. The microfluidic pattern was made using a photolithographic pattern on the negative photoresist. The fluidic system was capped with a polycarbonate disc that had been laminated with an optical quality pressure sensitive adhesive (3M 8142, 3M, Minneapolis, MN). Fig. 16.7 shows the fluidic pattern for this assay. This pattern is based on the structure developed in a similar approach used to demonstrate multiple enzymatic assays on CD.

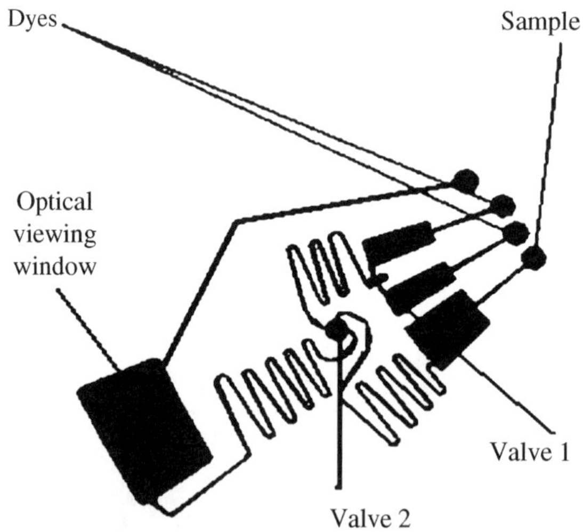

FIGURE 16.7 Microfluidic pattern for LIVE/DEAD®BacLight™ Bacterial Viability Assay. The dyes and sample are introduced into the reservoir chambers using a pipette. The dyes fill the chamber stopping at a capillary valve (valve 1). Similarly, the sample containing cells is introduced into the sample reservoir. The disc is rotated to a velocity of 800 rpm, the dyes are forced through the capillary valves and they are mixed as they flow through the switchback turns of the microfluidic channels. Simultaneously, the sample passes from the reservoir into a fluid channel where it meets the dye mixture at valve 2. The velocity of the disc is increased to 1600 rpm and the dye mixture and sample combine and mix in the switchback microfluidic path leading to the optical viewing window. (See color insert)

The dyes and sample were introduced into reservoir chambers using a pipette. The dyes fill the chamber stopping at a capillary valve (valve 1 in Fig. 16.7). Similarly, the sample containing cells was introduced into the sample reservoir. Upon rotation, the dyes were forced through the capillary valves and were mixed as they flowed through the switchback turns of the microfluidic channels. Simultaneously, the sample passed from its reservoir into a fluid channel where it met the dye mixture at valve 2 of Fig. 16.7. The velocity of the disc was increased and the dye mixture and sample combine and mix in the switchback microfluidic path leading to the optical viewing window. The dye-sample mixture is allowed to incubate in the dark at room temperature for 5 min. The optical viewing chamber was imaged twice, once with optics for the green signal and then with optics for the red signal. A typical fluorescence microscopy image of an overlay of the red and green images of stained *E. coli* is shown in Fig. 16.8.

The instrument for disc rotation and fluorescence imaging (Fig. 16.9) used a programmable rotational motor for various velocities and acceleration/deceleration rates. The use of standard microscope objectives enabled the selection of magnification. An automatic focusing system was used. The light source was a mercury lamp, which used standard low pass excitation filters for fluorescent excitation. A CCD camera was combined with standard emission filter cubes for imaging.

16.4.5 Integrated Nucleic Acid Sample Preparation and PCR Amplification

Nucleic acid analysis is often facilitated by polymerase chain reaction (PCR) and requires substantial sample preparation that, unless automated, is labor extensive. After the initial sample preparation step of cell lyses to release the DNA/RNA, a

FIGURE 16.8 Fluorescent microscopy over-laid images of red and green stained *E. coli* on CD from LIVE/DEAD®BacLight™ Bacterial Viability Assay. (See color insert)

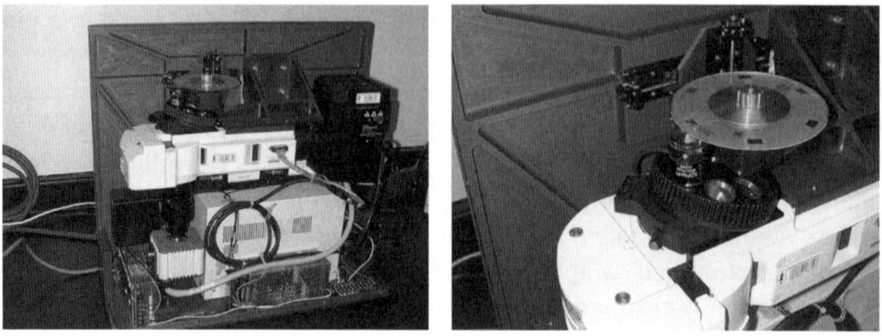

FIGURE 16.9 *Left*: Optical disc drive/imager with cover removed. Size of unit is made to fit in specific cargo bay of Space-Lab. *Right*: Zoom of microscope objectives and a disc loaded in the drive.

step must be taken to prevent PCR inhibitors, usually certain proteins such as hemoglobin, from entering into the PCR thermocyle reaction. This can be done by further purification methods such as precipitation and centrifugation, solid-phase extraction, or by denaturing the inhibitory proteins. Finally, the sample must be mixed with the PCR reagents followed by thermocyling, a process that presents difficulty in a microfluidic environment because of the relatively high temperatures (up to 95°C) required. In a small volume microfluidic reaction chamber, the liquid will be easily evaporated unless care is taken to prevent vapor from escaping.

Kellogg et al. combine sample preparation with PCR on the CD. The protocol involves the following steps: (1) mixing raw sample (5 μL of dilute whole bovine blood or *Escherichia coli* suspension) with 5 μL of 10 mM NaOH; (2) heating to 95 °C for 1–2 min (cell lyses and inhibitory protein denaturization); (3) neutralization of basic lysate by mixing with 5 μL of 16 mM tris–HCl (pH 7.5); (4) neutralized lysate is mixed with 8–10 μL of liquid PCR reagents and primers of interest; and (5) thermal cycling. The CD fluidic design is shown schematically in Fig. 16.10. Three mixing channels are used in series to mix small volumes. A spinning "platen" allows control of the temperature by positioning thermoelectric devices against the appropriate fluidic chambers. The CD contacts the PC board platen on the spindle of a rotary motor, with the correct angular alignment, which is connected by a slip ring to stationary power supplies and a temperature controller. Thermocouples are used for closed-loop temperature control and air sockets are used as insulators to isolate heating to reservoirs of interest. The thermoelectric at the PCR chamber both heats and cools and since the PCR reaction chamber is thin, 0.5 mm, fast thermocyling is achieved. Slew rates of ±2 Cs with fluid volumes of 25 μL and thermal gradients across the liquid of 0.5 °C are reported. It is important to note here that the PCR chambers were not sealed, vapor generated inside the PCR chamber is condensed on the cooler surfaces of the connecting microfluidic chamber and since the CD is rotating, the condensed drops are centrifuged back into the hot PCR chamber. This microcondensation apparatus is unique for the centrifugal CD platform. Details of the experimental parameters used can be found in the original reference, but to summarize, sample preparation and PCR

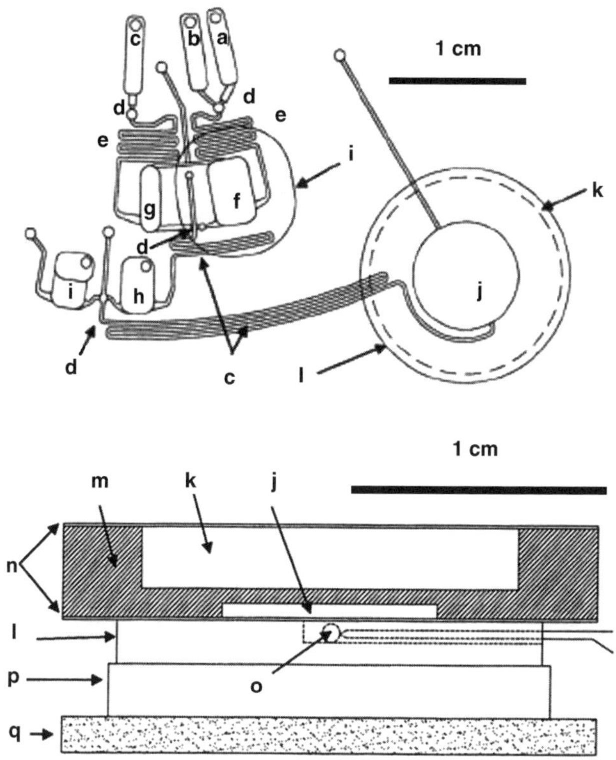

FIGURE 16.10 Schematic illustration of the CD microfluidic PCR structure. The center of the disc is above the figure. The elements are (**a**) sample, (**b**) NaOH, (**c**) Tris–HCL, (**d**) capillary valves, (**e**) mixing channels, (**f**) lysis chamber, (**g**) Tris–HCl holding chamber, (**h**) neutralization lysate holding chamber, (**i**) PCR reagents, (**j**) thermal cycling chamber, (**k**) air gap. Fluids loaded in (**a**), (**b**), and (**c**) are driven at a first RPM into reservoirs (**g**) and (**f**), at which time (**g**) is heated to 95 °C. The RPM is increased and the fluids are driven into (**h**). The RPM is increased and fluids in (**h**) and (**i**) flow into (**j**). On the right, the cross section shows the disc body (**m**), air gap (**k**), sealing layers (**n**), heat sink (**l**), thermoelectric (**p**), PC-board (**q**) and thermistor (**o**). (Micro Total Analysis Systems 2000©, Gregory J. Kellogg, Todd E. Arnold, Bruce L. Carvalho, David C. Duffy, Norman F. Sheppard, "Centrifugal Microfluidics: Applications," (2000) pp. 239–242, Academic Publishers.)

amplification for two types of samples, whole blood and *E. coli*, was demonstrated on the CD platform and shown to be comparable to conventional methods.

16.4.6 Sample Preparation for MALDI-MS Analysis

MALDI-MS peptide mapping is a commonly used method for protein identification. Correct identification and highly sensitive MS analysis require careful sample preparation. Manual sample preparation is quite tedious, time consuming, and can introduce errors common to multistep pippetting. MALDI-MS sample preparation protocols employ a protein digest followed by sample concentration, purification, and recrystallization with minimal loss of protein. Automation of the sample preparation process, without sample loss or contamination, has been enabled on the CD platform by the Gyrolab MALDI SP1 CD and the Gyrolab Workstation (Gyros AB, Sweden).

The Gyrolab MALDI SP1 sample preparation CD will process up to 96 samples simultaneously using separate microfluidic structures. Protein digest from gels or solutions are concentrated, desalted, and eluted with matrix onto a MALDI target area. The CD is then transferred to a MALDI instrument for analysis without the need for further transfer to a separate target plate. The CD fluidic structure contains functions for common reagent distribution, volume definition, valving, reverse-phase column (RPC) for concentration and desalting, washing, and target areas for external calibrants. Figure 16.11 shows the Gyrolab MALDI SP1 sample preparation CD. The CDs are loaded with reagents and processed in a completely automated, custom workstation capable of holding up to 5 microtiter plates containing samples and reagents and up to five CD microlaboratories. The reagents

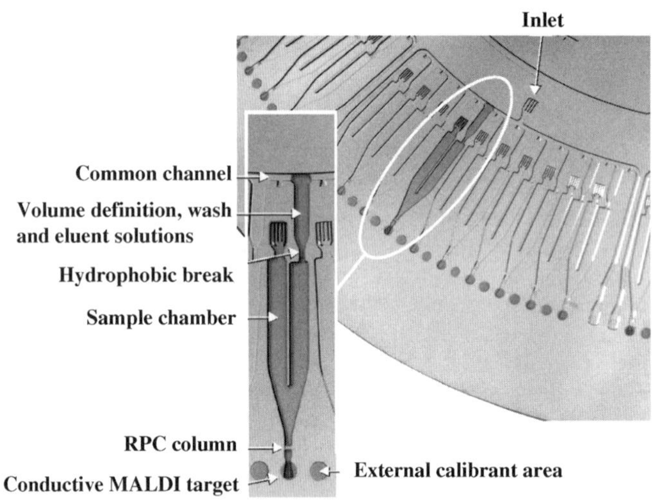

FIGURE 16.11 Image of Gyrolab MALDI SP1 sample preparation CD. The protein digest samples are loaded into the sample reservoir (see inset) by capillary action. Upon rotation, the sample passes through the RPC column. The peptides are bound to the column and the liquid goes out of the system into the waste. A wash buffer is loaded into the common distribution channel and volume definition chamber. The disc is rotated at a PRM that will empty the common distribution channel but not allow the wash solution to pass through the hydrophobic zone. A further increase of the RPM allows the well-defined volume of wash solution to pass the hydrophobic break and wash the RPC column then be discarded as waste. Next, a well-defined volume of the elution/matrix solution is loaded and passed through the column, taking the peptides to the MALDI target zone. The flow rate is controlled to optimize the evaporation of the solvent crystallization of the protein and matrix at the target zone. (Application Report 101, Gyrolab MALDA SP1, Gyros AB, Uppsala, Sweden.)

are taken from the microplates to the CD inlets using a precision robotic arm fitted with multiple needles, the liquid is drawn into specific inlets by capillary forces, and then the needles are cleaned by rinsing at a wash station. Samples are applied in aliquots from 200 nL up to 5 μL sequentially to each channel where it is contained using hydrophobic surface valves. The CD is then rotated, at an optimized rate, causing the sample to flow through an imbedded reverse phase chromatography column and liquid that passes through the column is collected in a waste container. Controlling the angular velocity dependent liquid flow rate maximizes protein binding to the column. A wash solution is introduced, by capillary action, into common distribution channels connected to groups of microstructures. The wash solution fills a volume definition chamber (200 nL) until it reaches a hydrophobic valve and the CD is rotated to clear the excess liquid in the distribution channel. Not until the rotational velocity is further increased is the defined wash volume able to pass through the hydrophobic valve and into the RPC column (SOURCE™15 RPC). The peptides are eluted from the column and directly on to the MALDI target area using a solution that contains α-cyano-4-hydroxycinnamic acid and acetonitrile using the same common distribution channel and defined volume as the previous wash step. Optimization of rotational velocity during elution enables maximum recovery and balances the rate of elution with the rate of solvent evaporation from the target surface. Areas in and around the targets are gold plated to prevent charging of the surface that would cause spectral mass shift and ensures uniform field strength. Well-defined matrix/peptide crystals form in the CD MALDI target area. Gyros reports high reproducibility, high sensitivity, and improved performances when compared to conventional pipette tip technologies. Data were shown that include: comparison of 23 identical samples, processed in parallel on the same CD, from a BSA tryptic digest and analysis of identical samples processed on different CDs, run on different days. Sensitivities were shown in the attomole to femtomole range indicating the ability for identification of low abundance proteins. The report attributed the superior performance of this platform to pretreatment of the CD surface to minimize nonspecific adsorption of peptides, reproducible wash volume and flow, and reproducible elution (volume, flow, and evaporation) and crystallization.

16.4.7 Modified Commercial CD/DVD Drives in Analytical Measurements

The commercial CD/DVD drive, commonly used for data storage and retrieval, can be thought of as a laser scanning imager. The CD drive retrieves optically generated electrical signals from the reflection of a highly focused laser light (spot size: full width at half maximum ~1 μm), from a 1.2 mm thick polycarbonate disc that contains a spiral optical track feature. The track is fabricated by injection molding and is composed of a series of pits that are 1–4 μm long, 0.15 μm deep, and about 0.5 μm wide. The upper surface of a CD is made reflective by gold or aluminum metallization and protected with a thin plastic coating. Information is generated, as the focused laser follows the spiral track, by converting the reflected light signal into digital information. A flat surface gives a value of zero; an edge of a pit gives a value of one. The data are retrieved at a constant acquisition rate and the serial values (0/1) are converted to data of different kinds for various applications (music, data, etc.). In addition to the code generated by the spacing of the pits, optical signals necessary for focusing, laser tracking of the spiral track, and radial position determination of the read head are monitored and used in feedback loops for proper CD operation. The laser is scanned in a radial direction toward the outer diameter of the disc with an elaborate servo that maintains both lateral tracking and vertical focusing.

Researchers have taken advantage of this low cost, high resolution optical platform in analytical DNA array applications. Barathur et al. (14) from Burstein Technologies (Irvine, CA), for example, have modified the normal CD drive for use as a sophisticated laser-scanning microscope for analysis of a Bio Compact Disk™ assay, where all analysis is carried out in microfluidic chambers on the CD. The assay is carried out concurrently with the normal optical scanning capabilities of a regular CD drive. The authors report on the application of this device for DNA micro spot-array hybridization assays and comment on its use in other diagnostic and clinical research applications. For the DNA spot-array application, arrays of captures probes for specific DNA sequences are immobilized on the surface of the CD in microfluidic chambers. Sample preparation and multiplexed PCR, using biotinylated primers, are carried out off-disc, then the biotinylated amplicons are introduced into the array chamber and hybridization occurs if amplicons with the correct sequence are present. Hybridization detection is achieved by monitoring the optical signal from the CD photo detector, while the CD is rotating. To generate an optical signal when hybridization has occurred a reporter is used, for the Bio Compact Disk™ assay the reporter is a streptavidin-labeled microsphere that will bind only to the array spots that have successfully captured biotinylated amplicons. The unbound microsphere reporters are removed from the array using simple centrifugation and no further rinsing is needed. As the laser is scanned across the CD surface, the microparticle scatters light that would have normally been reflected to the photodetector resulting in less light on the detector (bright field microscopy) and a distinctive electronic signal is generated. The electronic signal intensity data can be stored in memory then deconvoluted into an image. A 1 cm^2 microarray can be scanned in 20–30 s with a data reduction time of 5 min and custom algorithms that perform the interpretations in real time. Data were shown for identification of three different species of the *Brucella* coccobacilli on the CD platform. Human infection occurs by transmission from animals by ingestion of infected food products, contact with an infected animal, or inhalation of aerosols. Multiplex PCR-amplified DNA from all three species (common forward primers and specific reverse primers resulting in amplicons of different length for various species were used for verification of PCR on external gels) was incubated on arrays with species-specific capture probes. Removal of one of the species in the sample resulted in no probes present on that specific array spot verifying the specificity of the assay.

Alexandre et al. (15) at Advanced Array Technologies (Namur, Belgium), utilize the inner diameter area of a CD and standard servo optics for numerical information and operational control and employ a second scanning laser system to image DNA arrays on transparent surfaces at the outer perimeter of a CD. The second laser system, consisting of a laser diode module that illuminates a 50 μm spot on the CD surface, is scanned radially at a constant linear velocity of 20 mm/min while the CD is rotating. Each CD contains 15 arrays arranged in a single ring on the CD perimeter that extends in the radial direction for 15 mm. The arrays are rectangular and consist of 4 rows and 11 columns of 300 μm spots. The normal CD servo optics are located below the disc and the added imaging optics are above the disc. A photodiode head follows the imaging laser and the refracted light intensity is stored digitally at a high sampling rate. An image of each array on the disc is reconstructed by deconvolution of the light intensity data. The entire CD can be scanned in less than 1 min producing a total of 6 MB of information. Sample preparation and PCR amplification was carried out off-disc. Specific DNA capture probes were spotted on the surface of the CD using a custom arrayer that transfers the probes from a multiwell plate on to the surface of up to 12 discs using a robotic arm. Biotinylated amplicons are introduced on to the array chambers (one chamber for each array) and hybridization occurs if amplicons with the correct sequence are present. In order to get an optical signal that can be detected, after a rinse step, a solution of streptavidin-labeled colloidal gold particles is applied to the array followed by a Silver Blue solution (AAT, Namur, Belgium). The silver solution causes silver metal to grow on the gold particles thereby making the hybridization-positive micro-array spots refractive to the incident laser light. Results were shown for the detection of the five most common species of *Staphylococci* and an antibiotic resistant strain. The *fem A* and *mec A* genes of the various species of *Staphylococci* were amplified by primers common to all *Staphylococci* species then hybridized to a microarray containing spots with probes specific for the different *Staphylococci* species. The array also included a capture probe for the genus *Staphylococci* and a probe for the *mec A* gene that is associated with methicillin resistance of the *Staphylococci* species. The results were digitized and quantified with software that is part of the custom Bio-CD™ workstation. Signal-to-noise ratios were above 50 for all positive signals.

16.4.8 Microarray Hybridization for Molecular Diagnosis of Infectious Diseases

In recent years, microarrays have become important tools for nucleic acid analysis and gene expression profiling. The expression of thousands of genes can be monitored in a single experiment using this technology. A number of investigators have attempted to adapt this technology to rapidly detect infectious agents in clinical specimens for diagnostic purposes (16–20). However, such systems are still in their infancy and most of them require technologically complex biochips with integrated heating/cooling systems (21). Peytavi et al. (22) have reported a CD-based microfluidic platform for DNA microarray analysis of infectious disease, presenting an elegant solution to automate and speed up microarray hybridization. Staphiloccocal specific oligonucleotides were used as capture probes immobilized in 4 × 5 arrays of 125 μm spots on a standard 3 × 1 in. glass slide. The layout of the array is shown in Fig. 16.12a. A flow cell is designed to realize the self-contained hybridization process in the CD platform. As shown in Fig. 16.12b, the flow cell consists of a hybridization column 1, aligned with the DNA microarray on the glass slide, a sample chamber 2, and rinsing chambers 3 and 4. The reagent chambers are connected to the hybridization column with a microchannel which is 50 μm in width and 25 μm in depth. The flow cell is aligned with and adhered to the glass slide to form a DNA hybridization detection unit, up to 5 of which can be mounted into the CD platform fabricated from acrylic plastic using CNC machining (Fig. 16.12c). The reagents are positioned to be pumped through the hybridization column by centrifugal force in a sequence beginning with chamber 2 up to chamber 4 and this flow sequence is achieved by manipulating the balance between the capillary force and centrifugal pressure. The sample (chamber 2) is released first and flows over the 140 nL hybridization chamber (chamber 1) where the oligonucleotide capture probes are spotted onto the glass support. The rinsing buffers (chamber 3 and 4) are then released sequentially at a higher angular velocity and are used to wash the nonspecifically bound targets following the hybridization process.

This custom microarray hybridization microfluidic platform is easy-to-use, automated, and rapid. It uses standard glass slides that are compatible with commercial arrayers and standard commercial scanners found in most academic departments. In this removable microfluidic system, the hybridization chamber is composed of a low cost elastomeric material, PDMS using standard molding methods [i], engrafted with a microfluidic network. This elastomeric material reversibly sticks to the glass slide without any adhesives or chemical reactions, forming the microfluidic unit. Placed onto a plastic compact disc-like support, the microfluidic units are spun at different speeds to control fluid movements. To simplify hybridization experiments using this device, buffer compositions and capture probe sequences were optimized to be compatible with room temperature hybridizations to prevent the need for a heating device. Furthermore, this microfluidic system allows to drastically reduce the volume of reagents needed for microarray hybridizations and does not require a PCR amplicon purification step that may be time consuming.

In a passive hybridization system, a hybridization event requiring collision between a capture probe and the analyte relies solely on diffusion. In such systems, sensitivity is increased by using longer hybridization periods (23,24). One advantage of flow-through hybridization is that the probability of collision between the probe and the analyte is increased by the much shorter diffusion distance allowed by the shallow hybridization chamber, thereby accelerating the hybridization kinetics (25,26). In the study it was shown that for the same concentration of 15-mer oligonucleotides or 368 bp amplicons, a 5-min flow-through hybridization increased the kinetics of hybridization respectively by a factor of 2.5 and 7.5 in comparison with the passive hybridization. These results are in line with a previous study. Using a microfluidic system, Chung et al. have shown a sixfold rate increase between flow-through hybridization versus passive hybridization. However, this system required a 30 min hybridization step (27). Interestingly, the difference

(a)

(b)

(c)

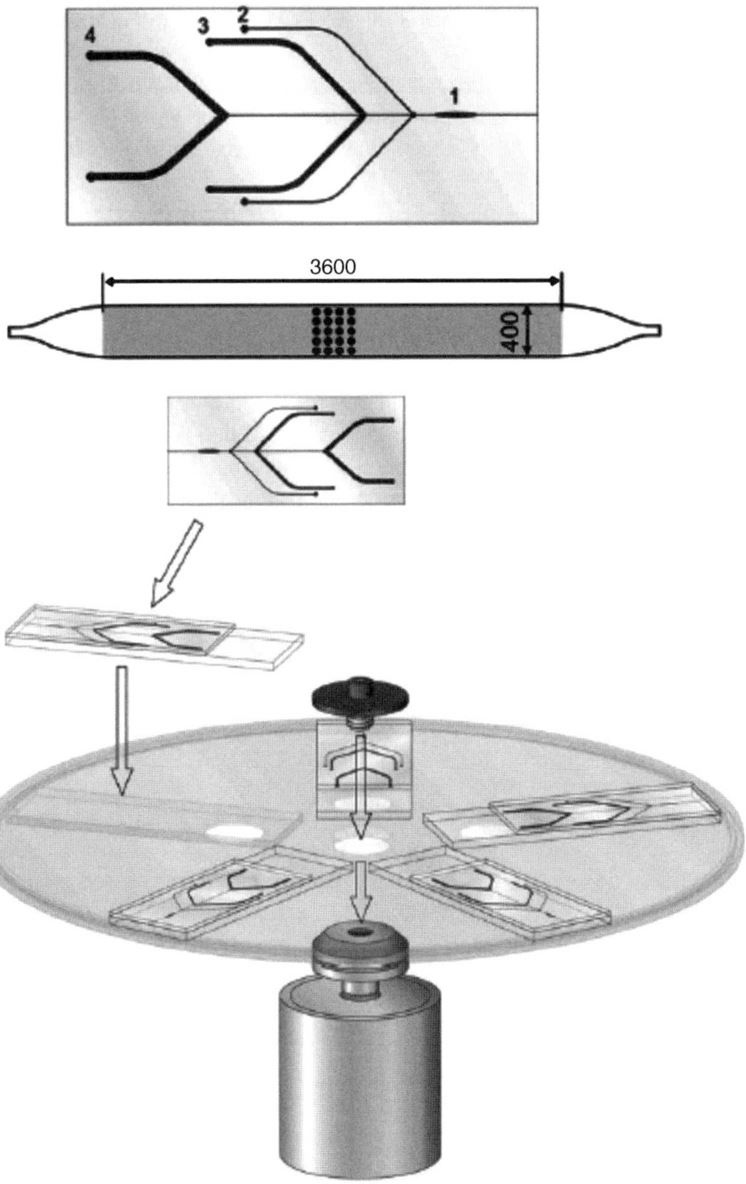

FIGURE 16.12 Schematic representation of the microfluidic system. (**a**) PDMS microfluidic unit: The test sample (chamber 2) is released first and flows over the hybridization chamber (chamber 1) where the oligonucleotide capture probes are spotted onto the glass support. The wash buffer in chamber 3 and the rinsing buffer in chamber 4 then start to flow at a higher angular velocity. (**b**) Schematic view of the hybridization chamber showing the dimension in micrometers and the area of the chamber (shaded section) that can accommodate up to 150 microarray spots. Layout of the staphylococcal microarray used in the present study is also showed (five capture probes for each species). (**c**) Engraved PDMS is applied to a glass slide on which are arrayed nucleic acid capture probes. The glass slide is placed on a compact disc support that can hold up to five slides.

between passive and flow-through hybridization was about three times more important for the amplicons as compared to the shorter 15-mer oligonucleotides. This could be explained by the higher diffusion coefficient of the smaller oligonucleotide molecules.

To be used for clinical applications, in addition to being rapid and inexpensive, a molecular test should be sensitive and specific. In 5 min of hybridization, the CD system showed a detection limit of 500 amol of amplified target. This result is comparable with results obtained with more complex microfluidic devices (28). One system, using chemiluminescence, shows a detection limit of 250 amol, but requires a 3-h hybridization time (29). To detect a significant fluorescent signal, an amplification step is required with microarray technology. The CD microfluidic system reported allows detection of amplicons amplified from 10 bacterial genome copies that is at least 1000 times more sensitive than results obtained by other groups showing microarray hybridization using microfluidic devices (30).

In terms of specificity, the CD system was able to discriminate four different *Staphylococcus* species using a post-PCR hybridization protocol of only 15 min The *S. aureus* probe designed with only one mismatch in the *S. epidermidis* amplicon sequence did not show any significant cross hybridization. This clearly demonstrates the possibility to discriminate one SNP using the CD system at room temperature and with only 10 μL of washing and rinsing buffer. This SNP discrimination capacity will allow a rapid identification of bacteria and their antibiotic resistance genes.

16.4.9 Cell Lysis on CD

There are many types of cell lysis methods used today that are based on mechanical (31), physical–chemical (32), chemical (33), and enzymatic, principles. The most commonly used methods in biology research labs rely on chemical and enzymatic principles. The main drawbacks of those procedures include intensive labor, adulteration of cell lysate, and the need for additional purification steps. To minimize the required steps for cell lysis, a rapid and reagentless cell lysis method would be very useful. Recently, cell lysis has been demonstrated by Kim et al. (34) on a microfluidic CD platform. In this purely mechanical lysis method, spherical particles (beads) in a lysis chamber microfabricated in a CD, cause disruption of mammalian (CHO-K1), bacterial (*E. coli*), and yeast (*Saccharomyces cerevisiae*) cells. Investigators took advantage of interactions between beads and cells generated in rimming flow (35,36) established inside a partially filled annular chamber in the CD rotating around a horizontal axis (Fig. 16.13). To maximize bead–cell interactions in the lysis chamber, the CD was spun forward and backward around this axis, using high acceleration for 5–7 min. Cell disruption efficiency was verified either through direct microscopic viewing or measurement of the DNA concentration after cell lysing. Lysis efficiency relative to a conventional lysis protocol was approximately 65%. Experiments identified the relative contribution of control parameters such as bead density, angular velocity, acceleration rate, and solid volume fraction.

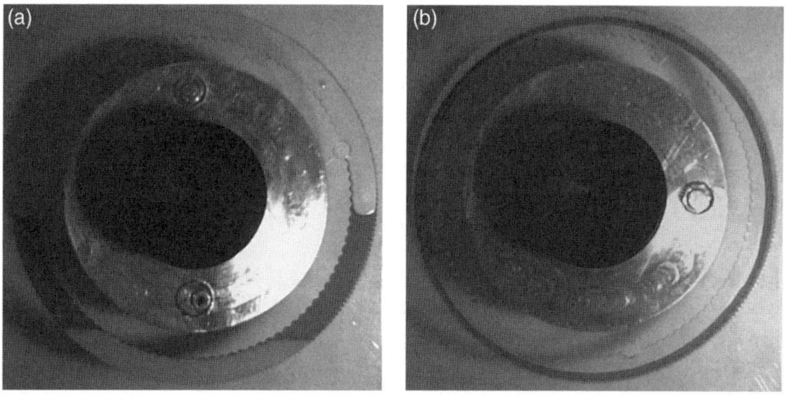

FIGURE 16.13 Flow patterns for two rotational states of the CD (**a**) At rest: beads are sedimented at the bottom of the annular chamber; (**b**) While spinning: two circumferential bands of beads (lighter) and liquid (darker) are observed. (See color insert)

More recent work (37) by the same investigators used the multiplexed lysis design shown in Fig. 16.14. Bead–cell interactions for lysing arise while the beads and cells are pushed back and forth (by switching the CD rotational direction) through a continuously narrowing chamber wall. This phenomenon is called the *key-stone effect*. There are two interaction forces associated with the key-stone effect: collision induced by the geometry and friction due to a velocity gradient set up along the chamber (i.e., fast in the core and slow around the wall). The investigators used real-time PCR to characterize the performance of this CD design and achieved 95% lysis efficiency of *B. globigii* spores.

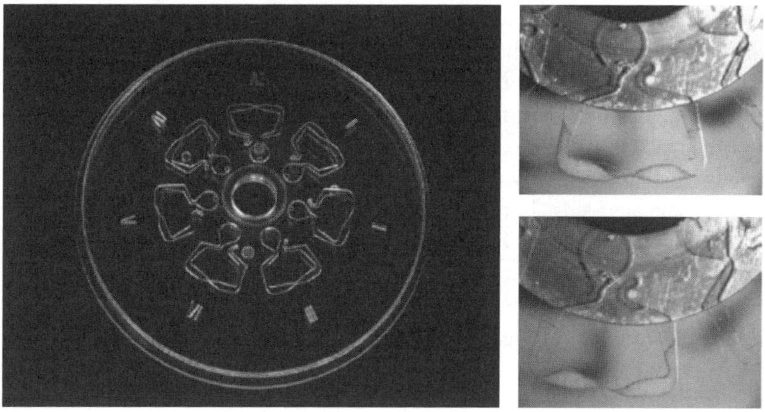

FIGURE 16.14 *Left*: photograph of CD. *Right*: Still images of a rotating CD (zirconia–silica beads and water loaded). (Upper right) More beads are observed on the left because of a rapid stop from a clockwise rotation. (Lower right) More beads are on the right because of a rapid stop from a counterclockwise rotation.

All prototype CDs in this work were fabricated using photolithography and PDMS molding. For the purpose of mechanical cell disruption, an ultra-thick SU-8 process was developed to fabricate a mold featuring high structures ($\sim$1 mm) so that sufficiently high lysing chambers could be formed in the PDMS.

16.5 CONCLUSIONS

In comparing miniaturized centrifugal fluidic platforms to other available microfluid propulsion methods we have demonstrated how CD-based centrifugal methods are advantageous in many analytical situations because of their versatility in handling a wide variety of sample types, ability to gate the flow of liquids (valving), simple rotational motor requirements, ease and economic fabrication methods, and large range of flow rates attainable. Most analytical functions required for a lab-on-a-disc, including; volume definition, dilution, mixing, calibration, separation have all been successfully demonstrated in the laboratory. Moreover, the possibility of maintaining simultaneous and identical flow rates, to perform identical volume additions, to establish identical incubation times, mixing dynamics, and detection in a multitude of parallel CD assay elements makes the CD an attractive platform for multiple parallel assays. The platform has been commercialized by Tecan Boston for high throughput screening, by Gyros AB for sample preparation techniques for MALDI and by Abaxis (in a somewhat larger and less integrated rotor format compared to the CD format) for human and veterinary diagnostic blood analysis. The Abaxis system for human and veterinary medicine uses only dry reagents, but for many diagnostic assays, requiring more fluidic steps, there are severe limitations in progressing toward the "lab-on-a-disc" goal as liquid storage on the disc becomes necessary. In high throughput screening situations, the CD platform is being coupled to automated liquid reagent loading systems and no liquids/reagents need to be stored on the disc. The latter made the commercial introduction of the CD platform for HTS somewhat simpler. There is an urgent need though for the development of methods for long-term reagent storage that incorporate both liquid and vapor barriers to enable the introduction of lab-on-a-disc platforms for a wide variety of fast diagnostic tests. One possible solution to this problem involves the use of lyophilized reagents with common hydration reservoir feeds, but the issue in this situation becomes the speed of the test as the time required for redissolving the lyophilized reagents is often substantial.

The CD platform is easily adapted to optical detection methods because it is manufactured with high optical quality plastics enabling absorption, fluorescence, and microscopy techniques. Additionally, the technology developed by the optical disc industry is being used to image the CD at the micron resolution and move to DVD and HD DVD will allow submicron resolution. The latter evolution will continue to open up new applications for the CD-based fluid platform. Whereas today the CD fluidic platform may be considered a smart microcentrifuge, we believe that in the future the integration of fluidics and informatics on the DVD and

HD DVD may lead to a merging of informatics and fluidics on the same disc. One can then envision making very sharp images of the bacteria under test and correlate both test and images with library data on the disc.

ACKNOWLEDGMENT

Funding provided by Genome Canada.

REFERENCES

1. Manz E, Verpoorte C, Effenhauser S, Burggraf N, Raymond DE, J Harrison DJ, Widmer HM. *J High Resolut Chromatogr* 1993;16:433–436.
2. Caliper Home Page:http://www.calipertech.com/.
3. Tecan-Boston Home Page: Look for LabCD-ADMET system:http://www.tecan-us.com/us-index.htm.
4. Abaxis Home Page:http://www.abaxis.com/.
5. Madou MJ. Fundamentals of Microfabrication. 2nd ed. New York: CRC Press; 2002.
6. Miyazaki S, Kawai T, Araragi M. Proceedings of IEEE Micro Electro Mechanical Systems (MEMS);1991. p283–288.
7. Jorgenson JW, Guthrie E. *J Chromatogr* 1983;255:335–348.
8. Harrison DJ, Fan Z, Fluri K, Seiler K. Solid State Sensor and Actuator WorkshopS. C; Hilton Head Island, SC;1994 p21–24.
9. Duffy DC, Gills HL, Lin J, Sheppard NF, Kellogg GJ. *Anal Chem* 1999;71:4669–4678.
10. Madou MJ, Kellogg GJ. *Syst Technol Clini Diagn Drug Discov* 1998;259:80–93.
11. Duffy DC, McDonald JC, Schueller OJA, Whitesides GM. *Anal Chem* 1998; 70:4974–4984.
12. Burbaum J. *Drug Discov Today* 1998;3:313–322.
13. Zoval JV, Boulanger R, Blackwell C, Borchers B, Flynn M, Smernoff D, Landheim R, Mancinelli R, Madou MJ. SETI Institute, Orbital Sciences, Dysyscon Inc., Stanford University and NASA Ames. Manuscript in preparation.
14. Barathur R, Bookout J, Sreevatsan S, Gordon J, Werner M, Thor G, Worthington M. *Psychiatric Genetics* 2002;12:193–206.
15. Alexandre I, Houbion Y, Collet J, Hamels S, Demarteau J, Gala J-L, Remacle J. *Biotechniques* 2002;33:435–439.
16. Wang Y, Vaidya B, Farquar HD, Stryjewski W, Hammer RP, McCarley RL, Soper SA, Cheng YW, Barany F. *Anal Chem* 2003;75:1130–1140.
17. Mikhailovich V, Lapa S, Gryadunov D, Sobolev A, Strizhkov B, Chernyh N, Skotnikova O, Irtuganova O, Moroz A, Litvinov V, Vladimirskii M, Perelman M, Chernousova L, Erokhin V, Zasedatelev A, Mirzabekov A. *J Clin Microbiol* 2001;39:2531–2540.
18. Bekal S, Brousseau R, Masson L, Prefontaine G, Fairbrother J, Harel J. *J Clin Microbiol* 2003;41:2113–2125.
19. Bavykin SG, Akowski JP, Zakhariev VM, Barsky VE, Perov AN, Mirzabekov AD. *Appl Environ Microbiol* 2001;67:922–928.

20. Westin L, Miller C, Vollmer D, Canter D, Radtkey R, Nerenberg M, O'Connell JP. *J Clin Microbiol* 2001;39:1097–1104.

21. Fan HZ, Mangru S, Granzow R, Ho PW, Dong Q, Kumar R. *Anal Chem* 1999;71:4851–4859.

22. Peytavi R, Raymond FR, Boissinot K, Picard FJ, Boissinot M, Bissonnette L, Ouellette M, Bergeron G, Jia G, Zoval J, Madou M. Anal Chem-2005, Submitted for publication.

23. Chan V, Graves DJ, McKenzie SE. *Biophys J* 1995;69:2243–2255.

24. McQuain MK, Seale K, Peek J, Fisher ST, Levy S, Stremler MA, Haselton FR. *Anal Biochem* 2004;325:215–226.

25. Bringuier E, Bourdon, *A Phys Rev E* 2003;67:011404.

26. Axelrod D, Wang MD. *Biophys J* 1994;66:588–600.

27. Chung YC, Lin YC, Shiu CZ. *Lab Chip* 2003;3:228–233.

28. Liu RH, Lenigk R, Druyor-Sanchez RL, Yang J, Grodzinski P. *Anal Chem* 2003;75:1911–1917.

29. Cheek BJ, Steel AB, Torres MP, Yu YY, Yang H. *Anal Chem* 2001;73:5777–5783.

30. Lenigk R, Liu RH, Athavale M, Chen Z, Ganser D, Yang J, Rauch C, Liu Y, Chan B, Yu H, M Ray R, R Marrero R, P Grodzinski P. *Anal Biochem* 2002;311:40–49.

31. Carlo DDI, Jeong K-H, Lee LP. *Lab Chip* 2003;3:287–291.

32. Lee W, Tai Y-C. *Sens Actuators A* 1999;73:74–79.

33. Sambrook J, Russell DW. Molecular Cloning. Cold Spring Harbor, NY: CSHL Press;2001.

34. Kim J, H Jang SH, Jia G, Zoval JV, Da Silva NA, Madou MJ. *Lab Chip* 2004;4:516–522.

35. Ruschak KJ, Scriven LE. *J Fluid Mech* 1976;76:113–125.

36. Thoroddsen ST, Mahadevan L. *Exp Fluids* 1997;23:1–13.

37. Kim J, Kido H, Zoval J, Peytavi R, Picard FJ, Gagn D, Bergeron MG. manuscript in preparation.

17

MICROFLUIDIC TECHNIQUES FOR POINT-OF-CARE *IN VITRO* DIAGNOSTICS

Emil P. Kartalov, Axel Scherer and Clive Taylor

17.1 INTRODUCTION

The ongoing revolution in biological sciences has generated high hopes for the advent of true personalized/preventive medicine. While the necessary biological tools are being developed at a fast pace, it has become clear that cost, operation, and manufacturability are equally challenging issues that must be appropriately addressed before these new methods can be widely accepted in medical practice. This is especially true in today's world of skyrocketing medical costs, shrinking insurance coverage, and rapid growth of the elderly and total population.

In the particular case of diagnostics, decentralized "near-patient" or "point-of-care" (PoC) testing has sought to provide rapid quantitative results at the bedside or in the clinic, thereby facilitating early detection and prevention, decreasing hospital stays, reducing transportation and administrative expenses, and decreasing errors from mishandling and miscommunication. While a few systems have been developed (e.g., the various glucose meters), the enormous potential for decentralized testing largely remains untapped, as the vast majority of medical diagnostic tests are still conducted in centralized hospital or clinical reference laboratories using high throughput methods and large equipments.

We are of the opinion that PoC diagnostic methods are unlikely to enter widespread use without new miniaturization solutions that produce accurate, reliable performance while simultaneously addressing logistic, economic, and

Biological Applications of Microfluidics, Edited by Frank A. Gomez

manufacturing issues. Accordingly, herein we describe two recent advances in elastomer microfluidics—high throughput multiantigen microfluidic fluorescence immunoassays (1) and microfluidic vias (2)—which are relevant to the practical development of PoC assay devices.

Accordingly, this chapter is divided into two major sections. First, we describe the current state-of-the-art commercial blood-test benchtop systems, outline the limitations of their approach, discuss miniaturization as an alternative, and offer microfluidic immunoassays (1) as a promising solution. We include a comprehensive discussion of the respective technological limitations, requirements for ubiquitous utilization, and challenges to industrialization. Second, we describe the new technique of microfluidic vias (2) and discuss some derivative devices and their potential application to biomedical diagnostics.

17.2 MICROFLUIDIC IMMUNOASSAYS

17.2.1 Benchtop Commercial Systems

Currently, large high throughput instruments (e.g., by Abbott, Roche, and Beckman) conduct the bulk of clinical blood tests and in essence constitute a "gold standard" for clinical laboratory assays. These instruments were engineered to maximize efficiency and economy in a centralized testing model. To a degree, their technical requirements and costs determine the organizational structure of clinical laboratory practice in developed countries.

17.2.1.1 Requirements and Costs For example, with the Roche Elecsys® system, 30 μL (for the 1010) and 20 μL (for the 2010) of sample are quoted as sufficient for one test of one patient. In practice, the amount used is 0.5–1 mL per test per patient, as some volume is lost in preparing serum from whole blood, each measurement uses about 250 μL, and many tests are done in duplicate to minimize the effect of potential errors. Thus, multiple tests on the same patient require a sample size on the order of 1–10 mL. Such volumes cannot be obtained by a simple finderprick. Instead, blood must be drawn from the vein. In doctors' offices, clinics, and hospitals, blood samples are obtained by a phlebotomist or another trained health care practitioner to ensure the quality of the sample and minimize the adverse effects of venipuncture (e.g., hematoma, infection). In many instances, the phlebotomy procedure necessitates extra delays (e.g., wait for the patient, an extra appointment, and extra transportation) incurring extra costs. In addition, the sample volume requirements pose particular problems for multiple blood tests of pediatric patients.

On the financial side, a large, freestanding or desktop clinical laboratory analyzer may cost upwards of $100,000 and require skilled technicians to maintain and operate on a daily basis. Consequently, its usage is restricted to centralized facilities, for example, hospital or commercial reference laboratories. Centralization means samples are collected elsewhere and then transported to the facility. This process incurs additional costs and delays, while allowance must be made for sample degradation prevention as well as the potential biohazard posed by some samples.

At the central laboratory, workflow is not uniformly distributed–heavy daily loads generate bottlenecks resulting in extra—expenses and delays, while instruments and staff operate below capacity when input is low, for example, during weekends and night shifts in hospitals. Further, a single poorly prepared serum sample may cause an obstruction in a large instrument and thus shut it down for maintenance. Delays and unused capacity add to the "per test" cost of the operation.

17.2.1.2 Throughput and Size Elecsys processing time could be as low as 9 min for a particular test in an emergency. However, the instrument cannot be operated cost effectively in this mode for more than a small fraction of the total test load. The practice of batch processing of samples for identical tests means a particular sample will not generally be processed until enough samples are collected to run the respective test at instrumental capacity. The 1010 carousel is designed for a panel of 6 tests and a batch of 66 samples, and produces 50 results per hour; the 2010 one handles 15 tests and outputs 85 results per hour. These technical and logistic factors, particularly transport, produce a mode of operation under which it is not uncommon to take a week or more for the results to become available. This amount of delay provokes anxiety in the patient, extends hospital stay for inpatients, and may interfere with the timeliness of medical intervention and care for outpatients.

The Elecsys is quite versatile, offering measurements of a list of about 50 analytes pertaining to infectious disease, cardiac condition, tumors, thyroid function, bone, and others. The Elecsys 1010 is hardly portable, weighing 110 kg and measuring 95 × 65 × 65 cm. At best, it could conceivably be installed in a vehicle but would require an extra power source supplying 110 V and 610 W, while shock resistance is low due to the glass tubes and mechanics of the robotic sample retrieval. Elecsys 2010 has similar limitations and is about 30% larger and heavier.

A further step toward higher throughput (and lower per test cost) is Roche E170. This instrument is quoted to require a 10–50 μL sample volume and offers a panel of 25 tests at a time among the same selection of about 50. The output is 170 results per hour. However, the E170 occupies 4 m^3, weighs 750 kg, and requires 4 kW power. The E170 may also be used as one of the modules in MODULAR *ANALYTICS*, which offers combined measurements of substrates, specific proteins, drugs, ions (ISE) among others with a throughput of several hundred or thousands of results per hour. If all four modules are E170s, the throughput is 680 results per hour. However, the total system then weighs 2850 kg, requires four times the space, and consumes 12 kW power. Other systems, for example, offered by Abbott and Beckman Coulter, have comparable characteristics and scale in a similar way.

Several recurrent themes can be discerned in reviewing modern clinical laboratory analyzers. In routine clinical practice, there are about 50 blood protein analytes that are measured through immunoassays. The claimed sample size is usually on the order of tens of microliters, but in practice the logistics of collection and processing require multiples of this volume. A linear increase in throughput and simultaneously loaded tests (batch size) requires a roughly linear increase in size, weight, power consumption, and cost of the instrument, although cost per test will decline if the instrument can be operated at optimal capacity. A strong desire for comprehensive integrated test panels is offset by a linear modularity resulting in very large systems. While the smallest of these systems may be used in smaller hospitals, clinics, and consolidated "doctors' offices," effective portable systems require different approaches as described below.

17.2.2 Improvements Offered by Miniaturization

17.2.2.1 Savings and Portability Decentralized testing requires small robust instruments that require less maintenance and can be easily operated by less skilled, nonlicensed personnel, while the accuracy and the reliability would have to compare well to the "golden standard" of the larger centralized testing process. Such instruments would allow full blood tests to be conducted at the hospital bedside or in every physician's office, eliminating transportation and handling requirements and their concomitant costs and delays. If the needed sample volume is greatly reduced, then a fingerprick could suffice, eliminating the phlebotomist appointments as well as making some blood tests far more accessible to pediatric patients.

Easier, faster, and less expensive testing would also facilitate earlier detection and diagnosis, resulting in potentially critical benefits for the health of the patient, and at the same time reducing or eliminating costs for treating of more advanced stages of the disease. As 20% of the insured use 80% of the resources in modern health care, especially through developed chronic conditions, the utility of early detection and prevention has become self-evident. In addition, if the devices for clinical testing are miniaturized, rendered simpler to use, and are more robust and shock resistant in operation, then rapid and accurate blood tests would become a reality in field conditions, such as in law enforcement or in military settings. These proposed improvements and their consequences are summarized in a block diagram in Fig. 17.1.

17.2.2.2 Commercial Examples One major development toward miniaturization of a limited range of blood tests is the iSTAT system. A cartridge is filled through a syringe or a pipette with a blood sample as small as 20 μL, and then inserted in a handheld device, which produces readout in 2 min. Each cartridge is hardwired to a specific set of several tests. The measurements are based on ion-selective electrode potentiometry (for ions, pH, urea), electrical conduction (for hematocrit), or generated current (for glucose). This system is already in widespread use in the hospital setting, especially for quick testing immediately before surgery. However, with the exception of a cardiac marker (cTnl), iSTAT currently does not offer measurements of blood proteins as do the large systems discussed above.

A small desktop system, Roche Diavant's Reflotron®, uses reflection to detect changes in color in the test strips. The system measures 15 analytes (including lipids, glucose, metabolites, and enzymes), produces the result within 3 min, outputs 18–30 results per hour, occupies about a cubic foot without the computer, and weighs 5.3 kg.

Reflotron® and iSTAT® are excellent examples of a physician's hopes and expectations for the future of *in vitro* diagnostics—multiple fast measurement capability in a small, inexpensive package, using small amounts of sample, with a per test cost that is acceptable considering the accrued advantages of PoC function. At present, neither system offers the standard extensive blood panels of their large robotic counterparts.

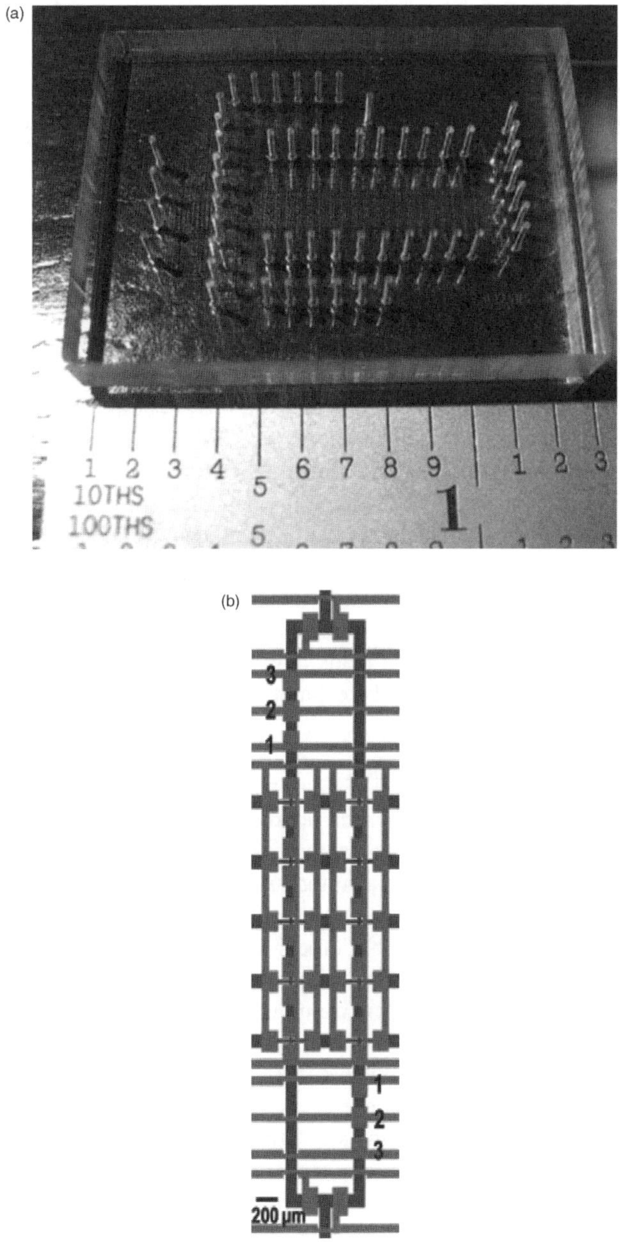

FIGURE 17.1 Multiplexed microfluidic fluorescence immunoassays. (**a**) A 100-chamber PDMS chip bound to an epoxide slide allows 5 tests of each of 10 samples, with 2 chambers per sample–test combination. The scale is in inches. (**b**) Functional diagram of an individual coliseum (the chip contains 10 such coliseum). Here, control channels and valves are drawn in red and flow channels in blue. Vertical and horizontal comb-like valve arrays enclose a pair of immunoassay chambers for each of the five tests. Valve arrays 1, 2, and 3 pump the sample in a circle along the coliseum (e.g. clockwise for actuation order {1, 2, 3}) with a lap time of 20 s. The fundamental technology (3,5,7) allows expansion of the test matrix to 50×50, in which case a desktop machine could reach instrumental throughput of 15,000 results per hour while using less than 1 μL per sample. (See color insert)

This limitation will only be overcome by taking essentially the immunoassay methods and reagents used in large laboratory platforms and "scaling them down" to work in true portable devices.

17.2.2.3 Challenges The major hurdle to overcome in achieving high throughput multiantigen immunoassays is the need for simultaneous handling of multiple samples for assays using multiple reagents, for which the function and detection methods are basically the same. This obstacle is one reason why the iSTAT does not perform multiplexed immunoassays but instead uses different methods (potentiometry, ohmmetry, or amperometry) to detect chemically orthogonal analytes, ensuring that diffusion and cross contamination are not the issues. By contrast, Elecsys-like systems measuring blood proteins through immunoassays have to use the same method for all samples and tests (antibodies and optical detection), while the available optical bandwidth in standard fluorescence/chemiluminescence is too small for large-scale multiplexing and cross contamination poses a significant technical challenge.

Presently, the only orthogonality in immunoassays is provided by antigen–antibody recognition methods, which require that tests are compartmentalized to prevent diffusion and to ensure unambiguous attribution of the detected signal. Such compartmentalization leads to extensive robotic fluid handling in the macroworld of large laboratory analyzers and is a major obstacle to miniaturization.

17.2.3 The Microfluidic Approach

17.2.3.1 The Argument If indeed compartmentalized fluid handling is a major factor contributing to the large size and complexity of current automated immunoassay instruments, it stands to reason that miniaturization of the fluid-handling systems (microfluidics) should lead to the miniaturization of the overall instrument as well. Extensive literature is available on microchannel devices in silicon and glass, utilizing etching and patterning techniques borrowed from the semiconductor industry. However, conducting multiplexed microimmunoassays in this environment requires the development of a means of microscale fluidic control (e.g., microvalves) to direct reagent flow into respective reaction chambers, while cross contamination (e.g., due to diffusion or mixing) is disallowed. The development of such a means remains elusive, although significant steps toward active mechanical devices in silicon have been made by researchers fabricating microelectromechanical systems (MEMS).

Another microscale solution is offered by the recent advent of PDMS polydimethylsiloxane (PDMS) microfluidics (3). PDMS is a soft, transparent, inert, elastomeric material that is produced by thermal curing of a monomer and a crosslinker in the presence of a platinum catalyst. Since PDMS is impermeable to water, it is widely used as a sealant, for example, for caulking bathtubs. Its biological inertness and nontoxicity have prompted its use in cosmetic implants and contact lenses.

17.2.3.2 Basic Techniques and Applications In PDMS microfluidics, microchannels are fabricated by pouring the mixed components over a soft-lithography mold, whose features are defined by photoresist mounds on a plain silicon wafer (4). After curing, the PDMS retains the reverse features. This replication molding is simple, rapid, and inexpensive. Separate layers of PDMS can be aligned on top of one another and bound together to produce a variety of configurations and devices (3). A microfluidic valve is constructed by crossing two channels traveling in two separate but bound layers (5). Applied pressure in one of the channels deflects the membrane separating the two channels to pinch off the other channel. If the pressure is released in the former, the membrane snaps back and the latter is open once again. Thus, the valve operation is very similar to a foot stepping on and off a garden hose.

The typical channel width is 100 μm, while the channel height and valve membrane thickness are 10–30 μm. Different valve dimensions produce different closing pressures (5) in a reproducible systematic fashion (6). Three valves acting on the same channel and activated in the proper sequence have the effect of a peristaltic pump (5). Channels, valves, and pumps can be arrayed to build complex large-scale integration (LSI) devices (7), which have been dubbed "BioLSI" when applied to biotechnology. Most recently, the advent of microfluidic "vias" (2) has allowed vertical connectivity between channels in different layers, offering further architectural flexibility and novel autoregulatory devices.

Thus, within a decade of its emergence, PDMS microfluidics has negotiated the long distance from the plain channel (4) to a plethora of specialized components organized by the thousands in large-scale integration devices (7). The now mature technology has been successfully used in a number of important applications, for example, protein crystallization (8), DNA sequencing (9), nanoliter PCR (10), cell sorting and cytometry (11), nucleic acids extraction and purification (12), and cell studies (13).

17.2.4 Multiplexed Microfluidic Fluorescence Immunoassays

17.2.4.1 Description Using PDMS microfluidic technology, we have developed a high throughput multiantigen fluorescence immunoassays system (1) to quantify blood analytes at their clinically relevant levels. The microfluidic chip (Fig. 17.1a) multiplexes sandwich immunoassays to allow 5 simultaneous tests for each of the 10 samples. Microfluidic valves (5) direct the pressure-driven reagent flow along a network of 10 μm tall channels. The network contains 5 rows and 20 columns. The columns are connected in pairs to form 10 circular paths (coliseums) (Fig. 17.1b). "Four-way" valving at each intersection of rows and columns forms a capture microchamber within which the immunoassay stack is built for a particular sample–test combination.

The chip uses 10–100 nL of each sample for all tests, thus enabling the development of a portable apparatus capable of conducting common blood tests by immunoassay methods using the volume obtained from a fingerprick. Simultaneously, the chip uses 300 nL (as low as 0.8 ng) of antibody per sample–test combination. In contrast, the state-of-the-art Elecsys® PSA kit from Roche uses 200 ng per sample–test or 250 times more.

17.2.4.2 Operation First, monoclonal antibody for each of the five different analytes is fed along its own row and bound to the glass substrate by epoxide chemistry. The unbound excess is flushed with Tris buffer in all rows in parallel, while unreacted epoxide moieties are passivated by binding Tris molecules. Similar passivation is conducted along columns to prevent direct binding of sample proteins. Immunoassay capture sites are thus formed at the intersections of rows and columns. Ten samples are fed along the columns to fill corresponding coliseums. Each sample is trapped in its coliseum and cycled over the respective capture sites by peristaltic pumping. Unbound material is flushed out along the columns. Biotinylated polyclonal antibodies are fed in parallel in their corresponding rows. Unbound excess is removed by parallel buffer feeds. Fluorescently tagged streptavidin is fed along all rows. Unbound excess is flushed with buffer. Fluorescence detection at each chamber quantifies the captured antigen. This quantity is divided by the processed sample volume to produce the respective analyte concentration for each sample–test combination.

17.2.4.3 Results and Discussion In initial studies, C-reactive protein (CRP), prostate-specific antigen (PSA), ferritin, and vascular endothelial growth factor (VEGF) were selected as representative of the types of analytes significant in medical diagnostics, as well as for the wide concentration range spanned by their clinically normal levels and for the availability of well-validated antigens and antibodies. The published measurements (1) utilized commercial antigens diluted in simple solutions of 0.1% bovine serum albumin (BSA) in PBS buffer. The signal-to-noise ratios at the lowest concentrations of antigen (10 pM) were 164 (CRP), 38 (PSA), 11 (ferritin), and 8 (VEGF). In yet unpublished measurements of the same antigens in human serum and plasma, we observed essentially the same behavior and calibration curves (1), without significant increases in background or noise, thus proving that the technique is applicable for work with processed blood samples.

Previously, serious concerns have been voiced in the microfluidic community concerning the possibility of significant nonspecific attachment of proteins to PDMS. Such attachment could produce false positive signal and interfere with the specificity of the detection system. We did not encounter such problems (1), quite possibly because of the treatment of the channels with BSA after the monoclonal antibody binding but before the feeding of sample, polyclonal antibodies, and fluorescence labels. In addition, the particular fluidic architecture of our device precludes a polyclonal antibody from being in contact with any capture sites other than the ones patterned with its corresponding monoclonal antibody. This restriction likely helps improve our results further, as it eliminates some of the possible pathways for cross talk and consequent false positives in printed microarray immunoassays.

17.2.5 Future Microfluidic Benchtop Systems

17.2.5.1 Benchtop Apparatus What are the potential engineering problems in building a microfluidic fluorescence immunoassays benchtop machine?

Electronic boxes and Lee valve arrays, which provide the pneumatic operation of the microfluidic control lines, are relatively easy to build and are even commercially available (e.g., by Fluidigm Corp). The subsystem can be fit in less than a quarter of a cubic foot and is controllable by an external computer running LabView functions.

The fluorescence microscope with its mercury lamp illuminator can be replaced with a bright LED of appropriate color. Since capture sites are separated spatially on the chip, it would be sufficient to use one fluorophore type throughout (1), simplifying the optical system. However, signal would still have to be extracted and fed to a sensitive detector. One possibility is an array of optical fibers aligned in a matrix to match the microfluidic chambers at one end, while the other end is fed to the detector, possibly through an optomechanical or acoustooptic transducer.

Another possibility is a mechanized mini stage that rasters a high numerical aperture objective under the chip, in analogy with modern scanners. The detector itself could be an inexpensive cooled CCD camera (1), a photomultiplier, or an avalanche photodiode. Thus, the engineering of such a benchtop system would be relatively straightforward, while its cost should be well below \$10 K, and thus significantly less than the current standard benchtop instruments.

17.2.5.2 Microfluidic Chip Limitations What would be the size of the test matrix for the system?

The optimal size of the chip matrix would be determined by targeted application, the capabilities of underlying technology, and the practical yield of industrial fabrication. For example, an individual physician's office might prefer one small disposable chip per patient, thus conducting about a dozen tests with multiple data points per test. For such tests, the chip would have perhaps a 10×5 matrix and fewer controls.

By contrast, a large pharmaceutical company conducting drug tests would likely prefer as large a matrix as technically possible to maximize throughput, minimize reagent costs, and make maximal use of the available experimental samples. The parallelism of the fundamental technology (7) allows a sizeable test-matrix increase with respect to the prototype, so a 50×50 chip, for instance, would be relatively straightforward to build. Moreover, the concomitant 10-fold increase in required sample and reagent volume is insignificant compared to the microliter or so spent in interfacing between the macro- and microworld, and thus would not increase the overall expenditure in terms of reagents.

However, if the coliseum geometry is retained, more complexity would be needed to control pumping across the larger distances. Also, larger distances would mean higher fluidic resistance if all other parameters are kept the same. Thus higher pressures would have to be applied to produce the same fluid velocities, while the average chip is unlikely to survive pressures exceeding 20 psi. Hence, the limitation on pressure containment may indirectly limit the chip size.

Finally, if quality control is gated to disallow a single failure, it might be challenging to produce economically acceptable fabrication yields of superchips. Most likely, the yield would be the strongest limitation among the above. The good news in this respect is that Fluidigm Corporation already offers relatively large devices for use with their Topaz system, while they have recently started marketing a 48×48 "dynamic array" chip as well.

17.2.5.3 Throughput and Operation What is the expected instrumental throughput of such a benchtop system?

In the reported prototype (1), most of the experimental time is spent taking pictures of each of the 100 chambers, manually, one at a time, using a fluorescence microscope. In a commercial benchtop system, this step can be improved by the hardware solutions discussed above. Software would then integrate the signals in the respective spots, compare to calibrations, and calculate the results. The system could be optimized to produce results within 10 min of sample loading, which would mean an instrumental throughput rate of up to 15,000 results per hour per chip using a 50×50 matrix.

Higher throughput and lower costs per test would make wider test panels more accessible to patients. Comprehensive measurements are likely to discover developing conditions that would otherwise be missed by the current symptom-based methodology. Such a capability would therefore move health care further on the road toward preventive medicine and early diagnostics, with concomitant benefits for the patients and their insurance companies.

The overall system would be significantly smaller, lighter, and less expensive than most current commercial benchtop instruments, as most of the fluid-handling hardware would be replaced by small disposable elastomeric chips that can cheaply be mass produced. In addition, the disposability of the chip offers a major operational advantage over current benchtop systems. As discussed above, a single badly prepared serum sample may clog an Elecsys-like machine and shut it down for maintenance, while incoming samples are backed up. Such a clog in a disposable chip would simply mean discarding the chip and moving to another one with the rest of the samples without delay.

The versatility of this system would be limited by the availability and specificity of antibodies, but this is the case with any immunoassay in general. Moreover, as new analytes (e.g. cancer markers) are identified, it would be straightforward to incorporate them in the test panel, while the panel size could be easily expanded due to the underlying technology. An orthogonal increase in versatility would be achieved by validating the system for interrogation of different types of samples, such as cerebrospinal fluid, synovial fluid, vaginal secretions, urine, saliva, suspended biopsy samples, and so on.

17.2.5.4 Industrialization The combination of high throughput, high parallelism, and small sample requirements would be particularly lucrative to pharmaceutical companies conducting large-scale screenings. As the cost of an experimental drug could reach \$1 million per gram of protein, the competitive advantages of microfluidic immunoassays become self-evident. Moreover, current chip prices are a result of this fact. Therefore, it is still economic to produce chips in small numbers and mostly by manual labor of technicians. However, the same price tag is unacceptable for individual health care of the average insured. Thus, the final major hurdle to overcome before ubiquity in medicine would be to reduce the price per chip, perhaps by two orders of magnitude. This could and should be accomplished by developing processes and building machines that mass produce microfluidic chips in strong analogy with the semiconductor industry.

17.2.6 Future Microfluidic Handheld Systems

There is little doubt that microfluidic immunoassay systems would be a valuable addition to the benchtop family of devices, but is it possible to produce a handheld based on the same technology?

While the chip itself is quite small and light, it would be challenging to miniaturize the electrical macrovalves that control it and the optical system that reads it. A better idea would probably be to move away from optical detection in favor of an electrical one (e.g. like iSTATs), while retaining the microfluidic compartmentalization that is critical for multiplexed immunoassays as discussed above. Miniaturizable methods of pneumatic valve actuation would also have to be developed. Notably, such a system would be highly shock resistant as it would contain no optical elements, and the only moving components would be highly durable elastomeric membranes.

17.2.7 Conclusions

The dream of personalized preventive medicine cannot be materialized without ubiquitous and affordable diagnostics. Decentralization and miniaturization are critical in achieving this goal. In this context, the recently reported high throughput multiantigen microfluidic fluorescence immunoassay chip is an important enabling step toward fast, inexpensive, portable, versatile systems capable of accurate protein-based test performance in a decentralized, ubiquitous setting, including over-the-counter (OTC) devices as well as scalability for high throughput screening in industrial settings.

17.3 MICROFLUIDIC VIAS AND DERIVATIVE APPLICATIONS

17.3.1 Technological Background

One of the chief limitations to the maximal complexity of microfluidic devices (and, consequently, their applications) has to do with design rules imposed by the fundamental technology and its fabrication techniques. Recently, we reported on a fundamental advance (2) that allows a simple and easy access to a large increase in the architectural complexity of microfluidic devices as well as opens new possibilities for technical developments and consequent applications. We dubbed the new device "via" in reference to its analog in modern semiconductor electronics.

Vias are vertical micropassages that connect channels fabricated in different layers of the same PDMS multilayer chip. The functional result is the 3D channels that lift the restrictions of the traditional architecture wherein channels could not leave their layer, and two channels within the same layer could not cross without mixing. These restrictions did not entirely prevent the emergence of expansive architectures (7), because particular applications could be shrewdly chosen to involve large-scale parallelization of simple identical operations, thereby requiring few controls and maximizing device density. However, as the field moves to functionally complex heterogeneous devices integrated on the same chip (14), laying out the respective circuitry would inevitably necessitate convenient, simple, and reliable vertical connectivity, just as it did in the semiconductor industry. Microfluidic vias provide this 3D connectivity, lift the above architectural restrictions, and facilitate new morphological and functional capabilities.

The pursuit of 3D devices is not new. Whitesides and colleagues at Harvard (15) developed an ingenious scheme wherein a complex system of multilayer photoresist molds, photoresist premasters, and PDMS masters were fabricated and then used in an involved many-step process to produce a 70-μm-thick PDMS layer housing 100-μm-wide vertical cylinders connecting 70-μm-tall channels fabricated in thick PDMS slabs. The resulting 3D technique was successfully used in protein and cell patterning (16). Jo et al. demonstrate a similar method involving physical clamping to control layer thickness (17). Whitesides and colleagues also developed a technique to produce 3D channels by mechanical deformation of straight channels (18). However, the challenging and labor-intensive fabrication required for the above devices has largely dissuaded researchers from further work with these methods.

By contrast, the fabrication of vias presented herein is about as simple, fast, and easy as the one of standard multilayer devices, thereby removing the practical obstacles to the wide use of 3D microfluidic architectures. In addition, vias can work with significantly smaller dimensions, for example, 7-μm-tall channels connected by 25-μm-wide vias. The ultimate limit in miniaturization is now set by the submicron resolution of optical lithography rather than the softness of PDMS masters.

17.3.2 Microfluidic Vias

17.3.2.1 Basic Technique The flow chart of via fabrication is drawn in Fig. 17.2. PDMS is spun onto a standard hybrid mold to a thickness smaller than the height of the taller features but larger than the height of the shorter features. As a result, the taller features protrude through the upper PDMS surface. After curing, a separately fabricated layer is assembled on top in such a way that the end of its channel matches the protruding mold feature.

After bonding, the device is peeled off and assembled on a substrate, for example, a glass slide. Stacking the two PDMS layers in this fashion produces a vertical connection (via) between channels fabricated in each layer. The functional result is a composite channel that switches from one layer to another as it winds its way through the chip. Fig. 17.3 contains photos of actual via devices.

17.3.2.2 Combinations If two such devices are combined, an overpass or an underpass is produced (Fig. 17.4). They allow two channels traveling in the same

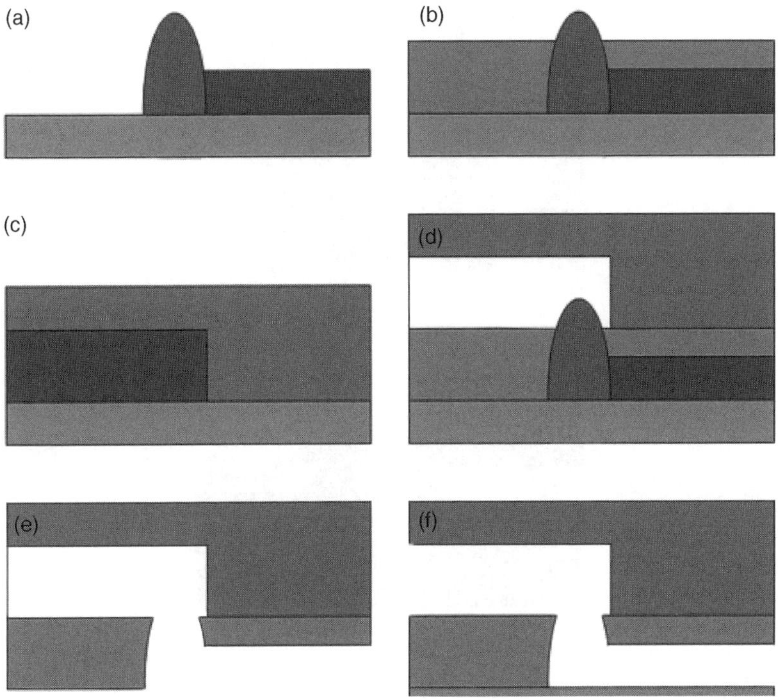

FIGURE 17.2 Via fabrication (drawing not to scale). (**a**) Standard hybrid mold is constructed using Su8 (violet) and Az50 (orange) photoresists spun at 7 and 30 μm heights, respectively. The mold is then baked at 200°C for 1 h to round the Az50 features. (**b**) 20:1 uncured PDMS (blue) is spun at 4000 rpm so that the high features protrude through and over the PDMS surface. (**c**) 5:1 uncured PDMS (pink) is poured over another mold. (**b**) and (**c**) are partially cured at 80°C for 30 min. (**d**) The (**c**) layer is peeled off and assembled onto the (**b**) layer. Curing is completed to bond the layers. The chip is peeled off the (**a**) mold (**e**) and assembled to a substrate of choice (green) (**f**). In the resulting device, the via connects channels fabricated in different layers.

layer to cross without mixing, as one of the channels switches over to an adjacent layer for the length of the crossing. Overpasses and underpasses can be arranged in series to enable 3D flexibility in the architectural layout of a two-layer chip.

Further enhancements are possible by increasing the number of stacked PDMS layers. Up to seven layers have been reported to be bound by optimized thermal curing procedures (5), while alternative binding schemes, for example, oxygen plasma treatment should allow even larger stacks. The multilayer stacking would be produced by binding a structure as in Fig. 17.2e structure as in Fig. 17.2b, peeling off the mold, and iterating. The final lowest elastomeric layer would be bound to a substrate. Multiple layers carrying channels interconnected with vias open an enormous phase space of architectural possibilities to explore in future devices and applications.

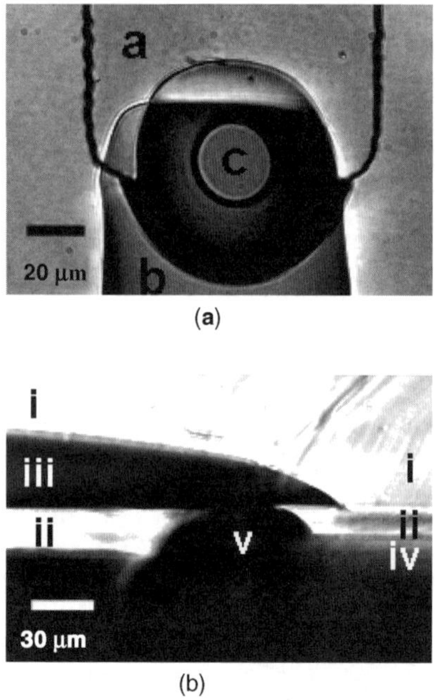

FIGURE 17.3 Rounded via devices. (**a**) Top view shows lower channel (a), upper channel (b), and via (c). The access annulus is clearly visible. (**b**) Side view cross-section shows upper layer (i), lower layer (ii), upper channel (iii), lower channel (iv), and via (v). The chip was peeled off the glass slide, cut with a razor, and laid on its side to take the photo. Compare to Fig. 17.2f.

17.3.2.3 Rounded Vias To determine the optimal via fabrication parameters, we created a two-layer matrix containing four independent sets of 3D channels assembled by 288 vias of lateral dimensions that systematically varied between 25 and 80 μm, in 5 μm steps. A number of chips at different thicknesses of the lower

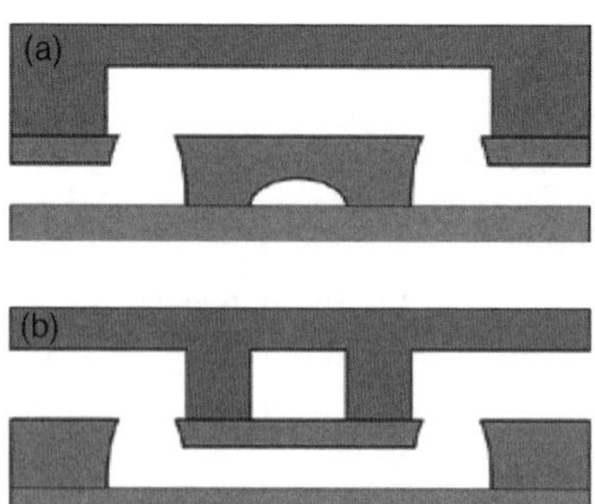

FIGURE 17.4 Two copies of the device from Fig. 17.2 can be combined to produce an overpass (**a**) or an underpass (**b**). They allow two channels traveling in the same layer to cross without mixing as one of the channels jumps to an adjacent layer for the length of the crossing. A series of such structures allows full 3D flexibility of architectural layout of the chip's channels accomplished by simple means.

PDMS layer were produced using different spin speeds. Optimal results were achieved for 4000 rpm (on a 3-inch wafer) and via lateral dimensions between 50 and 65 μm, where vias formed successfully and reproducibly.

For some devices of larger dimensions and/or lower spin speeds, surface tension of the uncured PDMS formed a hump over the tall mold features. This hump cured into an unbroken membrane, producing a defective via. At smaller lateral dimensions (and resulting larger ratios of height to width), melting the photoresist during the rounding process lowered the height of the tall mold features. Thus, they were too short to break through the subsequent PDMS layer.

17.3.2.4 Nonrounded Vias All vias discussed up to this point were made using rounded molds. Hence, we next fabricated devices using nonrounded (or square profile) molds of the same 288-via layout. Virtually, all vias of 25–80-μm lateral dimensions formed successfully at 2500 rpm on 3-inch. wafers. Fig. 17.5a and b show 25-μm square functional vias. The lack of melting preserves the height of the narrow features, while the sharp edges help break the PDMS surface tension, thereby explaining the greater success of this technique.

To demonstrate the correct workings of the chip, the four independent sets of channels were filled with solutions containing plain buffer, fluorescein, Cy3, or Cy5 dyes. One colorless image and the respective three fluorescence images were mixed in false color to produce Fig. 17.5c. Of the 288 sites in the chip, 286 formed vias and 2 formed septa.

17.3.2.5 Significance to PoC Diagnostics Even without their derivative compound devices, vias represent an important advance in microfluidic technology as applied to molecular biology and biomedical diagnostics. Vias save much "real estate" on the chip by making shorter connections and lifting design restrictions that ultimately stem from the previous inability of two channels in the same layer to cross without mixing. Thus vias allow a significant increase in functional complexity and device density without a comparable increase in fabrication difficulty or cost of the chip, thereby enabling higher level of feature integration and correspondingly smaller, more robust devices. Vias should also make it easier to construct efficient pumping architectures for larger complex chips, for example, for immunoassay arrays. Finally, the contributed ability of a channel to switch layers allows the integration of pushdown and push-up architectures within the same device, which is a critical feature in certain applications, e.g. on-chip cell expression analysis.

17.3.3 Microfluidic Septum

17.3.3.1 Technique The incomplete formation of vias at lower spin speeds and/or extreme dimensions suggested another device—a microfluidic septum. The flow chart of its fabrication mimics that of the via, except for spinning the PDMS slightly above all features of the hybrid mold. Another way to think about a septum is as a purposefully defective via.

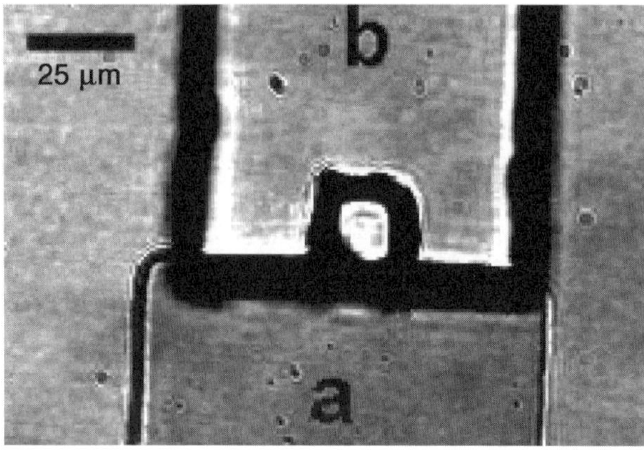

(a)

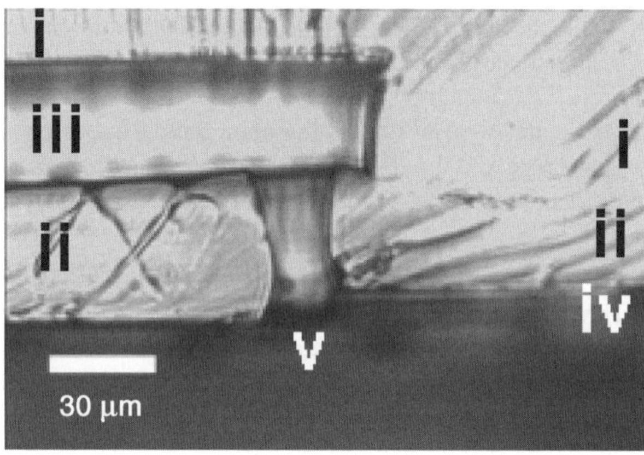

(b)

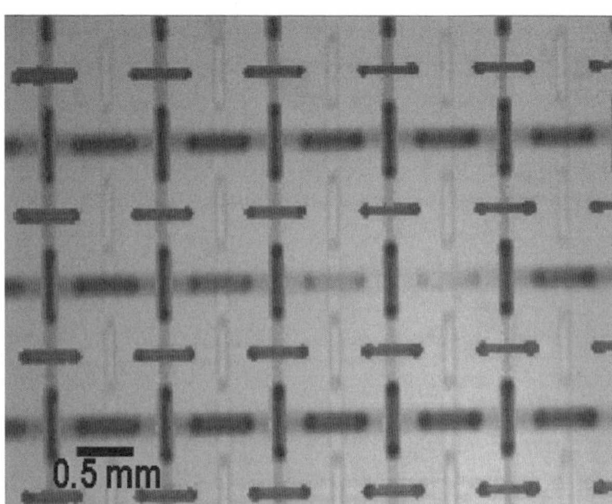

(c)

The septum has the useful property that it would breach at a characteristic pressure determined by its fabrication parameters. For example, an $80 \times 80\,\mu m$ septum at 3000 rpm on a 3-inch wafer over a 36μm tall rounded mold breaches at 5 psi. Therefore, a septum can be used to isolate a certain section of the chip, up to a specific applied pressure. Such "destructive engineering" has been demonstrated using voltage instead of pressure (19).

17.3.3.2 Applications The septum will likely prove to be a very useful device component in the industrial packaging of point-of-care devices because it could keep dust out, or reagents in, until the sealed section is to be accessed. Also, if the sealed section is connected to an exhaust, then the septum acts as an irreversible microfluidic fuse. When the applied pressure exceeds a certain hardwired value, the septum breaches, the fluid flows to the exhaust, and the pressure decreases. This scheme can be used to protect a sensitive section of the chip against excessive pressures.

If a system of septa tuned to different breaching pressures is built within the same chip, then the chip could be configured to varying final functionalities by applying respective pressures. This technique would allow mass production of identical chips that could later be finalized to suit specific needs. Each chip would then be operated at pressures too low to cause further architectural changes. In another embodiment of the same idea, the benchtop apparatus itself could be programmed to reconfigure the particular PoC chip through septa, in accordance with the physician's input based on a particular patient's needs.

17.3.4 Nested Bioarrays

17.3.4.1 Demonstration One of the applications of vias is printing of bioarrays (16), for example, protein or DNA arrays for immunoassays or hybridization expression analysis. As a particular example, we designed another chip containing four independent sets of 3D channels. Its primitive cell is a 4-plex (Fig. 17.6a), wherein each set accesses the substrate in one respective location.

As a direct example of the process, CRP and ferritin monoclonal antibodies were each fed in a separate channel set and bound to the respective epoxide locations. The unbound excess was washed away with buffer, the PDMS chip was peeled off, and the unreacted epoxide was passivated with Tris buffer (1). We pipetted a mixture of

FIGURE 17.5 Nonrounded vias form successfully over a wider spectrum of dimensions than the rounded vias. The smallest functional nonrounded vias made to date are 25-μm^2: (**a**) top view shows lower (a) and upper (b) channels. (**b**) Side-view cross-section shows upper layer (i), lower layer (ii), upper channel (iii), lower channel (iv), and via (v). The chip was peeled off the slide, cut with a razor, and laid on its side to take the photo. (**c**) A 4-plex design was executed with nonrounded vias at 2500 rpm PDMS spin speed. The four channel sets were filled with solutions containing plain buffer, fluorescein, Cy3, and Cy5 dyes; a colorless image and three false-color fluorescence images in blue, green, and red, respectively, were combined to produce the shown result. Of the 288 sites in the chip, 286 formed vias and 2 formed septa (seen here at the ends of what would have been a red segment, mid row, second from the right). (See color insert)

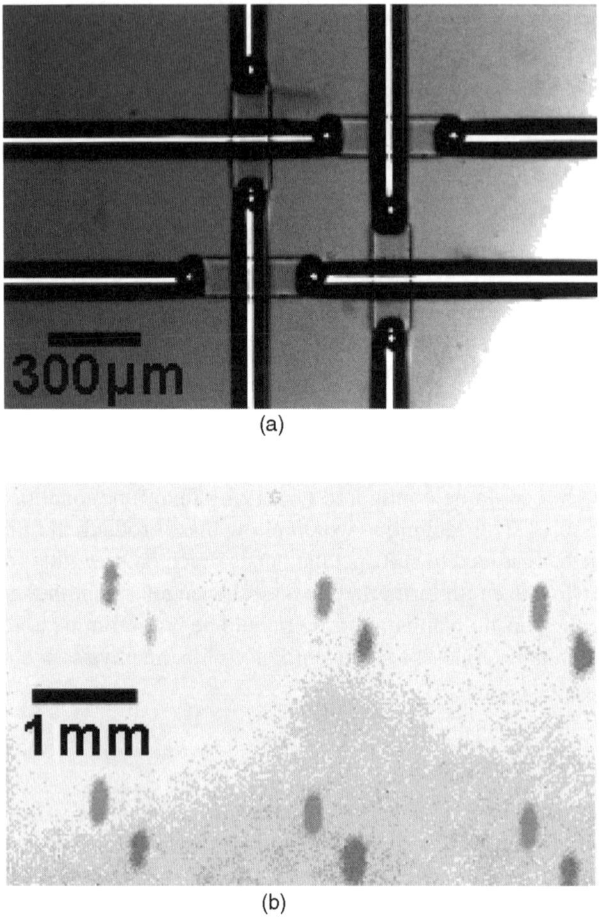

FIGURE 17.6 Printing of nested bioarrays. We built a modified chip containing a 6 × 6 array of nested arrays (one is shown in (**a**)), wherein four nonmixing sets of channels access the substrate surface at one location each. CRP and ferritin monoclonal antibodies were each fed into a separate channel set and bound to the epoxide substrate, while the third and fourth sets were not used here. After washing away the excess, the chip was peeled off and the rest of the epoxide was passivated with Tris buffer (1). A mixture of CRP and ferritin antigens and respective polyclonal antibodies tagged with Cy3 and Cy5 were incubated on the slide. After another wash, fluorescence detection produced a false-color image (**b**) demonstrating the correct assembly of the sandwich immunoassays of the nested bioarrays. If the initial substrate is diced after printing, this technique is a simple, fast, and inexpensive way to mass produce identical nanomedicine chips, each conducting multiple tests on a separate sample. In principle, nested bioarrays of arbitrary size can be printed with three-layer devices (see text). (See color insert)

CRP and ferritin antigens and fluorescently tagged polyclonal antibodies onto the slide. After incubation and washing, fluorescence detection led to the mixed false color image as in Fig. 17.6b. The green (red) spots are produced by the Cy3 (Cy5) tag on the CRP (ferritin) polyclonal antibody, while the locations of the spots match the substrate access of the CRP (ferritin) monoclonal antibody, respectively. The results indicate the correct completion of the corresponding sandwich immunoassays.

17.3.4.2 Limits How large can the superarray and nested array be, as allowed by the underlying via technology?

The 4-plex (Fig. 17.6) is a special case wherein underpasses double up as deposition access channels. If the number of individual reagents is increased, adjacent channels cannot jump over each other to create the superarray, unless they also access the substrate in extraneous locations.

To solve this problem, it is sufficient to increase the number of layers to three, wherein the first layer accesses the substrate, the second layer accommodates most of the length of the channels, and the third layer houses overpasses. This scheme can theoretically handle any number of individual reagents, and thus an arbitrary n-plex. The future should witness many important advances in patterning, as made possible by the presented microfluidic via technology.

17.3.4.3 Applications As Fig. 17.6b demonstrates, vias provide a simple, fast, and inexpensive way to construct arrays of bioarrays (or "nested bioarrays"). To borrow an idea from the semiconductor industry, a large substrate wafer could thus be derivatized with complete parallelism and subsequently diced to yield a large number of identical microchips, and each ready to be used for multiple tests on a separate sample. Such devices could form the centerpiece in a class of standardized inexpensive nanomedicine chips that could prove revolutionary in today's world of skyrocketing health care costs and mounting pressures for true personalized preventive medicine.

17.3.5 Autoregulatory Devices

17.3.5.1 Basic Principle and Architectures In the multilayer world of microfluidic valves (3,5,20), vias remove the distinction between controls and flow channels because the same 3D channel can now be a control channel in one section of the chip and a flow channel in another. Therefore, the same 3D channel must be able to act as a control channel upon itself. This observation forms the basis of new autoregulatory devices.

Previous work in microfluidic autoregulation utilized the non-Newtonian rheological properties of concentrated polymeric solutions (21,22). By contrast, the presented work is the first to show microfluidic autoregulation in Newtonian fluids, and thus in environments typical of most microfluidic applications.

Figure 17.7a and b shows the simplest device architectures using pushup valves. In the pushup configuration, pressure applied to the channel in the lower layer deflects the valve membrane upward to pinch off the channel in the upper layer (20).

Within the detour autoregulator (Fig. 17.7a), static pressure decreases from source to exhaust as fluid flows along the main channel. Meanwhile, static pressure remains constant along the dead-end detour channel leading to the valve. Therefore, the pushup valve experiences an effective pressure equal to the static pressure drop

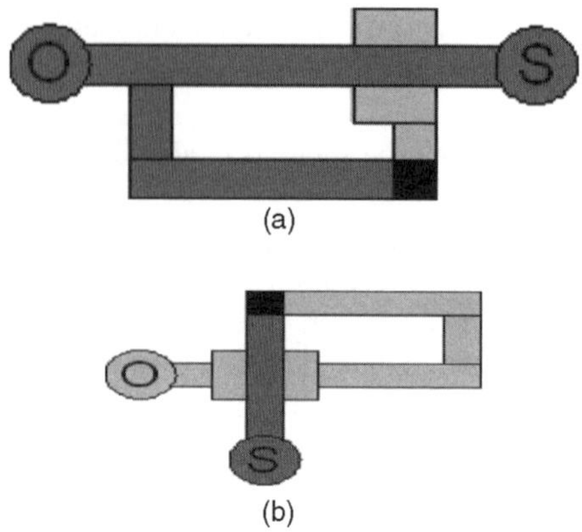

FIGURE 17.7 In a detour autoregulator (**a**), pressure is applied at origin O to produce fluid flow along the main channel to exhaust/sink S, while a detour channel is connected to a pushup valve through a via (black). Channels in the upper (lower) layer are drawn in red (blue), respectively. Static pressure decreases from origin to sink along the main channel due to the fluid flow, while the dead-end detour channel stays at the same static pressure as the split point. Thus the pushup valve (20) experiences a pressure difference equal to the respective pressure drop, which scales with applied pressure due to Poiseuille's law. At sufficiently large applied pressures, the valve starts constricting the main channel and the total resistance increases. Thus the device behaves nonlinearly with Newtonian fluids. A loop autoregulator (**b**) utilizes the same pressure drop idea. The loop eliminates the need for a detour channel but requires that the main channel returns to the starting point. Either architecture could be preferable in specific situations.

between the channel split and the main-channel segment above the valve. Because of the geometry and Poiseuille's law, this pressure drop scales with applied pressure. As the drop increases, the valve membrane deforms upward and constricts the main channel. Hence, total resistance increases with applied pressure and the device behaves nonlinearly with Newtonian fluids.

The alternative architecture (loop autoregulator) (Fig. 17.7b) utilizes the same pressure drop principle. The loop eliminates the need for a detour channel but requires that the main channel returns to the starting point. Either architecture could be preferable in particular situations.

17.3.5.2 Microfluidic Current Source We constructed a set of detour autoregulators on the same chip, where the only varying parameter was the main-channel distance X between split and valve. All valves were 100×100 μm, all lower channels were 7×100 μm, and all upper channels were 36×100 μm in lateral dimensions. The length of all main channels was $L = 14.2$ mm. The autoregulator set had $X/L = \{0.80, 0.68, 0.57, 0.46, 0.23\}$. The 60-μm-square vias were rounded and the PDMS was spun at 5000 rpm.

High purity water was flowed through each device. Throughput (Fig. 17.8a) was measured by timing the advance of the water meniscus in transparent tubing connected to the exhaust. Larger split-to-valve distances monotonically correspond to lower

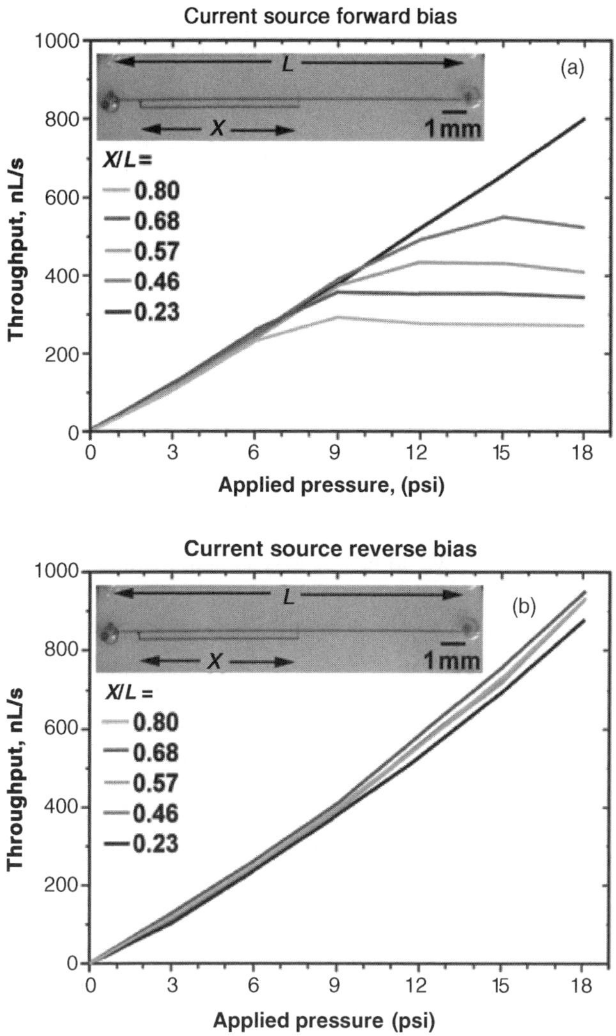

FIGURE 17.8 (**a**) Experimental results for the throughput as a function of applied pressure for a set of detour autoregulators (one is shown in the photo), where the varying parameter was the distance *X* between split and valve. Larger *X* produced larger pressure drops for the same applied pressure, valve dimensions, and total main-channel length *L*. Thus devices with larger *X*/*L* produced monotonically lower saturation points for throughput and pressure, demonstrating the predicted nonlinear behavior with Newtonian fluids. Hence, these devices can be used as microfluidic current sources with saturation characteristics that are tunable by architectural design. (**b**) The same devices were then reverse biased by exchanging the roles of origin and sink. In reverse bias, the devices function linearly as plain channels. The small curvature upward is due to slight dilation of the elastomeric channels at higher pressures and the third-power dependence of throughput on the smaller lateral dimension in Poiseuille's law.

saturation points, because identical valves experience larger percentages of the same total applied pressure. These experimental results offer a confirmation of the above qualitative predictions. They also demonstrate that the saturation throughput and saturation pressure of the device can be tuned by varying the split-to-valve length as a percentage of the main-channel length. Thus, in forward bias, a detour autoregulator can be used as a microfluidic current source when operated in its saturation regime.

Fig. 17.8b shows the experimental results when the same detour autoregulator devices were used in reverse bias. Then, the effective pressure acts on the valve in the opening rather than in the closing direction and the devices act like plain channels. The small curvature upward in both forward and reverse bias is due to slight dilation of the elastomeric channels at higher pressures and the third-power dependence of throughput on the smaller lateral dimension in Poiseuille's law.

17.3.5.3 Microfluidic Diodes Also, two such devices could be arranged back-to-back in series. Then each would act as a current source in its forward bias and as a plain channel in its reverse bias. An appropriately tuned design would ensure different saturation throughputs T in the different directions of applying pressure P. As such, the described compound device would be a new type of a microfluidic diode (23).

Curiously, unlike the detour case, a loop autoregulator in reverse bias would act as a current source rather than as a plain channel. For example, the device in Fig. 17.7b would see a pressure drop applied on a pushup valve in forward bias and a pushdown valve in reverse bias, leading to saturation behavior in both directions. Careful engineering of the involved dimensions would produce an asymmetry between the saturation points of the two biases, thereby ensuring that the loop autoregulator would act as yet another type of a microfluidic diode.

The above detour and loop diodes could be used as building blocks to produce more advanced devices, such as a microfluidic rectifier bridge, as is the case with their electronic counterparts. While the correspondence between electrical and fluidic circuitry is not complete, analogies would prove useful. The above devices would form the basis of novel microfluidic analog logic in Newtonian fluids.

17.3.5.4 Significance to PoC Diagnostics Autoregulatory devices provide new fluidic capabilities, which cannot be mimicked well by traditional on/off micro-valves. Further, these particular via-based autoregulators function using Newtonian fluids instead of concentrated solutions of exotic polymers (21,22), thereby making the former completely compatible with typical reagents for biomedical diagnostics.

Finally, autoregulators offer an alternative, passive way of fluidic control, in contrast to active control provided by on/off microvalves actuated by external control pressure input. Hence, autoregulators would prove useful in situations where the application and respective fluidic circuitry can utilize them instead of on/off valves, thereby decreasing the number of required independent external control lines and making the overall device smaller and simpler to operate. This is a major advantage since portability and ease of use are some of the chief concerns in PoC applications.

The above features make autoregulators a useful addition to the toolbox of micro-fluidic devices and techniques in general and microfluidic biomedical diagnostics in particular.

17.3.6 Conclusions

Microfluidic vias are an important advance enabling significant increases in functional complexity, device density, compactness, and fluidic handling capabilities of elastomer chips. Together with their logical derivatives, vias contribute new applications (e.g., nested bioarrays) as well as new devices featuring unique functionalities (e.g., autoregulators) to elastomer microfluidics in general and its application to PoC diagnostic devices in particular.

17.4 CONCLUSIONS

The macrofluidic robotic sampling approach of the state-of-the-art high throughput molecular-diagnostics clinical-laboratory analyzers results in their large size and unit cost, consigning them to specialized, centralized facilities. These facilities offset overhead and hidden costs by high throughput large-scale operation ("economies of scale") at a price of logistic complexity, transportation delays, and inconvenience to the patient. A good alternative would be fast, inexpensive, decentralized, point-of-care biomedical diagnostic instruments of comparable testing capabilities, but such apparatuses cannot be technically achieved without enabling advances in fluid-handling miniaturization.

Elastomer microfluidics promises to provide such miniaturization through its elegant polymer devices that feature intricate fluidic control at the microscale. The recent demonstration of high throughput microfluidic fluorescence immunoassays in particular is an important step to miniaturized protein-based diagnostics. In addition, new techniques and components, such as microfluidic vias and septa, contribute new capabilities by themselves and through their derivative devices, for example, nested bioarrays and autoregulators, thus further extending the range of addressable applications.

ACKNOWLEDGMENTS

Financial support for the respective research work (1,2) was provided by NIH 1RO1 HG002644-01A1, DARPA HR0011-04-1-0032, and Boeing's Multifunctional NanoSystem Technologies program. We also gratefully acknowledge the National Institutes of Health for Dr Kartalov's K99/R00 "Pathway to Independence" Career Award.

REFERENCES

1. Kartalov EP, Zhong JF, Scherer A, Quake SR, Taylor CR, Anderson WF. Biotechniques 2006;40:85–90.
2. Kartalov EP, Walker C, Taylor CR, Anderson WF, Scherer A, Proc Natl Acad Sci U S A 2006;103:12280–12284.

3. Kartalov EP, Anderson WF, Scherer A. J Nanosci Nanotech 2006;6:2265–2277.
4. Duffy DC, McDonald JC, Schueller OJA, Whitesides GM. Anal Chem 1998;70: 4974–4984.
5. Unger MA, Chou H-P, Thorsen T, Scherer A, Quake SR. Science 2000;288:113–116.
6. Kartalov EP, Scherer A, Quake SR, Taylor CR, Anderson WF. J Appl Phys 2007; 101:064505, Forthcoming.
7. Thorsen T, Maerkl SJ, Quake SR. Science 2002;298:580–584.
8. Hansen CL, Skordalakes E, Berger JM, Quake SR. Proc Natl Acad Sci U S A 2002;99: 16531–16536.
9. Kartalov EP, Quake SR. Nucleic Acids Res 2004;32:2873–2879.
10. Liu J, Hansen C, Quake SR, Anal Chem 2003;75:4718–4723.
11. Fu AY, Chou H-P, Spence C, Arnold FH, Quake SR. Anal Chem 2002;74:2451–2457.
12. Hong JH, Studer V, Hang G, Anderson WF, Quake SR. Nat Biotechnol 2004;22: 435–439.
13. Balagadde FK, You L, Hansen CL, Arnold FH, Quake SR. Science 2005;309:137–140.
14. Shaikh KA, et al. Proc Natl Acad Sci U S A 2005;102:9745–9750.
15. Anderson JR, Chiu DT, Jackson RJ, Cherniavskaya O, McDonald JC, Wu H, Whitesides SH, Whitesides GM, Anal Chem 2000;72:3158–3164.
16. Chiu DT, Jeon NL, Huang S, Kane RS, Wargo CJ, Choi IS, Ingber DE, Whitesides GM. Proc Natl Acad Sci U S A 2000;97:2409–2413.
17. Jo BH, Van Lerberghe LM, Motsegood KM, Beebe DJ, J Microelectromech Syst 2000; 9:76–81.
18. Wu HK, Odom TW, Chiu DT, Whitesides GM. J Am Chem Soc 2003;125:554–559.
19. MacDonald JC, Metallo SJ, Whitesides GM. Anal Chem 2001;73:5645–5650.
20. Studer V, Hang G, Pandolfi A, Ortiz M, Anderson WF, Quake SR. J Appl Phys 2004; 95:393–398.
21. Groisman A, Enzelberger M, Quake SR, Science 2003;300:955–958.
22. Groisman A, Quake SR, Phys Rev Lett 2004;92:094501/1–094501/4.
23. Adams ML, Johnson ML, Scherer A, Quake SR, J Micromech Microeng 2005;15: 1517–1521.

PART V

MICROFLUIDIC FABRICATION STUDIES

18

FABRICATION OF POLYMERIC MICROFLUIDIC DEVICES

EMAD EL-GIAR, EMANUEL WADDELL, MARQUITA JACOBS AND GLORIA THOMAS MAGEE

18.1 INTRODUCTION

One of the growing trends in chemical, biological, and clinical diagnostics is the development of miniaturized systems (also called microfluidic devices, micro-devices, or micrototal analysis systems, μ–TAS) (1–4). These devices are used to manipulate analytes within enclosed structures having dimensions in the 10–100 μm range. Miniaturization to this scale offers a wide range of advantages including reduced manufacturing costs, rapid analysis, high throughput, portable size dimensions, and minimal consumption of sample and reagents (5–12). Moreover, microfluidic devices are not limited in design to simple, linear columns, compared to other separation techniques and formats. Microfluidic structures can be integrated into modular and multiplexed devices able to perform many functions, such as centrifugation, reagent mixing, cell culture, sorting and/or lysis, PCR amplification, and do so with high throughput (13–19).

Two of the key elements in leveraging this potential in microfluidics are the substrate and microfabrication method used to develop microfluidic devices. Generally speaking, an ideal substrate material for microdevices (for biological applications) is one that is relatively inexpensive, mechanically strong, chemically and thermally stable, and biocompatible. Additionally, if electrophoresis, one of the most powerful tools in bioanalysis, is employed, the dielectric properties of the material must also be carefully considered. In terms of fabrication methods, several considerations must also be made, including the need for mass production, the

Biological Applications of Microfluidics, Edited by Frank A. Gomez

feature sizes desired, the need to rapidly change the desired pattern, and the substrate used.

During the early development of microfluidic devices, silicon-based substrates and fabrication technologies originally developed for microelectromechanical systems (MEMS) were borrowed from the semiconductor industry. Specifically, photolithography and wet chemical etching techniques were leveraged to produce microfluidic devices in glass and other silicon-based substrates. However, increasing demands for mass production of single-use, disposable clinical devices, combined with the high cost of the technology and substrate, have motivated researchers to seek other alternatives. Among these alternatives are plastics and polymer-based materials and replication methods of fabrication, which are both being exploited for microfluidics at increasing rates (20).

In this chapter, selected methods and techniques used for the fabrication of polymeric microfluidic devices used in biological applications are discussed. After a brief overview of the advantages and disadvantages of commonly substrate materials, microfabrication techniques are discussed as either direct write or replication techniques, with emphasis on methods used for production of plastic and polymeric devices. Specifically, mechanical machining, laser ablation, wire imprinting, and hot embossing are discussed. In closing, comments on the progress in use of these methods in biological applications are provided.

18.2 GLASS- AND SILICON-BASED MATERIALS

Since their introduction in the 1990s, microfluidic devices have been fabricated in glass, quartz, and other silicon-based materials (7,15,21–26). Several advantages supported this initial choice of substrate: The surface properties and chemistries of these materials are well evaluated and characterized, high purity single crystalline silicon wafers are commercially available in various orientations, and established fabrication techniques are available. These materials also boast high mechanical strength, high chemical resistance, low conductivity, and a wide optical transmission range. In addition, silicon-based materials contain negatively charged silanol groups that cause electroosmotic flow (EOF) under electrophoretic conditions, a benefit to applications that rely upon electrokinetic transport for fluidic transport. Microfabrication in these substrates was initially limited to isotropic wet chemical etching using hydrofluoric acid (HF) or KOH. Advanced techniques were quickly employed, included dry etching, deep reactive ion etching and plasma etching techniques. Using these microfabrication methods, glass devices have been successfully used for the analyses of biologically important analytes in a wide variety of applications spanning the range from small molecules to whole cells (27).

However, despite the advantages to using silicon-based materials and developments in fabrication techniques, many more challenges remain. Microfabrication of these materials often requires well-equipped clean room environments, extensive safety precautions for wet chemical etching, and drastically high voltages and temperatures for bonding lids to devices. Moreover, EOF is a detriment to some applications (e.g., DNA separations), and in silicon-based substrates, EOF is highly pH dependent, rendering it difficult to reproducibly control. In addition, silicon suffers from an opaque appearance, difficulty in integration with other materials, and noncompatibility with biological applications (28). However, the major disadvantage of silicon-based materials is the relatively high cost and limitations in mass production, which is of particular importance for medical and clinical diagnostic applications that require disposable tools to eliminate cross contamination.

18.3 PLASTICS AND POLYMERIC MATERIALS

The limitations and challenges of fabricating silicon-based microfluidic devices quickly led to the investigation of alternative substrates, such as ceramics, plastics, and polymer-based materials (29,30). Fortunately, a vast range of plastics and polymeric materials with diverse mechanical, chemical, electrical, and optical properties are commercially available. On the basis of their thermal behavior, these materials are classified as thermoplastics, elastomers, and duroplastics (31,32). *Thermoplastics* are polymers that melt and flow when heated. They are not highly crosslinked and have low melting and boiling points. Examples used for the fabrication of microfluidic devices are poly(methyl methacrylate) (PMMA), polycarbonate (PC), polypropylene (PP), polyvinylchloride (PVC), polyethylene (PE), polyethylenetetraphthalate glycol (PETG), and copolyesters. *Elastomers* are characterized by weakly crosslinked, rubber-like polymers. They do not melt before reaching their decomposition temperatures (17). The most well-known polymers in this group are polyurethane, poly(dimethylsiloxane) (PDMS), and epoxy. The last group, *duroplastics* (or thermosets) are polymers that do not melt but rather decompose when heated. They are highly crosslinked, hard, brittle, and thermally stable, thus they have not been exploited for microfluidic applications. Vinyl esters and phenol formaldehyde are typical examples of this group.

Advantages of microfabrication in thermoplastics and elastomers include the expanding availability and relatively low cost of fabrication technology and use of simple sealing methods that do not require clean rooms or high temperatures. However, possibly the most attractive feature to using polymer and plastic substrates is the ability to mass produce devices with high aspect ratios (33–36). Another important advantage of polymeric materials, compared with glassy and silicon materials, is a reduced electroosmotic flow during electrophoretic applications due to the absence of ionizable functional groups, which is of great importance, especially for improving the separation efficiency of biomolecules (37). Combining, these characteristics facilitate the production of disposable microdevices with a wide range of properties appropriate for diverse applications, including clinical, environmental, and biological analyses.

While clearly attractive, the use of polymer substrates also has limitations. One disadvantage is a smaller knowledge base concerning surface chemistries, compared to that which is documented for silicon-based materials. This makes it difficult to predict surface interactions, such as protein adsorption, which would result in poor efficiencies in separation applications. Surface modification techniques, such as covalent and dynamic coating, have been reported; however, these methods are substrate- and application specific. Another challenge is that polymer-based devices are not compatible with organic solvents and/or high temperatures. Many polymers dissolve in organic solvents (e.g., PDMS and PMMA), resulting in swelling and/or cracking (28,34). Nonetheless, polymers remain attractive for bioanalytical application and their use is growing.

18.4 APPROACHES TO MICROFABRICATION

Congruent with the increased use of plastics and polymeric materials in the fabrication of microfluidic devices, the last decade offered the emergence of a large number of microfabrication methods for the production of microfluidic devices in polymers with high precision and aspect ratios, several reviews of which have been written (38–40). Governing factors in the approach to microfabrication include the number of microdevices desired, the complexity of the design, needs for prototyping, and the time required for the technique (31). In general, the approaches can be translated into two classifications: direct write and replication techniques.

18.4.1 Direct Write Techniques

Direct write techniques refer to those techniques in which the desired pattern is translated directly into a substrate; they may be conceptualized as "pen-to-paper" techniques. In most instances, these techniques are used to "write" two-dimensional designs; however, three-dimensional structures have been demonstrated (24). A microdevice design may be machined into a stationary substrate by directing the movement of a writing instrument (any device that is capable of scribing micron-size channels in the substrate of interest) such that the pattern is machined directly in the substrate (41). Alternatively, some methods use a stationary writing instrument with respect to a substrate controlled by a motion stage. Writing instruments include, but are not limited to, lasers, ion beams, high pressure water or sand, and plasma. As such, direct write techniques are amenable to a wide variety of substrates.

Direct write techniques typically provide a rapid method of fabricating devices most appropriate for a research and development environment that demands the ability to prototype different designs. Advantages of this technique include rapid fabrication, flexibility in choice of substrates, virtually unlimited two-dimensional design constraints, and the ability to utilize a device without further processing. The primary disadvantages of this technique are that its serial nature (one device must be completely manufactured before another may be started) and that while the cost to fabricate individual devices may be relatively low and the amount of time required short, the technique does not compare favorably to injection molding and embossing when there is a need to produce hundreds or thousands of devices in a reasonable amount of time.

18.4.2 Replication Techniques

Replication methods are cost effective, and rapid techniques used for large-scale, high throughput production of microfluidic devices (31,42,43). These methods depend on the development of a mold (also called a master, stamp, or template) that contains the inverse of all of the features and microstructures of interest. Thus, the first step involves the generation of the mold design, followed by fabrication of the mold, and subsequent transferal of the mold, design to a polymeric substrate. Materials such as silicon, quartz, nickel, stainless steel, and brass are often used as molds due to their excellent surface qualities, high mechanical strength, good thermal conductivity, and easy mold releasc at the end of the replication step (7,31,44–46).

A mold can be produced by several techniques including mechanical micromilling, wet etching of silicon and glass, epoxy resin stamps, photolithography, LIGA, electroforming, electroplating, microthermoforming, computer numerical controlled (CNC) milling, and electrodischarge machining (14,38,45,47–53). Each of these techniques has unique characteristics with regard to the minimum feature size possible, aspect ratio of microstructures, reproducibility, and geometry and total surface area possible (53). For example, an SU-8 photoresist used to define the region of Ni growth in production of a Ni master has been described (52). In another approaches, micromachined PMMA have been used as a mold for the production of PDMS devices, as described in many references (54). The main advantage of this method is the elimination of the photoresist, thus reducing the total fabrication time and cost. Moreover, this method often offers deeper channels than those obtained with a photoresist. Generally speaking, the quality of the devices produced using replication methods depends mainly on three factors: the limitations of the technique used, the equipment, and the quality of the master mold (55). For example, the surface properties (e.g., surface roughness, optical properties, contact angle) of the imprints produced using the replication methods depend on the characteristics of the master as well as the conditions of the replication process.

The key advantage of replication methods, compared to direct write methods, is the reduction in total time and cost per device in mass production. The primary disadvantage to many replication methods is a time-consuming nature, which limits the investigation of new designs. In addition, the lifetime of a mold depends mainly on the replication method used (31,53).

18.5 SELECTED MICROFABRICATION TECHNIQUES

18.5.1 Wire Imprinting

Wire imprinting is quite possibly the cheapest, fastest, and most user-friendly direct write method available to produce microchannels in plastic materials (45,56,57). In this simple method, a hard metal wire (e.g., chromel, platinum, nicrome) with diameter in the micrometer range is used to imprint simple linear microfluidic channels in easily deformable plastic materials, such as PMMA. The metal wire is brought into conformal contact with plastic substrate and subjected to pressure and often heated to the glass transition temperature (T_g) of the substrate. The assembly is later removed and allowed to cool to room temperature before the clamps are removed. In some methods, the substrate and the wire are clamped placed between two plates of glass, steel, or alumina and the whole assembly is heated in a circulating air oven. In others, a heated hydraulic press is used, while imprinting at room temperature has also been described (43).

In a PMMA sheet having a thickness of 2 mm, two holes (3-mm diameter) were mechanically drilled through the sheet at an approximate distance from one another of 2 mm. A platinum wire of 300 μm diameter was then placed on top of the substrate and between the two wells, and then taped in place. The wire and PMMA were then clamped together between glass plates using large spring-loaded clamps. The entire assembly was heated at 115°C in a circulating air oven. After 30 min, the materials were allowed to cool before disassembly. As observed in the scanning electron microscope (SEM) image shown in Fig. 18.1, the resulting microchannel was roughly 400 μm wide and irregularly shaped with a rounded profile. Additionally, the substrate on either side of the microchannel was partially deformed. Upon visual inspection, it was observed that during the imprinting process, the platinum wire had been compressed from an initially rounded cross profile to one of a slightly elongated shape, which explains the slightly wider channel.

As illustrated, wire imprinting is very simple, cheap, and does not require the use of expensive equipment; however, the technique has obvious drawbacks. The disadvantages of this method include limitations to very simple geometries and a lack

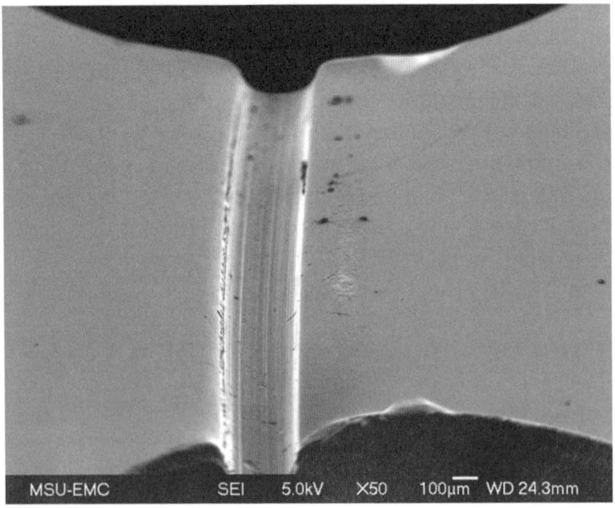

FIGURE 18.1 PMMA microchannel imprinted using a Pt wire (300-μm diameter). Microchannel is roughly 400-μm wide and slightly oval in cross-sectional shape as a result of the Pt wire becoming slightly flattened during the imprinting process.

of precision and reproducibility. Moreover, this method is limited to easily deformable plastics.

18.5.2 Laser Ablation

Laser ablation is a direct write technique where the writing instrument is the focused output of a laser, utilized to rapidly photochemically break bonds such that material is pressure ejected from the substrate (58). In some instances, the term is applied where the resultant microstructure is a result of thermal damage, rather than a chemical interaction between light and material. It is important to note that in the literature, the term ablation is used loosely in regard to microfluidics and bioanalytical devices. In a strict sense, ablation only occurs when chemical bonds are broken and there is a resultant pressure ejection of material (59). This distinction is important when one considers biological assays and chemical separations that are dependent upon surface chemistry.

In regard to bioanalytical microdevices, laser ablation is mainly used to fabricate devices in polymer substrates. Girault was one of the first researchers to fully describe the technique and its application to fabricating microchannels for microscale capillary electrophoresis (58). Since the majority of these bioanalytical devices are fabricated in polymeric materials, the laser wavelength is typically in the ultraviolet region of the spectra. Early work was based on methods developed in the microelectronics industry and primarily utilized excimer (XeCl (308 nm), KrF (248 nm), and ArF (193 nm)) lasers. As the technique became more established and was adopted by more research laboratories, Nd : YAG (532 nm, 266 nm) and Nitrogen (337 nm) gained in popularity. This is due to a wider availability in many research laboratories and a solid-state nature that results in more of a "turnkey" operation than excimer lasers, that require highly specialized and toxic gas mixtures. The aspect ratios and surface roughness of ablated microchannels are dependent on several parameters including the nature and properties of the polymer (e.g., molecular weight and thickness) and the experimental conditions (e.g., beam energy, wavelength and focusing, and local atmosphere at ablation site).

In this example, polycarbonate was selected as the substrate due to its compatibility with the 248-nm excimer laser used to produce regularly shaped channels. As previously described, the laser ablation system (LMT-4000, Potomac Photonics, Inc., Lanham, MD) included a laser light source, a circular aperture and other optical components, and a CNC programmed $X-Y$ motion stage with vacuum to hold the substrate in place (60). Nitrogen was swept over the ablation surface, while a vacuum nozzle positioned opposite was used to remove debris ejected during ablation. The polycarbonate substrate was moved with respect to the stationary laser at a linear rate of 0.5 mm/s for eight passes. The resulting microchannels exhibited deep "pits" at the intersections of channels, which is caused by the multiple passes of the laser beam during ablation. Additionally, a narrowing cross-sectional shape is observed as shown in Fig. 18.2b.

This technique has the primary advantages of direct write techniques for rapid prototyping as well as the ability to directly control surface chemistry during the

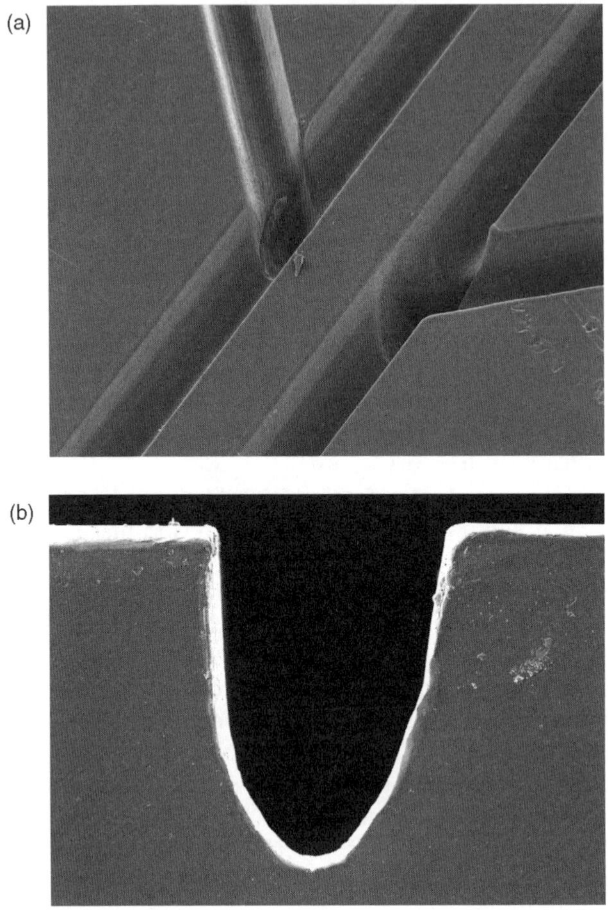

FIGURE 18.2 (**a**) SEM image of microchannels created in PC via laser ablation. "Pitting" is observed at the intersection of channels due to the multiple passes required to create intersecting channels. (**b**) Cross-sectional profile of a laser ablated polycarbonate microchannel by SEM.

fabrication process. Additionally, using laser ablation it is possible to write a wider variety of substrates, which include ceramics, glass, polymers, and metals. The CNC method of controlling movement of the substrate or laser is widely utilized, though there are a number of systems in which the laser or a focusing element is moved with respect to the substrate to write the desired design. In this case, the primary disadvantage is a limited size of the final product. Other disadvantages of the laser ablation method include technical expertise required to operate the laser system, which may range from simple to complex chemical surface changes that dependent upon the biological assay may need to be accounted for, and limitations of the particular laser wavelength that may limit one to a subset of the aforementioned materials.

18.5.3 Mechanical Machining

As the name implies, mechanical machining utilizes mechanical tools to transfer a design into a substrate, generally a polymeric or plastic substrate due to their overall to withstand the mechanical stresses imparted by the tool used. The types of tools used vary from diamond tipped lathes to micromills that use specialized metal tips created using focused ion beams. In mechanical machining, generally the movement of either the writing instrument or the substrate is computer numerically controlled. This method of control relies on translating the desired pattern into numerical $x-y$ coordinates and specifying engagement and disengagement of the surface with the tool (or vice versa dependent on whether the surface or the tool is being controlled). Mechanical machining methods can be used for direct write approaches; however, due to the initial investment and time required, it is typically employed in the fabrication of master masks and molds to be used in replication techniques, such as soft lithography and hot embossing.

An example of micromilled PMMA mold suitable for subsequent soft lithography to produce PDMS devices is shown in Fig. 18.3. In contrast, a micromilled mold for subsequent use in hot embossing is shown in Fig. 18.4. Both examples contain the typical twin-T (or offset T) separation channel with an offset injection zone. The

FIGURE 18.3 Bright field microscopy image of features milled in PMMA. The clear features observed are 100-μm wide and 100-μm tall with respect to the surrounding substrate. The circular patterns visible in the darker regions are artifacts of the mechanical milling process.

(a)

(b)

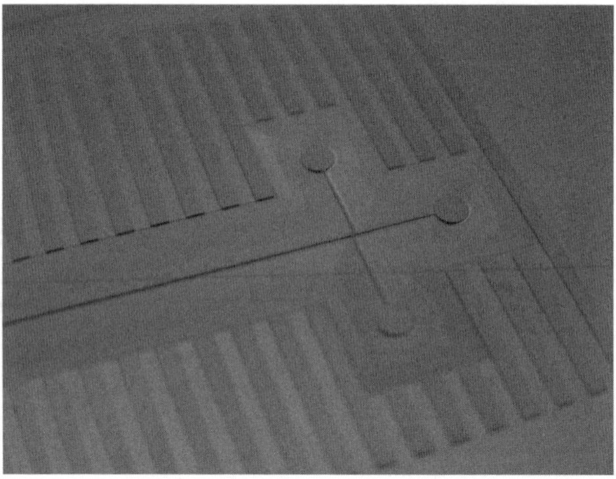

FIGURE 18.4 (**a**) Photo of brass mold created via micromilling. Three twin-T designs are shown, along with block-like "filler" structures. (**b**) Enlarged photo of the offset injection zone and fillers of a twin-T separation channel. (See color insert)

microstructures were created using a high precision micromilling machine (KERN MMP 2522, KERN Micro und Feinwerktechnik GmbH & Co. KG, Germany) equipped with a 50-mm solid carbide milling bit (Quality Tools, Hammond, LA) at a feed rate of 10–20 mm/min and 40,000 rpm (61).

In the PMMA example shown in Fig. 18.3, the clear features observed are 100-μm wide and 100-μm tall with respect to the surrounding substrate. The circular patterns visible in the darker regions are artifacts of the mechanical milling process used to "mill away" material surrounding the (taller) substrate left remaining. Noticeably, the inner corners are rounded, due to the rotating and circular drill bit used. In the brass mold shown in Fig. 18.4a and b, three designs are included for producing multiple devices in one embossing step. Also, block-like horizontal structures known as "fillers" are included in the design. These structures serve to protecting the tall and narrow (high aspect ratio) microstructure from mechanical stress and strain during hot embossing. If not designed properly, microstructures are prone to being sheared off during embossing (results not shown). The circular artifacts from the milling process are again observed in the lower lying areas surrounding the taller microstructures and fillers.

Advantages of mechanical machining include portability of commands from one instrument to another, a minimized need for highly specialized training, and elimination of clean-room environments. Disadvantages of this technique include the initial investment of the equipment, the time required, and slight restrictions in geometries dependent upon the tool dimensions.

18.5.4 Hot Embossing

Hot embossing is currently the most commonly used method for the replication of micro- and nanostructures in a wide range of thermoplastic polymeric materials in a precise and reproducible manner (46,53,62–65). This old, but low cost and high throughput method is very attractive for highly accurate mass production of microfluidic devices with high aspect ratios (10–50) (31,66). The process involves heating a thermoplastic polymeric material to its glass transition temperature (T_g) under pressure and in conformal contact with a mold containing a positive relief of the design to be transferred. Typically, a hydraulic press with heated platens is used. Under these conditions, the polymeric material becomes softened, allowing the features of the mold to be "stamped" into the polymeric material as an inverse replicate. The substrate's glass transition and melting temperatures and the thermal expansion coefficients of both the substrate and the mold material are important considerations for successful replication.

An example of a hot embossed microchannel is shown in Fig. 18.5a and b. Using a digitally programmable hydraulic press (Model AutoFour/30-3895, Carver Inc., Wabash, IN), the brass mold shown in Fig. 18.4 was brought into conformal contact with a PMMA sheet at a vertical force of 1100 lb and temperatures of 135 and 154°C for the upper and lower platens, respectively. After 30 min, the platens were cooled to room temperature while maintaining pressure over an additional 10 min. The mold and base were then mechanically separated. As observed in Fig. 18.5a, the microchannel of the mold is reproduced with extremely high accuracy, such that even the circular artifacts from the lower lying areas of the brass mold are embossed into the PMMA substrate. In the cross-sectional profile shown in Fig. 18.5b, the aspect ratio and square shape are precisely transferred into the PMMA substrate. The slight deformation in the upper-left corner of the microchannel and the slight cracks visible at the bottom corners are artifacts of the dicing process used to shear the microchannel across its cross-section.

FIGURE 18.5 (**a**) SEM image of a hot embossed PMMA microchannel (100 × 100 μm, W × H). The circular patterns observed on either side of the microchannel are artifacts resulting from the micromilling process. (**b**) SEM image of cross-sectional profile.

18.6 CONCLUSIONS

In this chapter, a brief overview of microfabrication methods was presented, with particular emphasis on selected techniques. In particular, the utility of direct write and replication methods for producing polymeric microdevices has been demonstrated. Unfortunately, the bioapplications for which these devices were designed is outside the scope of this work (67). It is expected that with the continued demonstration and use of microfluidic devices for utility in various fields of life science, development of novel materials, especially modified polymers and copolymers, and improved fabrication techniques are certain. New and interesting platforms are constantly appearing in the literature (68). It is also expected that the size of the channels will continue to reduce in size from the micrometer to the nanometer range in nanofluidic devices and that the use of disposable plastic-based microfluidic devices will bloom, especially for biomedical point-of-care diagnostics.

ACKNOWLEDGMENTS

The authors gratefully acknowledge Prof. Steven Soper (Department of Chemistry, Louisiana State University) for donation of mechanical micromilling services, Bill Monroe and Richard Kuklinski of the Mississippi State University Electron Microscope Center, and funding from the National Science Foundation (NSF #542336). Furthermore, the work of M. Jacobs was supported by the NSF Research Experience for Undergraduates (REU) Program (#453503) and EPSCoR.

REFERENCES

1. Laurell T, Nilsson J, Jensen K, Harrison DJ, Kutter JP editors. Micro total analysis systems 2004, Volume 1. Proceedings of mTAS 2004 8th International Conference on Miniaturized Systems for Chemistry and Life Sciences; 26–30 September 2004; Enschede, The Netherlands;2004.
2. Szumski M, Buszewski B. Crit Rev Anal Chem 2002;32:1–46.
3. Henry CS editor. Microchip capillary electrophoresis: Methods and protocols. In: Methods Molecular Biology (Totowa, NJ); 2006,338.
4. Gawron AJ, Martin RS, Lunte SM. Eur J Pharm Sci 2001;14:1–12.
5. Jia G, Madou MJ. MEMS Handbook, 2nd. 2006; 2, 3/1–3/214.
6. Breslauer DN, Lee PJ, Lee LP. Molec BioSyst 2006;2:97–112.
7. Harrison DJ, Fluri K, Seiler K, Fan Z, Effenhauser CS, Manz, A Science 1993; 261:895–897.
8. Marle L, Greenway GM. TrAC, Trends Anal Chem 2005;24:795–802.
9. Wang J. Anal Chim Acta 2004;507:3–10.
10. Zhang Y, Timperman AT. Analyst (Camb, UK) 2003;128:537–542.
11. Lion N, Reymond F, Girault HH, Rossier JS, Curr Opin Biotechnol 2003;15:31–37.
12. Kartalov EP, Quake SR. Nucleic Acids Res 2004;32:2873–2879.
13. Madou M, Zoval J, Jia G, Kido H, Kim J, Kim N. Ann Rev Biomed Eng 2006;8:601–628.
14. Jensen-McMullin C, Ng S, Bachman M, Sims C, Allbritton N, Li GP. Third IEEE/EMBS Special Topic Conference on Microtechnology in Medicine and Biology, Kahuku, HI, United States, May 12–15, 2005;2005. p 34–37.
15. Roman GT, Chen Y, Viberg P, Culbertson AH, Culbertson CT. Anal Bioanal Chem 2007;387:9–12.
16. Folch A, Tourovskaia A. BioMEMS Biomed Nanotechnol 2006;3:345–363.

17. Liu CN, Toriello NM, Mathies RA. Anal Chem 2006;78:5474–5479.
18. Kartalov Emil P, Zhong Jiang F, Scherer A, Quake Stephen R, Taylor Clive R, Anderson WF. Biotechniques 2006;40:85–90.
19. Dishinger JF, Kennedy RT. Anal Chem 2007;79:947–954.
20. Kamotani Y, Huh D, Futai N, Takayama S. BioMEMS Biomed Nanotechnol 2006;3:55–78.
21. Baldwin RP, Roussel Jr TJ, Crain MM, Bathlagunda V, Jackson DJ, Gullapalli J, Conklin JA, Pai R, Naber JF, Walsh KM, Keynton RS. Anal Chem 2002;74:3690–3697.
22. Waits CM, Morgan B, Kastantin M, Ghodssi R. Sens Actuators A: Physical 2005; A119:245–253.
23. Jacobson SC, Ramsey JM. Anal Chem 1998;146:613–633.
24. Giridhar Malalahalli S, Seong K, Schulzgen A, Khulbe P, Peyghambarian N, Mansuripur M. Appl Opt 2004;43:4584–4589.
25. Shintani T, Torimura M, Sato H, Tao H, Manabe T. Anal Sci 2005;21:57–60.
26. Jacobson SC, Moore AW, Ramsey JM. Anal Chem 1995;67:2059–2063.
27. Woolley AT, Lao K, Glazer AN, Mathies RA. Anal Chem 1998;70:684–688.
28. McDonald JC, Chabinyc ML, Matallo SJ, Anderson JR, Stroock AD, Whitesides GM. Anal Chem 2002;74:1537–1545.
29. Szekely L, Guttman A. Curr Anal Chem 2006;2:195–201.
30. Henry CS, Zhong M, Lunte SM, Kim M, Bau H, Santiago JJ. Anal Commun 1999;36:305–307.
31. Hu M, Watson CJ, Kennedy RT, Becker JB. Synapse (Hoboken) 2005;59:122–124.
32. Song S, Lee KY. Macromol Res 2006;14:121–128.
33. Hansen C, Quake Stephen R. Curr Opin Struct Bio 2003;13:538–544.
34. McDonald JC, Duffy DC, Anderson JR, Chiu DT, Wu H, Schueller OJ, Whitesides GM. Electrophoresis 2000;21:27–40.
35. Wang Y-C, Su P, Zhang X-X, Chang W-B. Anal Chem 2001;73:5616–5619.
36. Boone TD, Fan ZH, Hooper HH, Ricco AJ, Tan H, Williams SJ. Anal Chem 2002;74:78A–86A.
37. Soper SA, Ford SM, Qi S, McCarley RL, Kelly K, Murphy MC. Anal Chem 2000;72:643A–651A.
38. Fiorini GS, Chiu DT. Biotechniques 2005;38:429–446.
39. Verpoorte E. Lab Chip 2003;3:60N–68N.
40. Geschke O, Klank H, Telleman P, editors Microsystem Engineering of Lab-on-a-chip Devices. Weinheim: Wiley-VCH Verlag GmbH;2004.
41. Waddell EA, Locascio LE, Kramer GW. J Assoc Lab Automat 2002;7:78–82.
42. Heckele M, Guber AE, Truckenmueller R. Microsyst Technol 2006;12:1031–1035.
43. Xu JD, Locascio L, Gaitan M, Lee CS. Anal Chem 2000;72:1930–1933.
44. Zhong JF, Maltezos G, Sheriff Z, Burke K, Scherer A, Taylor CR, Anderson WF. Lett Drug Des Discov 2006;3:636–639.
45. Hong JW, Chen Y, Anderson WF, Quake SR. J Phys: Condensed Matter 2006;18: S691–S701.
46. Kimerling TE, Liu W, Kim BH, Yao D. Microsyst Technolo 2006;12:730–735.

47. Ehrfeld W, Lehr H, Michel F, Wolf A, Gruber H-P, Bertholds A. Proc SPIE—Int Soc Opt Eng 1996;2879:332–337.
48. Situma C, Wang Y, Hupert, M, Barany F, McCarley RL, Soper SA. Anal Biochem 2005;340:123–135.
49. Zhao C, Sinha JK, Wijayawardhana CA, Wittstock G. J Electroanal Chem 2004;561:83–91.
50. Moon SJ, Lee SS, Lee HS, Kwon TH. Microsyst Technol 2005;11:311–318.
51. Belligundu S, Shiakolas PS. J Microlith Microfab Microsyst 2006;5:021103/021101–021103/021108
52. Kim S-J, Yang H, Kim K, Lim YT, Pyo H-B. Electrophoresis 2006;27:3284–3296.
53. Rotting O, Ropke W, Becker H, Gartner C. Microsyst Technol 2002;8:32–36.
54. Briones MPP, Honda T, Yamaguchi Y, Miyazaki M, Nakamura H, Maeda H. J Chem Eng J 2006;39:1108–1114.
55. Bley P. Proc SPIE—Int Soc Opt Eng 1999;3876:172–184.
56. Martynova L, Locascio L, Gaitan M, Kramer G, Christiansen R, MacCrehan W. Anal Chem 1997;69:4783–4789.
57. Locascio LE, Perso CE, Lee CS, J Chromatogr A 1999;857:275–284.
58. Roberts MA, Rossier JS, Bercier P, Girault H. Anal Chem 1997;69:2035–2042.
59. Srinivasan R, Braren B. Chem Rev 1989;89:1303–1316.
60. Olsen KG, Ross DJ, Tarlov MJ. Anal Chem 2002;74:1436–1441.
61. Musyimi HK, Guy J, Narcisse DA, Soper SA, Murray KK. Electrophoresis 2005;26:4703–4710.
62. Datta P, Goettert J. Microsyst Technol 2007;13:265–270.
63. Becker H, Gartner C. Rev Mol Biotechnol 2001;82:89–99.
64. Becker H, Gartner C. Electrophoresis 2000;21:12–26.
65. Russo AP, Apoga D, Dowell N, Shain W, Turner AMP, Craighead HG, Hoch HC, Turner JN. Biomed Microdev 2002;4:277–283.
66. Worgull M, Heckele M, Schomburg WK. Microsyst Technol 2005;12:110–115.
67. Thomas G, El-Giar EM, Locascio LE, Tarlov MJ. In: Microfluidic Techniques: Reviews and Protocols. Totowa, NJ: Humana Press; 2005.
68. Gimm JA, Beebe DJ. BioMEMS Biomed Nanotechnol 2006;4:223–242.

19

NANO FOUNTAIN PEN: TOWARD INTEGRATED, PORTABLE, LAB-ON-CHIP DEVICES

LEVI A. GHEBER

19.1 INTRODUCTION

The pressing need to miniaturize biological and chemical assays has triggered a wealth of activity in the areas of chemical and biological sensing at scales never attempted before, ranging from few tens of micrometer down to a few tens of nanometer. Regardless of the prefixes and suffixes used in an attempt to define in one word the discipline to which various activities belong (nano-, micro-, -fluidics, bio-, -technology, -systems, etc.), the fact is that we are witnessing a worldwide inter- and cross- disciplinary effort of scientists to realize the miniaturization and integration on a small, portable platform of what nowadays is being done in a large laboratory equipped with expensive and advanced equipment and well-trained personnel: the "lab-on-a-chip" concept. When the original ideas is integrating on a chip fluid-related applications, such as separation, mixing, focusing, handling, amplification, among others with sensing (bio or chemical) and, moreover, with the powerful approach of (arrayed) parallel sensing of a multitude of targets, the power of such envisioned devices is tremendous and so is the complexity of the problems that require solutions.

Nanolithography of biological molecules has taken a path that is making a wide use of scanning probe microscopy (SPM) approaches. While microfluidics has borrowed very successfully approaches and methods from the electronics (semiconductor) world, biological molecules are hardly compatible with these techniques. Extreme pH, vacuum, UV light, elevated temperatures are the

Biological Applications of Microfluidics, Edited by Frank A. Gomez

characteristics to the processes for fabricating semiconductor devices and many microfluidic devices. All these are incompatible with biological molecules that need aqueous environment, neutral pH, and room temperature. SPM in their wide variety are able to work in this type of environment with nanometer (and better) precision and have the ability to modify substrates, in addition to imaging them.

The large majority of SPM-based methods use the probe of an atomic force microscope (AFM) as their working tool. It is used to either apply forces on the substrate (indenting, removing material, or inducing local perturbation) and deposit (print) molecules, or (when appropriately functionalized) perform local chemical reactions. Several examples are as follows.

Nanografting (1), NPWR (2), MFN (3) are the methods that use an AFM probe to mechanically remove a layer of protective self-assembled monolayer (SAM) while a different molecule is adsorbed to the substrate, replacing the removed molecules.

Direct deposition (printing) of molecules is realized using dip pen nanolithography (DPN) (4) and variations. DPN is a method that uses an AFM probe as a dip pen, where the probe is dipped in a molecular "ink" that is subsequently "written" on a surface by contacting the probe with the surface.

AFM assisted chemical and biochemical reactions include several examples. The extension of DPN, the electrochemical AFM Dip-pen (E-DPN) uses an (conductive) AFM probe to perform electrochemistry on nanometer scales, and thus write metallic nanowires (5). In another case, an AFM probe was used to perturb an ordered and closely packed phase of phospholipids, and this allowed the stereospecific enzyme PLA_2 to hydrolyze the lipids, thus creating nanochannels (6). A way to directly perform a biochemical reaction was shown using a protease immobilized to an AFM probe to controllably hydrolyze peptides adsorbed on the substrate (7). A similar approach, a tip immobilized enzyme, was used to digest a SAM of oligonucleotides, leading to the creation of nanochannels (8).

The relative ease of using a commercially available AFM probe is remarkable and has led to numerous demonstrations of printing nanostructures of DNA (3,9–13), various proteins (1,14–21), and peptides (22). Here, we would like to concentrate on another SPM-based tool used for nanobiolithography, fountain pen, or nanofountain pen (NFP).

19.2 NANO FOUNTAIN PEN

NFP consists of a glass or quartz capillary heat pulled to create a taper, rendering a sharp tip with an aperture with diameters ranging between a few tens and a few hundreds of nanometers.

The nanopipette is mounted as the probe of a scanning force microscope and controlled in the same fashion as an AFM probe is controlled (Fig. 19.1). The

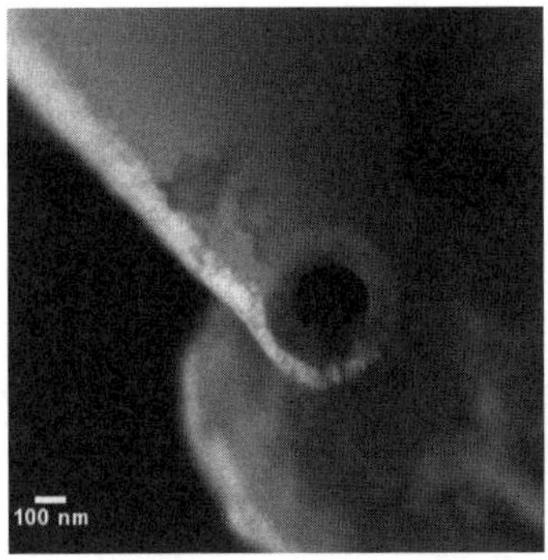

FIGURE 19.1 The end of a 200 nm aperture nanopipette.

nanopipette is filled with a fluid at its wide end, and the fluid is drawn to the tip by capillary forces, but it does not flow out due to surface tension. Deposition of the fluid occurs upon contacting the pipette with a substrate.

The use of micropipettes as probes for scanning probe microscopes was originally developed for near-field scanning optical microscopy (NSOM) as early as 1993 (23); however, originally the purpose of micropipettes use was to serve as light waveguides with subwavelength apertures. It was later that these nanopipettes were used for delivery of liquids to a surface (24,25), however, not in a biological context. The demonstration of the ability to print patterns of proteins of submicrometer dimensions followed in 2002 (26) and 2003 (27).

Although more difficult to manufacture and not amenable to mass production, like AFM probes, NFP is much more versatile and allows many more applications than AFM probe-based methods. In addition to delivering biological molecules (DNA (26,28), proteins (27,29), enzymes (30,31), etc.), it is possible with nanopipettes to deliver gases, acids (25), and polymers (24,32). In addition, NFP can write large amounts of "ink" without the need for repetitive dipping. Moreover, it should be possible to connect it with the output of chromatography equipment (such as HPLC) to consecutively write various proteins without pen lifting. NFP can be combined with electrical field to precisely control ejection of charged biomolecules, such as DNA (26) and proteins (28), down to single-molecular level.

We show here several examples from our work, describing the abilities of NFP and the ways its use may facilitate the large idea of portable, integrated, fluidic devices for diagnosis and monitoring.

19.3 PROTEIN PRINTING

Using the NFP, we have shown the ability to print green fluorescent protein (GFP) (27). The protein was printed on a layer of bovine serum albumin (BSA) that had previously been spread and dried (and hardened) on a glass support, providing a smooth surface to which the GFP associated. GFP was loaded in the pipette in phosphate-buffered saline solution (PBS) with a concentration of 1.5 μg/μL, providing the best environment to avoid its denaturation. We could print dots with 280 nm diameter and ~4 nm thickness, and lines longer than 70 μm with similar width and height (Fig. 19.2). The ~4 nm height is similar to the size of one protein molecule and indicates that we are printing a monomolecular layer of GFP.

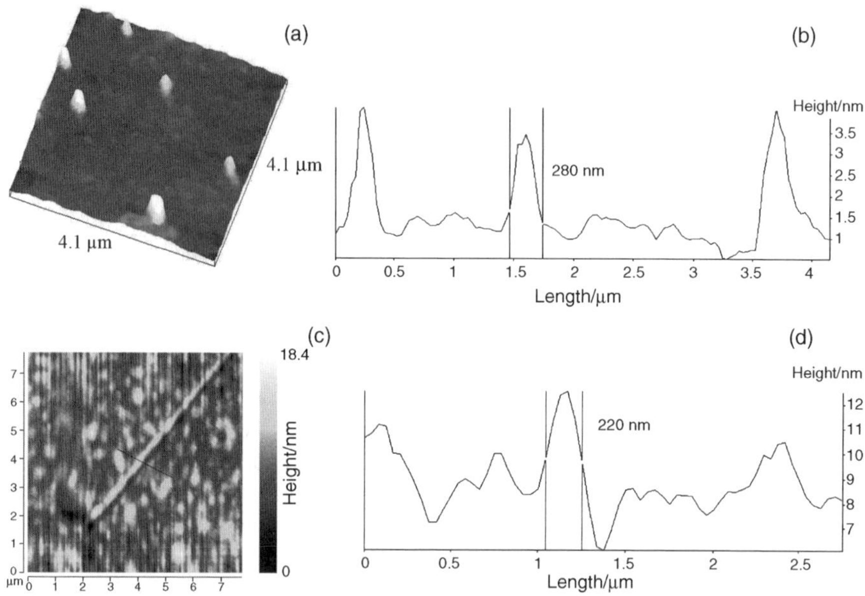

FIGURE 19.2 GFP printed on a hardened BSA substrate using NFP. (**a**) dots with 280 nm diameter and 4 nm height. (**b**) profile of three consecutive dots. (**c**) part of a line of GFP having a width of 220 nm and a height of 4 nm. (**d**) profile of the line in c.

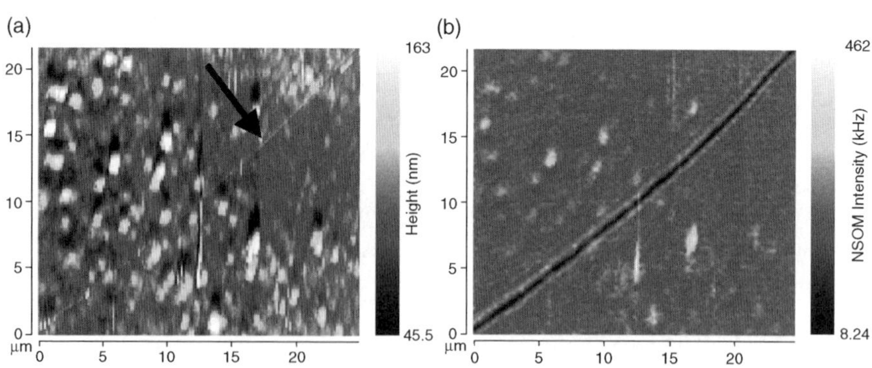

FIGURE 19.3 NSOM pair of images, acquired simultaneously. (**a**) topography. The GFP line is indicated with a black arrow, but is barely distinguishable on the background of numerous dust particles. (**b**) optical image obtained by illumination with 488 nm light and collection in reflection. The GFP line is the only feature appearing dark, since GFP strongly absorbs at this wavelength, while the rest of the sample reflects or scatters the light. (See color insert)

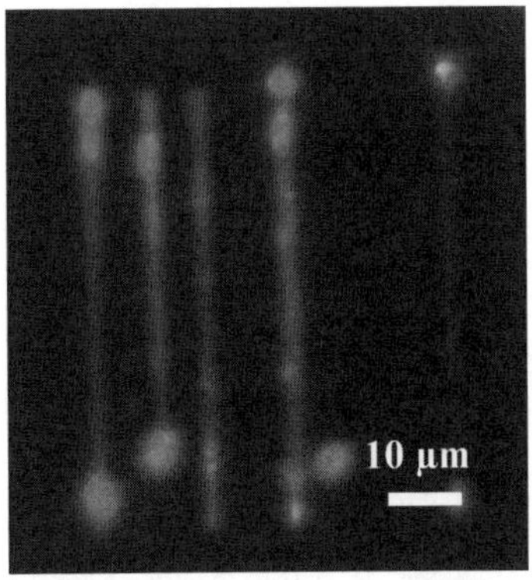

FIGURE 19.4 Lines of fluorescently labeled avidin printed with NFP on aldehyde slides and imaged in far field, with conventional fluorescence microscopy. (See color insert)

To prove that the protein maintains its activity (fluorescence), we used NSOM to image the printed features. GFP has an absorption maximum at 490 nm, so we used an Ar-ion laser 488nm as the light source. The images in Fig. 19.3 show simultaneously acquired topography (left) and reflection NSOM. The GFP line (black arrow) can be faintly observed in the topography image on the background of dust particles, however, is clearly seen as the dark stripe in the NSOM image as the only feature that absorbs the 488 nm light (the dust particles are scattering it).

We also showed that fluorescently labeled avidin can be printed in the same way on aldehyde slides, as shown in Fig. 19.3, and imaged with a conventional fluorescence microscope (Fig. 19.4). This opens the way of immobilizing virtually any (biotinylated) protein along the patterns of avidin, using the strong avidin–biotin affinity.

19.4 ENZYME LITHOGRAPHY

Proteolytic enzymes hydrolyze the peptide bond at very specific sites. For example, trypsin recognizes lysine and arginine and cleaves the amino acid sequence at these points. The high affinity of enzymes for their substrate offers the possibility to "encode" etching cues on a surface that will be recognized and followed by the etchant. This is a unique possibility offered by the biological world, which is absent in the inorganic world. We used this property to demonstrate enzyme-based nanolithography.

BSA contains 86 lysine and arginine amino acids; therefore, the protein is expected to collapse upon cleavage at all these points by trypsin. We used 1.3 mg trypsin (TPCK trypsin, bovine pancreas, Pierce) in 1 mL of 0.05 M sodium borate, pH 8.5 (BupH Borate buffer, Pierce). Holding the pipette in position for increasing durations gives rise to depressions of increasing volume. In Fig. 19.5a, we have showed three pairs of depressions obtained as a result of keeping the pipette in contact with the surface for 20, 60, and 120 s, as indicated. If the sample is moved

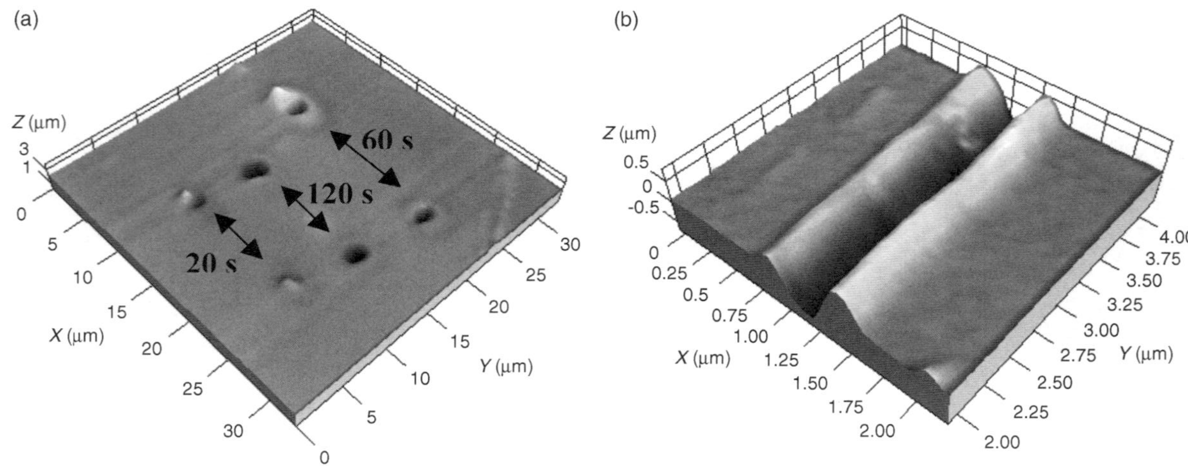

FIGURE 19.5 Depressions created in a BSA hardened substrate as a result of deposition of trypsin. (**a**) Depression volume increases with deposition time. (**b**) Trench created by moving the sample while the pipette is in contact. (See color insert)

while the pipette is in contact (under feedback control), a trench results, as shown in Fig. 19.5b.

To study the precise mechanism underlying the formation of depressions, we performed a control experiment, where the pipette was filled with only water. This resulted in protrusions, as shown in Fig. 19.6.

Interestingly, the size of the protrusions increases with the duration of contact between the pipette and the substrate. These protrusions are a result of swelling. Therefore, the formation of depressions must be a combination of swelling, with a faster collapse of cleaved proteins that ultimately results in depressions. To check this hypothesis, we measured the volume of protrusions as a function of contact time and the volumes of depressions as a function of contact time. These are plotted in Fig. 19.7. The graph labeled A describes the volume of the protrusions created with water only as a function of contact time. It is clear that the volume roughly increases linearly with time, and the solid line is a fit to the experimental points with $V = 0.0066\,t - 0.0056$. The line passes through the origin as expected.

The graph labeled B represents the volumes of the depressions (thus negative) as a function of contact time. The solid line is a fit to the experimental points with $V = -0.0188\,t + 0.1614$. Both from the equation and from the inspection of the graph, the line does not pass through the origin, but instead it crosses the $V = 0$ axis at ~8.6 s from the moment the pipette contacts the substrate. At $t = 0$, the moment of contact, the linear fit predicts a positive volume (swelling) of 0.16 μm^3. Therefore, the "competition" underlying the creation of depressions can be described with these data. The net etching of the substrate by trypsin alone, without the swelling, can be deduced from the rate of swelling (a) and the true rate of etching (b). The swelling occurs at a rate of ~0.007 μm^3/s and the resulting depressions are created at a rate of ~0.019 μm^3/s; therefore, the net etching rate of the trypsin is ~0.026 μm^3/s, as shown in Fig. 19.7c. The linear fit yields $V = -0.0255\,t + 0.174$, which crosses the $V = 0$ axis in ~6.8 s from the moment of contact.

Thus, on contact, the BSA starts swelling immediately at a rate of ~0.007 μm^3/s, a fact that exposes lysine and arginine to trypsin. Some 6.8 s later, trypsin starts cleaving the BSA at a rate of ~0.026 μm^3/s, while the BSA continues to swell. However, since the etching rate is faster than the swelling rate, approximately 2 s later (or 8.6 s from the moment of contact), the growing depression reaches the level

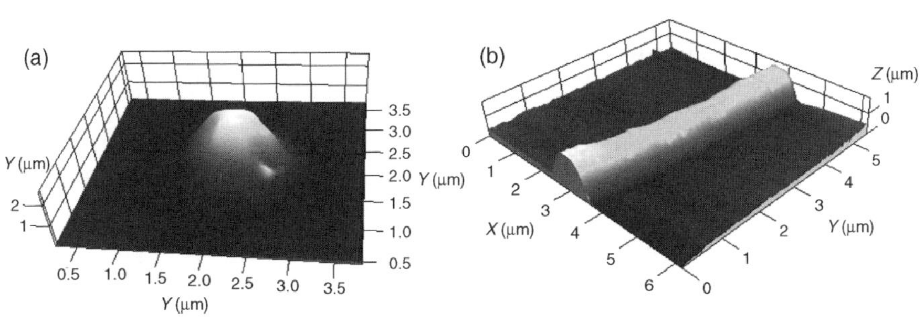

FIGURE 19.6 Protrusions created with a water-filled NFP.

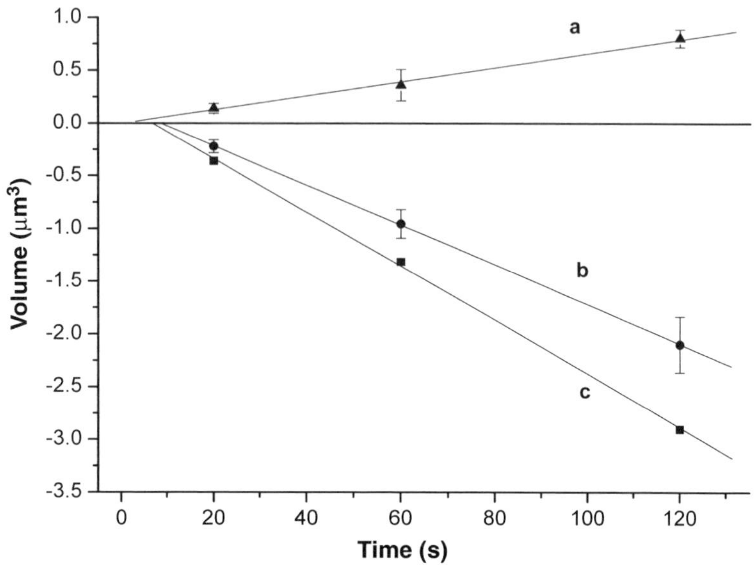

FIGURE 19.7 (**a**) Volumes of mounds are linear with time of water flow. Linear regression yields a swelling rate of ~0.007 μm^3/s. Intercept is zero, within experimental error. Error bars indicate variation of four different mounds for each time point. (**b**) Volumes of wells are linear with the time of trypsin flow. Linear regression yields an etching rate of 0.019 μm^3/s. Intercept predicts a protrusion of ~0.16 μm^3 at $t = 0$. Time point for which $V = 0$ is ~8.6 s. Error bars indicate the variation of three different wells for each time point. (**c**) Deduced net trypsin etching is calculated as the difference between the combined effect (**b**) and swelling (**a**). Linear regression yields an etching rate of ~0.026 μm^3/s. Time point for which $V = 0$ is ~6.8 s.

of intact (unswollen) substrate ($V = 0$). From this point on, the depression grows below this original level at a rate that is the combination between swelling and net etching rates, $-0.019 = 0.007 - 0.026$ μm^3/s.

Two additional control experiments further support this mechanism. The first is reducing the concentration of trypsin in the solution deposited. Assuming that the net etching rate of the trypsin is proportional to its concentration, reducing the concentration should lead to a lower etching rate, and possibly lower than the swelling rate, in which case we will end up with a protrusion, despite the fact that trypsin is present in the solution. Indeed, upon reducing the concentration by ~50% (0.6 mg/ml), a ridge is created (Fig. 19.8a), instead of a trench as in Fig. 19.5b. The combined effect results in a protrusion since the swelling rate in this case is faster than the etching rate. In addition, mixing the trypsin solution with soybean trypsin inhibitor—a natural highly specific inhibitor of trypsin—yields a ridge again (Fig. 19.8b), since the trypsin is rendered inactive, and the only observed effect is swelling.

We also studied the dynamics of channel formation using this enzyme lithography approach (31). When creating channels, additional parameters play an important role, such as the pipette aperture diameter and the velocity at which the sample is

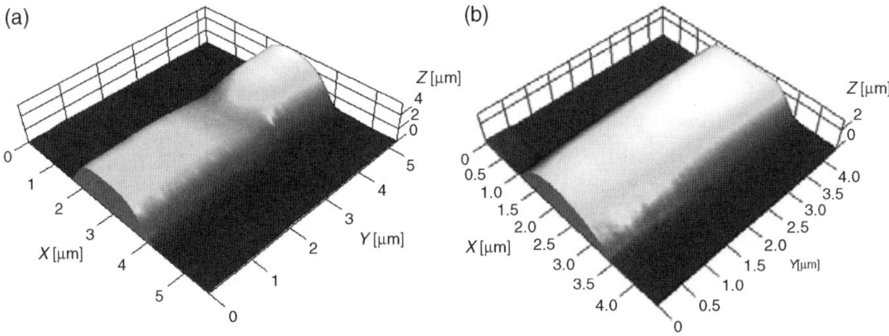

FIGURE 19.8 (**a**) Ridge resulting from writing with trypsin at ~50% concentration relative to Fig. 19.5b. (**b**) A ridge resulting from writing with trypsin at the same concentration as in Fig. 19.5b, but with soybean trypsin inhibitor mixed in the solution.

moved relative to the pipette. A comprehensive study shows that it is easy to control the cross-section of the channels, as shown in Fig. 19.9. The horizontal axis represents a product of pipette aperture area (A) and the dwell time between consecutive positions while drawing a line (Δt). The larger the Δt, the slower the velocity. This product is a measure of the volume of trypsin flowing to the substrate in

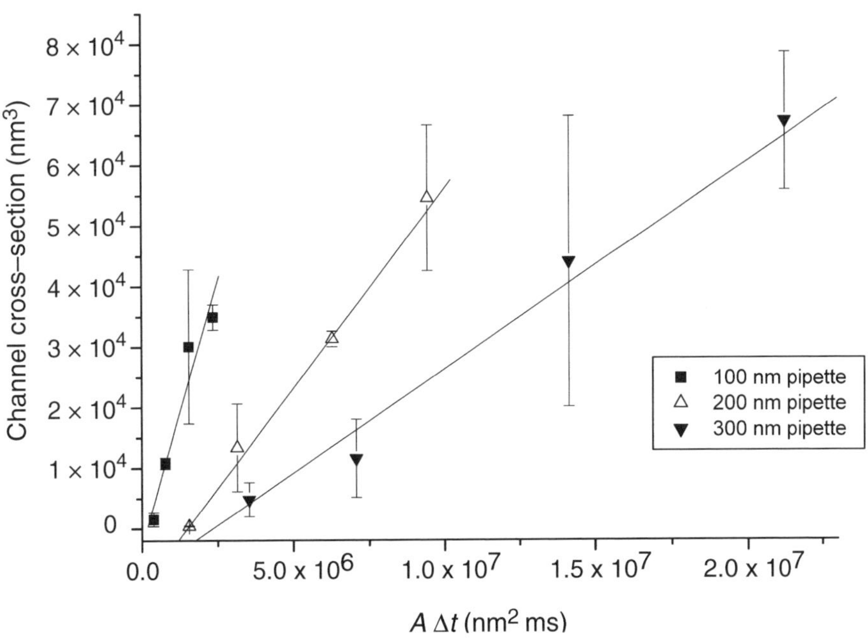

FIGURE 19.9 Channel cross-section as a function of the product of pipette aperture area (A) and the dwell time in a certain position along the channel (Δt).

a given position. When equating the volume of a unit length of channel with the volume of trypsin solution, it is observed that the cross-section of the channels created is proportional to the volume of trypsin and, interestingly, is different for pipettes with different diameters. The etching "efficiency" is higher with a 100 nm diameter pipette than with 200 nm and 300 nm.

Networks of channels can be created using this approach, as shown in Fig. 19.10b. The etching is additive in the sense that the depth at an intersection equals the sum of depths of the two channels that intersect. In the intersection imaged in Fig. 19.10a, the channel labeled 1 was created first and the channel labeled 2 was created later. From the profiles in Fig. 19.10c, it is visible that the depth of both channels is similar (profile a ~17 nm and b ~20 nm), while the intersection itself has a depth corresponding to the sum of depths of the intersecting channels (~36 nm).

We further showed that it is possible to flow liquid through such channels. In this experiment, we created channels in a biotinylated BSA (b-BSA) layer and subsequently deposited fluorescently labeled avidin with another pipette. The fluorescence image is shown in Fig. 19.11.

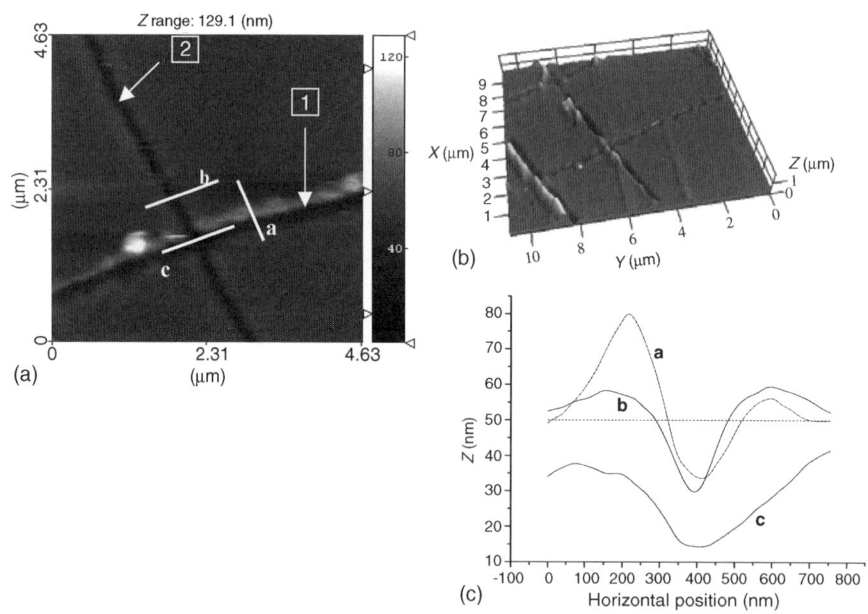

FIGURE 19.10 (**a**) An intersection of two channels. The channel labeled 1 was first created and the one labeled 2 was created later. (**b**) A network of channels of various width created by varying the velocity of the substrate with the same pipette. (**c**) Cross-sections along the lines labeled a, b, c in (**a**), showing that the depth of the channels 1 and 2 are similar (having been fabricated with the same parameters) and the depth of the intersection is the sum of the depths of each channel.

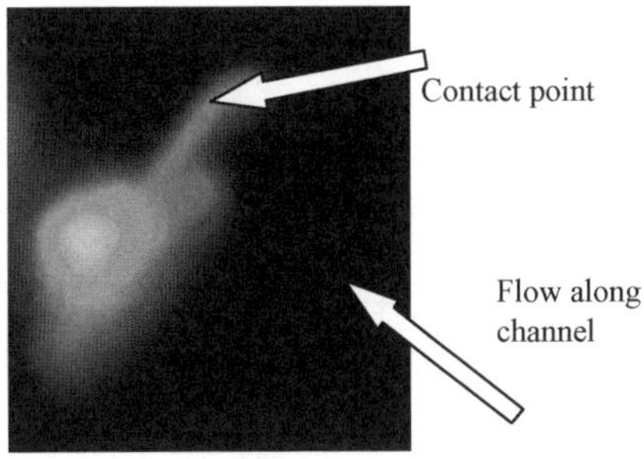

FIGURE 19.11 Fluorescently labeled Avidin was delivered with NFP at the position labeled "contact point" and flowed along a channel previously manufactured with a different pipette filled with trypsin solution on a biotinylated BSA surface.

19.5 POLYMER MICROLENSES

Owing to the versatility of the NFP, it is possible to deposit a monomer solution that can be subsequently polymerized using UV light. This can be used to manufacture microlenses one by one and with individual optical properties at very precise locations, a property that, in turn, can be very useful for integration of the optical reading system with the array of sensing elements and the microfluidics in the context of lab-on-a-chip devices.

Following deposition of drops and polymerization, they can be characterized with AFM and shown to have a spherical cap shape. Moreover, since the contact angle between the drop and the substrate is constant (determined by the wetting properties), depositing various volumes of monomer solution results in spherical caps with various diameters, which are precisely controlled (32). Fig. 19.12 shows an array of microlenses (Fig. 19.12a) manufactured this way, imaged with a white light microscope, in the focal plane of the lenses. In the inset the pipette shadow is visible,

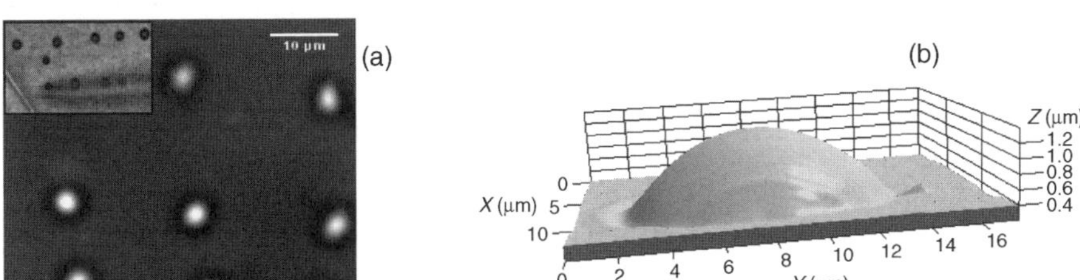

FIGURE 19.12 (**a**) White light microscope image of an array of microlenses in the focal plane of the lenses shows the concentration of light. The inset shows the pipette shadow, which appears while manufacturing the microlenses. (**b**) AFM image of one of the microlenses showing its topography.

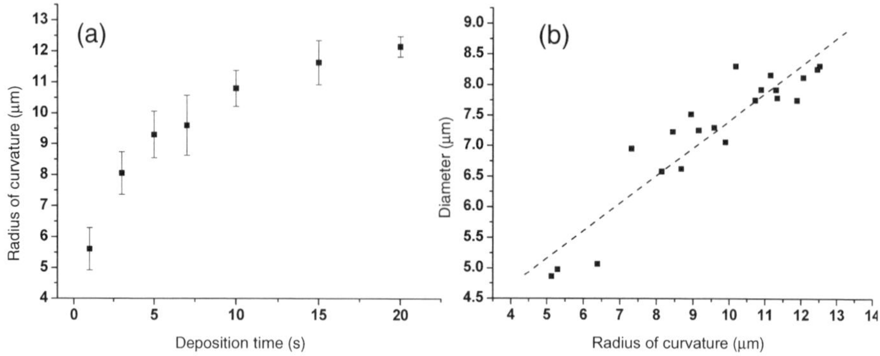

FIGURE 19.13 (**a**) Dependence of the radius of curvature on the deposition time. (**b**) The diameter and radius of curvature are related by a linear dependence, and the extracted contact angle in this case is 13°.

which appears while manufacturing the lenses. An AFM image of one of these microlenses is shown in Fig. 19.12b.

Since the diameter and the radius of curvature are related by $R = \frac{D}{2 sin\alpha}$, where R is the radius of curvature, D is the lens diameter and α is the contact angle, and the diameter depends on the deposition time (the volume of the drop), it is easy to control the radius of curvature of lens, as demonstrated in Fig. 19.13. D and R are measured from AFM images.

The radius of curvature, in turn, determines the focal length of the lens according to $f = R/(n_{\text{lens}} - 1)$, a fact that enables an easy way to control the lens power. This can be verified optically. The optical measurements were performed on the stage of an optical microscope with collimated white light. Series of images were taken at various distances from the plane of the lenses, along the optical axis, through the focal plane of the lenses. Such a series is shown in Fig. 19.14. The optical "slices" labeled 1 through 20 in Fig. 19.14a are taken at steps of 2 μm in the Z direction. The focal plane of the lens is between slices 14 and 15, as easily observed in Fig. 19.14c, corresponding to a focal length of 35 μm. The intensity in the focal plane across the width of the focused beam is shown in Fig. 19.14b and shows a $(1/e)^{0.5}$ width of 3.7 μm. The slices 1 through 20 were stacked, and a section through this stack along the optical axis is shown in Fig. 19.14d.

To prove that such microlenses can be used to concentrate the light emitted by fluorescent spots, we deposited a spot of fluorescein with a nanopipette on one side of a cover glass and a microlens with another nanopipette on the other side of the glass, as described in Fig. 19.15a. The white light image of this arrangement is shown in Fig. 19.15b. The fluorescence image of the fluorescein spot (in the plane labeled 2 in Fig. 19.15a) is shown in Fig. 19.15c. The image in Fig. 19.16d is taken in the focal plane of the lens (labeled 1 in Fig. 19.15a), and it shows the out-of-focus fluorescence from the fluorescein spot and the light concentrated by the lens in the middle. The

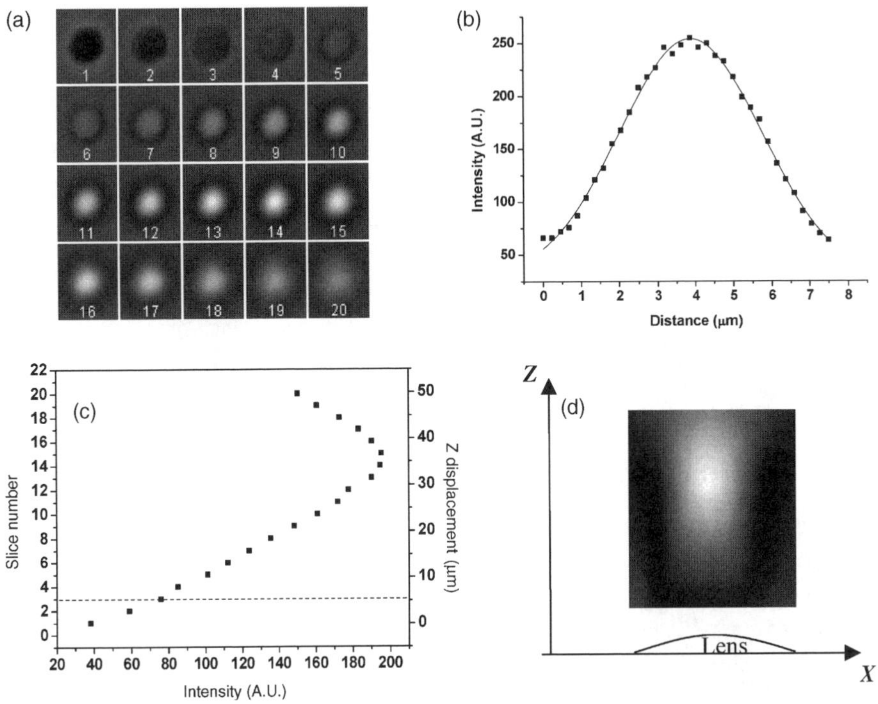

FIGURE 19.14 (**a**) A series of images of light focused through one microlens, starting below the plane of the lens (slice 1) through the focal plane (slices 14, 15) and above it (slices 16–20). (**b**) The intensity profile along the X axis (in the focal plane). The continuous line is a fitted Gaussian with a full width at $1/\sqrt{e}$ maximum of 3.7 μm. (**c**) The intensity profile along the Z axis (optical axis of the lens). The dashed line represents the plane of the lens. The focal plane is at $Z = 35$ μm, where the light intensity is maximal. (**d**) A stack formed of the slices 1–20 (in Fig. 19.4a) was sectioned along the $X–Z$ plane to obtain the light intensity distribution in this plane.

intensity profile along the horizontal line in Fig. 19.16d is plotted in Fig. 19.15e, where it is readily observed that the lens projects 1.5× more light in this plane than would have been collected without it.

Such arrangements should be possible on a more complex device. For example, an individual lens can be fabricated for each spot in a biological array, or individual lenses could be placed on top of micro/nanochannels to amplify the amount of light emitted by various fluorescent species. This could rid of one of the need for an objective (and the microscope) for light collection purposes and should greatly advance the portability of the device. By placing individual "objectives" where they are needed, the optical signal needs to be projected on a detector, such as a CCD array. Further integration of the light source should also be possible using laser diodes and embedded waveguides.

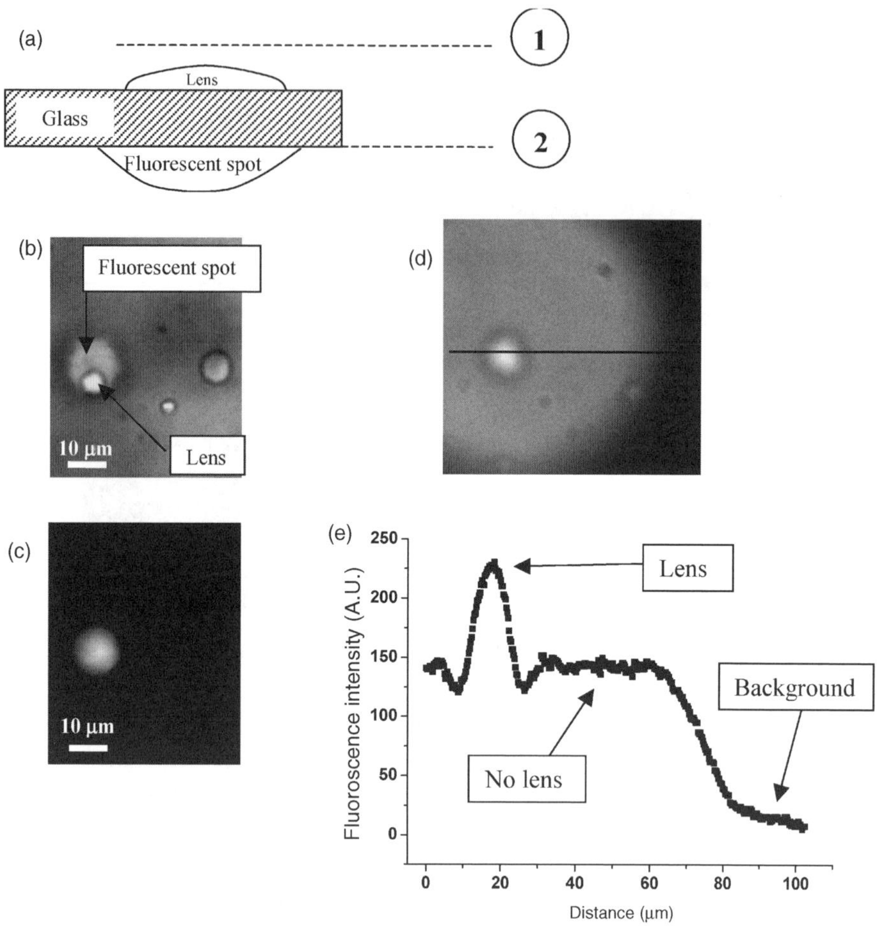

FIGURE 19.15 (**a**) Schematic representation of the experimental setup: a fluorescent spot and a microlens are deposited on the opposite sides of a coverglass (170 μm thickness) with two different nanopipettes. The focal plane of the microlens is labeled 1 and the plane of the fluorescent spot is labeled 2. (**b**) A bright-field image of the fluorescent spot and the lens taken in the focal plane of the lens (1). (**c**) A fluorescence image of the fluorescent spot taken in its plane (2). (**d**) A fluorescence image taken in the focal plane of the lens (1). The round halo is the fluorescent light emitted by the fluorescent spot, which is out of focus (in plane 2). The bright spot at the center of the halo is the image of the fluorescent spot concentrated by the microlens in its focal plane. (**e**) An intensity profile along the horizontal line in Fig. 19.16d. The light intensity produced by the microlens is 1.5 fold higher than the light intensity detected directly from the fluorescent spot.

19.6 CONCLUSIONS

Nano fountain pen is a very versatile SPM-based tool that can be used to fabricate most of the basic elements in a lab-on-a-chip device. It can be used to print biological arrays of proteins, DNA, among others with ~100 nm diameter and single-molecule thickness, or lines with similar dimensions and tens of micrometer length. It can deliver proteins without damaging them, as shown for the cases of GFP (fluorescence maintained) or enzyme lithography (function of trypsin maintained). NFP can be used to create nanochannels in biological substrates using "enzyme lithography." Using the high specificity of enzymes for their substrates, it should be possible to design complex architectures with very high resolution. Finally, NFP can be used to fabricate polymer microlenses to integrate the optical reading system with the fluid-handling systems and bioarrays.

REFERENCES

1. Wadu-Mesthrige K, Xu S, Amro NA, Liu GY. Langmuir 1999;15:8580–8583.
2. Amro NA, Xu S, Liu GY. Langmuir 2000;16:3006–3009.
3. Schwartz PV. Langmuir 2001;17:5971–5977.
4. Piner RD, Zhu J, Xu F, Hong SH, Mirkin CA. Science 1999;283:661–663.
5. Li Y, Maynor BW, Liu J. J Am Chem Soc 2001;123:2105–2106.
6. Clausen-Schaumann H, Grandbois M, Gaub HE. Adv Mater 1998;10:949–952.
7. Takeda S, Nakamura C, Miyamoto C, Nakamura N, Kageshima M, Tokumoto H, Miyake J. Nano Lett 2003;3:1471–1474.
8. Hyun J, Kim J, Craig SL, Chilkoti A. J Am Chem Soc 2004;126:4770–4771.
9. Liu MZ, Amro NA, Chow CS, Liu GY. Nano Lett 2002;2:863–867.
10. Wilson DL, Martin R, Hong S, Cronin-Golomb M, Mirkin CA, Kaplan DL. Proc Natl Acad Sci U S A 2001;98:13660–13664.
11. Demers LM, Ginger DS, Park SJ, Li Z, SChung, SW, Mirkin, CA. Science 2002;296:1836–1838.
12. Hyun J, Ahn SJ, Lee WK, Chilkoti A, Zauscher S. Nano Lett 2002;2:1203–1207.
13. Liu MZ, Liu GY. Langmuir 2005;21:1972–1978.
14. Wadu-Mesthrige K, Amro NA, Garno JC, Xu S, Liu GY. Biophys J 2001;80: 1891–1899.
15. Lee KB, Park SJ, Mirkin CA, Smith JC, Mrksich M. Science 2002;295:1702–1705.
16. Lee KB, Lim JH, Mirkin CA. J Am Chem Soc 2003;125:5588–5589.
17. Kenseth JR, Harnisch JA, Jones VW, Porter MD. Langmuir 2001;17:4105–4112.
18. Lim JH, Ginger DS, Lee KB, Heo J, Nam JM, Mirkin CA. Angew Chem Int Ed 2003;42:2309–2312.
19. Lee KB, Kim EY, Mirkin CA, Wolinsky SM. Nano Lett 2004;4:1869–1872.
20. Nam JM, Han SW, Lee KB, Liu XG, Ratner MA, Mirkin CA. Angew Chem Int Ed 2004;43:1246–1249.
21. Tang QL, Zhang YX, Chen LH, Yan FN, Wang R. Nanotechnology 2005;16: 1062–1068.
22. Agarwal G, Naik RR, Stone MO. J Am Chem Soc 2003;125:7408–7412.
23. Lieberman K, Lewis A. Appl Phys Lett 1993;62:1335–1357.

24. Hong MH, Kim KH, Bae J, Jhe W. Appl Phys Lett 2000;77:2604–2606.
25. Lewis A, Kheifetz Y, Shambrodt E, Radko A, Khatchatryan E, Sukenik C. Appl Phys Lett 1999;75:2689–2691.
26. Bruckbauer A, Ying LM, Rothery AM, Zhou DJ, Shevchuk AI, Abell C, Korchev YE, Klenerman D. J Am Chem Soc 2002;124:8810–8811.
27. Taha H, Marks RS, Gheber LA, Rousso I, Newman J, Sukenik C, Lewis A. Appl Phys Lett 2003;83:1041–1043.
28. Bruckbauer A, Zhou DJ, Ying LM, Korchev YE, Abell C, Klenerman D. J Am Chem Soc 2003;125:9834–9839.
29. Bruckbauer A, Zhou DJ, Kang DJ, Korchev YE, Abell C, Klenerman D. J Am Chem Soc 2004;126:6508–6509.
30. Ionescu RE, Marks RS, Gheber LA. Nano Lett 2003;3:1639–1642.
31. Ionescu RE, Marks RS, Gheber LA. Nano Lett 2005;5:821–827.
32. Sokuler M, Gheber LA. Nano Lett 2006;6:848–853.

20

SURFACE ENGINEERING OF MICROFLUIDIC DEVICES USING REACTIVE POLYMER COATINGS

Mutsumi Yoshida and Jörg Lahann

20.1 INTRODUCTION

Defined and stable surface properties along with the capability to immobilize active biomolecules onto a surface are key features for the development of miniaturized biodevices, such as micro total analysis systems (μ-TAS), microfabricated tools for genomics and proteomics, or cell-based assays (1). At the same time, the development of advanced surface modification protocols that are equally applicable to flat and three-dimensional surfaces have become more important with increasing miniaturization. Microfluidic systems have been applied to single-molecule level assays including studies of enzymes, protein folding kinetics, and DNA analysis (2). Other applications of microfluidic systems in biology include tissue engineering and the study of cell–biomaterial interactions (3,4), immunosensors (5,6), and other combinatorial assays using gradient or matrix of surface-immobilized ligands (6). In miniaturized systems, very large surface-to-volume ratios can occur. Under these conditions, even small chemical inhomogeneities can result in device malfunction. In addition, the range of materials, from which microfluidics devices have been fabricated, is quite diverse. Capitalizing on established fabrication methods from microelectronics and microelectromechanical systems, glass and silicon have been conventionally used for microfludics systems. For these substrates, silanization chemistry offers a wide range of possibilities with respect to surface modification (7).

Biological Applications of Microfluidics, Edited by Frank A. Gomez

Alternatively, microfluidics systems composed of polymeric materials have been pursued, because they can easily be microfabricated and mass produced, using techniques such as UV or carbon dioxide laser ablation (8,9), imprinting (10), replica molding (11,12), and injection molding (13). Polymers utilized in microfluidics systems include polyvinyl chloride, polystyrene, polyurethane, polycarbonate, poly (methylmethacrylate) (PMMA), and poly(dimethylsiloxane) (PDMS) (14) as well as a range of different hydrogels (15). In particular, PDMS has been widely fabricated via soft lithography, which enables rapid prototyping of microfluidic devices (11,12).

To fulfill the ample requirements of demanding biological applications, a series of surface modification techniques has been developed for PDMS material to control the interface between microfluidic device and biological sample, some of which have recently been summarized in an excellent review paper (14). While these polymers are promising for microfluidics applications, they are often hydrophobic and lead to undesired adsorption of analytes onto channel walls. Especially in systems that are designed to be in contact with biological layer, such as proteins or DNA, nonspecific protein adsorption is a major challenge and has been the focus of intense research (16–19). The continuous development of effective surface modification protocols that enable spatially and temporally controlled properties is essential to a range of different microfluidic applications. The latter especially applies to surfaces that are intended for immobilization or patterning of proteins, cells, and other biological molecules.

20.2 MICROFLUIDICS SURFACE MODIFICATION TECHNIQUES

Because of the high surface area-to-volume ratio in microfluidic systems, surface chemistry is of significant importance to miniaturized devices. For example, regulation of electroosmotic flow (EOF) and reduction of unintentional interaction of analytes at surface walls are critical issues that are governed through surface chemistry (7). The ability to spatially and temporally control surface chemistry and geometry of ligands that interact with cells is of particular importance for applications in which interaction of cells are explored. Some of the desirable surface properties for microfluidic devices have been summarized in Table 20.1.

In this chapter, methods used to modify microfluidics surfaces in the context of biomedical applications, specifically surface chemistries that enable subsequent control of protein immobilization and cell adhesion will be reviewed, with an emphasis on chemical vapor deposition polymerization technology.

20.2.1 Energy-Induced Surface Modifications

Treatment of microfluidics surfaces with energy sources has been mostly applied to hydrophobic polymers, in particular PDMS. Upon treatment, a PDMS surface becomes oxidized, rendering its surface more hydrophilic. Energy treatments include corona discharges (22), oxygen plasmas (11,23,24), and UV light (25).

TABLE 20.1 Properties of Ideal Microfluidics System

General design/electroosmotic flow properties	
	Controllable wetting properties
	Prevention of surface reconstruction
	Controllable of surface lubricity
Ligand immobilization	
	Density
	Spatial control
	Temporal control
	Ligand ratios
	Orientation
Manufacturability	
	Shelf life
	Sterilizability

Conversely, increase in PDMS surface hydrophobicity and roughness has been reported upon treatment with CF_4 plasma, as a result of surface fluorination (26). Although the details of changes that occur at the PDMS surface upon plasma or corona treatment are complex, formation of silica layer with increased oxygen content was shown by X-ray photon spectroscopy XPS (27), followed by recovery to hydrophobic state via displacement of silica-like structures by diffusion of low molecular weight noncrosslinked oligomers (23). Techniques used to study chemical composition of PDMS at surfaces include XPS (22,23,28,29), infrared spectroscopy (25), and chemical force microscopy (30). In addition, sum frequency generation spectroscopy has been used to characterize surface structure and kinetics of plasma modification of PDMS (31,32). Corona and oxygen plasma treatments of PDMS surface have been shown to increase EoF (30) and support electroosmotic pumping and filling with water (11). Recently, a method to generate a stable PDMS surface for microfluidics application has been reported. In this approach, unreacted oligomers are extracted from the PDMS using organic solvents followed by oxidation by air plasma to form a stable oxide layer (33). Collectively, these procedures tend to render PDMS microfluidics substrates more hydrophilic, at least for short times.

20.2.2 Surface Modifications Through Adsorbed Coatings

Coatings to modify microfluidics surfaces include physisorbed and chemisorbed materials. Physisorbed coatings are either dynamic (coating material incorporated into running buffer for electrophoresis) or static (coating material not in running buffer) (16). Alternatively, covalent binding of coating agents to substrates through functional groups or as permanent coatings has been successfully implemented. Materials used for coating of microfluidics channels are often charged or uncharged polymers, surfactants, or small molecules.

20.2.2.1 Surface Modification via Physisorption of Uncharged Molecules Physical adsorption has several potential advantages including ease of formation, coating regeneration, and decreased dependence on substrate surface chemistry for coating process (16). Polymers have been frequently used as dynamic coating of microfluidics devices. Dynamic coating of silica-based substrate with hydroxypropylmethylcellulose (34–36) as well as hydroxyethylcellulose (37) has been shown to decrease electroosmosis and protein adsorption. In addition, protein adsorption onto PMMA microchannel surfaces was improved by dynamic coating of nonionic detergent in combination with methylcellulose (38). However, uncharged hydrophilic polymers have been reported to fail to strongly adsorb to the substrate, causing the coating to be removed with time (39). Other polymeric coatings include poly(*N*,*N*-dimethylacrylamide) (PDMA), which decreases EOF on glass chips used for separation of proteins (40) or DNA (41). PDMA was used on PDMS and PDMS/glass surfaces to improve hydrophilicity (42). While hydrophobic polymers, such as PDMA or even poly(vinylpyrrolidone), can provide more stable coatings, they may reduce separation efficiencies because of interactions with hydrophobic analytes.

20.2.2.2 Surface Modification via Physisorption of Surfactants Surfactants have also been used for modification of microfluidics surfaces. Dynamic coating of a PMMA surface with nonionic surfactants Brij® 35, Brij® 78, or Brij® 76 increased the compatibility of PMMA surface with the medium used for separation of DNA (43). Similarly, dynamic modification of PDMS microchannels with Brij® 35 resulted in more hydrophilic surfaces with better protein separation (44). Poly (ethylene glycol) (PEG)-based surface coatings (Pluronics) on PDMS or UV-treated PDMS adsorbed differentially, leading to coating height differences due to the polymer conformation (45). The same phenomenon was also observed on hydrophobic PDMS and hydrophilic silica surfaces (46). Using a different approach, Lee and Voros adsorbed PEG, a polymer widely used for protein-resistant coatings, onto a PDMS surface (47). In this method, PEG copolymers with a poly-L-lysine backbone were electrostatically adsorbed to oxidized PDMS. A similar method has been adapted for PDMS microfluidics devices for geometrically controlled capture of specific proteins, while concurrently minimizing nonspecific protein adsorption (48).

Modification of PDMS with charged surfactants have also been reported. In particular, investigation of effects of anionic surfactants on PDMS microchips revealed that dynamic coatings with sodium dodecyl sulfate (SDS), phosphatidic acid, or deoxycholate increased EOF as well as separation rates of glucose and other compounds (49). Addition of SDS for electrophoresis of dye-labeled amino acids has also reduced adsorption of the dye into bulk PDMS and increased separation efficiency of dye-labeled amino acids through dynamic coating (50). Moreover, EOF properties of PDMS surfaces were altered by dynamic coating with tetrabutylammoniumchloride (TBACl) and SDS at acidic and basic pH (51). In this study, EOF was only slightly changed by TBACl, while the hydrophobic surfactant SDS adsorbed to the channel walls, and at lower pH significantly changed the EOF. Similarly, dynamic coating of PDMS channels with 2-morpholinoethanesulfonic acid in the presence of NaOH enhanced separation of glucose and amino acids by altering both zeta potential and EOF (52). Dynamic coatings with other charged molecules such as cationic ammonium starch derivatives have been shown to passivate adsorption of fluorescently labeled amino acids and peptide analytes on PMMA microchannel walls within a significant pH range (53).

20.2.2.3 Layer-by-Layer Films Another coating strategy based on adsorption is the layer-by-layer (LbL) technique (54). This method is based on electrostatic interactions or hydrogen bonding between a charged surface and a polyelectrolyte, followed by alternating adsorption of polyelectrolytes to form a multilayer assembly (55) (Fig. 20.1). LbL technique has been widely employed (56), as it can be applied to substrates of different sizes and topologies. Dynamic coating with successive multiple ionic layers of polybrene (cationic) and dextran sulfate (anionic) has been developed for generating negatively charged surfaces with decreased protein adsorption that can be operated within a wide range of pH (57,58). This technique has been extended to PDMS microchannels and resulted in stable, pH-dependent EOF (59).

The LbL technique has also been used to coat PDMS microfludics channels with silica nanoparticles (60). The channel surface was initially coated with poly (diallyldimethylammonium chloride), a positively charged polyelectrolyte containing quaternary ammonium ions, to alter EOF from cathodic to anodic, followed by negatively charged silica nanoparticles, which reversed EOF (53). As a result, the channel surface became more hydrophilic, as indicated by a decrease in contact angle from 113° for native PDMS to 64° after coating. Similarly, coating of PMMA or PDMS surfaces with alternating layers of poly(styrene sulfonate) and TiO_2 nanoparticles has been shown to significantly increase surface hydrophilicity and biocompatibility (61). Additionally, the LbL approach has been combined with surface-initiated grafting (62).

20.2.2.4 Surface Modification via Adsorption of Phospholipid Polymers Coatings of phospholipid polymers have been explored in microfluidics devices in efforts to decrease unfavorable analyte adsorption. While phospholipid coatings

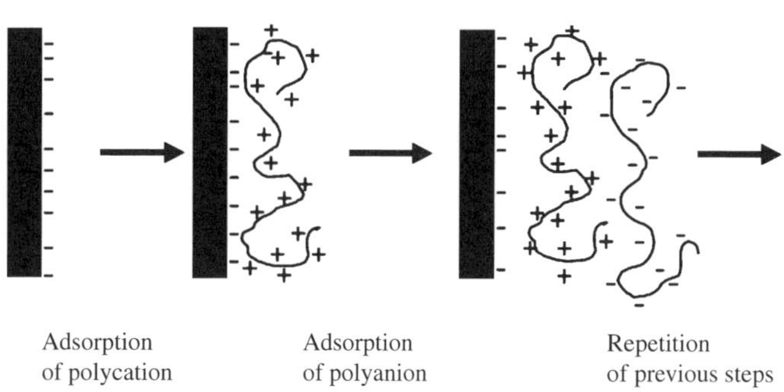

FIGURE 20.1 Alternating adsorption of polyelectrolytes in the layer-by-layer deposition technique. From Reference 56.

have been widely explored on glass, their use on PDMS is less frequently reported. Cremer et al. reported the kinetics of antibody/dinitrophenyl (DNP)-conjugated lipids binding to egg phosphatidyl choline bilayer-coated PDMS channels, as demonstrated based on total internal reflection fluorescence microscopy (63). In a later development, the same group reported the detection of glucose using immobilized streptavidin-conjugated enzyme via biotinylated phospholipid bilayers deposited at the luminal surface of PDMS microchannels (64). More recently, Phillips and Cheng systematically investigated the application of bilayer membranes on PDMS microchannels developed for immunoassays (65) and found that plasma-oxidized PDMS channels coated with phosphatidyl choline, but not first just plasma-treated PDMS channels, were stable against short-term hydrophobic recovery.

20.2.3 Chemical Surface Modifications

In contrast to coatings that depend on physisorption, which are typically less stable but relatively easy to prepare, chemical surface modifications may be more labor intensive, but are supported by covalent bonds, which tend to enhance their stability (7). Although PDMS is one of the most commonly used substrates for microfluidics devices, its hydrophobicity and chemical resistance have warranted the development of various chemical surface modification methods to improve its compatibility in microfluidics applications. In many of these techniques, the PDMS surface is initially oxidized via UV/ozone or plasma to render its surface more hydrophilic. This oxidative modification results in replacement of $Si{-}CH_3$ groups with $Si{-}OH$ groups (11,66,67).

20.2.3.1 UV-Induced Polymer Grafting One approach that has been successful in the modification of closed microfluidics channels is photochemical polymer grafting. In this method, radicals are created on a surface upon UV exposure and modifies only the surface without penentrating into the bulk (68–70). Although early modification methods for PDMS required photosensitizer prior to UV grafting (71), identification of improved reaction conditions has enabled a one-step modification on several polymeric surfaces (72–74). Allbritton et al. used a similar one-step approach to covalently link various hydrophilic polymers with different functional groups onto PDMS microchannels (75). This polymerization occurs in the presence of benzoyl alcohol that enhances grafting efficiency by facilitating polymerization chain transfer (72,73) and $NaIO_4$ that scavenges oxygen to minimize competition with monomer for free radicals on the surface (74). Properties of modified PDMS including surface chemistry and grafting density, hydrophobicity, and peptide adsorption were evaluated using attenuated total reflectance Fourier Transform infrared spectroscopy (ATR FTIR), contact angle measurements, and fluorescence microscopy, respectively. Relative to ungrafted PDMS, modified microchannels exhibited better separation efficiencies. The same research group has also reported grafting of mixtures of neutral and charged monomers to tailor surface properties—including electrophoretic mobility—by varying the concentration of the appropriate monomers (76). While these methods required modification of the channels prior to its assembly, PDMS channels may also be modified after its assembly by adsorbing the photoinitiator, benzophenone, onto the luminal surface of the microchannels (77). Various monomers were polymerized on PDMS channel surfaces using this approach. Modified channels retained stable electroosmotic mobility and allowed electrophoresis of model analytes. In addition, this method allows for patterning using a mask, and low resolution photopatterning was demonstrated. Using UV/ozone, a moderately hydrophilic and nonfouling PDMS microfluidic device was created via graft-copolymerization of PEG methyl ether methacrylate (78). Long-term stability of surface-modified PDMS for up to 56 days was confirmed based on (FTIR), contact angle measurements, and protein adsorption studies. This system also successfully sorted human sperms, demonstrating the applicability of this surface modification technique for cell-based systems. Fréchet et al. has also used a benzophenone-mediated UV-initiated photografting technique to modify several polymeric surfaces, including PDMS, for microfluidics applications (79). This method was used for grafting and/or patterning of various monomers.

20.2.3.2 Cerium-Catalyzed Radical Polymerization Reigner et al. reported surface modification of PDMS microchips via grafting and polymerization using cerium (IV) (80). Because cerium-initiated grafting requires presence of hydroxyl groups on the surface, PDMS as plasma-oxidized to appropriate modify its surface. Cerium, along with the monomers used for surface polymerization, was introduced to the system and reaction allowed to proceed for 30–60 min at room temperature. Several polymers including vinyl sulfonic acid, acrylic acid, 2-acrylamido-methylpropanesulfonic acid (AMPS), 4-styrenesulfonic acid, and stearylmethacrylate, were grafted onto PDMS surface using this approach. Electrophoretic performance of modified surfaces was evaluated by separation of fluorescence-labeled peptide mixtures, where AMPS-modified surface presented the best efficiency. In a subsequent step, the AMPS surface was further modified by addition of octadecylsilane, which reacts with free silanol groups present on PDMS surfaces after AMPS polymerization. After surface modification, a complex mixture of proteins was separated with high reproducibility.

20.2.3.3 Surface Modification via Plasma Polymerization To prepare a hydrophilic PDMS surface, Tabeling et al. plasma-polymerized and crosslinked acrylic acid coatings on the surface of PDMS microfluidics systems (81), (82). The process consists of a pretreatment step, in which surface is activated by argon plasma, a polymer deposition step, in which acrylic acid coating is deposited on pretreated PDMS via plasma of argon/acrylic acid, and a posttreatment step, in which the acrylic acid polymer films is crosslinked to suppress hydrophobic recovery of PDMS. Acrylic acid films of approximately 450 nm thickness were achieved with this method, and the coating was confirmed by XPS. Surface hydrophilicity was evaluated by contact angle measurements. Although the contact angle of the resultant surface increased from $<10°$ to $50°$-$60°$ upon rinsing with water, these measurements were lower than that of unmodified PDMS ($110°-120°$). In addition, using SEM, the coating was observed to remain stable.

20.2.3.4 Silanization Similarly, silanization approach has been employed to immobilize PEG using PEG di(trimethoxysilane) on PDMS after plasma-oxidation to increase surface wettability and protein resistivity (83). Using this approach, hydrophilic recovery of PDMS was delay, and fluorescent antibodies were patterned on to modified surfaces with high contrast and resolution. In contrast to plasma-oxidized PDMS surfaces that underwent hydrophobic recovery as early as 20 min after the oxidation, silanized surfaces remained hydrophilic and protein resistant for more than 3 days.

In an effort to decrease nonspecific protein adsorption, epoxy-modified hydrophilic polymers were grafted onto plasma-oxidized PDMS microchip silanized with 3-aminopropyl-triethoxysilanes (APTES) (84). Several polymers were prepared including glycidyl methacrylate (GMA) copolymerized with acrylamide (poly(AAM-co-GMA)) or dimethylacrylamide (poly(DMA-co-GMA)), and graft polymerized with polyvinylpyrrolidone (PVP-g-GMA) or polyvinylalcohol (PVA-g-GMA). Covalent binding to PDMS surface was confirmed using reflection absorption infrared spectroscopy, and this modification suppressed EOF in the PDMS microchannel. In addition, in comparison to nonmodified PDMS, protein adsorption was decreased on epoxy-modified surfaces.

20.2.3.5 Atom Transfer Radical Polymerization Atom transfer radical polymerization (ATRP) is a controlled free radical polymerization method, catalyzed by a transition metal (85,86). This method has also been applied to PDMS microchips to improve hydrophobic recovery. In their initial study, Wirth et al. modified PDMS surface with polyacrylamide using ATRP (87). In this report, PDMS surfaces were plasma-oxidized, and subsequently was reacted with trichlorosilane gas containing ATRP initiator (1-trichlorosilyl-2-*m*-*p*-chloromethylphenyl)ethane, followed by polymerization with acrylamide. Evaluation of polymerization kinetics using ATR FTIR revealed that polyacrylamide binding to PDMS was maximal 15 min of oxidation, indicative of maximal surface density of silanols at this time. In addition, as compared to unmodified PDMS, surfaces modified via ATRP of acrylamide showed enhanced resistance in protein adsorption and delayed hydrophobic recovery. In a more recent development, the same group evaluated ATRP of acrylamide onto PDMS microchannels (88). As in the previous study, PDMS channel surfaces were oxidized, followed by self-assembly of benzyl chloride initiator monolayer, onto which polyacrylamide layer was grown covalently by ATRP. The resultant channels exhibited hydrophilicity, resistance to protein adsorption, and successful electrophoresis of model proteins. Because the modification is only made to the inner channels of PDMS, there is no fluid leakage. While not on PDMS, Xu et al. has demonstrated ATRP to create a gradient of 2-hydroxyethyl methacrylate on a silicon wafer confined in a microchannel (89). In this method, initiator-functionalized self-assembled monolayers were formed on a wafer, onto which a solution of monomer and catalyst was pumped at a fixed rate. After the solution reached the outlet, the microchannel was disassembled and the silicon surface washed, thus creating a gradient of catalyst that was proportional to the distance from the outlet and equal to the time of surface exposure to the monomer solution. This method may be applied to PDMS surfaces, enabling investigation of composition gradients on electrophoresis, as well as subsequent immobilization of ligands, cells, proteins, and other biologically relevant molecules.

20.2.3.6 Reversible Addition–Fragmentation Chain Transfer Polymerization Reversible addition–fragmentation chain transfer (RAFT) polymerization is another living, controlled radical polymerization technique that may be used for surface modification of microfluidc devices (90). This polymerization is mediated by a dithioester chain transfer agent and is based on equilibrium between active radical and dormant chains. Recently, application of RAFT on silicon wafers, polymer films, and silica nanoparticles have been reported for the immobilization of polymers or polyelectrolyte brushes (91–94). In addition, RAFT polymerization on PDMS microfluidics systems was reported (62). In this study, PDMS surface was first treated with microwave plasma to introduce carboxylic acid groups via reaction with maleic anhydride, followed by hydrolysis, onto which the initiator was immobilized, and various polymers (*N*,*N*-dimethylacrylamide (nonionic), styrenesulfonate (SS) (negatively charged), and (*ar*-vinylbenzyl) trimethylammonium chloride (VBTAC) (positively charged)) were RAFT polymerized. In particular, PDMS surface immobilized with poly(VBTAC) was further modified by a LbL approach using defined poly(*p*-styrenesulfonate) poly-(PSS) and poly (VBTAC). Surface chemistry and morphology were evaluated using ATR FTIR and atomic force microscopy, respectively. While no difference in topology was observed among the different polymerized species, phase images indicated chemical differences. Water contact angle measurements revealed that surface modified with LbL deposition of poly-(VBTAC) and poly-(PSS) exhibited the lowest contact angle of 35.6°, lower than that of PDMS surface modified with poly-(VBTAC) (71°) or poly-(PSS) (72.5°) alone, due to lower surface density of the polymer. While protein adsorption and efficiency of these modified microfluidics channels in electrophoresis were not investigated, this study demonstrates the combination of surface modification techniques to enhance PDMS hydrophilicity.

20.2.4 Surface Modification via Chemical Vapor Deposition Polymerization

20.2.4.1 Reactive Coatings: Fundamentals Chemical vapor deposition (CVD) polymerization takes advantage of a unique class of polymers, poly(*p*-xylylenes), which can be prepared from ring-constrained aromatic molecules, [2.2]paracyclophanes (95–100). In many ways, [2.2]paracyclophanes show the behavior of ordinary aromatic compounds and can be converted in a wide range of differently substituted derivatives. When thermally activated, however, the bridges of [2.2]paracyclophanes can undergo homolytic cleavage. The resulting monomers tend to spontaneously polymerize as soon as they contact a sufficiently cold surface. The threshold temperature, below which polymerization occurs, varies with the functional groups of the [2.2]paracyclophane. The fact that polymerization occurs at cold

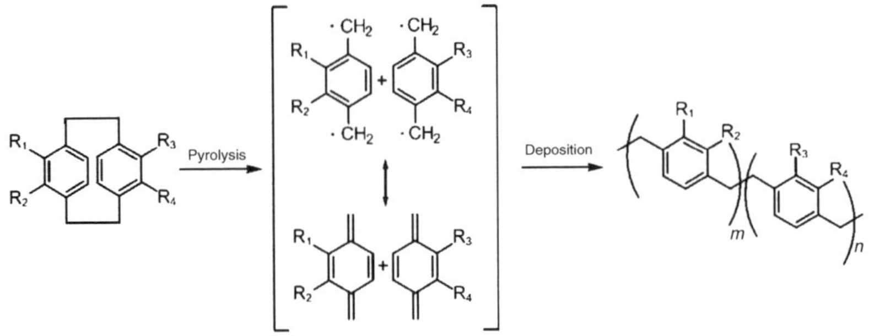

R_1	R_2	R_3	R_4	R_1	R_2	R_3	R_4
CHO	H	H	H	CH_2OH	H	H	H
$COCH_3$	H	H	H	CH_2OCH_3	H	H	H
COC_2H_5	H	H	H	CH_2OCOCH_3	H	H	H
$COCF_3$	H	H	H	$COOCH_3$	H	H	H
COC_2F_5	H	H	H	$COOCH_3$	$COOCH_3$	$COOCH_3$	$COOCH_3$
COPh	H	H	H	$CH_2OCOPpf$	H	H	H
COPh	H	COPh	H	CH_2OCOCF_3	H	H	H
NH_2	H	H	H	Cl	H	H	H
NH_2	H	NH_2	H	Br	H	H	H
CH_2NH_2	H	H	H	H	H	H	H
C≡CH	H	H	H				

FIGURE 20.2 CVD polymerization of functionalized [2.2]paracyclophanes From Reference 100.

temperatures, typically at or below room temperature, allows for coating of temperature-sensitive substrates and similar polymer-based microfluidics devices. As shown in Fig. 20.2, the CVD polymerization process is applicable to a range of functionalized [2.2]paracyclophanes and the choice of functional groups present in the polymer can be made within a considerable range (98). CVD polymerization is a controlled process that often proceeds with a minimum of side reactions, often resulting in beneficial biocompatibility (101). Functionalized poly(*p*-xylylenes) made by CVD polymerization are chemically well-defined polymer films and many show close to theoretical composition. Reactive CVD coatings can be functionalized with a wide range of different reactive groups, including amines, alcohols, active esters, ketones, aldehydes, and alkines. This provides versatility with respect to the type of surface chemistries that can be achieved. These functional groups have been used to immobilize various biomolecules including proteins, sugars, or DNA (100,102–106).

20.2.4.2 Chemical Vapor Deposition Coating of Preassembled Devices The use of vapor-deposited polymer coatings may provide an appealing approach for surface modification in open (107,108) and confined (109) microgeometries with both nonfunctionalized and functionalized poly(*p*-xylylenes). To create a general understanding of CVD polymerization within confined microgeometries, a recent study investigated CVD polymerization of seven different poly(*p*-xylylenes) within both removable and sealed microchannels (Fig. 20.3) (109). This study included five poly(*p*-xylylenes) with reactive side groups (so-called reactive coatings) as well as two commercially available poly(*p*-xylylenes). The PDMS microchannels used in this study were open at both ends and were 75 μm and 100 μm in height and width,

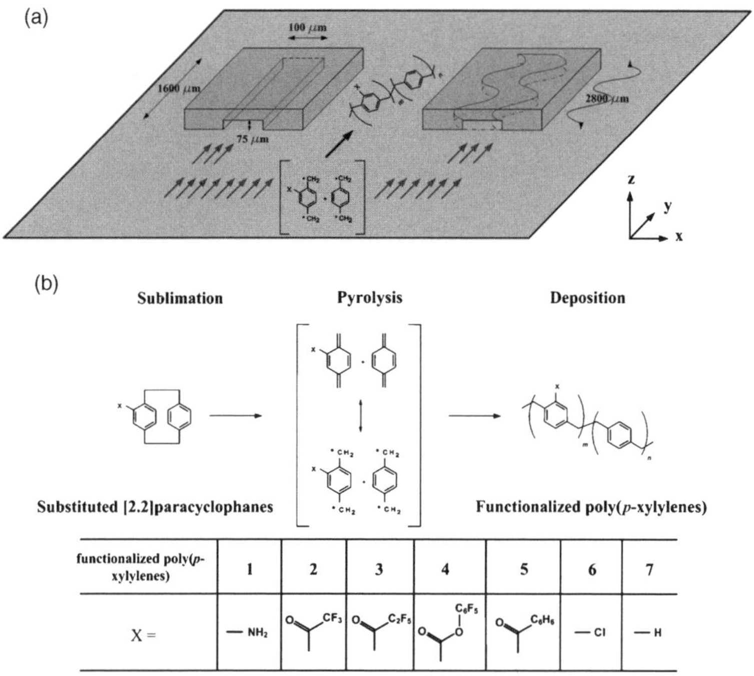

FIGURE 20.3 (**a**) Chemical vapor deposition polymerization within confined geometries of straight and meandering channels. (**b**) Functionalized poly(*p*-xylylenes) prepared via chemical vapor deposition polymerization used in this study. From Reference 109.

respectively. Both straight (1600 μm long) and meandering channel (2800 μm long) layouts were studied. After removal of the PDMS molds, the footprint of the deposited films remained on the silicon substrates as confirmed by optical microscopy. For all reactive coatings, homogenous surface coverage of different microgeometries featuring aspect ratio of up to 37 has been demonstrated based on optical microscopy and imaging XPS. The film thickness was mainly determined by the amount of a given [2.2]paracyclophane used for CVD polymerization. On the basis of thickness data, the degree of deposition, as defined as the ratio of film thicknesses measured at the center of the microchannel to that measured at the open substrate, was determined for each polymer. In essence, this ratio indicates how effective a CVD coating can be deposited with in confined geometry. The degree of deposition, will be 100% for a coating that is not limited by the confinement imposed by geometry. The degree of deposition was independent of film thicknesses, for a range of film thickness. Fig. 20.4 reveals strictly uncorrelated trends for both straight and meandering microchannels indicating the degree of deposition to be independent of the actual film thickness. Height profiles of deposited polymer footprints were obtained by atomic force microscopy, as well as imaging ellipsometry. Both methods indicated continuous transport and deposition throughout the entire microchannels. This behavior was independent of the channel geometry. Finally, the ability of

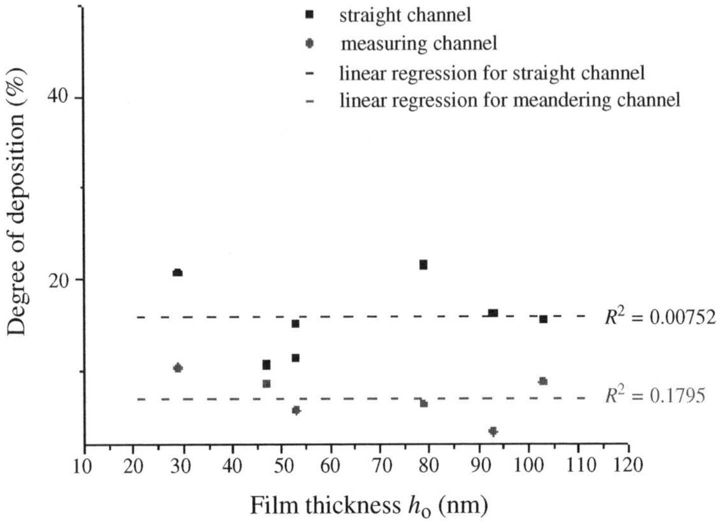

FIGURE 20.4 Degree of deposition versus film thickness for reactive coating 1. Linear regressions for straight ($R^2 = 0.00752$) and meandering ($R^2 = 0.1795$) channels revealed no correlation for the degree of deposition with varying film thickness.

reactive coatings to support chemical binding of biological ligands, when deposited in previously assembled microchannels, was demonstrated, verifying the usefulness of the CVD coatings for applications in micro/nanofluidics, where surface modifications with stable and designable biointerfaces are essential. Deposition of reactive CVD coatings within confined microgeometries bridges a critical technological gap toward surface-modified microfluidic and nanofluidic devices with use for biosensor and "BioMEMS" applications, bioseparation, design of high throughput screening platforms, or development of novel tissue constructs for regenerative medicine applications.

20.2.4.3 Spatially Directed Surface Modification via Photopatterning If the functional group of the poly(*p*-xylylene) is a benzoyl group, the resulting polymer (benzoyl-ppx) films become photoreactive (110). By recognizing the resemblance to benzophenone, a well-known photoinitiator, this chemical behavior can be understood. The carbonyl group in α-position to two phenyl rings generates free radicals upon excitation with light at 365 nm. Resulting free radicals undergo CH-abstraction resulting in a covalent linkage of whatever molecule has been in proximity to the surface. Insertion into NH bonds is favored over CH insertion, but the process is still fairly unspecific. In one example, the luminal surface of a microfluidic channel was coated with the benzoyl-ppx and the entire channel was then exposed to PEG solution. Illumination of the microchannel with light at 365 nm resulted in covalent attachment of the PEG. When the microchannel was exposed through a mask, spatially controlled immobilization was achieved (111). Unbound

PEG was removed in a subsequent washing step. Upon exposure to solutions containing proteins, such as albumin or fibrinogen, selective protein adsorption to areas that were not modified with the PEG (i.e., that were not illuminated during photoexposure) was observed. In contrast to fluid flow patterning, discontinuous patterns are equally accessible through his method. Similarly, this approach can be used to fabricate surface-tethered hydrogel elements with microchannels (110). In this case, the PEG is replaced by a photopolymerizable monomer, such as an acrylate or methacrylate, and exposure to light at 365 nm results in surface-initiated polymerization with the benzoyl-ppx acting as the initiator.

20.2.4.4 Covalent Immobilization of Biomolecules Within Microchannels While the light-directed immobilization of hydrogels or PEGs offers a simple approach to surface modification, biomolecules often require more specific immobilization schemes. Favorable orientation of a protein on a surface, that is, its presentation toward a potential binding partner, is often essential for its bioactivity. To achieve specific presentation of biomolecules on surfaces, we used CVD polymerization to create more selective anchor groups on the surface of the substrate. When immobilizing the antithrombotic protein r-hirudin, the N-terminal amino acid residues were located at a position determinative for its thrombin activity (112,113). In other words, the amino groups have a superior chemical reactivity compared to the remaining ε-amino groups (114). At the same time, free availability of this group is essential for thrombin inhibition. We coated the substrate with anhydride-ppx, which is characterized by a high reactivity of the functional groups toward amino groups (115). To immobilize r-hirudin via amino groups, an indirect immobilization strategy was chosen using an N-terminally protected r-hirudin derivative for immobilization. Deprotection of the surface-bound r-hirudin derivatives and intensive rinsing with SDS buffer delivered surfaces with covalently bound active r-hirudin. Surfaces prepared by covalent linkage of r-hirudin showed strongly enhanced prolongation of blood clotting due to inactivation of thrombin by surface-bound r-hirudin. These concepts are not limited to r-hirudin, but can be similarly applied to other proteins that may be of importance to microfluidic applications.

To selectively immobilize another biomolecules, heparin, on the polymer surface, an oligosaccharide immobilization strategy was chosen, since the heparin molecule contains oligosaccharide units. Functional groups of poly[4-trifluoroacetyl-*p*-xylylene-*co*-*p*-xylylene] (keto-ppx) were used to anchor the oligosaccharide molecules to the surface. This polymer has carbonyl functionality and the hydrazone reaction between hydrazide/hydrazine ($NH_2{-}NH{-}$) and carbonyl (C=O) groups was used for our surface reactions. To prove selectivity of the immobilization system, only certain areas of the substrate were activated, thereby establishing patterns. One hydrazide end of the dihydrazide, molecule reacts with the functional groups on the polymer, while the other end reacts with the saccharide molecule. This linker molecule covalently binds to heparin or glycoproteins.

A similar strategy was recently employed to demonstrate applicability to microfluidic devices. In this method, poly[*p*-xylylene carboxylic acid pentafluorophenolester-*co*-*p*-xylylene] was CVD polymerized onto PDMS to introduce functional groups onto which amino-terminated biotin ligands were immobilized on to the luminal surface of this device (116). The reactive coating was deposited as a thin film (~100 nm) and was strongly adherent to the PDMS surface, with uniform distribution of the pentafluorophenol ester groups within the microchannel inner surface, as visualized by fluorescence microscopy. Subsequently, this surface supported self-assembly of cell-capturing molecules including proteins and antibodies. Further, this microfluidics device was used for a model cell-based bioassay to monitor the effect of echistatin, a disintegrin, in a concentration-dependent manner, and demonstrates the applicability of this CVD-based surface modification approach to clinically relevant research including angiogenesis (116).

20.2.4.5 Immobilization of Multiple Biomolecules in Controlled Ratios via CVD Copolymerization For CVD copolymerization, mixtures of carefully purified dimmers 4-trifluoroacetyl [2.2]paracyclophane (keto-pcp) and 4-aminomethyl-[2.2]paracyclophane (amino-pcp) were initially sublimated under a reduced pressure of 0.56 mbar at temperatures between 90 and 100°C (117). Sublimated amino-pcp and keto-pcp were then transferred to the pyrolysis zone, which was heated to 670°C to ensure cleavage of the C–C bonds, resulting in corresponding quinodimethanes (monomers). In the last step, monomers were adsorbed on the substrate at temperature around 10°C and spontaneously polymerized. CVD copolymerization resulted in transparent and topologically uniform polymer films with thicknesses between 50 and 200 nm. The multifunctional coatings showed excellent chemical stability in a dry air environment. No significant change in composition or chemical behavior was found for samples stored in a dry air atmosphere for several weeks as compared to freshly prepared samples. All copolymers as well as the two individual polymers poly[4-aminomethyl-*p*-xylylene-*co*-*p*-xylylene] (amino-ppx) and poly[4-trifluoroacetyl-*p*-xylylene-*co*-*p*-xylylene] (keto-ppx) remained intact after rinsing with standard solvents, including water, chloroform, acetone, and ethanol.

Elemental composition of the copolymer poly[4-aminomethyl-*p*-xylylene-*co*-4-trifluoroacetyl-*p*-xylylene-*co*-*p*-xylylene] (amino/keto-ppx) was studied by XPS. XPS is capable of detecting atomic composition within a depth of about 10 nm. The copolymer amino/keto-ppx was compared to amino-ppx and keto-ppx. Characteristic chemical elements detected for the individual polymers, such as oxygen (keto-ppx), fluorine (keto-ppx), and nitrogen (amino-ppx), were simultaneously detected in the copolymer, indicating the presence of both functional groups on the surface. Because nitrogen is only present in the aminomethyl group of amino-ppx, while fluorine is a characteristic element for the ketone of keto-ppx, the ratio of the elemental composition of nitrogen and fluorine serves as a good indicator for the chemical quality of the copolymer films. Using XPS, nitrogen to fluorine ratio of 0.332 was determined which is in good accordance with the theoretically expected ratio of 0.333, suggesting that side reactions, such as decomposition of the

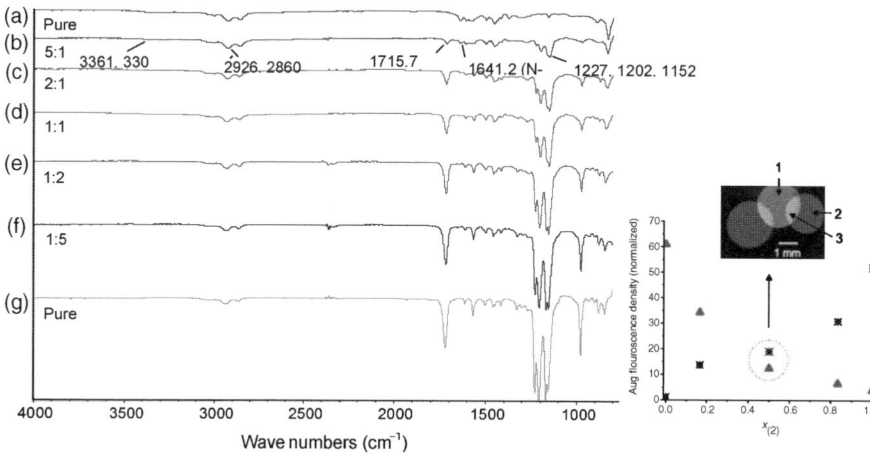

FIGURE 20.5 FTIR spectra of (3) with varying molar ratios of CH_2NH_2:$COCF_3$. (**a**) pure CH_2NH_2 (**b**) 5 : 1 (**c**) 2 : 1 (**d**) 1 : 1 (**e**) 1 : 2 (**f**) 1 : 5 (**g**) pure. The inset shows fluorescence reactivity versus specific binders with orthogonal reactivity for amines (red) or ketones (green) for a 50 : 50 composition. From Reference 117.

functional groups, were negligible. Furthermore, XPS revealed no signs of cross-reaction of the aminomethyl and keto groups under the conditions of CVD polymerization.

Finally, we verified that both functional groups contained in copolymer amino/keto-ppx are available for further surface modification by parallel immobilization of two model biomolecules. Moreover, by varying feed ratios of the two starting materials in the CVD polymerization, relative ratio of ligands bound to the surface was varied (Fig. 20.5). Relative increase of starting materials resulted in surfaces with increased amounts of amino-reactive ligands (red color), while a relative increase of amino-containing starting material used for CVD polymerization resulted in an increased fluorescence signal due to the keto-reactive dye. The availability of the functional groups for further surface reaction is essential for immobilization of two different types of (bio)molecules.

20.3 CONCLUSIONS

Continuous miniaturization of microfluidics systems for biological applications requires precisely defined and stable surface modifications for subsequent immobilization of biologically relevant molecules. As surface-to-volume ratios are amplified, even slight deviations from homogeneity in the surface properties may lead to misguided results.

While microfluidics systems have been composed of various base materials, many of which are associated with specialized surface modification protocols, this chapter focused on modification strategies for polymers, specifically PDMS. As mentioned,

PDMS provides numerous advantages including its adaptability for mass production and fabrication methods, but is still hindered by issues associated with wettability and analyte adsorption. In this regard, several techniques that modify the PDMS surface, especially as a microfluidics device substrate have been presented herein. Some of these techniques inherently change the surface chemistry via application of energy, while others take advantage of this altered surface property for further chemical modification to immobilize particular moieties or ligands. Together, these techniques serve to enhance general design and EOF properties of microfluidics systems, as well as provide precise control over ligand immobilization, all of which are essential in fulfilling the demanding requirements in biological applications. Future work on microfluidics substrate modification strategies may include controlled immobilization of multiple biologically-relevant molecules as well as stimuli-responsive materials for the development of advanced cell-based devices.

REFERENCES

1. Manz A, Graber N, Widmer HM. Sens Actuators B: Chemical 1990;1:244–248.
2. Dittrich PS, Manz A. Nat Rev Drug Discov 2006;5:210–218
3. Shin H. Biomaterials 2007;28:126–133
4. Khademhosseini A. Proc Natl Acad Sci U S A 2006;103:2480–2487
5. Bange A, Halsall HB, Heineman WR. Biosens Bioelectron 2005;20:2488–2503.
6. Delamarche E, Juncker D, Schmid H. Adv Mater 2005;17:2911–2933.
7. Belder D, Ludwig M. Electrophoresis 2003;24:3595–3606.
8. Roberts MA, Rossier JS, Bercier P, Girault H. Anal Chem 1997;69:2035–2042.
9. Wang SC, Lee CY, Chen HP. J Chromatogr A 2006;1111:252–257.
10. Martynova L, Locascio LE, Gaitan M, Kramer GW, Christensen RG, MacCrehan WA. Anal Chem 1997;69:4783–4789.
11. Duffy DC, McDonald JC, Schueller OJA, Whitesides GM. Anal Chem 1998;70: 4974–4984.
12. Qin D, Xia YN, Whitesides GM. Adv Mater 1996;8:917–919.
13. McCormick RM, Nelson RJ, AM G, Benvegnu J, Hooper HH. Anal Chem 1997;69: 2626–2630.
14. Makamba H, Kim JH, Lim K, Park N, Hahn JH. Electrophoresis 2003;24:2607–3619.
15. Peppas NA, Hilt JZ, Khademhosseini A, Langer R. Adv Mater 2006;18:1345–1360.
16. Doherty EA, Meagher RJ, Albarghouthi MN, Barron AE. Electrophoresis 2003;24: 34–54.
17. Liu J, Lee ML. Electrophoresis 2006;27:3533–3546.
18. Huang TT, Mosier NS, Ladisch MR. J Sep Sci 2006;29:1733–1742.
19. Dolnik V. Electrophoresis 2004;25:3589–3601.
20. Falconnet D, Csucs G, Grandin HM, Textor M. Biomaterials 2006;27:3044–3063
21. Kane RS, Takayama S, Ostuni E, Ingber DE, Whitesides GM. Biomaterials 1999;20: 2363–2376.

22. Hillborg H, Gedde UW. Polymer 1998;39:1991–1998.
23. Fritz JL, Owen MJ. J Adhesion 1995;54:33–45.
24. Lai JY, Lin YY, Denq YL, Chen JK. J Adhes Sci Technol 1996;10:231–242.
25. Efimenko K, Wallace WE, Genzer J. J Colloid Interf Sci 2002;254:306–315.
26. Yan YH, Chan-Park MB, Yue CY. Langmuir 2005;21:8905–8912.
27. Toth A, Bertoti IBlazso M, Banhegyi G, Bognar A, Szaplonczay P. J Appl Polym Sci 1994;52:1293–1307.
28. Owen MJ, Smith PJ. J Adhes Sci Technol 1994;8:1063–1075.
29. Hettlich HJ, Otterbach F, Mittermayer C, Kaufmann R. Biomaterials 1991;12: 521–524.
30. Wang B, Abdulali-Kanji Z, Dodwell E, Horton JH, Oleschuk RD. Electrophoresis 2003;24:1442–1450.
31. Chen C, Wang J, Chen Z. Langmuir 2004;20:10186–10193.
32. Ye H, Gu Z, Gracias DH. Langmuir 2006;22:1863–1868.
33. Vickers JA, Caulum MM, Henry CS. Anal Chem 2006;78:7446–7452.
34. Gilges M, Kleemiss MH, Schomburg G. Anal Chem 1994;66:2038–3046.
35. Lindner H, Helliger W, Dirschlmayer A, Jaquemar M, Puschendorf B. Biochem J 1992;283:467–471.
36. Bean SR, Lookhart GL. Electrophoresis 1998;19:3190–3198.
37. Kleemiss MH, Gilges M, Schomburg G. Electrophoresis 1993;14:515–522.
38. Mohamadi MR, Mahmaoudian L, Kaji N, Tokeshi M, Baba Y. Electrophoresis, 2007; 28:830–836.
39. Mandabhushi RS. Electrophoresis 1998;19:224–230.
40. Bousse L, Mouradian S, Minalla A, Yee H, Williams K, Dubrow R. Anal Chem 20021, 73:1207–1212.
41. Emrich CA, Tian HJ, Medintz IL, Mathies RA. Anal Chem 2002;74:5076–5083.
42. Chen L, Ren J, Bi R. Electrophoresis 2004;25:914–921.
43. Song L, Fang D, Kobos RK, Pace SJ, Chu B. Electrophoresis 1999;20:2847–2855.
44. Dou YH, Bao N, Xu JJ, Meng F, Chen HY. Electrophoresis 2004;25:3024–3031.
45. Hellmich W, Regtmeijer J, Duong T, Ros R, Anselmetti D, Ros A. Langmuir 2005; 21:7551–7557.
46. Barnes TJ, Prestidge CA. Langmuir 2000;16:4116–4121.
47. Lee S, Voros J. Langmuir 2005;21:11957–11962.
48. Marie R, Beech JP, Voros J, Tegenfeldt JO, Hook F. Langmuir 2006;22:10103–10108.
49. Garcia CD, Dressen BM, Henderson A, Henry CS. Electrophoresis 2005;26:703–709.
50. Roman GT, McDaniel K, Culbertson CT. Analyst 2006;131:194–201.
51. Ocvirk G, Munroe M, Tang T, Oleschuk R, Westra K, Harrison DJ. Electrophoresis 2000;21:107–115
52. Dou YH, Bao N, Xu JJ, Chen HY. Electrophoresis 2002;23:3558–3566.
53. Kato M, Gyoten Y, Sakai-Kato K, Nakajima T, Toyo'oka T. Anal Chem 2004;76: 6792–6796.
54. Deche G, Schlenoff JB. Multilayer Thin Films—Sequential Assembly of Nanocomposite Materials. Wiley-VCH;2002.

55. Schwarz S, Eichhorn K-J, Wischerhoff E, Laschewsky A. Colloids Surf A 1999;159: 491–501.
56. Groth T, Lendlein A. Angew Chem Int Ed 2004;43:926–928.
57. Katayama H, Ishihama Y, Asakawa N. Anal Chem 1998;70:5272–5277.
58. Katayama H, Ishihama Y, Asakawa N. Anal Sci 1998;14:407–408.
59. Liu Y, Fanguy JC, Bledsoe JM, Henry CS. Anal Chem 2000;72:5939–5944.
60. Wang W, Zhao L, Zhang JR, Wang XM, Zhu JJ, Chen HY. J Chromatogr A 2006;1136: 111–117.
61. Kommireddy DS, Patel AA, Shutava TG, Mills DK, Lvov YM. J Nanosci Nanotechnol 2005;5:1081–1087.
62. Bae WS, Convertine AJ, McCormick CL, Urban MW. Langmuir 2007;23:667–672.
63. Yang T, Jung S-Y, Mao H, Cremer PS. Anal Chem 2001;73:165–169.
64. Mao H, Yang T, Cremer PS. Anal Chem 2002;74:379–385.
65. Phillips KS, Cheng Q. Anal Chem 2005;77:327–334.
66. Chaudhury MK, Whitesides GM. Langmuir 1991;7:1013–1025.
67. Xia YN, Whitesides GM. Angew Chem Int Ed 1998;37:551–575.
68. Chan CM, Polymer Surface Modification and Characterization. Cincinatti, OH: Hanser/Gardner Publications; 1994.
69. Garbassi F, Morra M, Occhiello E, Polymer Surfaces. New York: John Wiley and Sons; 1998.
70. Jagur-Grodzinski J, Heterogenous Modification of Polymers. New York: John Wiley and Sons;1997.
71. Inoue H, Kohama S. Appl Polym Sci 1984;29:877–889.
72. Uchida E, Uyama Y, Ikada Y. J Polym Sci A1 1989;27:527–537.
73. Richey T, Iwata H, Oowaki H, Uchida E, Matsuda S, Ikada Y. Biomaterials 2000;21:1057–1065.
74. Uchida E, Uyama Y, Ikada Y. J Appl Polym Sci 1990;41:677–687.
75. Hu SW, Ren XQ, Bachman M, Sims CE, Li GP, Allbritton NL. Anal Chem 2002;74: 4117–4123
76. Hu SW, Ren XQ, Bachman M, Sims CE, Li GP, Allbritton NL. Langmuir 2004;20: 5569–5574.
77. Hu SW, Ren XQ, Bachman M, Sims CE, Li GP, Allbritton NL. Anal Chem 2004;76:1865–1870.
78. Wu J, Chung Y, Belford K, Smith G, Takayama S, Lahann J. Biomed Microdevices 2006;8:99–107.
79. Rohr T, Ogeletree DF, Svec F, Fréchet JMJ. Adv Funct Mater 2003;13:264–270.
80. Slentz BE, Penner NA, Regnier FE. J Chromatogr A 2002;948:225–233.
81. Barbier V, Totoulian M, Li H, Arefi-Khonsari F, Ajdari A, Tabeling P. Langmuir 2006;22:5230–5232.
82. Arefi F, Andre V, Montazer-Rahmati P, Amouroux J. Pure, Appl Chem 1992;64: 715–723.
83. Papra A, Bernard A, Juncker D, Larsen NB, Michele B, Delamarche E. Langmuir 2001;17:4090–4095.

84. Wu D, Zhao B, Dai Z, Qin J. Lab Chip 2006;6:942–947.
85. Kato M, Kamigaito M, Sawamoto M, Higashimura T. Macromolecules 1995;28:1721–1723.
86. Wang JS, Matyjaszewski K. J Am Chem Soc 1995;117:5614–5615.
87. Xiao DQ, Zhang HW, Wirth MJ. Langmuir 2002;18:9971–9976.
88. Xiao DQ, Van Le T, Wirth MJ. Anal Chem 2004;76:2055–2061.
89. Xu C, Wu T, Drain CM, Batteas JD, Beers KL. Macromolecules 2005;38:6–8.
90. Chiefari J, Chong YK, Ercole FKJ, Jeffery JLe TPT, Mayadunne RTA, Meijs GF, Moad CL, Moad G, Rizzardo E, Thang SH. Macromolecules 1998;31: 5559–5562.
91. Baum M, Brittain WJ. Macromolecules 2002;35:610–615.
92. Zhai GQ, Yu WH, Kang ET, Neoh KG, Huang CC, Liaw DJ. Ind Eng Chem Res 2004;43:1673–1680.
93. Yoshikawa C, Goto AT, Fukuda T, Yamamoto K, Kishida A. Macromolecules 2005;38:4604–4610.
94. Li CZ, Benicewicz BC. Macromolecules 2005;38:5929–5936.
95. Lahann J, Klee D, Hocker H. Macromol Rapid Commun 1998;19:441–444.
96. Lahann J, Hocker H, Langer R. Angew Chem Int Ed 2001;40:2947.
97. Lahann J, Choi IS, Lee J, Jensen KF, Langer R. Angew Chem Int Ed 2001;40:31666–33169
98. Lahann J, Langer R. Macromolecules 2002;35:4380–4386.
99. Nandivada H, Chen HY, Bondarenko L, Lahann J. Angew Chem Int Ed 2006;45:3360–3363.
100. Lahann J. Polym Int 2006;55:1361–1370.
101. Schurmann K, Lahann J, Niggemann P, Klosterhalfen B, Meyer J. Kulisch A, Klee D, Gunther RW, Vokwerk D. Radiology 2004;230:151–162.
102. Lahann J, Plüster W, Gattner H-G, Klee D, Höcker H. J Mater Sci Mater Med 2001;12:807–810.
103. Lahann J, Balcells M, Rodon T, Lee J, Choi JS, Jensen KF. Langmuir 2002;18:3632–3638.
104. Nandivada H, Chen H-Y, Lahann J. Macromol Rapid Commun 2005;26:1794–1799.
105. Lahann J. Chem Eng Commun 2006;193:1457–1468.
106. Yoshida M, Langer R, Lendlein A, Lahann J. Polymer Rev 2006;46:347–375.
107. Lahann J, Balcells M, Lu H, Radon T, Langer R, Jensen KF. Micro Total Anal Syst 2002;1:443–445.
108. Suh KL, Lahann J. Appl Phys Lett 2003;83:4250–4252.
109. Chen H-Y, Elkasabi Y, Lahann J. J Am Chem Soc 2006;128:374–380.
110. Suh KY, Langer R, Lahann J. Adv Mater 2004;16:1401–1405.
111. Chen H-Y, Lahann J. Anal Chem 2005;77:6909–6914.
112. Chang JY. FEBS Lett 1983;164:307–313.
113. Dodt J, Muller HP, Seemuller U, Chang JY. FEBS Lett 1984;165:180–184.
114. Tulinsky A. Semin Thromb Hemost 1966;22:117–124.

115. Lahann J, Pluster W, Rodon T, Fabry M, Klee D, Gattner HG, Hocker H. Macromol Biosci 2002;2:82–87.

116. Lahann J, Balcells M, Lu H, Radon T, Kensen KF, Langer R. Anal Chem 2003;75:2117–2122.

117. Elkasabi Y, Chen H-Y, Lahann J. Adv Mater 2006;18:1521–1526.

21

MICROCHIPS CONTAINING *IN SITU* PATTERNED POLYMERIC MEDIA FOR BIOCHEMICAL ANALYSIS

Robert J. Meagher and Anup K. Singh

21.1 INTRODUCTION AND SCOPE

Emerging research areas in biology, such as genomics, proteomics, and clinical analysis, increasingly require larger numbers of experiments performed in a smaller amount of time using a limiting amount of starting biological sample. This requires scaling down of the analysis methods, and, analogous to the integrated microelectronic-chip revolution, "microfluidic chips" are starting to transform the field of biochemical analysis by integrating many functions that could not easily be integrated at a conventional scale. To realize these "micrototal analysis systems" that go from a biological sample to complete analysis in a single chip, many bio/chemical processes such as mixing, dilution, concentration, transportation, separation, and reaction need to be integrated in a single chip. Integration of these fluidic manipulations, however, requires that suitable materials such as a dialysis membrane or chromatographic media be strategically placed in different locations of a chip.

Many different approaches can be used to integrate functional materials in a chip, and one of the easiest and the most versatile methods of forming materials *in situ* relies on photopolymerization to cast micro- and nanoporous materials in channels of a microchip. These polymeric monoliths, crosslinked gels, or membranes are formed in minutes and can easily be tailored to obtain desired pore size (10 nm–1 μm), surface chemistry, and function. The use of light allows exquisite control over spatial patterning of these materials in a microchip, analogous to photolithography, using a

Biological Applications of Microfluidics, Edited by Frank A. Gomez

mask and a UV source. The patterned material could be mobile and used as a valve or could be fixed in a channel to act as separation media, a filter, or immobilized reactor. It is also possible to make many of these elements in the same chip to allow the integration of the corresponding functions. For example, a microchip containing a photopolymerized size-exclusion membrane upstream of a photopolymerized polyacrylamide gel can be used to perform sample clean up or concentration prior to sodium dodecyl sulfate- polyacrylamide gel electrophoresis (SDS-PAGE). Microfluidic chips containing functional polymers have found widespread applications and have been used to adapt many workhorse biochemical techniques, such as SDS-PAGE, chromatography, immunoassays, sample preconcentration, and dialysis.

This review specifically focuses on functional materials (a) that are patterned and formed *in situ* within microchannels and (b) that directly facilitate biochemical analyses. Functional media can be synthesized off-chip and introduced into the chip, as is the case with entangled polymer sieving matrices used for electrophoresis (1,2), or silica particles or sol gel precursors used for solid-phase extraction or chromatography (3,4). Additionally, separation media can, in some cases, be incorporated directly as part of device fabrication, for example, arrays of micro- or nanofabricated posts used to facilitate electrophoretic separation (5), or nanofluidic channels used for electrokinetic trapping (6). Microchannel coatings are an active area of research and facilitate all types of biological analyses (7,8), and can be selectively patterned *in situ* (9,10). Finally, *in situ* patterning of polymers, hydrogels, and monoliths has widely been used to form structural elements (barriers, frits, and even channel walls), as well as control elements (e.g., valves, micromixers, and pumps) of microfluidic devices (11–16). There are certainly points of overlap between all of these topics, and integrated devices might contain several of these elements. A review on solid supports by Peterson (17) and a book chapter by Fintschenko et al. (18) cover some of these applications. For the remainder of this chapter, we will limit our scope specifically to the case of patterned materials that enable biochemical analyses.

Microchip-based biochemical analyses using *in situ* formed polymeric materials are listed in Table 21.1, and many of the them are described in detail later in this chapter. The list in Table 21.1 and the selected references are in no way exhaustive but are intended to provide illustrative examples of the wide array of uses for these materials.

21.2 GENERAL INFORMATION ABOUT PATTERNED MATERIALS

21.2.1 Types of Patterned Materials

Broadly speaking, polymeric materials can be divided into two categories: gels (or hydrogels), which are polymer networks swollen with water (water content may be >90%) and solid polymeric structures with low water or solvent content. Both types of materials are sometimes termed "monoliths," although this term is most frequently applied to solid polymer structures prepared by phase-separation polymerization. The most familiar hydrogel encountered in microfluidics is crosslinked

TABLE 21.1 Examples of Microfluidic Devices with Cast-in-Place Polymeric Materials

Target Analyte	Chip Application	Cast-in-Place Material	References
Proteins/peptides	SDS-PAGE	Polyacrylamide gel	(25,27)
	Isoelectric focusing	Polyacrylamide gel	(27)
	Native PAGE	Polyacrylamide gel	(28,31)
	Immunoassay	Polyacrylamide gel	(28,31)
	Electrophoretic preconcentration	Polyacrylamide gel membrane	(29,31)
		Zwitterionic acrylate monolith membrane	(30)
	Electrochromatography (CEC)	Porous acrylate polymer monolith	(21,34)
	Liquid chromatography (HPLC)	Porous acrylate polymer monolith	(33)
	Dialysis	Zwitterionic acrylate monolith membrane	(23)
DNA	dsDNA sizing	Polyacrylamide gel	(37,40)
	ssDNA sizing	Polyacrylamide gel	(37)
	dsDNA stretching	Polyacrylamide gel	(42)
	Solid-phase extraction (SPE)	Silica particles in methacrylate polymer monolith	(44)
	Sequence-specific capture	Polyacrylamide gel with grafted oligonucleotides	(46–48)
	Electrophoretic preconcentration	HEMA and acrylic acid copolymer hydrogels	(51)
Cells	Bioreactors/biosensors	PEG hydrogels	(52–54)
	Cell trapping and culture	Assorted biocompatible hydrogels	(58–60)
	Magnetic capture and sorting	PEG hydrogel	(57)
	Immobilized surface gradients	PEG and polyacrylamide hydrogels	(62,65)
Small molecules/ other analytes	Immobilized enzyme biosensors	PEG hydrogels	(66–69)
		Porous polyacrylamide monolith	(70)
	Immobilized enzyme bioreactors	Porous monoliths of acrylamide, methacrylates	(71)
	SPE and chromatography	Porous acrylate polymer monolith	(72,73)

polyacrylamide. Monomer concentrations of roughly 5–20% can be used in combination with the crosslinker *N,N′*-methylenebisacrylamide (BIS) to form a nanoporous gel that is both chemically and mechanically stable, with good sieving properties for DNA and protein electrophoresis. Other hydrophilic acrylamide derivatives, poly(ethylene glycol) (PEG), and copolymers with hydrophilic groups can also be used to form hydrogels. A variety of hydrogels that respond to environmental factors such as temperature and pH have been described (19), as well as materials that undergo chemically or thermally triggered phase transitions, such as Ca^{2+}-triggered gelation of alginate. Hydrogels appear homogeneous at the microscale, but a closer look would reveal a structure of interconnected gel fibers with a network of nanoscale pores with a complex size distribution.

Solid polymeric monoliths can be prepared in microchannels by phase-separation polymerisation, that is, polymerization of monomers in a porogenic solvent. Growing polymer chains become less soluble in the solvent, and phase-separate or precipitate, leaving a porous solid structure. A variety of acrylamide, acrylate, and methacrylate monomers and crosslinkers can be used, for example, butyl acrylate and 1,3-butanediol diacrylate; a more complete list of common monomers can be found in the review on monolithic media by Ro et al. (20). The pore-size distribution, mechanical properties, and surface properties (e.g., zeta potential) can be tailored by an appropriate choice of monomers and solvent. Both nanoporous and microporous structures can be produced by phase-separation polymerization. For glass microdevices, old or degraded polymer monoliths can be removed by thermal oxidation (21).

In situ polymerization may be chemically, thermally or photochemically initiated. Chemical initiation is not particularly suitable for chip application because polymer is formed in the entire fluid path including openings of the chips used to connect to the outside world. In theory, one can remove the polymer from places where it is not desired, but the process is cumbersome, imprecise, and irreproducible. Thermal initiation also suffers from the problem of spatial control of polymerization and inhomogeneities resulting from transients during heating and cooling, although microwaves can be used to achieve polymerization more rapidly (22). Photoinitiation can occur at room temperature and is easier to control (the duration of the reaction is defined by the exposure to the light source). Exposure to light can be localized to desired areas of the chip using a photomask or other patterning techniques.

21.2.2 Advantages of *In Situ* Patterned Materials

An advantage that miniaturization offers in general is the ability to integrate different biochemical processes and the components required to perform them. A complex microfluidic chip contains multiple fluidic channels and materials to perform such diverse tasks as filtering, pumping, valving, dialysis, separation, detection, and the like. Compared to larger fluid-handling and analysis systems, an integrated microfluidic chip performs these tasks much faster using smaller amounts of reagents and has the potential to be significantly cheaper if mass produced. Integrating functions at the microscale also greatly reduces sample loss and dilution, potentially allowing detection of amounts that are not possible at the larger-scale operation.

The primary advantage of patterning functional materials within microchannels is the ability to localize and immobilize the desired material in a specific location within a chip. Immobilization of a rigid material, such as a crosslinked gel or a polymer monolith, effectively prevents mixing of functional materials from different sections of a microdevice. Although this advantage may not be immediately obvious for simple devices, for example, simple electrophoretic analysis with a cross injection chip, the ability to create distinct patterned zones can enable complex, multistep operations that would be difficult to achieve with a freely flowing material such as a simple buffer or a polymer solution. Photopatterning can enable, for example, the localization of the gel to one stretch of channel while leaving other portions of the channel open for injection, detection, or valving structures. Multiple zones of gels with distinct monomer composition or concentration and even gels with concentration gradients can be patterned within a single channel. Patterning of rigid materials can simplify the production of integrated devices with multiple operations, without requiring complex valving structures to isolate different zones of the chip. A wide variety of well-characterized photoinitiated chemistries is available, allowing a great deal of flexibility over the formation of microstructures. Other types of *in situ* patterning are possible as well, including flow patterning, offering other possibilities for unique microstructures.

21.2.3 Challenges Associated with *In Situ* Patterned Materials

Major challenges associated with *in situ* patterned polymers are (a) reproducibility of material formation from chip to chip and (b) limited shelf-life resulting from evaporation of liquid or degradation of polymer. These problems are compounded by the difficulty in characterizing materials using traditional analytical techniques. Other than observation of gross defects, such as bubbles in a polyacrylamide gel, it can be difficult to assess the reliability of material formation on a chip-by-chip basis. The most sensitive analytical techniques for determining the microstructure of a monolith or gel, such as electron microscopy, necessarily involve destruction of the device and may raise difficult questions about appropriate sample preparation techniques. At any rate, once a device has been destroyed to analyze the materials inside, the results are not necessarily representative of the properties of the next device fabricated, and thus quality control is difficult. However, if the process of photopolymerization can be standardized and automated (analogous to photolithography used in semiconductor fabrication), the yield and reproducibility can be vastly improved. Another limitation of chips containing polymers is that the polymers need to be hydrated at all times. If the polymer dries out, it is practically impossible to rehydrate it to its original condition. Moreover, shrinkage resulting from drying can lead to polymer detaching from the walls of the channels. Another factor reducing shelf-life can be degradation of polymer. For example, polyacrylamide-based hydrogels are prone to hydrolysis, and storage at 4°C near neutral or acidic pH is required to extend the life of the gel.

Rigid, immobile materials are, by definition, fixed in place, and thus can limit the reusability of microdevices. For certain types of analysis, a single crosslinked gel or polymer monolith might be used dozens or even hundreds of times with no difficulties. But for other analyses, particularly where buildup or carryover of contaminants from previous samples is a problem, replacement of the gel after each sample is necessary. The trend in lab-on-a-chip devices seems to be toward low-cost, single-use, disposable devices, and thus reusability of the device may be a moot point.

While the polymers used are very inexpensive, the true cost of *in situ* fabrication of materials requires a careful analysis. Many of the patterning techniques described in the literature are fairly slow, labor-intensive processes with multiple steps, carried out one chip at a time by a skilled researcher. Moreover, most researchers would admit that their success rate is certainly less than 100% in device preparation. Many patterning techniques are surely amenable to increased automation, which would go along with an increase in device reliability and decreased cost per device. Hence, for commercial applications, it is important to choose processes that are amenable to mass production, without leaving any difficult, exotic, or temperamental patterning techniques to be implemented by the end user.

21.2.4 Approaches to Photopatterning of Materials

The optical transparency of microfluidic chips enables photoinitiation of polymerization, which can be as simple as replacing chemical/thermal initiators (e.g., ammonium persulfate and TEMED) with an appropriate UV photoinitiator such as 2,2′-Azobis[2-methyl-*N*-(2-hydroxyethyl)propionamide (VA-086). Several techniques are available for defining regions of illumination when photopatterning microdevices. Ultraviolet patterning is most common, as effective photoinitiators are available that strongly absorb at wavelengths available from common light sources, although photoinitiators are also available for visible, infrared, and deep UV wavelengths. The simplest approach to UV patterning is perhaps to position the region of interest below (or above) the objective of a microscope and expose the region to focused illumination from a UV-rich light source such as a mercury lamp. The resulting light intensity is non-uniform, both vertically and laterally, and this approach is only useful for simple applications where precise control over polymer properties or boundaries is not required. More precise patterns can be defined by contact lithography using flood illumination through a photomask. A chrome mask pattern on glass is effective, but a simple laser printer transparency, electrical tape, or other UV-blocking material may be used as well. Proper alignment of the mask prior to light exposure is important. Diffraction effects limit the resolution of features that can be defined using projection of incoherent light through a photomask, particularly when the microchannels are defined in comparatively thick glass or plastic substrates (thicknesses ranging from 0.5 to 1.5 mm are common). Photopolymerization using flood illumination typically occurs over time scales of $\sim$10 min, and thus diffusion of activated initiator during the course of polymerization can result in diffuse edges of the polymeric elements. UV lamps typically generate a significant amount of heat, and cooling fans or other measures for heat removal are useful when the chip is placed in close proximity to the light source as polymerization kinetics are strongly affected by temperature. Sealing liquid reservoirs with plastic tape or film may also be useful in presenting evaporation during extended polymerizations.

For the highest resolution and the fastest polymerization, illumination can be performed using shaped laser light. Small, solid-state frequency-tripled Nd:YAG lasers ($\lambda = 355$ nm) with output of several milliwatts are readily available for modest cost and can be used to photopattern polymers with illumination times as brief as a few seconds. Projection of collimated laser light through a slit ensures that the light is fairly uniform spatially and cropped at the edges to eliminate the standard Gaussian drop-off in intensity. Scanning mirrors or a translating stage can potentially be used to define patterns more complex than a single spot, and cylindrical optics can be used to define a laser sheet or line. Spatial resolutions of ~10–15 microns are readily achieved using a simple optical setup; attention should, of course, be paid to using high quality components with appropriate coatings and optical properties for ultraviolet wavelengths, and appropriate laser safety practices should always be observed, particularly with invisible wavelengths.

When performing photopatterning, attention to certain experimental details is important. Monomer solutions should be well degassed immediately prior to introduction into microchannels (sonication under vacuum with gentle agitation for several minutes is an effective technique for small volumes). For the highest resolution of features, the fluid should be totally quiescent within the microchannels, and it may be necessary to allow the device to sit undisturbed for several minutes to ensure that any flow driven by slight differences in liquid level in wells has settled down. This comes with the caveat that during the equilibration time, oxygen diffuses into the carefully degassed solution. The addition of radical scavengers such as hydroquinone at part-per-million levels can help to minimize stray polymerization caused by diffusion of radical species outside of the region of illumination (23), although too much scavenger can inhibit the polymerization reaction in the illuminated region.

Optimal results are typically obtained when the gel or the monolith is covalently attached to the microchannel wall, as there is often some shrinkage associated with polymerization and detachment can occur at the microchannel wall. For glass and some plastics, covalent attachment is frequently achieved by coating the microchannel wall using a bifunctional reagent such as 3-methacryloxypropyltrimethoxysilane (MAPS), which reacts directly with polymer chains during polymerization. Another common approach is to graft a linear polymer onto the microchannel wall and then crosslink the photopatterned gel to this polymer "interlayer."

Silanization of glass surfaces is a field unto itself with a large body of sometimes contradictory literature as to the optimal conditions to produce optimal coatings. The interested reader is directed to a study by Munro et al. in which the subject of silanization for microchannel coating is treated in some depth (24). Our experience has been that when silanization is followed by the patterning of a crosslinked polymer, formation of a "perfect," molecularly smooth, densely packed monolayer is probably not required, and a fairly robust procedure, as outlined by Kirby et al. (10) and by Herr and Singh (25), is sufficient. Attention to fine details such as the humidity of the air in the laboratory is not critical to the long-term stability of the gel, although keeping the silane dry and rinsing microchannels thoroughly between steps of the procedure are both important.

21.3 PHOTOPATTERNED MATERIALS FOR PROTEIN ANALYSIS

Microfluidic techniques for protein analysis are the subject of widespread research and development, and microdevices geared toward proteomic analysis were recently reviewed by Freire and Wheeler (26). To date the largest commercial application for microfluidic protein analysis has been SDS-protein electrophoresis using an entangled polymer sieving matrix (e.g., the Agilent Bioanalyzer or competing products from Caliper and Bio-Rad). Replaceable polymer matrices are effective for this relatively simple application, but photopatterned materials enable numerous other microdevices for more complex protein analysis.

21.3.1 Crosslinked Sieving Gels for Protein Separations

SDS-PAGE is one of the most commonly used techniques for researchers working with proteins. It separates proteins based on their size and is typically performed in slab gels. It is used to determine protein purity, and molecular weight, and also to identify proteins after a processing step such as affinity separation. It is also the second step in two-dimensional gel electrophoresis. The current slab-gel devices require manual loading of samples using pipettes, and a typical run is completed in 45 min to an hour. After SDS-PAGE, the gel is stained with a dye, destained to remove excess dye, and then imaged in a scanner. These steps take several additional hours to complete. If proteins need to be analyzed further, then extra steps of gel excision and extraction are required.

The adaptation of SDS-PAGE to a microfluidic format integrates and automates most of the steps outlined above. More importantly, the entire process of protein loading, SDS-PAGE, and detection can be completed in less than a minute per sample. Photopatterned crosslinked acrylamide gels were developed for sieving electrophoresis of both native and SDS-denatured proteins. In a microfluidic approach directly analogous to slab gel SDS-PAGE, Singh and coworkers at Sandia National Laboratories demonstrated high resolution protein sizing using commercial SDS-protein standards as well as clinically relevant cytokines and biotoxins (25,27). The gel (4–12% total monomer) was patterned in the separation channel of a simple offset-T chip while masking off the loading area during photopolymerization. Leaving the loading zone open speeds injection of analytes and relieves injection bias that can occur when analytes have to migrate through a sieving polymer to get to the injection zone, although the interface between the gel and the open channel can also be the site of nonideal sieving and size exclusion effects (28). Chips can also be fabricated with all channels filled with a single gel concentration or by a two-step photopatterning technique with a low concentration (~3.5% T) "loading zone" and high concentration (~8% T) "separation zone."

In other work, the group at Sandia has used photpatterned gels for isoelectric focusing (27), and native protein electrophoresis applied to problems ranging from immunoassays for biotoxins to detection of dilute biomarkers of dental disease in saliva. Native gel electrophoresis was used to detect the C-fragment of tetanus toxin (TTC) and anti-TTC in buffer or serum using an immunoassay format, with LIF enabling a detection limit for anti-TTC of ~680 pM (28).

21.3.2 Integrated Preconcentration-Protein Analysis

In biological and clinical applications, there is an ever-increasing need to analyze the low abundant species and to this end integrated polyacrylamide membranes have been developed by Hatch et al. for electrophoretic preconcentration of dilute proteins (29). A two-step photopatterning procedure is used. First, the channel is filled with a concentrated (~20% T, 5% C) monomer solution, and then a laser is used to rapidly polymerize a small (~20 μm wide) dense polyacrylamide "membrane" near the injection zone of a chip. The remaining monomer solution is removed and replaced with a lower concentration monomer solution (~5–8% T, 5% C), forming a gel tailored to the size range of the analytes of interest. The high concentration polyacrylamide plug acts as a size-exclusion membrane. By applying an electric field across the membrane, proteins and other large macromolecules can be made to stack at the gel interface. Concentration factors of several orders of magnitude can be achieved in this manner. The entire protein content of a long channel can be electrophoretically stacked into a narrow (~5–25 μm) zone at the membrane interface, with subsequent injection into the separation channel enabled by simply changing the direction of the applied field. Successive preconcentration steps can allow the mixing of multiple reagents, for example, proteins and antibody, within a narrow zone, thus allowing immunoassays to be carried out on-chip.

Electrophoretic preconcentration of proteins was also demonstrated using an acrylate-based polymer monolith membrane (30). The phase-separation polymerization chemistry employed is not compatible with polyacrylamide gel electrophoresis, but it would be a useful choice for preconcentration of proteins for open-channel or capillary zone electrophoresis (CZE).

Herr et al. recently demonstrated an immunoassay for the enzyme MMP-8, a biomarker in saliva, for periodontal disease (31). Integrated preconcentration membranes enabled detection of MMP-8 present in saliva with sensitivity comparable to an ELISA gold standard in an automated chip-based procedure requiring <10 min. The photopatterned membrane and the separation gel in this device enable multiple sample preparation steps to be performed on-chip, making this device an example of a true "sample in, answer out" device capable of handling real clinical samples. The layout and operation of the device is illustrated schematically in Fig. 21.1, along with representative electrophoretic immunoassay data.

21.3.3 Porous Polymer Monoliths for Chromatography

Chromatography (e.g., HPLC), because of its outstanding separation power and versatility, is one of the most common analysis methods for proteins and peptides, and its miniaturization holds substantial promise for rapid analysis of complex biological samples. A major roadblock in adaptation of chromatography to microchips has been

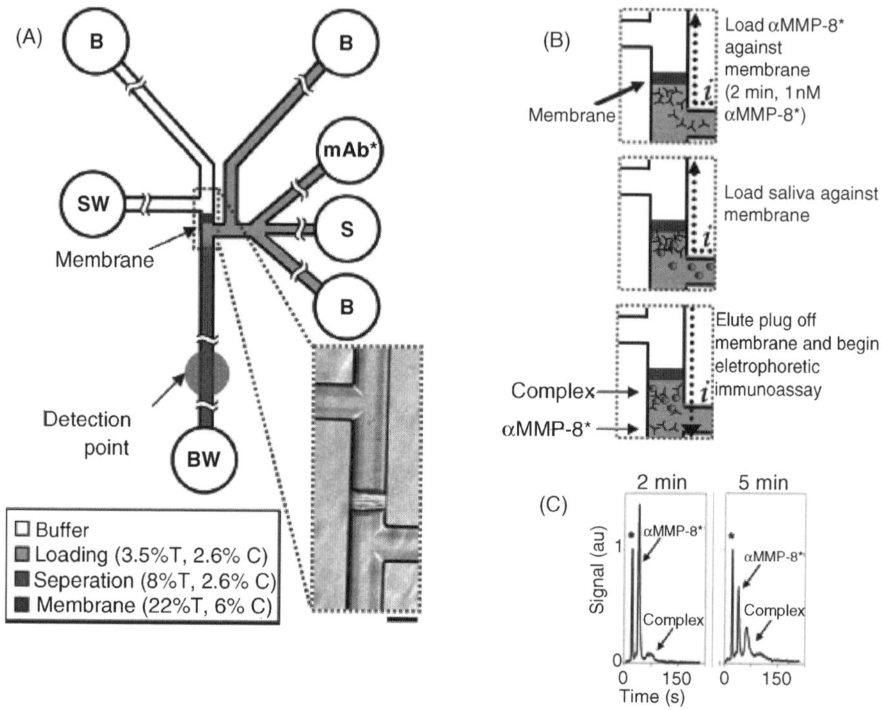

FIGURE 21.1 (**a**) Schematic of chip for electrophoretic immunoassay with integrated preconcentration membrane. The different zones (loading, separation, membrane, and buffer) are formed by successively filling the chip with different monomer compositions and photopolymerizing specific regions. Designations of wells are B = buffer, S = sample, mAb* = fluorescently labeled monoclonal antibody, SW = sample waste, and BW = buffer waste. (**b**) Schematic of electrophoretic preconcentration and mixing of reagents at integrated membrane. Current flow is indicated by arrow labeled *i*. (**c**) Electropherograms showing the formation of substantial immune complex with 5 min of preconcentration time. The peak labeled (*) is a protein internal standard. Figures copyright 2007 by The National Academy of Sciences of the United States of America.

the issue of placement of chromatographic media in the channels. Uniform packing of channels with particles is irreproducible, hard to implement, and requires fabrication of frits and side channels. Following the methodology developed by Svec, Frechet, and coworkers (16,32), Throckmorton et al. used UV-iniated polymerization to cast porous polymer monoliths of acrylate derivatives (21). The monomer mixture included some lauryl acrylate to give the packing a hydrophobic (C12) character. The monoliths were cast *in situ* in less than 10 min, were very reproducible with respect to separation characteristics, and allowed easy manipulation of separation parameters such as charge, hydrophobicity, and pore size. Moreover, the solvent used to cast the polymer enabled electroosmotic flow, allowing the separation channel to be conditioned without the need for high pressure pumps. The microchip chromatography offered significant improvements over capillary-based and HPLC separations, as they were an order of magnitude faster (6 peptides in 45 s; 9 min for HPLC separation on a C18), efficient (up to 600,000 plates/m), and reproducible (run-to-run variability <3%). Micrographs of the photopatterned monolith, including an SEM showing the porous structure of the monolith, are shown in Fig. 21.2, along with chromatographic data obtained from the separation of a peptide mixture.

Similar acrylate monoliths were also used as supports for on-chip high pressure liquid chromatography (33). Performing HPLC on a chip, at pressures of >400 psi, presents numerous technical challenges with flow control, injection, and world-to-chip interface. The researchers solved the problem of injection using a novel photopolymerized, pneumatically actuated fluoropolymer valve and used custom high pressure fittings for world-to-chip connections.

CEC with porous polymer monliths has been an enabling technology for coupling chip separations to electrospray mass spectrometry. Lazar et al. used photopatterning

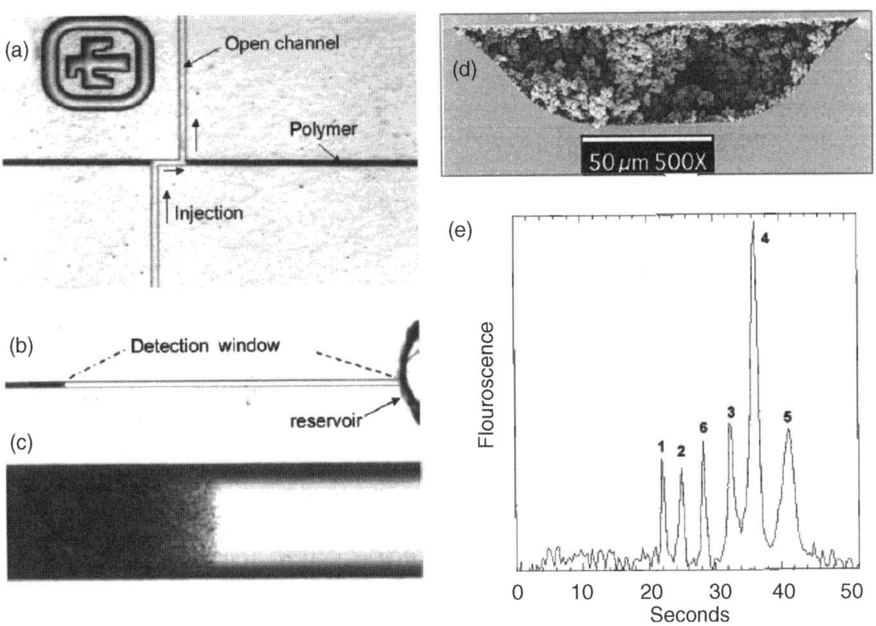

FIGURE 21.2 Photopatterning of polymer in channels for optimization of different functions. The polymer was patterned using a mask and a UV lamp. (**a**) and (**b**) Patterning of monolith to keep injection arms and detection window void of polymer in an offset double-T chip to allow for fast injection and lower background for detection. (**c**) Top view of the edge of polymer monolith. The sharpness of the interface indicates good resolution of the patterning process. (**d**) SEM micrograph of channel filled with porous polymer monolith, with 1 μm average pore size. (**e**) Reversed-phase electrochromatography of NDA-labeled peptide standards in a microchip with negatively charged lauryl acrylate monolith. The field strength was 770 V/cm, with a mobile phase consisting of 30:70, ACN/25 mM borate, pH 8.2 containing with mM octane sulfonate. The peptides are (1) papain inhibitor, (2) proctolin, (3) opioid peptide (R-casein fragment 90–95), (4) Ileangiotensin III, (5) angiotensin III, and (6) GGG. Reprinted with permission from Reference (21). Copyright 2002, American Chemical Society

to create porous polymer monoliths on a chip using a mixture of glycidyl methacrylate and methyl methacrylate monomers (34); polymerization was followed by derivatization with a secondary amine to render a positively charged, weakly hydrophobic (C4) surface. CEC of a peptide digest was performed in acidic conditions, allowing ESI–MS in positive ion mode. Electroosmotic flow generated by the positively charged surface provided a steady stream of liquid flow, although the design of the ESI source was challenging and the process of injection momentarily disrupted the electrospray.

21.3.4 Polymer Membranes for Dialysis

Laser patterning has been used to fabricate porous zwitterionic polymer dialysis membranes by phase-separation polymerization of 2-(*N*-3-sulfopropyl-*N*,*N*-dimethylammonium)ethyl methacrylate (23). Song and coworkers used shaped laser light to pattern a long porous polymer membrane down the center of a microchannel, effectively separating the channel into two halves. The membrane was ~35 μm wide and 5 mm long, and was fabricated piecewise in ~600 μm segments. The membrane was supported on glass "posts" fabricated at the center of the channel as part of device fabrication, which improved the mechanical strength of the membrane. Different solvent mixtures were used to tailor the pore size of the membrane; increasing the amount of 2-methoxyethanol (a good solvent for the polymer) relative to water (a poor solvent) delays the onset of phase separation, resulting in larger pores. The device was amenable to countercurrent dialysis, as opposed to the majority of laminar-flow microfluidic devices that operate only in a cocurrent mode. The device facilitated the partial removal of an organic dye from an aqueous stream (using a low molecular weight cutoff membrane), as well as a protein (using a high molecular weight cutoff membrane). Complete removal would require further optimization of the device, possibly including a longer channel, higher channel aspect ratio, or lower sample flow rate operation. Micrographs showing the device layout and details of the membrane are shown in Fig. 21.3a and b, and fluorescence micrographs showing removal of a fluorescent dye by dialysis are shown in Fig. 21.3c.

21.4 DNA PURIFICATION AND ANALYSIS

21.4.1 Sieving Media for Electrophoretic Analysis

Analytical separations of DNA fragments according to chain length are of critical importance in modern biology. Electrophoretic separation of single-stranded DNA ladders generated by the Sanger cycle sequencing reaction is the key analytical step in conventional DNA sequencing technology. A variety of other genotyping or DNA fingerprinting analyses, including short tandem repeat (STR) sizing, single nucleotide polymorphism (SNP) analysis, single strand conformation polymorphism (SSCP) and heteroduplex (HA) analysis, and single-base extension (SBE) assays also rely upon high resolution electrophoretic separation of single- or double-stranded

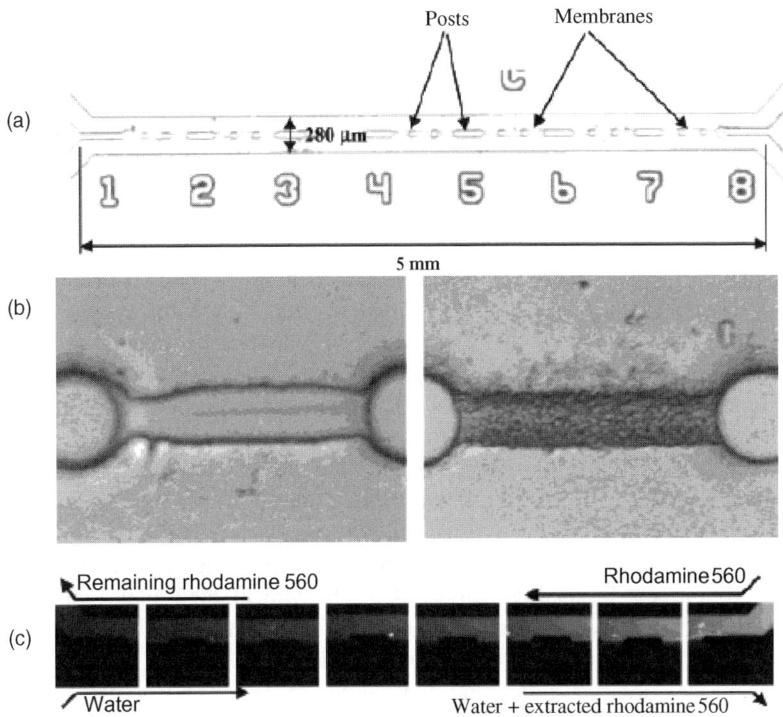

FIGURE 21.3 Porous polymer monolith membrane formed by laser patterning of zwitterionic acrylate monomers. (**a**) Channel layout showing position of membrane down the center of a long microchannel, with microfabricated "posts" for support. The membrane was patterned between posts in segments of ~600 μm at a time. (**b**) Optical micrographs of membrane segments between posts. The image on the left is of a low MWCO membrane (<5700 Da), while the image on the right shows a higher MWCO membrane. Pore sizes were controlled by varying the compatibility of the solvent used in polymerization. (**c**) Fluorescence micrograph showing removal of rhodamine 560 from a flowing stream by countercurrent dialysis across the microfabricated membrane. Reprinted with permission from Reference (23). Copyright 2004, American Chemical Society.

DNA. In addition, the sizing of PCR fragments is a common analytical tool in the everyday operations of a molecular biology lab.

Many of the analytical techniques for DNA were originally developed in crosslinked slab gels and over time were adapted for use in capillary electrophoresis. Many of the CE methods have since been adapted for use in microfluidic chips. In analytical DNA separations, especially for demanding applications such as DNA sequencing, the properties of the sieving matrix are of utmost importance. *In situ* polymerized polyacrylamide gels were never particularly successful for DNA sequencing by capillary electrophoresis for a variety of reasons including bubble formation during polymerization, sample carryover between analyses, and gel deterioration with repeated use, and thus *in situ* polymerization was largely abandoned with the advent of replaceable, moderate viscosity entangled polymer sieving matrices such as linear polyacrylamide. The practical difficulties associated with forcing viscous polymer solutions into chip microchannels have led to the development of improved sieving matrices with viscosities more appropriate for use on chips. Off-chip production affords the opportunity for rigorous dialysis and purification of the polymer, as well as detailed analytical characterization of the molar mass distribution, which is important for high quality, reproducible performance (35). For an in-depth treatment of microfluidic DNA analyses using replaceable polymer matrices, the reader is directed to the 2004 review by Kan et al. (36).

Despite the prevalence of noncrosslinked sieving matrices for chip electrophoresis, several researches have used *in situ* polymerized polyacrylamide gels for DNA electrophoresis on chips (37–40). Ugaz et al., in particular, compared a variety of sieving matrices including linear polyacrylamide, Pluronic F-127, agarose, and photocrosslinked polyacrylamide for the electrophoretic separation of 20-bp and 100-bp dsDNA ladders, and found improved resolution for the crosslinked polyacrylamide versus linear polyacrylamide matrices (37). The authors attributed the improved performance to reduced diffusion of the DNA molecules in the crosslinked versus noncrosslinked matrix (38). In comparing performance, it is important to point out that the authors used *in situ* polymerized linear polyacrylamide, and the resulting polymer was likely inferior in quality compared to the typical high molar mass linear polyacrylamide synthesized off-chip, which requires careful control of polymerization conditions to achieve the desired molar mass distribution. With respect to issues surrounding deteriorating gel quality and buildup of large DNA contaminants over multiple runs, Ugaz et al. point out that their device is easily disassembled and cleaned, and a fresh gel cast, allowing the device to be reused an indefinite number of times.

The advantages of crosslinked gels are most pronounced in complex, integrated devices such as that demonstrated by Pal et al. for influenza virus genotyping (40). This device, illustrated schematically in Fig. 21.4a, includes on-chip PCR, restriction digestion, and electrophoretic analysis. The use of a photocrosslinked gel allows a simple means of isolating the gel from the rest of the chip without additional valving, although this device also features integrated phase-change valves to isolate regions of the chip. A sample electropherogram obtained from on-chip genotyping of influenza virus is shown in Fig. 21.4b, with good resolution of bands obtained in 135 s. The authors project that similar devices could be mass produced for less than $1 per chip, which can be suitable for single-use applications.

A photopatterned crosslinked gel is also advantageous in devices designed for the recovery of specific bands of DNA from an electrophoretic separation, such as the device reported by Lin et al. (41). With a fixed gel patterned in the main separation channel only, specific DNA bands can be eluted into a side channel or well containing only buffer. This effectively eliminates contamination of the recovered DNA with high molecular weight sieving polymer, which could be of concern depending on the downstream application intended for the recovered DNA.

The ability to precisely position a crosslinked gel in a microchannel was central to a recent study in DNA chain dynamics (42), in which Randall and coworkers used a hyperbolic channel contraction to create an electric field gradient to stretch T4 DNA. The device is illustrated schematically in Fig. 21.5a. The degree of stretching

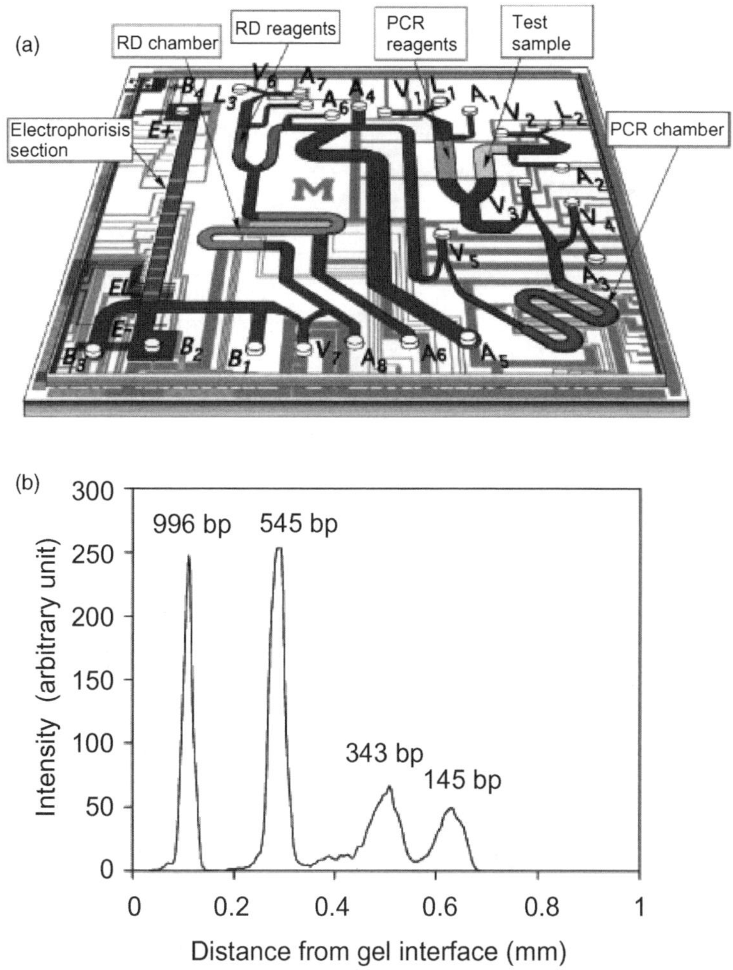

FIGURE 21.4 (**a**) Schematic of integrated chip for influenza strain genotyping showing photopatterned gel electrophoresis zone integrated with on-chip PCR and restriction digest (RD) along with integrated valves, electrodes, and heaters. (**b**) Gel electrophoresis analysis for on-chip PCR-RFLP strain identification of influenza virus. The electropherogram represents the intensity profile of a gel scan following 135 s of electrophoresis. The 343 and 996 bp bands are markers, and the 145 and 545 bp bands are HpaI restriction fragments from PCR-amplified influenza DNA. Reprinted with permission from Reference (40). Copyright 2005, Royal Society of Chemistry. (See color insert)

depends on the configuration of the 169 kbp DNA chain as it enters the contraction, resulting in nonuniform behavior in the contraction (a consequence of "molecular individualism"), as illustrated with two different DNA molecules in Fig. 21.5b. To achieve uniform stretching of all DNA molecules, the authors biased the configuration of the DNA by photopatterning a short plug of polyacrylamide gel immediately upstream from the contraction using UV illumination through a

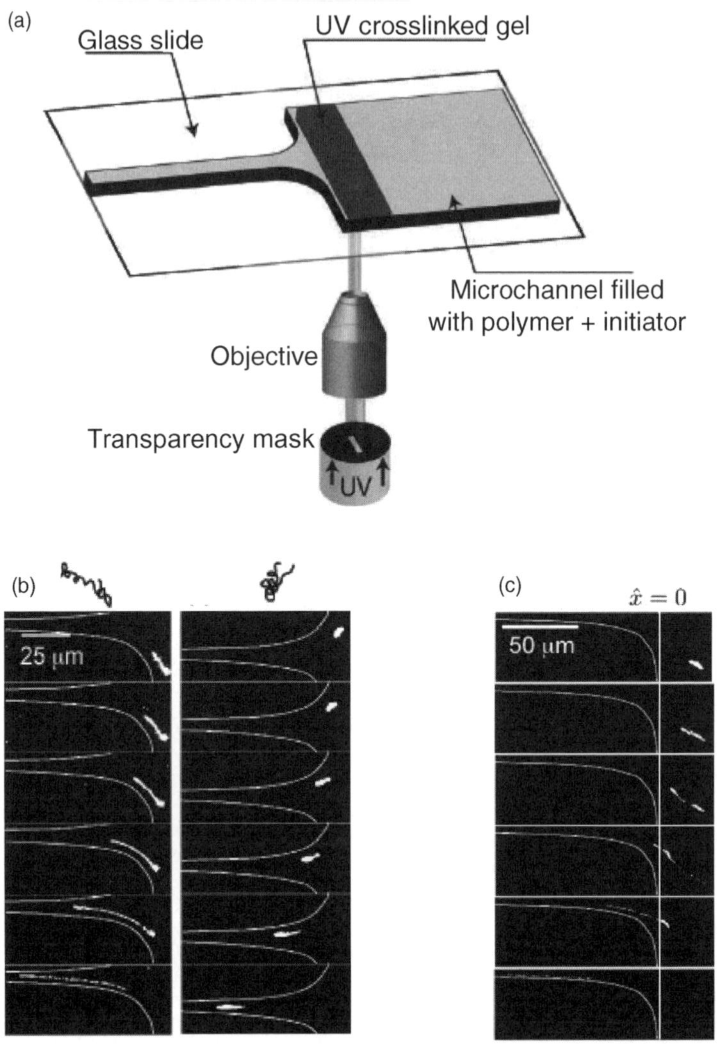

FIGURE 21.5 (**a**) Schematic of microdevice with a hyperbolic contraction for creating an electric field gradient for stretching of DNA. A polyacrylamide gel plug is photopatterned immediately upstream of the contraction to enhance stretching behavior. (**b**) Fluorescence micrographs of individual fluorescently labeled T4 DNA molecules entering the contraction with no gel upstream. The degree of stretching in the contraction depends upon the initial conformation of the DNA molecule as it enters. (**c**) Fluorescence micrographs of a single fluorescenlty labeled T4 DNA molecule entering the contraction after passing through the photopatterned polyacrylamide gel plug. As the DNA molecule emerges from the gel into the contraction, the gel acts as a "tether," restraining the portion of the DNA molecule that is still in the gel, while the portion of the molecule that has exited the gel plug stretches.
Reprinted with permission from Reference (42). Copyright 2006, Royal Society of Chemistry.

photomask. The process of DNA reptation through the gel results in stretching or in alignment of the chain as it exits the gel plug, and this perturbation in the configuration of the DNA is sufficient to ensure complete stretching of the chain in the contraction. This is illustrated for a single DNA molecule in Fig. 21.5c. The precise placement of the gel plug is important, with a well-defined boundary between the gel and the contraction defining two different regimes of DNA migration. Photopatterning in this case provides a straightforward way to achieve a function that could not be achieved with freely flowing polymer solutions.

To address the problem of characterization of *in situ* polyacrylamide production, Wang and Ugaz recently described oscillatory shear rheometry measurements of polyacrylamide gel properties during UV photopolymerization between closely spaced plates, simulating *in situ* polymerization in a microchannel (43). The authors measured the effects of UV intensity, polymerization temperature, and monomer and crosslinker concentrations on the gel storage modulus *G*', and used the results to predict gel pore size, which was correlated to pore size estimates based on DNA mobility. These results should be of interest for determining *a priori* the polymerization conditions required to produce a gel with the desired properties, either for DNA separations or for protein separations.

21.4.2 Photopatterned Materials for DNA Sample Pretreatment

Solid-phase extraction using silica beads or sol–gel monliths has been described by several groups for on-chip isolation and concentration of DNA from biological samples (3,4). This purification technique is based on the reversible, salt and pH-dependent adsorption of DNA to silica surfaces and is similar to the familiar bench-scale DNA purification kits sold by Qiagen and other vendors. In microchannels, silica beads may simply be packed against microfabricated or monolithic frits, but in at least one instance, photopatterning of a porous polymer monolith was used to immobilize the solid silica support within a Zeonor microchannel (44). Following the methodology developed by Frechet and coworkers for the formation of porous monoliths in plastic channels (16,45), Bhattacharyya and Klapperich began by grafting a thin, noncrosslinked polymer coating to the Zeonor channels to act as an interlayer to improve the adhesion of the polymer monolith to the channel walls. Next, a mixture of silica particles was suspended in a mixture of monomers (ethylene dimethacrylate and butyl methacrylate), with porogenic solvents and photoinitiator. Polymerization resulted in a porous structure covalently attached to the microchannel walls (*via* the polymer interlayer), and the silica bed was successfully used for DNA extraction from a model system. The silica particles tended to settle out of the precursor solution during polymerization, leading to nonuniform distribution of silica, and the authors found that simply inverting the chip partway through polymerization resulted in a more uniform distribution. The monolith slowly broke down over time, although stability was sufficient for a low cost, disposable plastic device.

A highly sequence-specific approach to both DNA identification and purification in microchannels has been developed based on polyacrylamide hydrogels grafted with "capture" DNA oligonucleotides. DNA oligos can readily be incorporated into polyacrylamide hydrogels by adding a 5′-acrylate modification (e.g., Acrydite™) during solid-phase synthesis. Hydrogels with an appropriate DNA oligonucleotide grafted throughout the polyacrylamide gel can serve as a "capture matrix," selectively trapping DNA with complementary sequence as it migrates through the gel during electrophoresis (46). This arrangement offers efficient mass transfer and capture of low concentrations of input DNA, and hybridization is reversible upon heating. Olsen and coworkers demonstrate DNA concentration enhancement in excess of 100-fold with a "capture gel" photopatterned within a polycarbonate device. By selectively patterning multiple zones with different oligos, an array or bar code device can be created with different regions specifically capturing different target sequences. In the initial implementation, gel plugs tended to break down or delaminate from channel walls within 25 min of continuous operation, possibly due to electroosmosis through microscopic pathways near the channel wall and possibly exacerbated by substantial body forces exerted on the charged gel plugs during electrophoresis. Adhesion of the gel plugs to the channel walls could be improved by oxidizing first the (plastic) microchannel surface with UV/O_3 treatment, followed by the attachment of an acrylate-modified anchor to the wall (47). This pretreatment resulted in a 2.5-fold increase in gel lifetime. With the improved plug formation procedure, Zangmeister and coworkers used the DNA-containing hydrogels to implement a DNA detection assay based on displacement of a fluorescently labeled DNA strand, with a detection limit of 0.1 μM using a simple fluorescence microscope and CCD camera for imaging (48).

With lifetimes of ~1 h, these DNA-containing hydrogels are significantly less stable than uncharged polyacrylamide hydrogels that have been demonstrated in glass microchannels, although it is unclear whether this is due to the presence of the DNA in the hydrogel, or less robust attachment to plastic versus glass microchannels, or other factors affecting the polymerization. The Mathies group has used a similar "capture matrix" approach for on-chip purification of DNA sequencing reactions prior to electrophoretic analysis (49,50), except that this group incorporated the "capture" oligonucleotide in a noncrosslinked or sparsely crosslinked polymer matrix fabricated off-chip, which could be replaced between runs. Since long-term stability of photocrosslinked capture gels was problematic, the replaceable capture matrices may have advantages, provided that appropriate measures (e.g., valving) are taken to isolate the capture matrix from other portions of the chip.

Photopatterned highly crosslinked hydrogel plugs have been used for online preconcentration of single-stranded DNA (51), in a manner similar to the polyacrylamide membranes used for protein preconcentration by Hatch and coworkers (29,31). Dhopeshwarkar used UV polymerization to pattern plugs of hydroxyethylmethacrylate (HEMA) and copolymers of HEMA and acrylic acid (AA) in PDMS microchannels. The poly(HEMA) hydrogels were neutral, while the poly(HEMA-co-AA) hydrogels carried a negative charge. Although impressive preconcentration factors were observed, the process of subsequent electrophoretic analysis is not as simple. With the neutral hydrogel plugs, the DNA partially enters the hydrogel during preconcentration, and some remains trapped in the gel upon switching voltages, resulting in a sample loss and a spread out injection. With the negatively charged hydrogel plugs, charge exclusion prevents the DNA from entering the gel, but the gel generates electroosmotic flow that reduces the effectiveness of preconcentration and "smears" the concentrated zone. Additional optimization is needed to minimize the effect of electroosmotic flow (EOF) generated by the negatively charged hydrogel plug.

21.5 PATTERNED MATERIALS FOR CELL CULTURE AND ANALYSIS

Applications of patterned materials to cells have been less common than to protein and DNA analysis, perhaps because microfluidic tools for manipulating and analyzing cells are a more recent development. Most of the patterned materials are hydrogels, which is logical, since hydrogels can be biocompatible, with complex 3D architectures, and modified with cell signaling molecules. Several authors report the immobilization of bacterial or mammalian cells in hydrogels as bioreactors or biosensors (52–54). Photopatterning is common, although one study reports using a microheater to locally switch the state of a thermoresponsive hydrogel, encapsulating cells following manipulation by optical tweezers (55). One report by Moorthy and Beebe describes the formation of a porous polymer monolith to act as a simple filter for size-selective cell separations, in this case to separate blood cells from serum, although buildup of cells eventually impedes flow through the filter (56).

In a demonstration of passive cell sorting on a chip (as opposed to "active" cell-by-cell techniques such as FACS), Pregibon et al. used a combination of magnetic fields and photopatterning to create arrays of PEG-encapsulated magnetic beads (Dynabeads) (57). A magnetic field was used to hold the beads in place on the chip surface, and photopatterning was used to encapsulate individual clusters or regions of beads. The authors demonstrated the capture of B-cells from a flowing stream on an immobilized array of B-cell-specific (anti-CD19) beads. The authors also demonstrated magnetically mediated sorting of B-cells from T-cells in an indirect approach, with off-chip incubation of a cell mixture with anti-CD19 magnetic beads, followed by on-chip magnetic capture of bead-bearing B-cells on a patterned array of magnetic beads.

Several reports describe laminar flow patterning approaches to form hydrogels *in situ* to trap cells or provide 3D culture environments (58–60). Flowing streams containing gel precursors are brought into contact in microchannels, triggering gel formation and trapping cells. In an example of flow-patterned gel formation, diffusion of Ca^{2+} ions is used to trigger gelation of alginate in a device with two inputs (58) similar to an "H-filter" or "T-sensor" (61). Yeast cells are immobilized in a region defined by the diffusion of Ca^{2+} ions into the alginate stream; the cells can be released by reducing the Ca^{2+} concentration in the inlet stream. Gel formation is fairly slow ($\sim$20 min), but it is reversible and does not interfere with cell proliferation. A more complex chip layout allows trapping of single cells by limiting the area of contact between the Ca^{2+} and the alginate-containing streams.

Several other authors report using a combination of microfluidic flow patterning and photopolymerization to create unique environments to study cell behavior.

Burdick and coworkers report photopolymerization to create an immobilized gradient of RGDS, an adhesive ligand for endothelial cells (62). The spatial gradient of the acryloyl-modified ligand in PEG-acrylate macromers was generated using a flow gradient generator similar to that first described by Jeon and coworkers in the Whitesides group (63). Photopolymerization was initiated downstream of the gradient generator, creating an approximately linear immobilized gradient of the ligand. Endothelial cell adhesion to the gradient hydrogel varied as a function of position along with the concentration of the adhesion ligand. The formation of a smooth, linear gradient required the careful tuning of the flow rates into the gradient generator; the authors propose that more complex gradient shapes could also be generated using previously described techniques (64).

In a related approach, Zaari and coworkers report creation of a hydrogel with a gradient in elastic modulus to measure the response of cells to the stiffness of the hydrogel substrate (65). These authors again used a microfluidic gradient generator to establish a flowing gradient followed by photopolymerization; in this case the two inputs to the gradient generator were acrylamide solutions with different concentrations of crosslinker. The elastic modulus (measured by AFM) varied by a factor of about 20 from one side of the gel to the other, a distance of 2.8 mm. The hydrogel was modified with adhesive proteins and incubated with smooth muscle cells or fibrolasts. Cell spreading was more extensive on stiffer regions of the gel, with apparent differences in cytoskeletal structure. Interestingly, this study appeared to reveal a threshold value of elastic modulus required for cell spreading.

In both of these flow gradient patterning studies, the immobilized gradients were formed with lateral dimensions of 1–3 mm, whereas typical cells might range from 10 to 100 μm in size. Particularly for spreading cells, the width of the cell may be several percent of the width of the gradient, meaning that individual cells can experience significant variation in chemical or mechanical environment across the width of the cell. Depending on your point of view, this may either confound interpretation of experimental data or may accurately simulate *in vivo* environments, which are frequently characterized by steep gradients. Formation of such steep, immobilized gradients in small domains is feasible using the combination of microfluidic and photopatterning techniques, but would be quite difficult using common approaches for creating macroscopic gradients.

21.6 OTHER BIOMOLECULES

The primary focus of development for microfluidic chips has thus far been to enable or miniaturize analyses of protein, DNA, and increasingly of cells. However, a number of researchers have applied the techniques of microfluidics, including photopatterned materials, to other types of biological analyses, notably techniques involving immobilized enzymes. Several researchers report immobilization of enzymes and reagents within photopatterned hydrogel plugs for use as biosensors of glucose or other analytes (66–69). A patterned acrylamide monolith with immobilized enzyme has also found use as a glucose biosensor (70). Several researchers also report immobilization of trypsin in porous polymer monoliths to enable protein digestion on-chip prior to chromatography or mass spectrometry (71). Additionally, techniques for creation of porous polymer monoliths for CEC or SPE of proteins have also been applied to CEC and SPE of small-molecule analytes such as drugs, with integrated interface to mass spectrometry (72,73).

21.7 CONCLUSIONS

Development of first-generation microfluidic chips focused on performing individual analyses such as DNA or protein separation. In these devices, while function such as injection was integrated, the chips generally did not have a high level of integration such as those required to perform multidimensional analyses or total analysis (sample in–answer out) requiring integrated sample pretreatment. The next-generation chips, like their microelectronics counterparts, involve a higher level of integration in terms of both scale and functionality. Increased complexity of integrated functionalities requires integration of diverse functional materials in a microdevice. These materials could include inorganic and organic porous gels, monoliths, and membranes required to perform diverse functions including separation, reaction, sample cleanup, valving, pumping and so on. Polymerization initiated by UV light is a facile way of making many of these functional elements and has extensively been used to form separation media for biochemical separations as reviewed in preceding sections. This method allows formation of desired materials in premade microchips and hence does not increase the cost of fabrication significantly. The use of light also permits exquisite control over spatial placements of these materials in a chip and opens the path of producing these chips reproducibly and in very large volumes using standard microfabrication tools. It is important to realize that while not all required materials can be made by photopolymerization, this technique has significantly improved our ability to integrate functional materials in microfluidic chips.

REFERENCES

1. Chiesl TN, Shi W, Barron AE. Anal Chem 2005;77:772–779.
2. Tian HJ, Landers JP. Anal Biochem 2002;309:212–223.
3. Breadmore MC, Wolfe KA, Arcibal IG, Leung WK, Dickson D, Giordano BC, Power ME, Ferrance JP, Feldman SH, Norris PM, Landers JP. Anal Chem 2003;75:1880–1886.
4. Wu QR, Bienvenue JM, Hassan BJ, Kwok YC, Giordano BC, Norris PM, Landers JP, Ferrance JP. Anal Chem 2006;78:5704–5710.
5. Huang LR, Tegenfeldt JO, Kraeft JJ, Sturm JC, Austin RH, Cox EC. Nat Biotechnol 2002;20:1048–1051.
6. Wang YC, Stevens AL, Han JY. Anal Chem 2005;77:4293–4299.

7. Doherty EAS, Meagher RJ, Albarghouthi MN, Barron AE. Electrophoresis 2003;24:34–54.
8. Dolnik V. Electrophoresis 2004;25:3589–3601.
9. Chen HY, Lahann J. Anal Chem 2005;77:6909–6914.
10. Kirby BJ, Wheeler AR, Zare RN, Fruetel JA, Shepodd TJ. Lab Chip 2003;3:5–10.
11. Kim, D, Beebe, DJ. Lab Chip 2007;7:193–198.
12. Agarwal AK, Sridharamurthy SS, Beebe DJ, Jiang HR. J Microelectromech Syst 2005;14:1409–1421.
13. Beebe DJ, Moore JS, Yu Q, Liu RH, Kraft ML, Jo BH, Devadoss C. Proc Nat Acad Sci USA 2000;97:13488–13493.
14. Beebe DJ, Moore JS, Bauer JM, Yu Q, Liu RH, Devadoss C, Jo BH. Nature 2000;404:588–590.
15. Wang XC, Yang XH, Zrang XM. Anal Sci 2006;22:1099–1104.
16. Rohr T, Yu C, Davey MH, Svec F, Frechet JMJ. Electrophoresis 2001;22:3959–3967.
17. Peterson DS. Lab Chip 2005;5:132–139.
18. Fintschenko Y, Kirby BJ, Hasselbrink EF, Singh AK, Shepodd TJ In: Svec F, Tennikova TB, Deyl Z, editors. Monolithic Materials: Preparation, Properties and Applications. Amsterdam: Elsevier; 2003. p659–686.
19. Roy I, Gupta MN. Chem Biol 2003;10:1161–1171.
20. Ro KW, Nayalk R, Knapp DR. Electrophoresis 2006;27:3547–3558.
21. Throckmorton DJ, Shepodd TJ, Singh AK. Anal Chem 2002;74:784–789.
22. Moschou EA, Nicholson AD, Jia GY, Zoval JV, Madou MJ, Bachas LG, Daunert S. Anal Bioanal Chem 2006;385:596–605.
23. Song S, Singh AK, Shepodd TJ, Kirby BJ. Anal Chem 2004;76:2367–2373.
24. Munro NJ, Huhmer AFR, Landers JP. Anal Chem 2001;73:1784–1794.
25. Herr AE, Singh AK. Anal Chem 2004;76:4727–4733.
26. Freire SLS, Wheeler AR. Lab Chip 2006;6:1415–1423.
27. Han J, Singh AK. J Chromatogr A 2004;1049:205–209.
28. Herr AE, Throckmorton DJ, Davenport AA, Singh AK. Anal Chem 2005;77:585–590.
29. Hatch AV, Herr AE, Throckmorton DJ, Brennan JS, Singh AK. Anal Chem 2006;78:4976–4984.
30. Song S, Singh AK, Kirby BJ. Anal Chem 2004;76:4589–4592.
31. Herr AE, Hatch AV, Throckmorton DJ, Tran HM, Brennan JS, Giannobile WV, Singh AK. Proc Nat Acad Sci USA 2007;104:5268–5273.
32. Yu C, Davey MH, Svec F, Frechet JMJ. Anal Chem 2001;73:5088–5096.
33. Reichmuth DS, Shepodd TJ, Kirby BJ. Anal Chem 2005;77:2997–3000.
34. Lazar IM, Li LJ, Yang Y, Karger BL. Electrophoresis 2003;24:3655–3662.
35. Buchholz BA, Barron AE. Electrophoresis 2001;22:4118–4128.
36. Kan CW, Fredlake CP, Doherty EAS, Barron AE. Electrophoresis 2004;25:3564–3588.
37. Ugaz VM, Lin RS, Srivastava N, Burke DT, Burns MA. Electrophoresis 2003;24:151–157.
38. Ugaz VM, Burke DT, Burns MA. Electrophoresis 2002;23:2777–2787.
39. Chen XJ, Ugaz VM. Electrophoresis 2006;27:387–393.

40. Pal R, Yang M, Lin R, Johnson BN, Srivastava N, Razzacki SZ, Chomistek KJ, Heldsinger DC, Haque RM, Ugaz VM, Thwar PK, Chen Z, Alfano K, Yim MB, Krishnan M, Fuller AO, Larson RG, Burke DT, Burns MA. Lab Chip 2005;5:1024–1032.
41. Lin RS, Burke DT, Burns MA. J Chromatogr A 2003;1010:255–268.
42. Randall GC, Schultz KM, Doyle PS. Lab Chip 2006;6:516–525.
43. Wang J, Ugaz VM. Electrophoresis 2006;27:3349–3358.
44. Bhattacharyya A, Klapperich CM. Anal Chem 2006;78:788–792.
45. Stachowiak TB, Rohr T, Hilder EF, Peterson DS, Yi MQ, Svec F, Frechet JMJ. Electrophoresis 2003;24:3689–3693.
46. Olsen KG, Ross DJ, Tarlov MJ. Anal Chem 2002;74:1436–1441.
47. Zangmeister RA, Tarlov MJ. Langmuir 2003;19:6901–6904.
48. Zangmeister RA, Tarlov MJ. Anal Chem 2004;76:3655–3659.
49. Paegel BM, Yeung SHI, Mathies RA. Anal Chem 2002;74:5092–5098.
50. Blazej RG, Kumaresan P, Mathies RA. Proc Nat Acad Sci USA 2006;103:7240–7245.
51. Dhopeshwarkar R, Li SA, Crooks RM. Lab Chip 2005;5:1148–1154.
52. Zguris JC, Itle LJ, Hayes D, Pishko MV. Biomed Microdevices 2005;7:117–125.
53. Heo J, Thomas KJ, Seong GH, Crooks RM. Anal Chem 2003;75:22–26.
54. Koh WG, Pishko MV. Anal Bioanal Chem 2006;385:1389–1397.
55. Arai F, Ng C, Maruyama H, Ichikawa A, El-Shimy H, Fukuda T. Lab Chip 2005;5: 1399–1403.
56. Moorthy J, Beebe DJ. Lab Chip 2003;3:62–66.
57. Pregibon DC, Toner M, Doyle PS. Langmuir 2006;22:5122–5128.
58. Braschler T, Johann R, Heule M, Metref L, Renaud P. Lab Chip 2005;5:553–559.
59. Toh YC, Ng S, Khong YM, Samper V, Yu H. Assay Drug Dev Technol 2005;3:169–176.
60. Kim MS, Yeon JH, Park JK. Biomed Microdevices 2007;9:25–34.
61. Hatch A, Kamholz AE, Hawkins KR, Munson MS, Schilling EA, Weigl BH, Yager P. Nat Biotechnol 2001;19:461–465.
62. Burdick JA, Khademhosseini A, Langer R. Langmuir 2004;20:5153–5156.
63. Jeon NL, Dertinger SKW, Chiu DT, Choi IS, Stroock AD, Whitesides GM. Langmuir 2000;16:8311–8316.
64. Dertinger SKW, Chiu DT, Jeon NL, Whitesides GM. Anal Chem 2001;73:1240–1246.
65. Zaari N, Rajagopalan P, Kim SK, Engler AJ, Wong JY. Adv Mater 2004;16:2133–2137.
66. Zhan W, Seong GH, Crooks RM. Anal Chem 2002;74:4647–4652.
67. Heo J, Crooks RM. Anal Chem 2005;77:6843–6851.
68. Koh WG, Pishko M. Sens Actuators B, Chemical 2005;106:335–342.
69. Lee NY, Jung YK, Park HG. Biochem Eng J 2006;29:103–108.
70. Mersal GAM, Bilitewski U. Electrophoresis 2005;26:2303–2312.
71. Peterson DS, Rohr T, Svec F, Frechet JMJ. Anal Chem 2002;74:4081–4088.
72. Yang YN, Li C, Lee KH, Craighead HG. Electrophoresis 2005;26:3622–3630.
73. Yang YN, Li C, Kameoka J, Lee KH, Craighead HG. Lab Chip 2005;5:869–876.

PART VI

HYBRID MICROFLUIDIC APPLICATIONS

22

COUPLING ELECTROCHEMISTRY TO MICROFLUIDICS

JONATHAN A. VICKERS AND CHARLES S. HENRY

22.1 INTRODUCTION

One of the most significant and important trends in analytical chemistry in the past 30 years has been the development of miniaturized instrumentation for the analysis of complex biological and environmental samples. The drive toward smaller instruments and detection systems is driven by the ability to reduce the sample size down to sizes as small as a single cell, while also reducing the consumption of often expensive reagents. The early focus in this field was on the development of micro- and ultramicroelectrodes to probe neurotransmission processes in cellular and tissue samples (1–8). Since this early work, miniaturization has become a dominant trend in many other fields of analytical chemistry ranging from mass spectrometry to optical sensors. Arguably, the field that has been affected the most, however, is the field of chemical separations. In the early 1990s, Widmer's group at Ciba Giegy presented the first microfabricated separation device that provided both high speed and high efficiency separations. This initial work has subsequently led to the development of a whole new field of chemical instrumentation referred to by many names, including lab-on-a-chip, miniaturized total analysis systems, and microfluidics (9–11). The microfluidics field holds significant promise for the development of truly integrated portable analysis systems capable of analyzing very complex samples at the point of measurement.

To achieve the goals of total integration and portability, all aspects of the microchip must be miniaturized, including detection and pumping. While this seems

Biological Applications of Microfluidics, Edited by Frank A. Gomez

a trivial aspect of development, it has been a significant stumbling block in the realization of the full potential of these methods. Many modern microfluidic systems use small microchips coupled to large pumps and optical detection systems for analysis, preventing their use at the point-of-measurement. Significant progress is being made in the field of microfluidic pumping and valving through the development of three-dimensional microchip structures (12,13). The development of portable detection systems has, however, lagged behind the fluid manipulation methods in large part because most detection systems used to date are optical. Electrochemistry provides a viable detection option that is already available in miniaturized formats. Electrochemistry generally is considered to have better sensitivity and selectivity than absorbance methods but lower sensitivity than laser-induced fluorescence methods. Electrochemistry also has the advantage of being able to detect many compounds without derivatization, simplifying the overall analysis as well as reducing the cost.

The goal of this chapter is to provide an overview of methods used to couple electrochemistry to microfluidic devices. The chapter is divided into four sections. The first section provides a brief overview of electrochemical methods that are useful for microfluidic systems. The second section provides an overview of the adaptation of electrochemical detection to microfluidic devices. The third section provides an overview of applications for microfluidic devices that use electrochemistry as the detection mode. The final section is a discussion of the current limitations and future opportunities in field.

22.2 ELECTROCHEMICAL METHODS OF ANALYSIS

A quick look at any basic book on electrochemistry will present the reader with a myriad of electrochemical methods that have been developed for different applications. For coupling to microfluidic systems, there are generally three methods that prove useful, amperometry, voltammetry, and conductivity. The two most common methods applied to microfluidic systems are amperometry and conductivity methods (14–19). In addition, potentiometry (measurement of potential at zero current) and voltammetry (measurement of current as a function of potential) have been demonstrated and hold significant promise, particularly for microfluidic sensors (20,21). For each detection mode, the benefits of small, portable instrumentation are still maintained.

22.2.1 Amperometry

Amperometry is the most common mode of electrochemical detection coupled with microfluidic devices (18,22,23). Amperometric signals are obtained by measuring current as a function of time at either a constant (DC amperometry) or a pulsed (pulsed amperometry, PAD) potential (Fig. 22.1). The choice of DC amperometry versus PAD is dependent on the specific application of the scientist. Amperometry works well for compounds that are easily oxidized or reduced such as catecholamine-

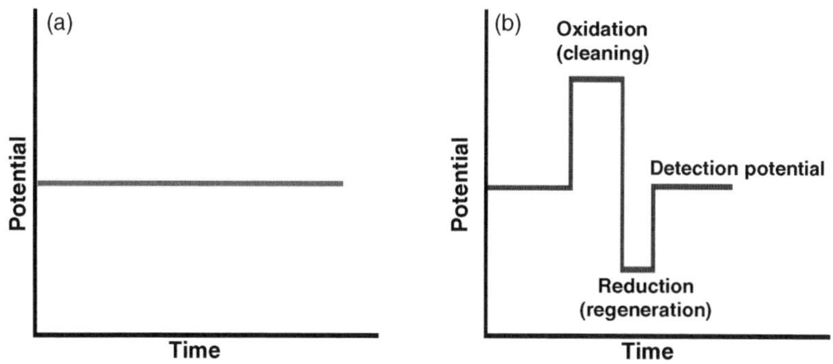

FIGURE 22.1 (**a**) Amperometric detection. A constant potential is applied to the working electrode for detection. (**b**) Pulsed amperometric detection (PAD). A potential waveform is applied to the working electrode for detection.

based neurotransmitters. Amperometry is also the most commonly used mode of electrochemical detection for microfluidic biosensors because the product of the enzymatic reactions are normally easily oxidized or reduced. PAD works better for compounds that either foul the electrode after oxidation or require the presence of a surface oxide to catalyze the reaction (24). Compounds that are particularly suited for measurement by PAD include alcohols, carbohydrates, thiols, thioethers, and amines. With many of these analytes, direct detection by any other method is challenging to impossible. In both DC amperometry and PAD, significant degrees of selectivity can be achieved by controlling the potential of the working electrode. This advantage has been used to selectively detect one group of analytes in the presence of many other electrochemically active analytes.

22.2.2 Conductivity

The second most common electrochemical detection method is conductivity (15). Conductivity detection is applied primarily as a detection mode coupled to microchip separation devices such as capillary electrophoresis (CE) and works by measuring the change in solution resistance between two electrodes as an analyte band passes through the separation channel. Two modes of conductivity detection have been reported, contactless and contact, with contactless being the more common of the two (14,15,19,25). Contact detection is the simpler of the two modes. In this mode, the electrodes are placed in direct contact with the solution and a low frequency (1–10 kHz) is applied between the electrodes. Very low detection limits have been achieved using contact conductivity. Galloway et al. reported detection limits below 10 nM for amino acids using a poly(methylmethacrylate) microchip (14). The major limitation to contact detection is the requirement that the detection electrodes be outside of any major separation voltage. To overcome this problem, contactless conductivity detection has been developed (15). In contactless detection, the detection electrodes are isolated from the separation channel by an insulator layer (19). Measurements of solution resistance are achieved by capacitively coupling the electrodes through the insulator to the solution. Capacitive coupling requires higher excitation frequencies (10^5-10^6 Hz) and improved shielding from outside current sources to achieve detection. The primary advantage, however, is the ability to place the electrodes anywhere along the separation channel, making device construction much less complicated. Contactless detection has not achieved as low of detection limits as contact, with most investigators reporting detection limits of 1–10 μM.

22.2.3 Potentiometry and Voltammetry

Potentiometry and voltammetry are less common than the other two modes of detection. Voltammetry, particularly in the form of cyclic voltammetry has been used to provide qualitative information about samples (26). The major limitation to the use of voltammetry in microfluidics is sensitivity. Detection limits for voltammetry are normally 100 μM or higher, making it less than ideal for the analysis of low abundance analytes in biological samples. The major benefit of voltammetry is the ability to determine peak purity for samples that are closely eluting. Potentiometry, on the contrary, has the potential to offer a highly selective and sensitive method for the detection of a wide variety of ionic compounds. In potentiometry, ions differentially migrate through a selective membrane, causing a potential difference to arise between two electrodes. Selectivity is gained based on the selectivity of the membrane. Potentiometric methods coupled to microfluidic devices have been slow to develop, however, because of the difficulty of coupling an ion-selective electrode to the microchip (27).

22.3 MICROFLUIDIC DEVICES

Electrochemistry has been coupled to a number of different types of microfluidic devices because of the inherent advantages of small size, analytical selectivity, and low detection limits. The most common microfluidic device used with electrochemistry is the electrophoresis microchip. The development and use of hydrodynamic-based flow devices is, however, gaining ground in the microfluidics world. This section contains a review of the different types of microfluidic devices coupled with electrochemistry is provided.

22.3.1 Electrophoresis Microchips

Far and away, the most common microfluidic devices coupled with electrochemistry are the electrophoresis microchips or microchip capillary electrophoresis (MCE). MCE as a separation technique has become quite popular for fast separations of complex mixtures. MCE can also analyze very small volumes (nL–pL) and has the potential for on-chip sample processing prior to separation for further improvement in the assay selectivity. MCE is beginning to gain ground in industrial applications as more standardized instrumentation and applications have begun to appear. The main consideration in coupling electrochemical detectors with MCE is the arrangement of the electrodes. Different designs are required for amperometric detection than conductivity detection. In both cases, however, isolation of the detection current from the separation current is a major concern because the separation current (μA) is normally several orders of magnitude greater than the detection signal (nA).

22.3.2 Amperometric Detectors

A wide variety of electrode configurations for amperometric detectors have been described in the literature by a growing group of scientists. Early work in this field, reported by Ewing's group, used electrode arrays situated at the end of an open separation channel to perform amperometric detection (28). Injection was achieved using a conventional fused silica capillary that rastered across the opening of the separation channel. This pioneering work was soon followed by the report of electrochemical detection using a more traditional T-shaped MCE device by Mathies' group (29). Subsequently, a large number of designs have been reported. These designs are summarized in Fig. 22.2 (18,29–31). The first consideration in designing the electrode system is to decide whether on-chip or off-chip electrodes are to be used. On-chip electrode systems are fabricated as part of the microchip device, normally using thin-film micromachining methods. Electrode materials are normally metallic with Au and Pt being the most common. On-chip electrode configurations have the advantages of being able to incorporate multiple detection electrodes (adding selectivity) and simple electrode alignment as a result of their direct inclusion in the microchip. The main problem with on-chip electrode systems is a lack of electrode stability under normal electrophoresis operating conditions. Off-chip electrode configurations use an electrode that is physically separate from the microchip. In this configuration a wide variety of electrodes including screen printed, disk, and carbon paste can be coupled to the microchip. These electrodes tend to be much more stable than microfabricated electrodes and can be easily chemically modified prior to analysis. The major limiting factors with off-chip electrodes are the inability to incorporate multiple detection electrodes on a single device and the need to manually align electrodes with the separation channel. A hybrid device that incorporates solid metal microwires in an on-chip configuration has been reported by the author's laboratory (16,32). In this system, the stability of off-chip electrodes as well as the ability to incorporate multiple electrodes is achieved in a configuration that self-aligns the working electrodes with the separation channel. Fig. 22.3 shows a schematic of this system as well as a photograph of a device containing multiple working electrodes.

The second consideration in the design of electrochemical detectors for MCE is whether or not to include a current decoupler in the system. As mentioned above, the major limiting factor for coupling electrochemistry to MCE is the need to isolate the detection current from the separation current. Early devices achieved this by placing the working electrode just outside the exit of the capillary into the waste reservoir, a configuration referred to as end column detection. In this way, the current was sufficiently reduced so that it does not affect the electrochemistry. The main problem with this approach is that it allows the eluting band to diffuse into solution before

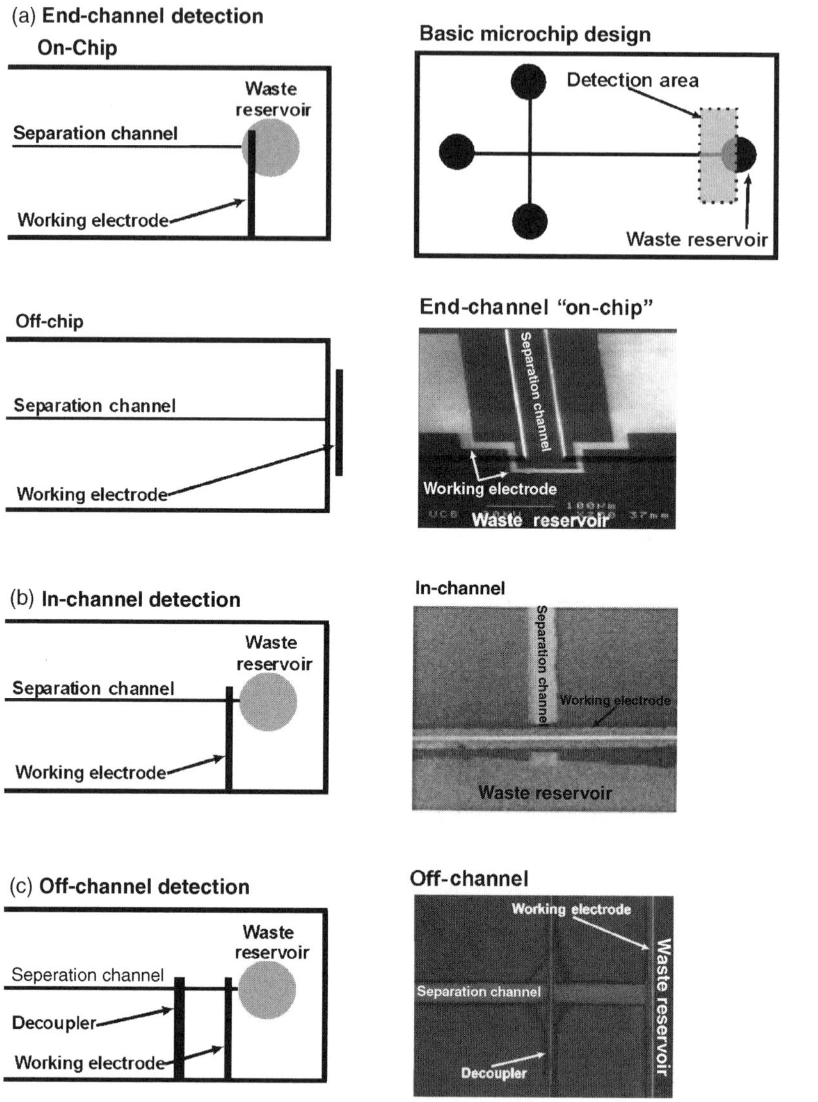

FIGURE 22.2 Four ways of aligning working electrodes on a microchip CE–EC system. (**a**) End-channel detection, both on-and off-chip. (**b**) In-channel detection. (**c**) Off-channel detection. Pictures are from the Mathies, Lunte, and Henry groups (top to bottom) (29–31).

detection resulting in detection limits that are in the low micromolar range. Since the early days of electrochemical detection with conventional CE, current decouplers have been used to improve the detection performance. A current decoupler is simply a physical device that permits the current to be grounded prior to but still in the

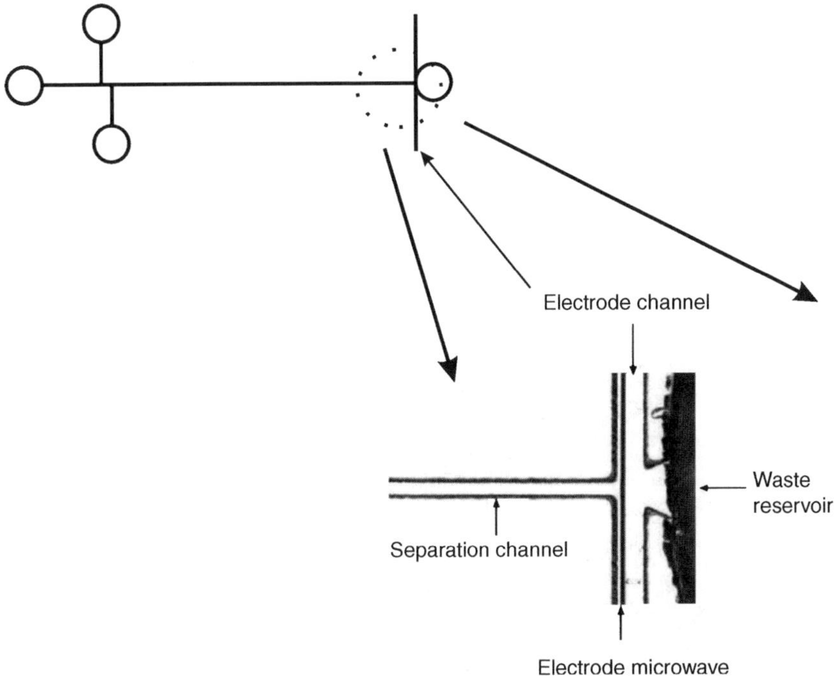

FIGURE 22.3 Microwire electrode for use with electrochemical detection on microchip CE device. *Top*: Schematic of the microchip showing placement of the electrode alignment channel. *Bottom*: Photograph showing electrode alignment in a completed microchip. Microchip design from the Henry group (16).

capillary. This configuration is often called off-channel detection. The result is a detection limit that is 2 to 3 orders of magnitude lower than can be achieved in end-column systems. Microchips enabled with decouplers have been generally slow to develop in large part as a result of the lack of a suitable way to create the grounding point prior to the detection electrode. To date, the most successful designs have incorporated a Pd electrode in the channel prior to the working electrode as shown in Fig. 22.4. Pd is unique in its ability to absorb and remove large volumes of H_2 gas (33). H_2 gas is generated at the cathode in any electrophoretic separation and, if not removed, can cause the channel to become blocked. Pt can also absorb H_2 but not as efficiently and thus lower separation potentials can be used with these systems. Lunte's group was the first to develop this type of a system in a MCE device. Subsequent reports by other groups have extended their early work to achieve detection limits in the range of 1–10 nM (31–34).

22.3.3 Conductivity

The type and number of electrode configurations used with conductivity detection is more limited than amperometric detectors. The main consideration for

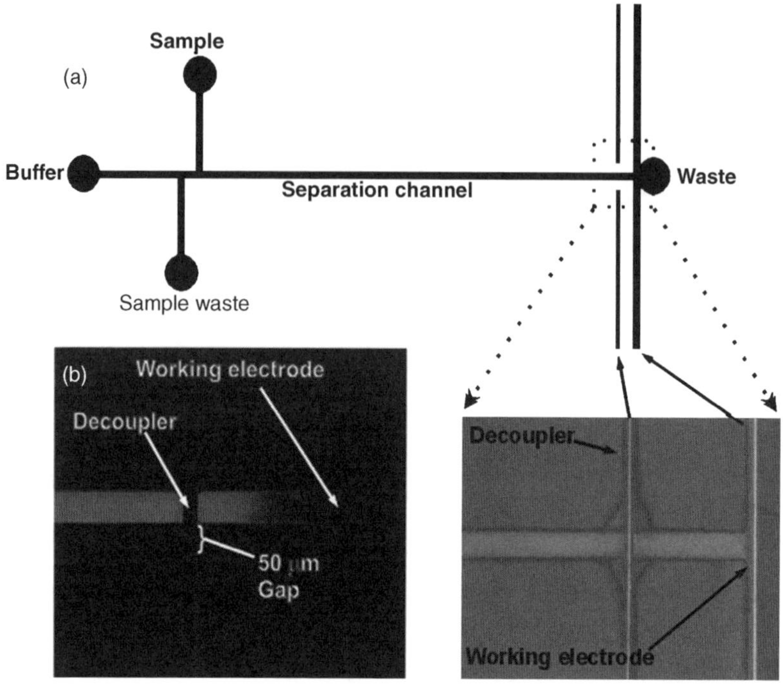

FIGURE 22.4 (**a**) Schematic of the microchip showing placement of the electrode alignment channels. The lower right is a photograph showing electrode alignment in a completed microchip. Left electrode is Pd microwire while the right electrode is the Pt working electrode. (**b**) Decoupled microchip. Fluorescent image of 1 mM FITC as it passes the decoupler. No sample leakage was observed around the Pd microwire (31).

conductivity detection is whether the detection system will be contact or contactless mode as described above. In both cases, the solution resistance between two electrodes is measured as a function of time. For contact conductivity, the electrodes can be placed directly opposite each other in a parallel configuration as in the work of Galloway et al. or one after another in a serial configuration (Fig. 22.5a and b) (14,35). Contactless electrodes are normally placed in a serial configuration as shown in Fig. 22.5c and can be made from any of a number of materials including copper, Au, and aluminum foil (25). One additional consideration for contactless detectors is the inclusion of a ground electrode between the two working electrodes. Some groups have suggested that the inclusion of a ground electrode makes the overall system more sensitive while other groups have suggested that it makes little difference.

22.3.4 Microfluidic Biosensors

In addition to MCE devices, electrochemistry has been coupled to microfluidic devices in the form of biosensors. Electrochemical biosensors have a long history as

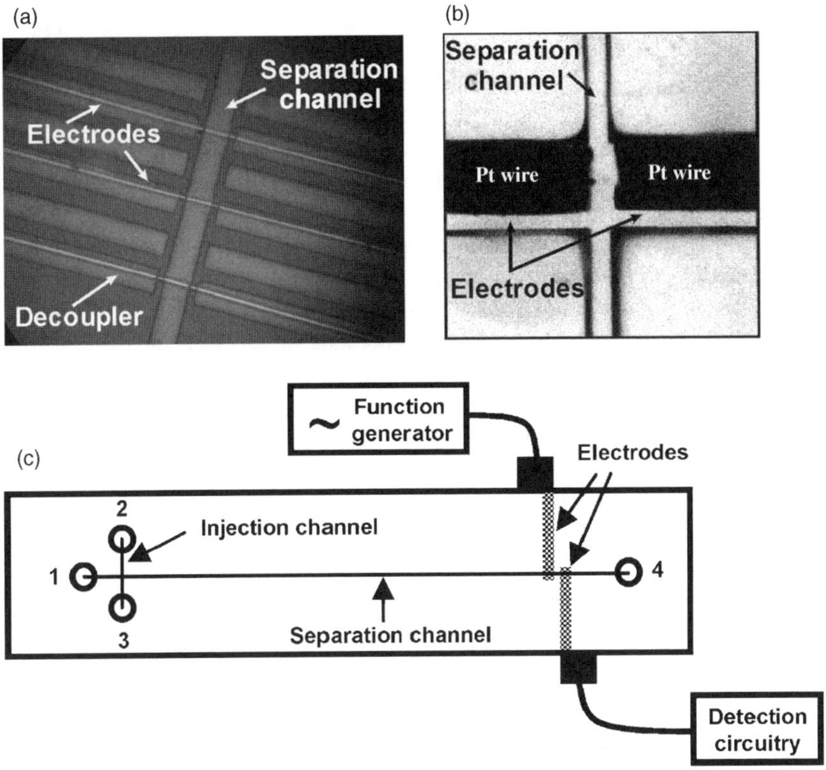

FIGURE 22.5 Electrode alignment for conductivity detection on a microchip. (**a**) Contact conductivity microchip with microwire electrodes in a serial conformation from the Henry group. (35) (**b**) Contact conductivity microchip with platinum electrodes in a parallel conformation from the Soper group (14). (**c**) Diagram of a contactless conductivity microchip in a serial configuration from the Hauser group (25).

simple, cheap analytical devices capable of detecting a wide range of species. The most common electrochemical biosensor is the wired enzyme sensor used in glucometers (36). The incorporation of electrochemical biosensors in microfluidic devices may seem to be pointless on first inspection because biosensors are generally used directly in the sample. On further inspection, there are real benefits to be gained by incorporating biosensors in microfluidic devices because of the ability to incorporate multiple biosensors with different selectivities (37). This can be achieved through either the incorporation of multiple electrodes or the incorporation of multiple catalytic (enzymatic) elements coupled with a single detector. Both approaches have their benefits and pitfalls. In the multiple biosensor approach, it is easier to block interferences from affecting the analytical signal but the fabrication is generally more complicated. The incorporation of multiple catalytic elements with a single electrode has the advantage of reducing the number of electrodes and therefore potentiostats that are needed for a single-sensing scheme. The main challenge with this approach is that it is more difficult to block the electrode surface from possible interfering compounds.

In addition to enzymatic biosensors such as those used for glucose, significant progress has been made in the development of DNA detection schemes using electrochemical biosensors (37–39). DNA detection is important for a wide range of biological and biosecurity applications ranging from the detection of pathogenic microorganisms in drinking water to detection of known single nucleotide polymorphisms that are known to be associated with cancer (40). DNA detection biosensors work on one of two formats, intercalating electrophores or enzymatic sensors. With intercalating electrophores, an electrode is first modified with a complimentary strand of DNA for the desired target. After exposure to and hybridization with DNA from a sample matrix, the electrochemically active molecule binds to resulting double stranded DNA. Upon application of a suitable voltammetric or amperometric method, it is possible to quantify the amount of DNA present in the sample. Selectivity is gained in this mode by the DNA hybridization. For enzymatically labeled DNA biosensors, sample DNA is first captured at a chemically modified electrode surface. After this initial hybridization event, a second strand of DNA labeled with an enzyme such as alkaline phosphatase is hybridized to the target DNA and a substrate is added. This approach provides an easy signal amplification strategy but requires multiple sample processing steps and the additional of a substrate.

22.4 APPLICATIONS

A wide variety of applications have been reported for electrochemical detectors coupled to microfluidic devices. These applications can largely be broken down into the areas of biological and environmental analysis. The following section is a review of some of the general trends in this field but is not ment to be completely inclusive. For more comprehensive coverage of all applications, readers should refer to one of the many excellent reviews published in *Electrophoresis*, *Electroanalysis*, and *Analytical Chemistry* (28,37,39–42).

22.4.1 Biological Applications

The majority of applications coupling electrochemistry with microfluidic devices fall in the category of biological. Microfluidic devices offer many benefits for the analysis of biological samples due to the small volume of sample and solution, the ability to precisely control flow volumes and rates at the nL and pL level, and the ability to integrate both analytical function as well as sample preparation and cell culturing to name a few. Again, the types of applications can be divided by the type of microfluidic device that is being used, electrophoresis or biosensor.

22.4.1.1 Electrophoresis Microchips Electrophoretic microchips coupled with electrochemical detection have been used for a wide range of analysis functions. One of the more common application areas for amperometric detection has been and continues to be the analysis of neurotransmitters and particularly catecholamines. Catecholamines such as dopamine and epinephrine are important signaling molecules in the neurotransmission pathways of the brain and other tissues. Abnormal levels have been widely associated with many different neurodegenerative diseases including Parkinson's and Alzheimer's (43). Catecholamines are also easily oxidized as a result of their basic quinine ring structure making them attractive targets to begin electrochemical detection studies. Finally, catecholamines are relatively well-behaved analytes, causing minimal electrode fouling and possessing excellent reproducibility. MCE coupled with electrochemical detection has been widely used to measure catecholamines in both control samples and biological systems.

Beyond catecholamines, a number of investigators have reported the ability to separate and detect many different classes of small biological molecules. Wang's group reported the ability to separate and detect labeled amino acids (44). In this example, the native amino acids could not be detected with DC amperometry and were thus labeled with naphthalene-2,3-dicarboxyaldehyde to generate easily oxidized compounds. A number of other groups, including Chen's group, Hauser's group, and Baldwin's group have shown the ability to detect amino acids and carbohydrates using Cu working electrodes instead of Au, Pt, or C (45–47). Cu electrodes react with primary amines and diol-containing molecules in strong basic conditions through an oxide-mediated process to give analytical signals for these compounds. The ability to perform direct detection of amino acids and carbohydrates is powerful because it elements the need for analyte derivatization and the associated loss of selectivity. Thiols are another class of compounds with wide significance for bioanalytical chemistry. Lunte's and Martin's groups have both published novel methods for the detection of this class of compounds using DC amperometry (48,49). In Lunte's work, Au/Hg amalgam electrodes were generated and used for detection of homocysteine and other thiols (49). Au/Hg amalgam electrodes are known to have a high specificity for thiol groups and work extremely well at low pH. Since not all thiols can be readily separated at low pH, Martin developed a chemically modified electrode system using cobalt pthalocyanine in C paste to achieve detection of thiols (48). In both cases, good sensitivity and specificity was achieved.

The author's group has largely focused on the development of PAD methods for detection of a wide range of analytes that DC amperometry cannot readily achieve. As discussed above, PAD can be used for the detection of alcohols, carbohydrates, amines, thiols, and thioethers. In the first report of coupling PAD with MCE, the separation and detection of a series of carbohydrates, amino acids, and thioethers-based antibiotics was demonstrated (32). Subsequently optimization of the detection process, including different pulse waveforms was presented. Sensitive and selective detection of a wide range of carbohydrates and amines was demonstrated using MCE–PAD. As an additional application of this work, creatinine was measured in urine samples as a simple method to estimate renal function.

Conductivity detection has also proven to be quite useful for the detection of small biological molecules when coupled with MCE. Conductivity does not have the specificity of amperometric methods and therefore more sample preparation is required. Despite this, a number of groups have shown the promise of the technique through the detection of a range of biological analytes including inorganic cations (Li^+, Na^+, and K^+), peptides, organic acids, and bioactive amines to name a few. In each case, conductivity detection offered the advantages of label-free detection and universality. An example separation of Li^+ from serum from van den Berg's group is shown in Fig. 22.6 (50). Li is an important phychoactive drug that requires close monitoring for effective treatment. A combination of moving boundary electrophoresis and capillary electrophoresis with conductivity detection was used to detect Li^+ from diluted blood. A detection limit of 0.1 mM in a 140 mM Na^+ background was achieved, demonstrating the potential for direct application of MCE to blood analysis.

In addition to small molecules, proteins have also been detected by MCE using both amperometric and conductivity detection. For amperometric detection, Wang's group used immunoaffinity labels coupled with enzymes to generate a measurable signal from an otherwise undetectable protein to measure insulin (51). This assay was coupled with additional enzymes to simultaneously detect creatine and creatinine. Conductivity detection has also been used with MCE to detect macromolecules. Hauser's group reported the separation and detection of IgM using an electrophoretic immunoassay (52). While IgM is not a high priority clinical analyte, the ability to separate and detect proteins and immunocomplexes offers great

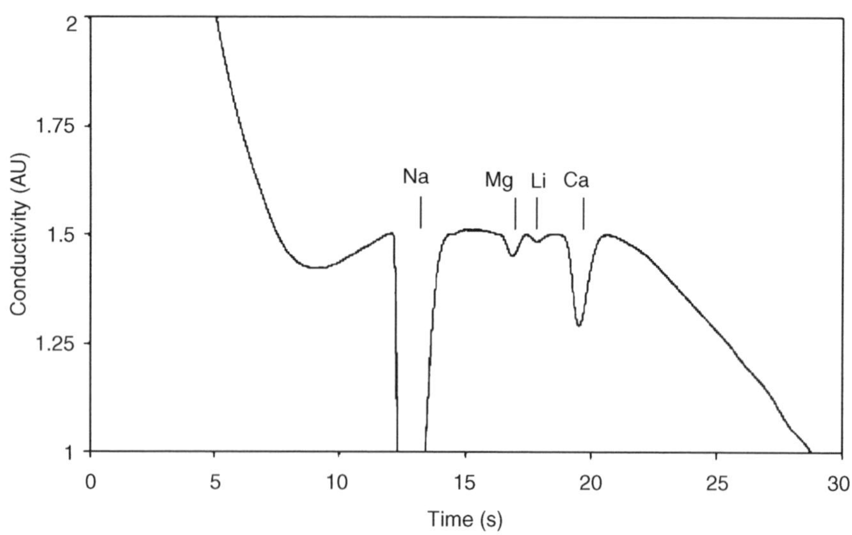

FIGURE 22.6 Electropherogram of a serum sample with 144 mmol/L sodium, 0.74 mmol/L lithium, and 2.59 mmol/L calcium. Work done by van den Berg group using conductivity detection (50).

promise for extending this technique to the analysis of analytes of more clinical and biological significance.

22.4.1.2 Microfluidic Biosensors A number of applications have been reported for microfluidic biosensors as well. The biosensors can largely be broken down into the sensors designed for DNA and sensors based on enzymatic reactions. In both cases, a wide range of biological applications has been reported for these types analytical measurement devices and readers are encouraged to refer to one of the many excellent reviews published in this field additional information on applications. DNA biosensors have largely focused on the development of multianalyte detectors coupled with microfluidics for sample pretreatment, whether it is for clinical or biological applications. An example of this type of sensor was published by the Crooks' group (53). In this application, complimentary DNA strands were immobilized on the walls of microfluidic channel. A second complimentary strand labeled with alkaline phosphatase was allowed to hybridize with the target DNA. The enzyme substrate was subsequently added to the channel and an electrochemically active product, *p*-aminophenol (PAP) was generated. The concentration of PAP was directly related to the amount of target DNA present in the original sample. Detection of 1 nM DNA was reported for a total solution volume of 20 nL corresponding to the detection of 20 mole of DNA.

Enzymatic biosensors have also been widely reported for microfluidic devices. Most of these devices use enzyme systems for analytes that result in the production of an easily measured product such as H_2O_2. An example of this type of enzyme includes the large family of oxidases such as glucose oxidase. H_2O_2 can be easily detected at a number of different electrode types and thus is quite convenient as a by-product of enzymatic reactions for use in electrochemical biosensors. An example of this type of sensor was recently published by Moser et al. (54). In this microfluidic system, sensors for glucose, lactate, glutamate, and glutamine were generated inside of a microfluidic channel. The resulting sensor had a total volume between 2.1 and 6 μL depending on the configuration and was used to measure the target analytes in a series of sample matrixes ranging from blood to fermentation broth. This biosensor highlight one of the main advantages of coupling microfluidics with biosensors, the ability to multiplex detectors with integrated fluid handling.

22.4.2 Environmental Applications

The number and range of environmental applications using microfluidics is not as extensive as the biological applications, yet this area has the potential to be greatly impacted by the development of specialized microfluidic devices. Again, the majority of applications published to date have used MCE as the microfluidic device. For amperometric detection, the primary focus has been on phenolic species as these are priority pollutants for the Environmental Protection Agency. An example application was recently published by Garcia's group (55). In their work, they demonstrated the ability to separate and detect three different phenolic compounds with detection limits of approximately 1 μM. In another vein of environmental research, the author's group recently published the analysis of ambient aerosols for levoglucosan (56). Levoglucosan is produced during the burning of wood and is used as a marker for tracking the impact of forest fires regional atmosphere conditions. MCE coupled with PAD was used to separate levoglucosan from glucose and galactosan in the analysis of laboratory generated aerosols. The primary advantages of this methodology are the small sample size and short analysis times relative to the chromatographic methods normally used.

22.5 CONCLUSIONS AND FUTURE DIRECTIONS

Microfluidic devices hold the potential to significantly change the way biological measurements are done. One area of continued development that is important for point-of-measurement applications is miniaturization of the detection. While optical detection currently dominates in microfluidic devices, electrochemistry can provide an important alternative that is easier to miniaturize and less expensive than many optical methods. Electrochemistry has the advantage of providing a range of selectivities from nearly universal for conductivity to highly selective with amperometry. Applications ranging from fundamental biology to clinical diagnostics to environmental monitoring have been achieved using microfluidics coupled with electrochemistry.

While significant development has been shown in this field, there are many more challenges and opportunities for future development. For example, there are currently no examples of microfluidic biosensors for environmental analysis. This field could greatly benefit from better and cheaper diagnostic tools. Another area of challenge is to increase the number of detection elements for both microfluidic and MCE devices. Increasing the number of detection elements will increase the flexibility of the analysis device for more applications of even more complex samples. Finally, one area that could still use significant development is the incorporation of ion selective electrodes with microfluidic devices of all kinds. Ion selective electrodes offer excellent sensitivity and selectivity in their present form and would fill an important niche in the point-of-measurement field.

REFERENCES

1. Bond AM. Analyst 1994;119:R1–R11
2. Cheng HY, Schenk J, Huff R, Adams RN. J. Electroanal Chem 1979;100:23–31.
3. Dayton MA, Geier GE, Wightman RM. Life Sci 1979;24:917–924.
4. Feng JX, Brazell M, Renner K, Kasser R, Adams RN. Anal Chem 1987;59: 1863–1867.
5. Gerhardt GA, Oke AF, Nagy G, Moghaddam B, Adams RN. Brain Res 1984;290: 390–395.
6. Leszczyszyn DJ, Jankowski JA, Viveros OH, Diliberto EJ, Near JA, Wightman RM. J Neurochem 1991;56:1855–1863.

7. Rice ME, Adams RN. Abstracts of Papers of the American Chemical Society 1982;184:126.
8. Schenk JO, Miller E, Rice ME, Adams RN. Brain Res 1983;277:1–8.
9. Manz A, Graber N, Widmer HM. Sens Actuators B: Chemical 1990; B1244–248
10. Manz A, Harrison DJ, Verpoorte EMJ, Fettinger JC, Luedi H, Widmer HM. Chimia 1991;45:103–105
11. Manz A, Harrison DJ, Verpoorte EMJ, Fettinger JC, Paulus A, Luedi H, Widmer HM. J Chromatogr 1992;593:253–263.
12. Eddington DT, Beebe DJ. J Microelectromech Syst 2004;13:586–593.
13. Karlinsey JM, Monahan J, Marchiarullo DJ, Ferrance JP, Landers JP. Anal Chem 2005;77:3637–3643.
14. Galloway M, Stryjewski W, Henry A, Ford SM, Llopis S, McCarley RL, Soper SA. Anal Chem 2002;74:2407–2415
15. Guijt RM, Evenhuis CJ, Macka M, Haddad PR. Electrophoresis 2004;25:4032–4089.
16. Liu Y, Vickers JA, Henry CS. Anal Chem 2004;76:1513–1517
17. Liu Y, Wipf DO, Henry CS. Analyst 2001;126:1248–1251.
18. Vandaveer WRIV, Pasas-Farmer SA, Fischer DJ, Frankenfeld CN, Lunte SM. Electrophoresis 2004;25:3528–3549.
19. Zemann AJ. Electrophoresis 2003;24:2125–2137.
20. Johnson RD, Bachas LG. Anal Bioanal Chem 2003;376:328–341.
21. Yunus K, Fisher AC. Electroanalysis 2003;15:1782–1786.
22. Lacher NA, Garrison KE, Martin RS, Lunte SM. Electrophoresis 2001;22:2526–2536
23. Vandaveer WRIV, Pasas SA, Martin RS, Lunte SM. Electrophoresis 2002;23: 3667–3677.
24. LaCourse WR. Pulsed Electrochemical Detection in High Performance Liquid Chromatography. New York: Wiley;1997.
25. Abad-Villar EM, Kuban P, Hauser PC. Electrophoresis 2005;26:3609–3614.
26. Wang J, Polsky R, Tian B, Chatrathi MP. Anal Chem 2000;72:5285–5289
27. Ibanez-Garcia N, Mercader MB, da Rocha ZM, Seabra CA, Gongora-Rubio MR. Chamarro, J. A. Anal Chem 2006;78:2985–2992.
28. Gavin PF, Ewing AG. Anal Chem 1997;69:3838–3845.
29. Woolley AT, Lao K, Glazer AN, Mathies RA. Anal Chem 1998;70:684–688.
30. Martin RS, Ratzlaff KL, Huynh BH, Lunte SM, Anal Chem 2002;74:1136–1143.
31. Vickers JA, Henry CS. Electrophoresis 2005;26:4641–4647.
32. Garcia CD, Henry CS. Anal Chem 2003;75:4778–4783.
33. Kok WT, Sahin Y. Anal Chem 1993;65:2497.
34. Lacher NA, Lunte SM, Martin RS. Anal Chem 2004;76:2482–2491.
35. Liu Y, MacDonald DA, Yu XY, Hering SV, Collett JL, Henry CS. Analyst 2006;131:1226–1231.
36. Heller A. Acc Chem Res 1990;23:128–134.
37. Bange A, Halsall HB, Heineman WR. Biosens & Bioelectron 2005;20:2488–2503.
38. Goral VN, Zaytseva NV, Baeumner AJ. Lab Chip 2006;6:414–421.
39. Wang J. Biosens Bioelectron 2006;21:1887–1892.

40. Soper SA, Brown K, Ellington A, Frazier B, Garcia-Manero G, Gau V, Gutman SI, Hayes DF, Korte B, Landers JL, Larson D, Ligler F, Majumdar A, Mascini M, Nolte D, Rosenzweig Z, Wang J, Wilson D. Biosens Bioelectron 2006;21:1932–1942.
41. Dittrich PS, Tachikawa K, Manz A. Anal. Chem. 2006;78:3887–3908.
42. Wang J. Talanta 2002;56:223–231.
43. DeKosky ST, Marek K. Science 2003;302:830–834.
44. Wang J, Chen G, Pumera M. Electroanalysis 2003;15:862–865.
45. Dou YH, Bao N, Xu JJ, Chen HY. Electrophoresis 2002;23:3558–3566.
46. Schwarz MA, Galliker B, Fluri K, Kappes T, Hauser PC. Analyst 2001;126:147–151
47. Voegel PD, Baldwin RP. Electroanalysis 1997;9:1145–1151.
48. Kuhnline CD, Gangel MG, Hulvey MK, Martin RS. Analyst 2006;131:202–207.
49. O'Shea TJ, Lunte SM. Anal Chem 1993;65:247–250.
50. Vrouwe EX, Luttge R, Vermes I, van den Berg A. Clin Chem 2007;53:117–123.
51. Wang J, Zhang X. Anal Chem 2001;73:844–847.
52. Abad-Villar EM, Tanyanyiwa J, Fernandez-Abedul MT, Costa-Garcia A, Hauser PC. Anal. Chem. 2004;76:1282–1288.
53. Seong GH, Zhan W, Crooks RM. Anal Chem 2002;74:3372–3377.
54. Moser I, Jobst G, Urban GA. Biosens Bioelectron 2002;17:297–302.
55. Ding YS, Mora MF, Garcia CD. Anal Chim Acta 2006;561:126–132.
56. Garcia CD, Engling G, Herckes P, Collett Jr, JL, Henry CS. Environ Sci Technol 200539:618–623.

23

MANIPULATING MASS-LIMITED SAMPLES USING HYBRID MICROFLUIDIC/NANOFLUIDIC NETWORKS

Jamie M. Iannacone, Bo Young Kim, Travis L. King, Paul W. Bohn and Jonathan V. Sweedler

23.1 INTRODUCTION

The considerable interest in developing micro total analysis systems (μ-TAS) (1) is fueled by the promise of miniaturized analytical platforms that can perform comprehensive laboratory analyses in a single compact device. In addition to the added benefits of decreased analysis times and increased sensitivity compared to traditional macroscale schemes, the required sample volumes are often significantly reduced and the potential for applications involving mass-limited samples may be realized. The use of such microfluidic-based systems can be expanded further by incorporating nanofluidic components to create hybrid networks with increased spatial and temporal control of fluid transport, in particular, extending the utility of microfluidic networks to interconnected three-dimensional (3D) architectures.

Analysis of mass-limited samples is often challenging due to the careful handling required to preserve analyte integrity. Many samples, such as pheromones or neurotransmitters, are naturally found in small quantities and are, therefore, limited in their availability. These samples must be handled in a way that prevents their loss during procedures that often involve several steps. Other types of samples are limited by mass due to some attribute of the analyte molecule. Toxins, for example, must be safely handled in small quantities to minimize potential harm to the experimentalist.

Biological Applications of Microfluidics, Edited by Frank A. Gomez

In addition to their detection, many toxins must also be classified and distinguished from other similarly structured molecules. μ-TAS systems, ideal for the characterization of such availability- and attribute-limited low mass samples, offer the potential to perform sequential analytical steps (i.e., separation, isolation, and detection) and precise fluidic manipulations in confined nanometer- to micrometer-sized structures.

23.2 NANOFLUIDICS

While microfluidic technology enables many otherwise difficult measurements in biological systems, the incorporation of nanofluidic elements, characterized by structures with length scales in the range of $10 < d < 100$ nm, into such architectures confers additional functionality. For example, hybrid microfluidic/nanofluidic networks have enabled such important manipulations as preconcentration schemes (2), as well as the gated injection and collection of attomole quantities of analyte (3). The unique features provided by nanofluidics include new transport phenomena that are not observed in larger scale systems, significantly increased surface-area-to-volume ratios, and volumes small enough that diffusion can be exploited for molecular transport and reactive mixing. Furthermore, the investigation of fundamental properties governing flow through these nanoscale components can lead to a greater understanding of phenomena that occur at this length scale, including single-molecule dynamics and chemical reaction mechanisms.

As early as 1965, theoretical investigations by Rice and Whitehead (4) focused on electrolyte flow in nanoscale cylindrical capillaries of infinite length as a function of pressure and/or potential. In the limit of small zeta potential, these authors expressed the potential φ at a radial distance r given by the Poisson–Boltzmann equation

$$\frac{1}{r}\frac{d}{dr}\left(r\frac{d\varphi}{dr}\right) = \kappa^2\varphi \tag{23.1}$$

where κ is the inverse Debye length, or double-layer thickness of an ionic solution given by,

$$\kappa = \sqrt{\frac{8\pi ne^2}{\varepsilon kT}} \tag{23.2}$$

and defined as a function of the number ion density n, dielectric constant ε, the Boltzmann constant k, and temperature T. Equation 23.1 can be modified to describe the velocity of fluid under a combined pressure and electrical-driven flow,

$$\frac{1}{r}\frac{d}{dr}\left(r\frac{dv_z}{dr}\right) = \frac{1}{\eta}\frac{dp}{dz} - \frac{F_z}{\eta} \tag{23.3}$$

where dp/dz is the pressure gradient, F_z is the body force per unit volume, and η is the coefficient of viscosity. At conditions where the radial distance r is comparable to the capillary radius a, and the necessary boundary conditions are applied, the net charge density of the solution can be expressed in terms of the Poisson equation and used to derive the expression for electroosmotic velocity in the absence of applied pressure

$$v_z(r) = \frac{\varepsilon\varphi_0}{4\pi\eta} E_z \left[1 - \frac{I_0(\kappa r)}{I_0(\kappa a)}\right] \tag{23.4}$$

where the E_z is the axially applied electric field and I_0 is the zero-order modified Bessel function of the first kind. From Equation 23.4, we learn that electrokinetic flow in a narrow, cylindrical capillary is dependent on the κa product (4). More specifically, when $\kappa a >> 1$, as observed with microfluidic channels, traditional electrokinetic profiles are established. Nanoscale capillaries, on the contrary, display an inner diameter similar to the Debye length, κ^{-1}, and two different flow regimes may be realized. As illustrated in Fig. 23.1a, a densely packed electrical double layer is observed in cases when $\kappa a >> 1$, where a relatively small Debye length is obtained with high ionic strength solutions (4–6). In this classical example, ion migration dominates molecular transport. However, a more diffuse electrical double layer is generated in nanofluidic elements filled with low ionic strength solutions, where $\kappa a \leq 1$, as illustrated in Fig. 23.1b. In this case, electroosmosis governs fluid

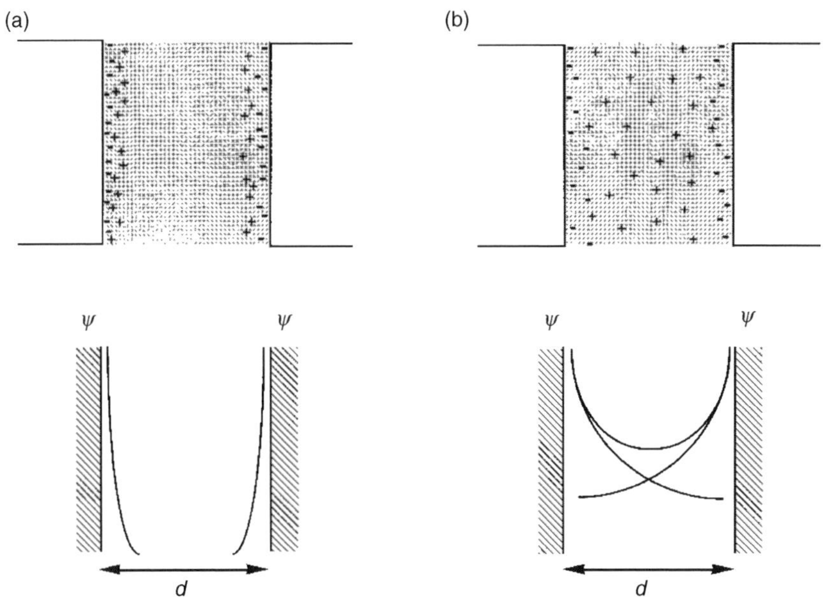

FIGURE 23.1 Schematic diagram representing the electrical double layer structures and potential profiles within nanopores at the extreme conditions where (**a**) $\kappa a \gg 1$ and (**b**) $\kappa a \leq 1$. Reproduced with permission from Reference 5. Copyright © 1998 American Chemical Society.

flow (4–6). When such a double-layer overlap exists, theory predicts that pressure-induced streaming potentials will decrease and pore conductivity will increase (7). Also, nanochannels with charged surfaces become unipolar solutions of counterions at bulk solution concentrations that neutralize the surface charge, essentially excluding co-ions from the channel (7–9). The influence of surface charge density on ionic transport is a function of the enhanced surface area-to-volume ratio inherent to nanoscale structures. Much research has focused on supporting theory, as well as the experimental verification of such surface-related phenomena (6,8–12). For example, the conductivity inside nanopores was theoretically shown to be dominated by the surface charge density and, therefore, is enhanced relative to that observed in the micrometer-scale (9,12). Ku and Stroeve (10) experimentally demonstrated the reduced transport of protonated and deprotonated proteins through surface-charged nanopores as a function of the electrostatic interactions that occur at low ionic strength. More recently, Lopez and Petsev (11) developed theoretical models for calculating the electrostatic potential, as well as the electroosmotic fluid flow profiles of symmetric electrolytes in capillaries with charged walls.

23.3 HYBRID MICROFLUIDIC/NANOFLUIDIC SYSTEMS

The unique characteristics of fluid flow in nanoscale structures can be exploited to expand the capabilities of traditional μ-TAS. Integrated systems with both microfluidic and nanofluidic elements fall into two categories, planar devices or 3D fluidic architectures. In planar devices, microfluidic channels are connected via nanochannels, or gradually narrowed to nanoscale dimensions. To access the third dimension, membranes that contain nanometer diameter pores can be incorporated between vertically separated microfluidic layers. There is significant interest in each system type, as both nanofluidic phenomena and the inherent capabilities available at the nanoscale are explored.

The distinction between microfluidic and nanofluidic flow regimes has been demonstrated in numerous experiments. Planar hybrid devices fabricated in Si by interferometric lithography (13) have been shown to separate dye molecules over short distances (<1 mm) of 35–200 nm diameter nanochannels due to different electroosmotic flow profiles that result as a function of analyte and surface charge (14). More specifically, the negatively charged nanochannel walls contribute to a greater electroosmotic velocity of similarly charged dye molecules with their localization toward the channel center, while neutral dyes adsorb onto the channel walls and exhibit slower flow profiles. In micrometer-scale channels, however, the opposite is true because the negatively charged dyes are slowed by electrophoretic drag. Theoretical models also support these contrasting electroosmotic flow behaviors (14). Han and Craighead (15–17) report the use of nanochannels as entropic traps for complex separations. Part of an alternating pattern of microfluidic and nanofluidic regions, nanoscale channels with depths smaller than the radius of gyration, R_o, of analyte molecules were fabricated in Si by photolithography to serve as molecular sieves (17). Because large, coiled analytes such as DNA

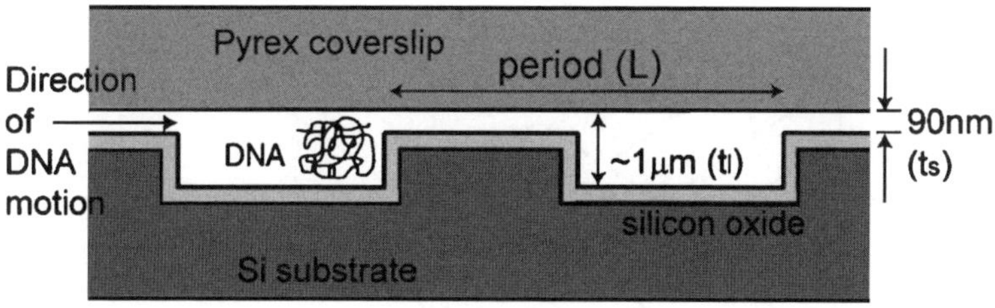

FIGURE 23.2 Schematic diagram of the entropic trapping device. The microfluidic regions are as deep as 1 μm, while the nanofluidic regions are 90 nm, which are much smaller than the R_0 of the double-stranded DNA molecules that are trapped in the microfluidic regions and entropically hindered from entering the nanochannels. Reprinted figure with permission from Reference 17. Copyright © 1999 by the American Physical Society.

are electrophoretically driven through the planar device depicted in Fig. 23.2, they undergo conformational changes that cost entropic free energy. As a result, molecules are temporarily trapped at the entrance of the nanofluidic regions until significant surface contact is made, followed by their introduction into the high field channels, and finally, their escape from the trap. Separations are realized because analytes of differing physical properties spend differing amounts of time in the "trapped" configuration prior to entering the nanofluidic manifold. The longer and/or larger the R_o of the analyte molecules, the more surface contact is made with the nanochannel boundaries, the more rapid the escape times, and the higher the overall mobility (15).

Cao and coworkers (18) have also investigated the entropic barriers that naturally exist at the boundary of nanofluidic environments. Using diffraction gradient lithography to create continuous spatial gradient structures, entropic barriers were reduced with a gradual narrowing of volume cross-sections from the microfluidic to nanofluidic regime. These integrated systems show potential for uniformly stretching DNA, a phenomenon that occurs in structures smaller than the persistence length of double-stranded DNA, about 50–70 nm (18,19). Hybrid systems containing nanochannels were also used to demonstrate the ion-enrichment and ion-depletion effects that occur at the ends of channels that exhibit double–layer overlap (20). Han et al. (2) utilized the same ion-depletion phenomenon to trap molecules and concentrate samples into pL–nL volumes, an application described in further detail below.

The second class of hybrid microfluidic/nanofluidic networks features nanoporous membranes to create layered, or 3D, architectures. Microchannels are most often fabricated using rapid prototyping protocols for poly(dimethylsiloxane) (PDMS) (21) and then positioned perpendicularly on each side of a commercially available polymeric nanocapillary array membrane (NCAM) (22–27). The result is a three-tiered, stacked device with an array of nanopore interconnects that join two vertically separated microfluidic environments, as depicted in Fig. 23.3.

Fluid transport is via pressure-driven or electrokinetic flow with improved spatial and temporal control of fluids compared to the planar, or two-dimensional networks.

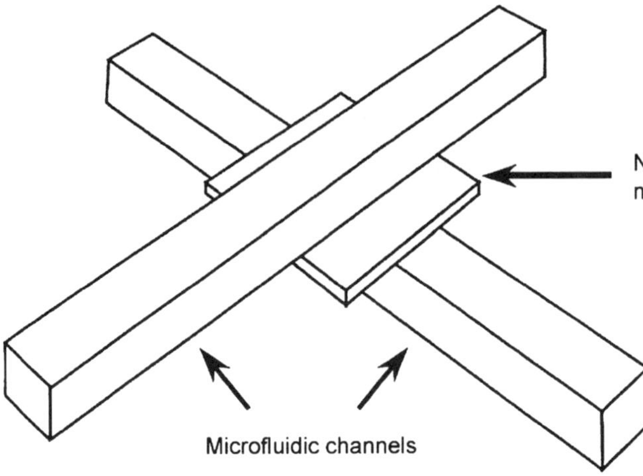

FIGURE 23.3 Schematic diagram of the molecular gate assembly where crossed microfluidic channels are separated by a nanochannel-containing membrane. Reprinted from Reference 25. Copyright © 2003, with permission from Elsevier.

The remainder of this chapter focuses on these 3D architectures because they show great promise for manipulating mass-limited samples. Our laboratories, for example, have employed these hybrid devices, equipped with polycarbonate NCAMs, to carry out analytical operations such as gated injections, electrophoretic separations with the subsequent collection of select bands, reactive mixing at rapid time scales, and sample treatment prior to online mass spectrometric detection (3,22,24,28,29). Lee and Whitesides (23,27) have reported the notable use of similar constructs for chiral separations and combinatorial microchip reactions, respectively. Several promising applications of the 3D hybrid architectures for mass-limited samples exist with innovative fabrication and/or modification schemes that are currently under investigation. One recent variation, for example, reports the use of reactive ion-etched poly(methylmethacrylate) (PMMA) microchannels to form multilayered networks for fluid manipulation across several spatial planes (30). Other future advances include the incorporation of functionalized NCAMs, an effort pioneered by Martin et al. (31–33) and adapted by many others. These membranes present selective transport properties based on the physicochemical interactions between the functionalized nanopores and analyte molecules, as well as molecular recognition when modified with molecules (e.g., antibodies) that bond only to specific moieties.

23.3.1 Digital Fluidics

The integration of nanofluidic membranes within microfluidic devices opens up new possibilities for fluid control and manipulation. Clearly, the interplay between microfluidic and nanofluidic flow phenomena determines the ultimate transport behavior in hybrid systems. For example, Kuo et al. (25) carried out experiments in devices composed of a poly(vinylpyrrolidone) (PVP)-coated polycarbonate nuclear

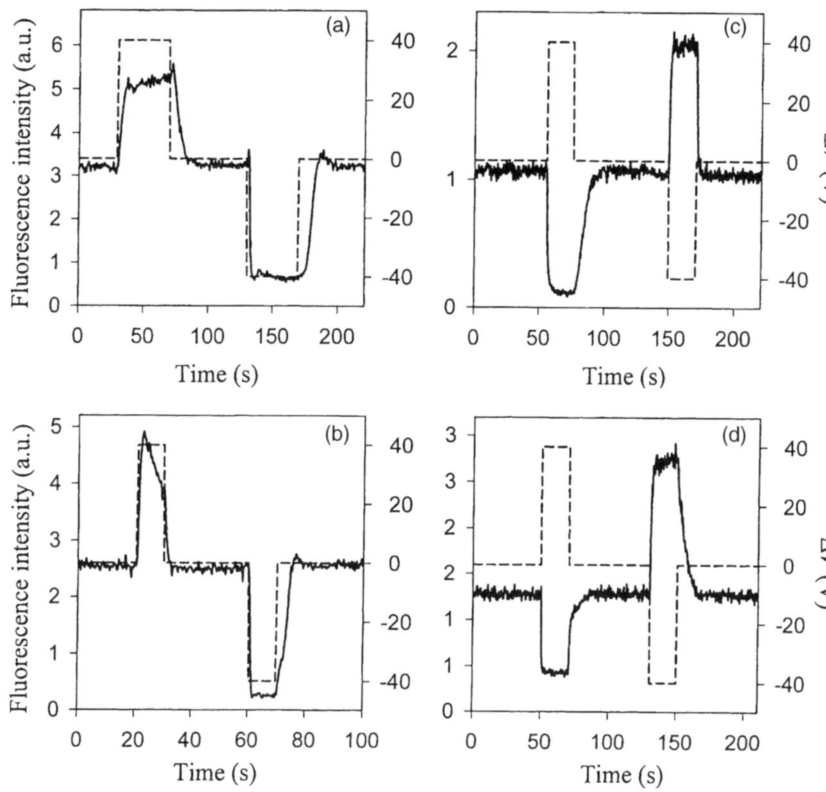

FIGURE 23.4 Fluorescence intensity (left ordinate) and applied bias (right ordinate) versus time showing transport of 170 nM fluorescein in 5 mM phosphate buffer (pH 8) across PCTE membranes with various pore diameters: (**a**) 15 nm, (**b**) 30 nm, (**c**) 100 nm, (**d**) 200 nm. Reprinted from Reference 25. Copyright © 2003, with permission from Elsevier.

track etched (PCTE) membrane embedded between two vertically separated poly (dimethylsiloxane) microchannels. In their system, PDMS exhibits a negative surface charge and PVP a positive surface charge at pH 8, resulting in the preferred electroosmotic flow vectors for the microchannels and nanochannels in opposite directions. In their trans-NCAM electrokinetic transport experiments, the net-transport direction is determined by NCAM pore size (Fig. 23.4). To transport molecules from a fluorescein solution-filled channel to a buffer-filled channel, the applied potentials required in systems containing 15 and 30 nm pores are the opposite polarity of those containing 100 and 200 nm pores. This switch in flow characteristics clearly demonstrates the shift between microchannel- and nanochannel-dominated transport. Analysis of the relative electrical impedances of the micro- and nanochannels reveals increasing nanochannel resistance as pore size decreases, leading to increased relative electric field strengths in the nanochannels and, ultimately, causing the reversal in the net transport direction (25).

The properties of the hybrid microfluidic/nanofluidic systems described above enable a variety of unit analytical operations such as gated injections (22,24), fraction collection (3,24), sample-stream mixing (29), and analyte preconcentration (2,29,34,35) to be carried out in a single, integrated device. Specifically, the trinary nature of fluid flow (reverse, off, and forward) across integrated nanofluidic elements allows for the externally controllable transport of fluidic bands throughout 3D networks. The forward and reverse states have clear applications in gated injections and fraction collection. The off state is equally important in that it enables chemically disparate environments to be maintained within an interconnected device (36).

The fluid transport properties of NCAMs integrated between microfluidic channels make them ideal gated sample injectors. Transfer of fluidic voxels between the separated microchannels can be effected by application of an axial electrical potential across the NCAM. The amount of analyte injected is a function of both injection time and applied potential, with minimum injection volumes characterized by the volume of the intersection between the NCAM and microchannels, 125 pL for 50 μm by 50 μm channels (22). Post injection separations of FITC-labeled arginine and tryptophan mixtures by Cannon et al. (22) yielded separation reproducibilities of 1%. In addition, gated injections through NCAMs also resulted in differential transport of charged sample species, which has been studied both theoretically and experimentally (28,37). Experimentally, Iannacone et al. (28) injected a salty peptide solution through an NCAM and into a buffer-filled microchannel prior to electrospray ionization toward an ion trap-mass spectrometer. Differential transport of species across the NCAM discriminated against small cation transport and reduced the number of Na^+ ions available for adduct formation, limiting the number of unwanted Na-adducts in the resulting mass spectra.

In addition to functioning as gated injectors, NCAM-separated microfluidic channels can also be utilized as gateable fraction collectors. For example, Tulock et al. (3) reported automated peak detection and fraction collection of separation products in a hybrid microfluidic/nanofluidic device. In their work, electrophoretic separation is followed by a preparative band isolation step in which picoliter fluid voxels (containing attomoles of analyte) are transferred from the separation channel to a vertically separated microchannel via application of an electrical bias across an NCAM interconnect. Results with mixtures of FITC-labeled amino acids show selective, nearly quantitative band transfer (Fig. 23.5).

While the capability of hybrid nanofluidic/microfluidic systems to control the flow of fluids are of interest, the converse ability of NCAMs to keep disparate chemical environments separate also allows complex sample handling schemes by enabling the creation of zones specifically optimized for each unit in a series of analytical operations. For example, an electrophoretic separation could be carried out under acidic conditions, the selected analyte band transferred to another environment, and then a derivatization reaction carried out under basic conditions. The separation of microfluidic environments with different pH levels, which often represents a challenge due to the very high mobility of H^+ ions, has been demonstrated in hybrid microfluidic/nanofluidic systems. Fa et al. (36) investigated proton transport across NCAMs via confocal fluorescence microscopy to develop

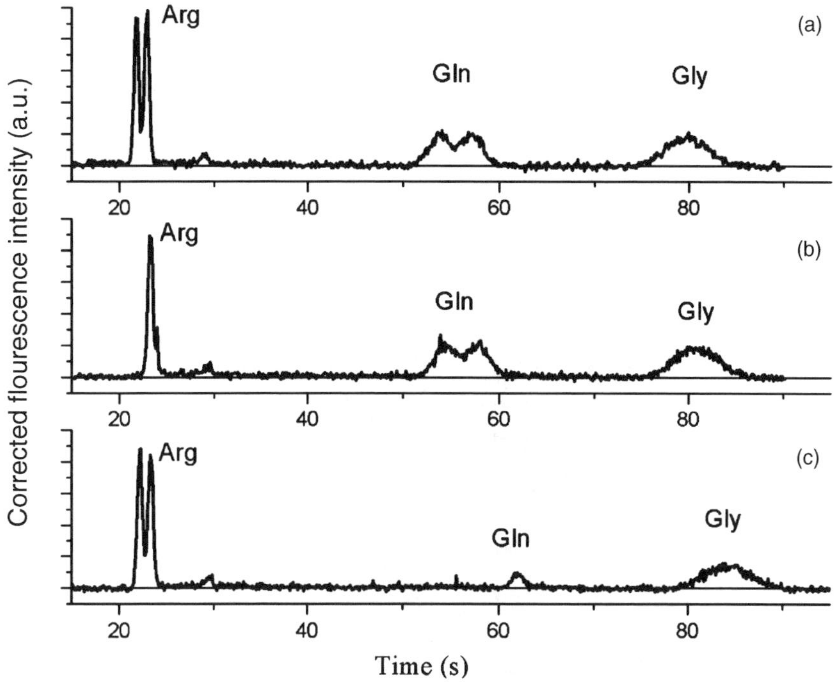

FIGURE 23.5 Electropherograms of FITC-labeled Arg, Gln, and Gly: (**a**) no collection; (**b**) collection of the first component of the Arg double peak; (**c**) collection of the Gln band. Separation conditions: 5 mM phosphate buffer (pH 8) at a field strength of 625 V/cm. Reproduced with permission from Reference 3. Copyright © 2004 American Chemical Society.

spatiotemporal maps of pH under both diffusive and electrokinetic transport. These studies demonstrate that in regimes where $\kappa a < 1$ (i.e., the electrical double-layer overlaps), greater than 1000-fold proton concentration differences can be maintained under conditions of electrokinetic flow. Practically, this demonstrates the ability to maintain separate chemical environments during sample injection or fraction collection in hybrid nanofluidic/microfluidic devices.

Mixing streams of fluid has traditionally been a challenge in microfluidic devices. Low Reynolds numbers lead to laminar flow and thus, diffusive mixing, which is slow on microfluidic length scales. A variety of methods have been developed to decrease mixing times in planar microfluidic devices, including hydrodynamic focusing (38) and flow lamination (39). Hybrid microfluidic/nanofluidic devices, however, allow new mixing strategies to be developed. For example, Jeon et al. (40) developed a passive nanofluidic mixer consisting of thousands of lithographically fabricated 3D nanoscale features integrated into a microchannel, causing the fluid streams to split and recombine. Also, Kuo et al. (29) have demonstrated an active mixer consisting of an NCAM sandwiched between two perpendicular, vertically separated microchannels. In their work, a trans-NCAM potential is applied, transporting fluid through the NCAM. The characteristic distance between the NCAM pores is ≤1 μm, meaning that diffusive mixing in the plane parallel to the NCAM occurs with a characteristic time scale on the order of milliseconds. Additionally, mixing perpendicular to the NCAM plane was also complete within the same time window, resulting in homogeneous mixing across the receiving microchannel (Fig. 23.6). Demonstrating the utility of this technique, Kuo et al. (29) measured the kinetics of a fluorescent reaction. Glycine was injected into a channel containing *o*-phthaldialdehyde and 2-mercaptoethanol, resulting in a

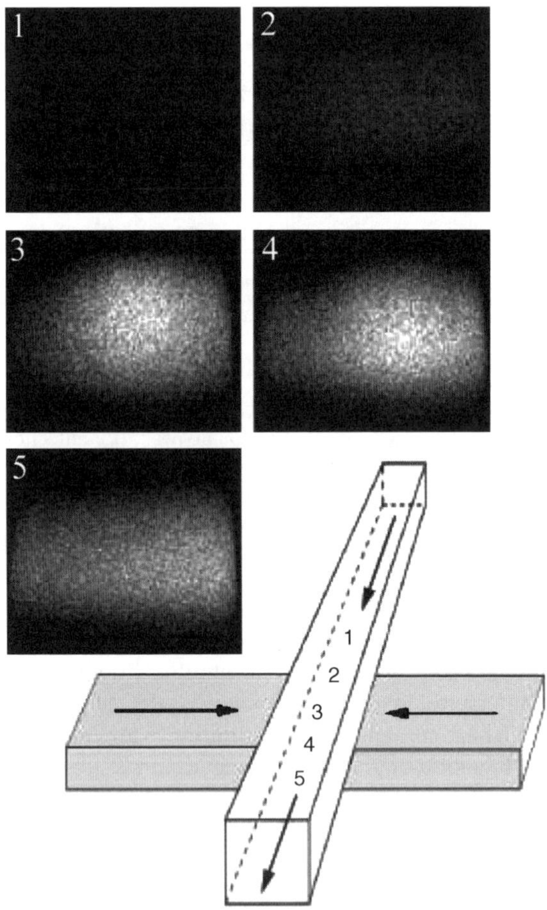

FIGURE 23.6 Cross-sectional fluorescence images normal to the flow axis at positions upstream (1) through downstream (5), of the 90 μm wide nanocapillary array intersection. The fluorescein solution in the horizontal channel was injected through the nanocapillary array into the receiving (observation) channel. Positions (2), (3), and (4) are at the upstream edge, in the middle, and at the downstream edge of the intersection, respectively. From Reference 29. Copyright © 2004 by John Wiley & Sons, Inc. Reprinted by permission of John Wiley & Sons, Inc.

fluorescent isoindole. A constant flow rate was maintained in the receiving channel, thus allowing an image of fluorescence intensity versus distance to be equated to concentration versus time.

The efficient handling of externally acquired samples and the practicalities of loading solutions onto microchip-based devices require, at a minimum, microliters of sample solution. Additionally, analyte molecules of interest (e.g., signaling molecules) often occur in intrinsically small quantities, as previously mentioned. Thus, the samples introduced into microfluidic devices are frequently quite dilute, necessitating preconcentration to enhance separation and detection efficiency. The properties of nanofluidic elements integrated into microfluidic devices have been exploited to achieve active concentration of many analyte species (2,25,34,35). Khandurina and coworkers (34) may have been the first to exploit transport differences between in-plane nanofluidic and microfluidic structures to concentrate analyte molecules. Kuo et al. (25) reported sample stacking of fluorescein in devices consisting of two vertically separated microfluidic channels connected by an NCAM with 15 nm pores when a trans-NCAM electrical potential is applied. In addition, Zhang and Timperman (35) achieved concentration of fluorescein and small anionic peptides. In their work, a PVP-coated PCTE membrane with 10 nm pores was placed in a microfluidic device so that the nanopores formed a continuous fluidic circuit with a single microfluidic channel. When an electrical potential was applied to the device, anionic analyte molecules were concentrated prior to the nanofluidic membrane, while neutral and cationic molecules passed through. Concentration of both fluorescein and small peptides were demonstrated, achieving concentration factors of 300 for fluorescein. Finally, Wang et al. (2) have used nanofluidic elements to create a space-charged, layer-based trap for peptides and proteins. In their work, devices in which nanochannels (ca. 40 nm) connect two microchannels were fabricated in silicon. Electrical potentials applied across the nanochannels generated a charge-depleted region in the microchannel, directly above the nanochannels. When a second potential was applied to the microchannel (tangential to the trans-nanochannel potential), peptides and proteins (e.g., green fluorescent protein) concentrated at the boundary of the charge-depleted layer, yielding concentration factors of 10^6–10^8.

23.4 FUNCTIONALIZED NCAMs

The unique properties arising from the fundamental nanofluidics of NCAMs in 3D hybrid microfluidic/nanofluidic networks can be further extended when the NCAM is chemically functionalized. The advantages of the incorporation of such NCAMs within the hybrid network are that the active (on) and rest (off) states of the NCAMs can be manipulated by external control at will, and mass-limited sample handling is facilitated due to the limited volume of the integrated device (41). The ability to control switching between rest and active states can be accomplished by exploiting the physicochemical properties of the immobilized molecules on NCAMs. Thus, the external stimuli may be electromagnetic radiation, applied electric field, or

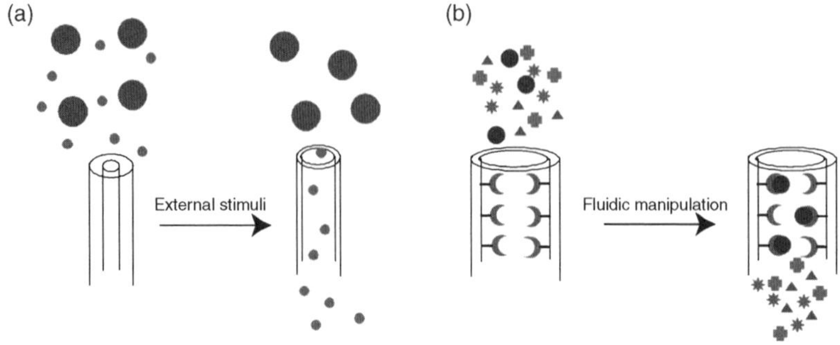

FIGURE 23.7 Concept of functionalized NCAMs. (**a**) depicts the concept of size-dependent molecular flow switching of the NCAM by external stimuli; (**b**) depicts the sensing of a specific analyte from a complex mixture.

temperature. These states can also be controlled by fluidic manipulation to change the chemical environment within the NCAM, such as pH, ionic strength, or hydrophobicity, to exploit the non-covalent interactions of the chemical species (41). Fig. 23.7 illustrates the concept of functionalized NCAMs where (a) shows a size selection and (b) shows a molecular recognition on a single nanopore. Many of the well-characterized, tailored nanotube membranes can be potentially incorporated into the 3D hybrid microfluidic/nanofluidic network to perform differential chemical transport, separation (42–44), sensing (45,46), or selective capture and release of mass-limited analytes. Examples of such membranes include, but are not limited to, an ordered array of nanoporous polymer membranes, for example, polycarbonate (42), PMMA (47,48), and poly(ethylene terephthalate) (PET) (42).

Each nanopore of the NCAM in the 3D hybrid microfluidic/nanofluidic networks can be considered as an intelligent attoliter-volume chemical reaction chamber where loading and unloading of the reagents can be externally controlled by fluidic manipulation. Confining mass-limited reactants within a nanopore significantly increases their molar concentration. As the pore diameter decreases, the effective concentration of a single molecule in an individual nanopore increases, as listed in Table 23.1. Therefore, the utility of these attoliter-volume chemical reaction chambers can expedite the rate of reaction in cases where the kinetics are dependent

TABLE 23.1 Molar Concentration of a Single Molecule in a Single Nanopore

Pore Diameter (nm)	Single Pore Volume[a] (L)	Single Molecule Concentration (M)
200	3.14×10^{-17}	5.29×10^{-9}
100	7.85×10^{-18}	2.11×10^{-8}
30	7.07×10^{-19}	2.35×10^{-7}
10	7.85×10^{-20}	2.11×10^{-6}

[a]Thickness of the NCAM is generally 6–10 μm depending on the pore size; 10 μm thickness was used to calculate the volume of a single pore.

on the molar concentration of the reagents (49) and/or those that are diffusion limited (50). For example, most biological sample handling involves limited amounts of samples with small diffusion coefficients. Usually, the diffusion coefficients (D) of biological molecules are on the order of 10^{-10} m^2/s $< D < 10^{-9}$ m^2/s (51,52), where the upper limit corresponds to a molecular transport ($\sqrt{D\,t}$) of about 30 nm in 1 μs (53).

23.4.1 Methods for Nanopore Modification

To introduce functionality to the nanoporous membranes, immobilization of molecules on the NCAM surface, including the pore walls, is necessary. Surface modification of NCAMs can be accomplished with various immobilization chemistries, many of which were developed for biological sensor fabrication. Although many surface immobilization strategies exist, NCAM functionalization is most often realized by gold-thiol chemistry because it offers a simple and straightforward strategy to modify the surface with desired molecules (42,54,55). The immobilized monolayer can be extended via various linker chemistries (54) by using amine terminated thiol, carboxylic acid terminated thiol, and thiolated polymer precursors. The utilization of gold-thiol chemistry requires a gold surface and molecules with accessible sulfhydryl or disulfide functional groups. Gold coating on NCAMs is most often carried out by thermal metal evaporation (56), sputter coating (57), or electroless plating (58), all of which can be implemented on nonconductive surfaces, unlike electrodeposition (56,58). Thermal metal evaporation produces a smooth, uniform surface that offers unambiguous control over the surface modification chemistry (56). However, gold deposition is restricted to the planar NCAM surface and the mouth of the nanopore wall (56). Although electroless plating (58) produces a rather rough surface compared to thermal metal evaporation, it allows deposition of gold throughout the length of the nanopores, enhancing the effective surface area for functional molecule immobilization (46).

Physical entrapment of functionally active molecules within a polymer network is another way to modify the NCAM surface. For example, sol–gel chemistry provides a convenient way to prepare nanoporous thin films (42,59), which are formed by hydrolysis–condensation of inorganic ($M(OR)_n$, $M = Si$, Ti, Zn, W, etc.) or hybrid ($R\text{-}Si(OR)_3$) precursors (42,59,60). Various molecules with specific functions (60), including biologically active molecules such as enzymes, can be encapsulated within the network (61). Another example of physical entrapment is modifying the surface with an electropolymerizable polymer, such as polypyrrole, to entrap enzymes within the polymer network, as well as within the nanopores (57).

Finally, functionalized nanoporous substrates can be prepared by doping the polymer substrate with the appropriate molecules before the nanopores are synthesized. For instance, Lee et al. (47) reported a fluorescence sensor based on an electronic energy transfer system. In their work, the fluorescence of perylene (donor) doped PMMA nanopores is reversibly quenched by the passage of disodium fluorescein (acceptor) through the pores. The nanocapillary array system described was synthesized by using a 200 nm Al_2O_3 anodic filter as a template (47). However, the nanopores can also be directly fabricated on thin (~10 µm) PMMA films by a focused ion beam, which produces NCAMs that are suitable and more likely to be incorporated into 3D hybrid microfluidic/nanofluidic networks (48).

23.4.2 Transport Control in Tailored NCAMs

Size-selective transport on the molecular scale can be achieved by tailoring NCAMs with phase transition polymers. There are two types of phase-transition polymers whose volume transition is responsive to external stimuli. The first type is exemplified by poly(acrylic acid), a polyelectrolyte that is responsive to electric field and/or pH, while poly(*N*-isopropylacrylamide) (PNIPAAm) and cellulose derivatives that are responsive to temperature are characteristic of the second type (62). The electric field-responsive (voltage-tunable) volume transition polymer can be a hydrogel that can hold large volumes of water without dissolution since it consists of an elastic crosslinked network. For example, Lokuge and Bohn (63) reported a voltage-tunable phase transition in a copolymer comprised of 2-hydroxyethyl methacrylate and ethylene glycol dimethacrylate on a gold surface and addressed the potential applications for surface-modified NCAMs. Changes in the amount of water contained within the network in response to applied potentials of opposite polarities either lead to swelling or shrinkage of the voltage-tunable polymer, modulating the effective nanopore diameter results (63). Compared to other stimuli, manipulating voltage is both facile and rapid, making it an attractive candidate for switching between active and rest states of the functionalized NCAMs for size selection. This study also described a PNIPAAm phase-transition polymer grafted on a gold NCAM where temperature controls the switching for nanopore size selection (56). PNIPAAm is a negatively charged polymer where temperature-induced contraction occurs when the polymer is heated above the lower critical solution temperature (LCST ~32°C). Therefore, grafted PNIPAAm on gold NCAMs undergoes a temperature-induced volume change, which leads to an effective pore diameter change (56). A schematic illustration of the temperature and corresponding volume change of the grafted polymer is shown in Fig. 23.8 (56). Whereas previously fabricated stimulus-responsive nanopore constructs had demonstrated simple on–off flow switching, careful control over molecular weight by atom-transfer radical polymerization produced a composite PNIPAAm-NCAM structure that showed reversible, size-selective flow-switching behavior effected by temperature changes.

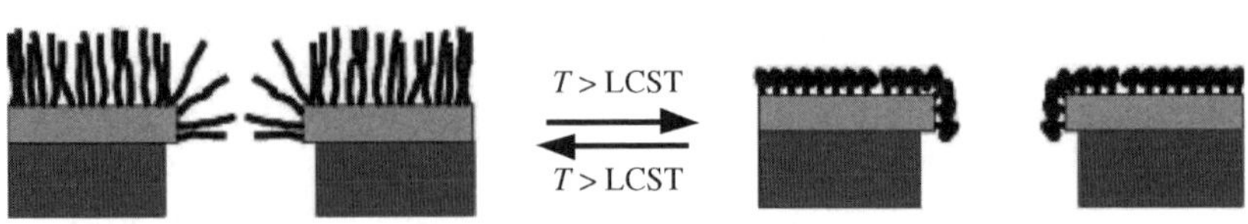

FIGURE 23.8 Schematic illustration of the temperature-dependent response of the surface-grafted polymer to perform size selection. Reproduced with permission from Reference 56. Copyright © 2007 American Chemical Society.

Harrell et al. (64) reported DNA-functionalized gold nanopores with asymmetric entrance and exit apertures to create artificial ion channels. The authors demonstrated blocking and unblocking of the nanopore entries dependent on the length of the immobilized DNA strand and the applied electrical bias. The asymmetric nanochannels were prepared by anisotropic etching of PCTE membranes to form 12 μm long pores with a large diameter of 5 μm and a small diameter of 60 nm (42,64). Depending on the applied electrical bias, the immobilized DNA on the asymmetric nanopores may lie down (on) or stand up (off), which leads to switching of the molecular transport. The ion current across the nanopores, which is indicative of molecular transport, as well as a schematic illustration of the modified asymmetric nanopore and the DNA response dependent on the applied electric field, are depicted in Fig. 23.9 (64).

Alkyl thiol-modified gold NCAMs show transport selectivity depending on the hydrophobicity of the immobilized thiol (31,32,65). Jirage et al. (32) investigated the selective transport of toluene, its homologous series, and pyridine through

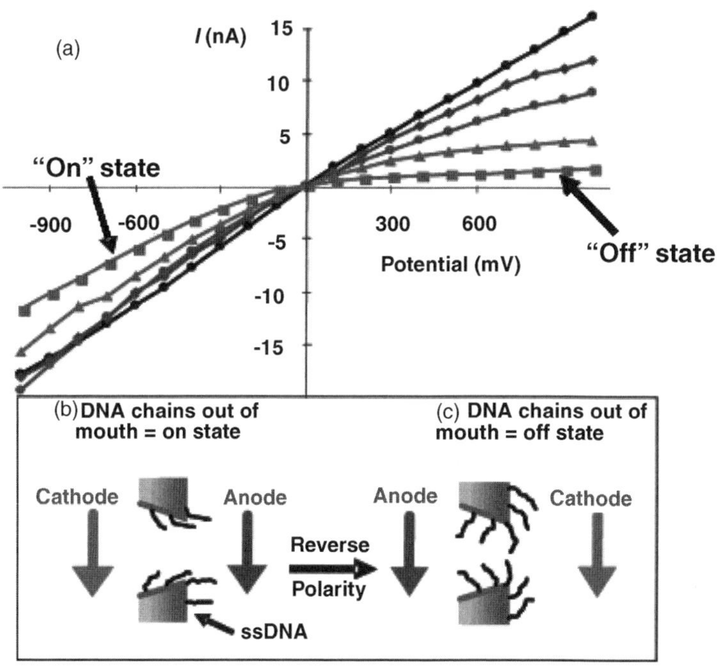

FIGURE 23.9 (**a**) shows the ion current curve dependent on the applied potential across the nanopores, where the smaller pore diameter is 40 nm. The black curve corresponds to asymmetric, unmodified nanopores; shaded diamonds, circles, triangles and squares correspond to the modification with 12-mer, 15-mer, 30-mer, and 45-mer of DNA, respectively. (**b**) and (**c**) show schematic diagrams of the behavior of the immobilized DNA when electric field is applied. Reproduced with permission from Reference 64. Copyright © 2004 American Chemical Society.

$HS\text{-}C_{16}H_{33}$ (C_{16} thiol) and $HS\text{-}C_2H_4OH$ (C_2H_4OH thiol) immobilized NCAMs, as well as the corresponding transport mechanisms. Modification of the gold NCAMs with these thiols modifies the interior environment of the nanopores, which leads to the preferential transport of molecules depending on their hydrophobic interaction. In this case, transport through the nanopores is dominated by the partition term (K) of the total flux (J) rather than by diffusion (32). The flux (J) is defined as

$$J = -DKC_{aq}/I \tag{23.5}$$

where D is the diffusion coefficient within the NCAM, K is the partition coefficient between the aqueous solution of the permeant molecules and the NCAM, C_{aq} is the concentration of the permeant molecule, and I is the thickness of the NCAM, in a model analogous to that used in reverse-phase liquid chromatography (RPLC). The C_{16} thiol-modified gold NCAMs showed enhanced transport of hydrophobic molecules, such as toluene, while the C_2H_4OH thiol preferentially transported more hydrophilic molecules, such as pyridine (32). In a two-molecule permeation experiment, C_{16} thiol-modified gold NCAMs showed 165 times higher flux of toluene than those of pyridine (32). Although the study highlighted the passive control of molecular flux, the substantial changes in molecular transport across the nanopores lead the way toward active switching of electrical bias to perform separations in a manner similar to RPLC.

When gold NCAMs are modified with ω-carboxylated thiol (66) or cysteine (67), they show selective ion transport, which can be modulated by changing the pH of the solution. Hou et al. (66) investigated the transport properties of benzenesulfonate (BS) across $HS(CH_2)_{10}COOH$-modified gold NCAMs. When the pH was low enough to protonate the surface-immobilized acid, the BS flux was reportedly high, while the opposite was true for deprotonated surfaces at high pH (66). Lee et al. (67) reported a cysteine-immobilized gold NCAM that showed a switch in the preferential transport of ions with a pH change. Cation transport was enhanced at high pH (pH 12) when both the amine group and the carboxylic acid group were unprotonated. However, anion transport was enhanced at low pH (pH 2.0) when both of the functional groups were protonated. Note that these are exactly the results one would expect from an exclusion of coions from the interior of the nanopore, in favor of mobile counterions. At pH 6.0, the flux of both cations and anions showed intermediate values, which was interpreted as the isoelelectric point (pI) of the chemisorbed cysteine (67). However, the pI of the cysteine in solution is 5, which is lower than that of other common amino acids and reflects the contribution of the −SH moiety in the molecule (67). Considering the observed result, the authors of both studies pointed out that the pI of the chemisorbed chemical species cannot be precisely measured (66,67). Again, these pH-responsive selective transport studies were performed with passive flux observations, but the transport of the fluid across the membrane could also be effected electrokinetically.

Selective transport can be realized by surfactant-modulated nanopores as well. Schmuhl et al. (41) presented a tunable transport nanoporous membrane where the selective transport behavior is modulated by reversible adsorption of surfactants on the internal nanochannel surface of an alumina-supported membrane. Thus, the transport of chemical species across the nanopores is controlled by the fluidic manipulation of the surfactants. The surfactants used for surface modification were cetyltrimethylammonium bromide (CTAB), sodium dodecyl sulfate (SDS), and triethylhexylammonium bromide (C6TEA), which have tails of 16, 12, and 6 carbon atoms, respectively. The molecular transport is determined by coulombic and/or steric interactions and, therefore, responsive to the nature (charge and carbon chain length) and concentration of the surfactant. When the nanopores were modified with a low concentration (10^{-5} M) of SDS, fluorescein (Fl^{2-}) transport was suppressed compared to methyl viologen (MV^{2+}) transport. Furthermore, cationic molecule, MV^{2+}, transport was effectively suppressed with CTAB, relative to C6TEA (41). However, the nanopores were blocked for both anion and cation transport at sufficiently high concentration of surfactant in the order C6TEA $<$ SDS $<$ CTAB. The influence of surfactant concentration was also shown to be reversible.

23.4.3 Molecular Recognition in Tailored NCAMs

Tailored NCAMs can sense specific molecules when molecular recognition molecules are immobilized on the NCAMs; many are adopted from biological systems such as nucleic acids, enzymes, and antibodies. In addition, not only can these types of platforms sense the targeted molecules, but they also have the potential to concentrate or selectively capture and release the mass-limited analytes on demand.

Wernette et al. (46) reported DNAzyme-modified gold NCAMs with an internal standard for Pb(II) sensing. An enzyme strand (HS-17EDy) was immobilized via gold-thiol chemistry with mercaptohexanol (MCH) to displace inactive DNA species at the gold surface. The cleavable substrate strand labeled with a fluorophore (17DS-Fl) was then hybridized to the immobilized enzyme strand to form a DNAzyme complex. When immobilized DNAzyme reacts with Pb(II), the cleavable strand is released and the fluorescence is detected in solution (46). The prepared DNAzyme-modified gold NCAM has long-term stability in ambient conditions and can be regenerated when the immobilized enzyme strand is rehybridized with a cleavable substrate strand (46). A schematic depiction of the DNAzyme immobilization procedure and the simplified Pb(II) recognition scheme are illustrated in Fig. 23.10 (68).

Kohli et al. (43) showed single-base mismatch recognition of a DNA-functionalized NCAM by immobilizing DNA hairpins on gold NCAMs. DNA hairpins have complementary base sequences at each end of the molecule so that they hybridize and form a loop structure under proper conditions. The DNA hairpin-immobilized gold NCAM showed expedited selective transport of a perfect complementary sequence DNA (PC-DNA) compared to the gold NCAM without an immobilized DNA hairpin (43). Also, the DNA hairpin-immobilized gold NCAM transported the PC-DNA across the membrane more efficiently than single-base mismatch DNA (1MM). The selectivity coefficients (α), defined as the ratio of the flux of PC-DNA to 1MM on the DNA hairpin-immobilized gold NCAM, were $\alpha_{HP,\ PC/1MM} = 3$ for 1MM and $\alpha_{HP,\ PC/7MM} = 7$ for 7MM (43).

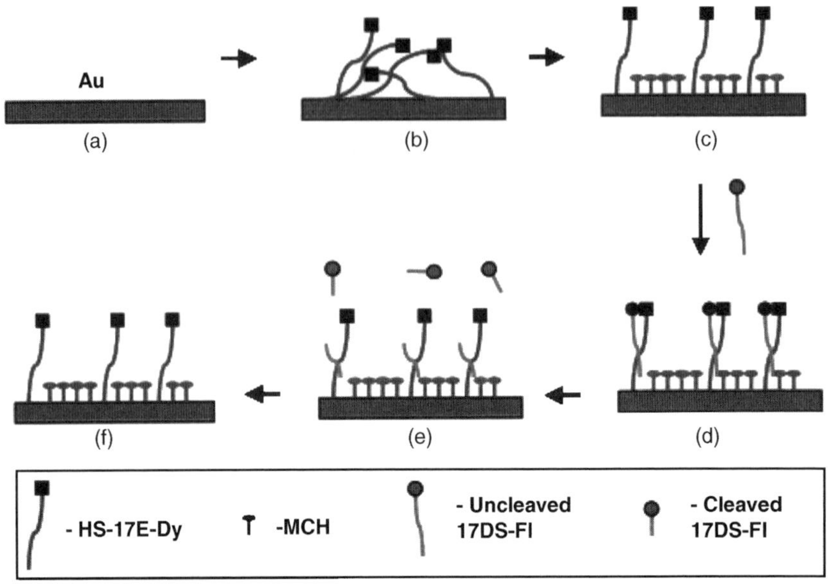

FIGURE 23.10 Schematic illustration of the gold surface-modification procedure with Pb (II)-specific DNAzyme and simplified working scheme of the sensor. (**a**) Clean gold surface; (**b**) HS-17Edy-modified surface before MCH was immobilized; (**c**) HS-17Edy-modified surface after MCH was immobilized; (**d**) HS-17Edy-modified surface after hybridized with 17DS-FI; (**e**) sensor surface releasing cleaved 17DS-FI after recognized Pb(II); (**f**) sensor surface ready to be regenerated by rehybridization. Reproduced with permission from Reference 68. Copyright © 2005 American Chemical Society.

Enzyme-modified NCAMs show specificity to the substrate of the immobilized enzyme. Lakshmi et al. (57) reported alcohol dehydrogenase apoenzyme (apo-ADH), apoaldehyde dehydrogenase (apo-AdDH), and D-amino acid oxidase apoenzyme (apo-D-AAO) immobilized NCAMs for a facilitated transport (substrate selective and enantioselective) without cofactors to prevent undesirable chemical conversion during the transport of specific molecules. Polypyrrole was electropolymerized on gold-coated NCAMs prepared by sputter coating to entrap the enzyme within the nanopores, as well as within the polymer network. Enzyme capture occurred because polypyrrole is permeable to small molecules (substrate) and impermeable to large molecules such as proteins (enzymes) (57). Each enzyme-loaded NCAM showed selective transport for its own substrate, for example, apo-AHD-NCAM showed selective transport for ethanol compared to phenol; apo-AdDH-NCAM showed selective transport for vanillin compared to the NCAM not loaded with enzyme; and apo-D-AAO-NCAM showed enantioselectivity to D-phenylalalnine with the selectivity coefficient of 3.3 when a 400 nm pore diameter membrane was used, and 4.9 when a 30 nm pore diameter membrane was used (57).

Glucose oxidase (GOx)-modified NCAMs (69,70), horseradish peroxidase (HRP)-modified NCAMs (71) and bienzyme HRP-GOx-modified NCAMs (72) have also been reported. The enzymes were immobilized on gold NCAMs prepared by electroless plating by forming a self-assembled-monolayer of anchor molecules, such as mercaptoethylamine or mercpatopropionic acid, and using a glutaraldehyde or carbodiimide coupling reaction (69–72). Although these enzyme-modified NCAMs were originally created as electrochemical sensors for their corresponding substrates, glucose and H_2O_2, they certainly have the potential to be incorporated into 3D hybrid microfluidic/nanofluidic networks.

Other molecular recognition systems adopted to tailor NCAMs include biotin/streptavidin, protein G/immunoglobulin G (IgG), and antibody/antigen. Siwy et al. (45) reported a single, conical gold nanotube protein biosensor on a 12 μm thick PET membrane. The large pore entry diameter was about 600 nm and the small pore entry diameter was about 30 nm (45). The diameter of the small pore entry used for the experiment was $5 < a < 9$ nm, comparable to the size of the immobilized biomolecules after gold was deposited for modification. Thus, when the immobilized molecular recognition molecule captured the specific analyte, the nanopore was blocked and the transmembrane ion current was not observed (45). Kim et al. (73) reported the fragment of antigen binding of antibody (Fab′-SH; antihuman insulin)-modified gold NCAMs and showed selective capture and release of the specific antigen (human insulin) from a complex sample mixture. The performance of the Fab′-tailored gold NCAM was assessed by matrix-assisted laser desorption/ionization mass spectrometry (73).

23.5 THE FUTURE

The unique properties of hybrid microfluidic/nanofluidic networks enable traditionally planar μ-TAS devices and operations to be extended to 3D architectures, as well as enabling new capabilities, such as the ability to maintain separate, chemically unique environments within a single, interconnected device and to transfer fluid between these environments at will. In many prior studies, devices have been limited to two microfluidic layers and a single layer of nanofluidic interconnects, with this limitation imposed by the practicalities of rapid prototyping using PDMS. In recent work from our laboratory, Flachsbart et al. (30) developed a new series of hybrid microfluidic/nanofluidic devices that allow an arbitrary number of layers. In such devices, PMMA microfluidic channels are coupled with PCTE NCAMs via adhesive contact printing. These systems have advantages, not only in their ability to enable more complex fluidic designs but also in the increased long-term electroosmotic flow stability afforded by PMMA (30). The increased number of available layers allows greater flexibility and enhanced functionality in sample processing and manipulation, especially because multiple and distinct NCAMs can be incorporated into a single device. As an example, one could imagine performing a gated injection from a microchannel filled with a complex sample mixture into another channel where a preparative electrophoretic separation would be carried out. A specific analyte band could then be collected from this separation, transferred to yet another spatial plane and into a channel filled with a chiral selector where, after another separation, a specific enantiomer could be transferred to a final microchannel interfaced to a mass spectrometer for detection. In another example, an antibody-modified NCAM could capture a specific analyte from a complex sample mixture, thus performing a combined preparatory separation and *de facto* preconcentration, prior to an on-demand release by fluidic manipulation for further analytic steps. Clearly, the extended capabilities provided by the integration of nanofluidic elements within more traditional microfluidic architectures expands the ability to perform complex analytical operations on mass-limited samples while retaining sample integrity.

ACKNOWLEDGMENTS

This study, which is based on work from the authors' laboratories, is supported by the National Science Foundation under award no. DMI-0328162 and cooperative agreement CTS-0120978, the Department of Energy DE FG02 07ER15851, and the Strategic Environmental Research and Development Program under award no. W9132T-05-2-0028.

REFERENCES

1. Manz A, Graber N, Widmer HM. Sens Actuators B, Chemical 1990;1:244.
2. Wang YC, Stevens AL, Han J. Anal Chem 2005;77:4293–4299.
3. Tulock JJ, Shannon MA, Bohn PW, Sweedler JV. Anal Chem 2004;76:6419–6425.
4. Rice CL, Whitehead R. J Phys Chem 1965;69:4017–4024.
5. Kemery PJ, Steehler JK, Bohn PW. Langmuir 1998;14:2884–2889.
6. Kuo TC, Sloan LA, Sweedler JV, Bohn PW. Langmuir 2001;17:6298–6303.
7. Fievet P, Szymczyk A, Aoubiza B, Pagetti J. J Membr Sci 2000;168:87.
8. Daiguji H, Yang P, Szeri AJ, Majumdar A. Nano Lett 2004;4:2315–2321.
9. Stein D, Kruithof M, Dekker C. Phys Rev Lett 2004;93:035901.
10. Ku JR, Stroeve P. Langmuir 2004;20:20030–20032.
11. Petsev DN, Lopez GP. J Colloid Interface Sci 2006;294:492–498.
12. Qiao R, Aluru NR. Phys Rev Lett 2004;92:198301.
13. O'Brien MJ, Bisong P, Ista LK, Rabinovich EM, Garcia AL, Sibbett SS, Lopez GP, Brueck SRJ. J Vac Sci Tech B 2003;21:2941–2945.
14. Garcia AL, Ista LK, Petsev DN, O'Brien MJ, Bisong P, Mammoli AA, Brueck SR, Lopez GP. Lab Chip 2005;5:1271–1276.
15. Han J, Craighead HG. Science 2000;288:1026–1029.
16. Han J, Craighead HG. Anal Chem 2002;74:394–401.
17. Han J, Turner SW, Craighead HG. Phys Rev Lett 1999;83:1688.
18. Cao H, Tegenfeldt JO, Austin RH, Chou SY. Appl Phys Lett 2002;81:3058–3060.
19. Bakajin OB, Duke TAJ, Chou CF, Chan SS, Austin RH, Cox EC. Phys Rev Lett 1998;80:2737.

20. Pu Q, Yun J, Temkin H, Liu S. Nano Lett 2004;4:1099–1103.
21. Duffy DC, McDonald JC, Schueller OJA, Whitesides GM. Anal Chem 1998;70:4974–4984.
22. Cannon Jr DM, Kuo TC, Bohn PW, Sweedler JV. Anal Chem 2003;75:2224–2230.
23. Ismagilov RF, Ng JM, Kenis PJ, Whitesides GM. Anal Chem 2001;73:5207–5213.
24. Kuo TC, Cannon Jr DM, Chen Y, Tulock JJ, Shannon MA, Sweedler JV, Bohn PW. Anal Chem 2003;75:1861–1867.
25. Kuo TC, Cannon DM, Shannon MA, Bohn PW, Sweedler JV. Sens Actuators A, Physical 2003;102:223–233.
26. Long Z, Liu D, Ye N, Qin J, Lin B. Electrophoresis 2006;27:4927–4934.
27. Wang PC, Gao J, Lee CS. J Chromatogr A 2002;942:115–122.
28. Iannacone JM, Jakubowski JA, Bohn PW, Sweedler JV. Electrophoresis 2005;26:4684–4690.
29. Kuo TC, Kim HK, Cannon DM, Shannon MA, Sweedler JV, Bohn PW. Angew Chem Int Ed 2004;43:1862–1865.
30. Flachsbart BR, Wong K, Iannacone JM, Abante EN, Vlach RL, Rauchfuss PA, Bohn PW, Sweedler JV, Shannon MA. Lab Chip 2006;6:667–674.
31. Hulteen JC, Jirage KB, Martin CR. J Am Chem Soc 1998;120:6603–6604.
32. Jirage KB, Hulteen JC, Martin CR. Anal Chem 1999;71:4913–4918.
33. Nishizawa M, Menon VP, Martin CR. Science 1995;268:700–702.
34. Khandurina J, Jacobson SC, Waters LC, Foote RS, Ramsey JM. Anal Chem 1999;71:1815–1819.
35. Zhang Y, Timperman AT. Analyst 2003;128:537–542.
36. Fa K, Tulock JJ, Sweedler JV, Bohn PW. J Am Chem Soc 2005;127:13928–13933.
37. Chatterjee AN, Cannon DM, Gatimu EN, Sweedler JV, Aluru NR, Bohn PW. J Nanoparticle Res 2005;7:507–516.
38. Knight JB, Vishwanath A, Brody JP, Austin RH. Phys Rev Lett 1998;80:3863.
39. Bessoth FG, deMello AJ, Manz A. Anal Commun 1999;36:213–215.
40. Jeon S, Malyarchuk V, White JO, Rogers JA. Nano Lett 2005;5:1351–1356.
41. Schmuhl R, van den Berg A, Blank DHA, ten Elshof JE. Angew Chem Int Ed 2006;45:3341–3345.
42. Baker LA, Jin P, Martin CR. Crit Rev Solid State Mater Sci 2005;30:183–205.
43. Kohli P, Harrell CC, Cao Z, Gasparac R, Tan W, Martin CR. Science 2004;305:984–986.
44. Wirtz M, Yu S, Martin CR. Analyst 2002;127:871–879.
45. Siwy Z, Trofin L, Kohli P, Baker LA, Trautmann C, Martin CR. J Am Chem Soc 2005;127:5000–5001.
46. Wernette DP, Swearingen CB, Cropek DM, Lu Y, Sweedler JV, Bohn PW. Analyst 2006;131:41–47.
47. Lee S, Muller AM, Al-Kaysi R, Bardeen CJ. Nano Lett 2006;6:1420–1424.
48. Cannon DM, Flachsbart BR, Shannon MA, Sweedler JV, Bohn PW. Appl Phys Lett 2004;85:1241–1243.

49. Carroll FA. Methods of studying organic reactions. In:Perspective on Structure and Mechanism in Organic Chemistry. Pacific Grove, CA: Brooks/Cole Publishing Company;1998.
50. Cooper JW, Chen J, Li Y, Lee CS. Anal Chem 2003;75:1067–1074.
51. Terazima M. Phys Chem Chem Phys 2006;8:545–557.
52. Liu MK, Li P, Giddings JC. Protein Sci 1993;2:1520–1531.
53. Bard AJ, Faulkner LR. Basic potential step methods. In:Electrochemical Methods, Fundamentals and Applications. 2nd ed. New York: Wiley;2001.
54. Love JC, Estroff LA, Kriebel JK, Nuzzo RG, Whitesides GM. Chem Rev 2005;105:1103–1169.
55. Kohli P, Wirtz M, Martin CR. Electroanalysis 2004;16:9–18.
56. Lokuge I, Wang X, Bohn PW. Langmuir 2007;23:305–311.
57. Lakshmi BB, Martin CR. Nature 1997;388:758–760.
58. Menon VP, Martin CR. Anal Chem 1995;67:1920–1928.
59. Hulteen JC, Martin CR. J Mater Chem 1997;7:1075–1087.
60. Valle K, Belleville P, Pereira F, Sanchez C. Nat Mater 2006;5:107–111.
61. Miller JM, Dunn B, Valentine JS, Zink JI. J Non-Crystalline Solids 1996;202:279–289.
62. Sutani K, Kaetsu I, Uchida K, Matsubara Y. Rad Phys Chem 2002;64:331–336.
63. Lokuge IS, Bohn PW. Langmuir 2005;21:1979–1985.
64. Harrell CC, Kohli P, Siwy Z, Martin CR. J Am Chem Soc 2004;126:15646–15647.
65. Lee SB, Martin CR. Chem Mater 2001;13:3236–3244.
66. Hou ZZ, Abbott NL, Stroeve P. Langmuir 2000;16:2401–2404.
67. Lee SB, Martin CR. Anal Chem 2001;73:768–775.
68. Swearingen CB, Wernette DP, Cropek DM, Lu Y, Sweedler JV, Bohn PW. Anal Chem 2005;77:442–448.
69. Delvaux M, Demoustier-Champagne S. Biosens Bioelecton 2003;18:943–951.
70. Delvaux M, Demoustier-Champagne S, Walcarius A. Electroanalysis 2004;16:190–198.
71. Delvaux M, Walcarius A, Demoustier-Champagne S. Anal Chim Acta 2004;525: 221–230.
72. Delvaux M, Walcarius A, Demoustier-Champagne S. Biosens Bioelecton 2005;20: 1587–1594.
73. Kim BY, Swearingen CB, Ho J-A, Romanova EV, Bohn PW, Sweedler JV. J Am Chem Soc; 2007;129:7620–7626.

24

MAGNETIC BEAD-BASED METHODS TO STUDY THE INTERACTION OF TEICOPLANIN WITH PEPTIDES AND BACTERIA

Menake E. Piyasena and Frank A. Gomez

24.1 INTRODUCTION

Microbead-based analysis, utilized in many biochemical studies, is advantageous as it provides for easy separation of bound complexes from unbound species without the use of sophisticated techniques and instrumentation. The high surface-area-to-volume ratio of beads allows for sample concentration, thereby, increasing the limits of detection (LOD).

Integration of microbeads into microfluidic devices can be advantageous in some respects. In microfluidics, fluid has its characteristic laminar flow profile due to the low Reynolds (Re) number (1). Laminar flow is an issue where rapid mixing and reaction is required due to the slow diffusive mixing, the most prolific method of mixing in laminar flow. Larger the molecule size and higher the diffusion distance, longer it takes to diffuse. Even at micron scale the diffusion can be a slow process. Presence of beads in microfluidic systems can facilitate the mixing thus enhance many chemical reactions (2).

Beads have been used in microfluidic devices for many types of biochemical applications including immunoassays, protein purification, separation, and others (3–6). Bioassays on a microfluidic format allow for rapid analysis with small reagent volumes that is important when expensive and/or hazardous compounds are involved. Inclusion of nonmagnetic beads in microfluidic devices requires a retaining

Biological Applications of Microfluidics, Edited by Frank A. Gomez

mechanism to trap beads. This is often acquired by making internal structures like pillars and dams. Magnetic beads can be packed into microchannels using permanent magnets or electromagnets. The absence of pillars or dam-like structures to hold beads permanently and the ability to release beads just by simply removing the magnetic field will ensure these devices are reusable.

Separations based on magnetic properties are common in biotechnology (7) reasons being the myriad of magnetic beads commercially available and the variety of diameters and surface chemistries possible (8). The variety of chemistries is critical and provides for facile immobilization of many biological receptors. Magnetic beads have been used in many biological applications including thermotherapy for cancer cells (9), immunoassays (10–12), cell separation (13–15), DNA hybridization (16), protein and peptide separation (17), and pathogen detection (18,19).

Glycopeptide antibiotics have been used to treat bacterial infections extensively. They inhibit the growth of gram-positive bacteria by suspending the synthesis of the bacteria cell wall (20–22). Vancomycin from *Streptomyces orientalis* has historically been used as the last option of treatment for many bacterial infections due to the increased resistance for many other types of antibiotic agents (13). Teicoplanin (teic) from *Actinoplanes teichomyceticus* (Fig. 24.1) is a glycopeptide antibiotic of the vancomycin family (23,24). Teicoplanin binds to the D-Ala-D-Ala terminus on the cell wall of gram-positive bacterium via hydrogen bonds thereby preventing the enzyme mediated crosslinking of peptidoglycan and eventual cell death (14–16). Teic is more active against *Staphylococcus* species than vancomycin, including methicillin-resistant strains, and it has less side effects and exhibits a longer half-life in humans (18,25–27). The growth of resistance to antibacterial agents is an ever-

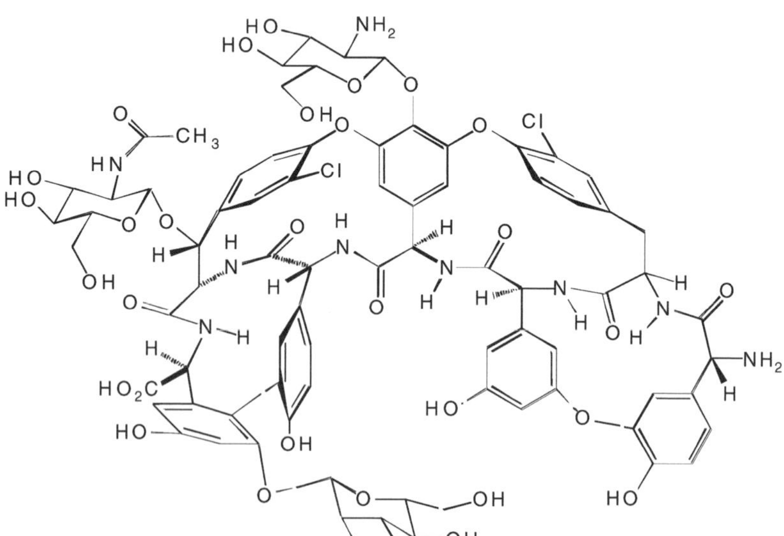

FIGURE 24.1 The structure of teicoplanin, a glycopeptide antibiotic of the vancomycin family.

increasing worldwide problem that threatens the chemical effectiveness of drugs used in the treatment of many infectious diseases. The development of new, fast, and sensitive techniques to analyze bacteria–antibiotic interaction is important for several reasons. In the case of clinical testing low concentrations of bacteria require several steps (incubation, enrichment, and amplification). There is growing concern, that in the future, pathogens (foodborne, waterborne, or airborne) will be used as weapons in terrorist activities. Hence, sensitive and instant detection of pathogens are warranted.

The development of novel bead-based methods to study the interaction of antibiotics with specific ligands compliments current analytical techniques by providing an alternative technique to estimating affinity parameters between two species. There have been previous publications where magnetic particles have been used to analyze antibiotic binding to bacteria. Lin et al. (28) used MALDI-MS techniques to analyze the binding of three different gram-positive bacteria to vancomycin-tethered magnetic nanoparticles. The lowest concentration they could detect was $\sim 7 \times 10^4$ cfu/mL. Gu et al. (19) could detect 10^1-10^2 cfu/mL level of gram-positive and gram-negative bacteria using vancomycin-tethered nanoparticles via optical and scanning electron microscopy.

In this chapter we mainly present a fluorescent-based magnetic microparticle technique to examine bacteria interactions with the antibiotic teicoplanin. We first demonstrate the concept by using a model system comprising the fluorescent-labeled peptide (5-carboxyfluorescein-D-Ala-D-Ala-D-Ala (5-FAM$(DA)_3$) and other related peptides. Bacterial binding to magnetic particles at ultralow concentrations is demonstrated. The binding of 5-FAM$(DA)_3$ to teic-beads is monitored via spectrofluorometry and flow cytometry. The binding of *Staphylococcus aureus* was analyzed via spectrofluorometry. Furthermore, we explore the feasibility of developing a microfluidic chip to investigate the binding of teicoplanin to 5-FAM$(DA)_3$ peptides.

24.2 EXPERIMENTAL

24.2.1 Derivatization of Magnetic Beads with Antibiotics

Carboxylic acid terminated magnetic beads (~2.7 μm diameter) (Invitrogen, Carlsbad, CA) were derivatized with the antibiotic, teicoplanin (Advanced Separation Technologies Whippary, NJ) (teic-beads). A solution (100 μL) of magnetic beads from the stock solution was taken into a microcentrifuge tube and a rare earth magnet was used externally to separate the beads from the solution. The supernatant was discarded and the beads were washed (5×) with 2-(*N*-morpholino)ethanesulfonic acid (MES) buffer (25 mM, pH 5). A solution (60 μL) of teicoplanin (10 mg/mL) in MES (25 mM, pH 5) was added to the beads and incubated for 30 min. A solution (30 μL) of freshly prepared *N*-(3-dimethylaminopropyl)-*N'*-ethylcarbodiimide hydrochloride (EDC) (Sigma, MO) solution (100 mg/mL in cold water) and MES buffer (10 μL) were added to the bead solution. The mixture was incubated (6 h) at room temperature with slight shaking. The solution was discarded and beads were suspended in Tris

(50 mM, pH 7.4) buffer for 15 min. Beads were washed with PBS (100 mM, pH 7.4, 150 mM NaCl, 0.1% BSA) (5×) and then resuspended in PBS (1 mL) (100 mM, pH 7.4, 150 mM NaCl). A hemocytometer count gave the bead concentration as 1.1×10^8 beads/mL.

24.2.2 Preparation of Bacteria

S. aureus strain RN4220 was obtained from the laboratory of Richard Novick (29). Four to five *S. aureus* colonies from a freshly streaked Luria–Bertani (LB) agar plate were inoculated into LB broth in an Erlenmeyer flask and grown for approximately 2 h at 37 °C. The culture was centrifuged at 6000 rpm for 30 min. The supernatant was removed and the pellet was resuspended in phosphate buffered saline (pH 7.4). The optical density of the culture in PBS was measured at 600 nm. The culture was then serially diluted in PBS from 10^{-4} to 10^{-8} OD_{600} and the diluted cell suspensions were used in the bead-binding assay. The *S. aureus* live cell counts (colony forming units) per unit volume for these cell suspensions were determined by 10-fold serial dilution and plating 100 μL of appropriately diluted samples onto an LB agar plate, three replicate plates per dilution. After an overnight incubation of the plates at 37 °C, a set of three plates with appropriate numbers of colonies (between 30 and 300) were counted and cell counts (cfu/mL) for the original cell suspension were calculated.

24.2.3 Peptide Binding Studies with Teic-Bearing Magnetic Beads

Binding of $FAM(DA)_3$ (Anaspec, San Jose, CA, Custom synthesized) to teic-beads in the absence and presence of two competing peptides, D-Ala-D-Ala ($(DA)_2$) and *N*,*N*-diacetyl-lys-D-Ala-D-Ala ($NNL(DA)_2$) (Sigma, St. Louis, MO), were investigated. Flow cytometric, fluorescent imaging, and microfluidic techniques were employed to monitor direct binding of teic-beads $FAM(DA)_3$. Spectrofluorometry was used to study the peptide binding to the beads in suspensions.

24.2.3.1 Spectrofluorometry

24.2.3.1.1 Suspension-Based Binding Assays Spectrofluorometric measurements were performed in constant wavelength mode on a Fluoromax-3 spectrofluorometer. The sample was excited at 488 nm and the emission monitored at 520 nm. As shown in Fig. 24.2, suspension-based methods were utilized to analyze the binding of labeled peptide to teic-beads. Samples containing $\sim 2.2 \times 10^6$ beads/mL (teic-beads and nonderivatized beads) were incubated in different concentrations (1 nM to 10 mM depending on the type of experiment) of 5-$FAM(DA)_3$ (160 μL) in Eppendorf tubes with slight shaking. The beads were separated from the supernatant using an external magnet. The supernatant 5-$FAM(DA)_3$ was measured for its fluorescence intensity. To determine the reaction rate of ligand with derivatized beads, 5-$FAM(DA)_3$ (5 nM) was incubated with teic-beads. The fluorescence intensity decrease was monitored at different time intervals (0–30 min).

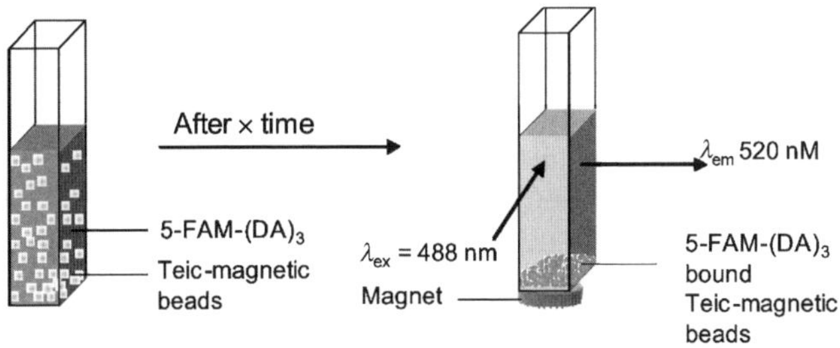

FIGURE 24.2 Suspension-based assays for spectrofluorometry. Magnetic beads are initially suspended in $FAM(DA)_3$. An external magnet was used to deposit the beads. Fluorescent intensity of the supernatant is measured before and after the incubation with beads at 520 nm.

24.2.3.1.2 Analysis of Fluorescent Peptide Binding to Teic-Beads A sample containing $\sim 2.2 \times 10^6$ beads/mL were incubated in ligand (0.1–100 μM) in Eppendorf tubes for 15 minutes. The supernatant intensities of 5-FAM$(DA)_3$ were measured as previously described. The measured intensities were used to determine the concentration of ligand ($[L_b]$) associated with the beads as shown in Eq. 24.1. In Eq. 24.1, I_0 is the 5-FAM$(DA)_3$ intensity before reacting with teic-beads and $[L]_0$ is the initial concentration of ligand. The number of 5-FAM$(DA)_3$ molecules per bead at a particular initial concentration was calculated using Eq. 24.2. Here, A and n are Avagadro's number and concentration of beads, respectively.

$$[L]_b = \frac{I_0 - I_t}{I_0}[L]_0 \quad (24.1)$$

$$\text{Sites} = \frac{[L]_b A}{n} \quad (24.2)$$

24.2.3.1.3 Competitive Binding Assays with $(DA)_2$ and NNL$(DA)_2$. A solution (80 μL) of NNL$(DA)_2$ and $(DA)_2$ (5 nM–0.5 μM) was separately mixed with a solution (80 μL, 10 nM) of 5-FAM$(DA)_3$. Binding of 5-FAM$(DA)_3$ was measured as described above. Normalized intensity, as described in Eq. 24.3, was used to demonstrate the competitive binding of 5-FAM$(DA)_3$ with beads in the presence of two similar peptides. The derived data were also used to demonstrate the different affinities of NNL$(DA)_2$ and $(DA)_2$ toward teic-beads.

$$\textit{Normalized intensity} = \frac{I_0 - I_t}{I_0} \quad (24.3)$$

24.2.3.1.4 Stability Studies of Derivatized Beads Beads (1.1×10^8 /mL) were generally stored in PBS (100 mM, pH 7.4, 150 mM NaCl) at 5°C for long-term use. The reactivity of beads of this stock solution was monitored by reacting portions of the stock with 5-FAM$(DA)_3$ periodically for 1 month. The fluorescent intensity of reacted 5-FAM$(DA)_3$ supernatant was normalized to the intensity of the original peptide solution of each day as in Eq. 24.3.

24.2.3.1.5 Bacteria Binding to the Teic-Beads A solution (160 μL) of a nonpathogenic strain of *S. aureus* was incubated with teic-beads for 30 min at room temperature. Beads were separated from the bacteria solution and reincubated with 5-FAM$(DA)_3$ (5 nM). The fluorescent intensity of the supernatant 5-FAM$(DA)_3$ was measured after 5 min. A control experiment was performed by incubating teic-beads with 5-FAM$(DA)_3$ (160 μL, 5 nM). Another control was performed with nonderivatized beads incubated in bacteria and subsequently with 5-FAM$(DA)_3$ to monitor nonspecific binding of reagents to beads.

24.2.3.2 Fluorescent Imaging The teic- and blank beads reacted with 5-FAM$(DA)_3$ (10 μM) and were washed (5×) with PBS buffer and used for fluorescent imaging purposes. An inverted epi-fluorescence microscope (Nikon Eclipse-TE 2000U) equipped with SPOT INSIGHT™ color CCD camera and SPOT imaging software (Diagnostic Instruments, MI) were used to capture the fluorescent images of beads.

24.2.3.3 Flow Cytometry The flow cytometric analysis used a Becton–Dickinson FACScan flow cytometer (Sunnyvale, CA) equipped with an argon-ion laser. The laser output was fixed at 488 nm. Samples (20 μL) of teic- and nonderivatized beads (control) were incubated with different concentrations of 5-FAM$(DA)_3$ in micro-centrifuge tubes for 15 min. The bead samples were washed (3×) with PBS buffer and resuspended in 0.5 mL for flow cytometry analysis.

24.2.3.4 Microfluidics

24.2.3.4.1 Fabrication of Microfluidic Chip A microfluidic channel (Fig. 24.3a) was prepared by conventional soft lithographic techniques. Briefly, a mold was created on a 3 inch silicon wafer using photolithography (30). A solution (2 mL) of degassed PDMS prepolymer solution was spin coated onto the mold at 2000 rpm for 30 s to obtain a thin layer of PDMS. The sample was baked for 15 min at 70 °C. Two 2 mm diameter rare earth magnets were aligned with the microchannel with the help of a "holding" magnet (Fig. 24.3b). Degassed PDMS prepolymer solution was poured onto the assembly and baked for 2 h at 70 °C to obtain a thicker layer of PDMS on top of the thin layer. The combined PDMS layers were peeled off from the wafer. Holes were punched at either end of the microchannel for the liquid connections. The PDMS chip was irreversibly sealed onto a glass slide (Fig. 24.3b) using an argon plasma (30).

A peristaltic pump, connected to the chip via silicon tubing, was used to introduce beads and reagents into the chip. The chip was first rinsed with PBS buffer containing 0.1% BSA to minimize the nonspecific binding of compounds to the PDMS surface. A known amount of derivatized beads was injected at the inlet end of the device by using a peristaltic pump. Continuous flow was maintained by connecting silicone tubing (0.64 mm i.d.) filled with buffer. The beads were packed

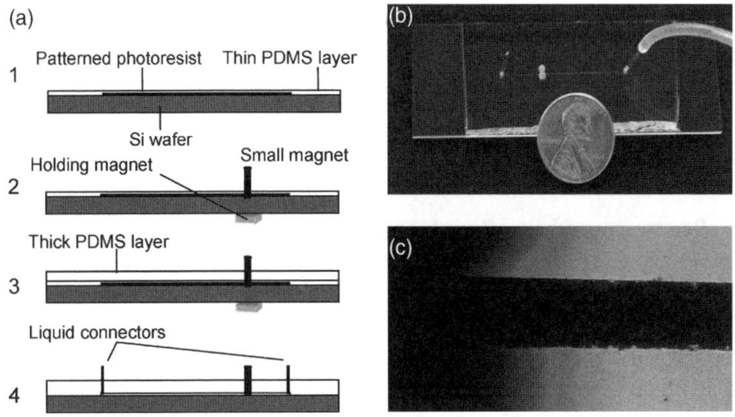

FIGURE 24.3 Fabrication of the microfluidic chip. (**a**) Steps involved in preparing rare earth magnets embedded PDMS chip. (**b**) Image of final device. (**c**) Optical micrograph of a magnetic bead-packed microchannel. Shaded areas are where the magnets are embedded.

at the two magnets (Fig. 24.3c). The packed bead segment was washed (5 min) with PBS buffer.

24.2.3.4.2 Sample Injection and Detection in the Chip The peristaltic pump was stopped and the silicone tubing containing the running buffer at the inlet end was removed from the chip. A 5 μL plug of 5-FAM$(DA)_3$ was manipulated into the tube and connected to the microchannel. The sample plug was manipulated through the microfluidic channel at a rate of 0.4 μL/min using a peristaltic pump. The fluorescence intensity was measured periodically and, when the labeled compound reached the bead segment, the flow was stopped for 10 min. After the beads were incubated with the labeled compound, flow was restarted to remove the excess reagent from the device while monitoring the fluorescent intensity. Control experiments were performed with 5-FAM$(DA)_3$ and nonderivatized beads.

24.3 RESULTS AND DISCUSSION

24.3.1 Derivatization of Magnetic Beads with Antibiotics

The primary amine in teicoplanin (Fig. 24.1a) can be coupled to carboxylic acid groups on the surface of the magnetic beads via amide bonding mediated by carbodiimide activation. Although self-polymerization of teicoplanin is possible due to the terminal carboxylic group, the use of excess ligand in the reaction prevents this from occurring. Since beads can be easily separated from the reactant solutions, excess reagents can be removed by thorough washing of the beads. By incubating derivatized beads with Tris buffer, unreacted activated carboxylic acid groups on the magnetic beads are quenched. Derivatized beads were washed with PBS buffer containing 0.1% BSA to minimize the nonspecific binding of ligands. For long-term storage, beads were suspended in a known amount of PBS buffer. Bead concentration was calculated by hemocytometer counting.

24.3.2 Suspension Assays of Fluorescent Peptides Binding to Teic-Bearing Beads

24.3.2.1 Reactivity and Stability Assays with Teic-Beads Direct monitoring of fluorescent particles in steady state spectrofluorometry can lead to erroneous results due to the scattering effects from the particles themselves. In such a situation, indirect methods can be used to monitor the binding of fluorescent ligands to receptors on beads. When a solution of 5-FAM$(DA)_3$ is incubated with teic-beads, the binding of fluorescent ligands to receptors can be monitored via the decrease in fluorescence intensity of the supernatant. The reaction rate between teicoplanin and D-Ala peptides in solution is in the range of a few milliseconds. Fig. 24.4a shows direct monitoring of binding interactions between labeled ligand and receptors on beads. The initial intensity of the supernatant is decreased by 50% within the first minute indicative of the beads reactivity. We did not find any appreciable decrease in affinity of derivatized teicoplanin to D-Ala peptides. After approximately 10 min, the change in the decrease in intensity is not significant and an overall decrease of 73% is reached after 30 min of incubation. The long shelf life of beads enables the continuous use of a batch of beads for many assays without repetitive preparations. Fig. 24.4b demonstrates that the beads are reasonably stable even after 1 month. The intensities are normalized according to Eq. 24.3. The reactivity decreases by 13%–15% after the first week. This loss may be attributed to the dissociation of some teicoplanin molecules from the surface of the bead. Overall, we have observed that an intensity decrease above 5 nM of 5-FAM$(DA)_3$ after reacting with freshly prepared teic-beads is 73%–84%.

24.3.2.2 Equilibrium Binding of 5-FAM$(DA)_3$ to Teic-Bearing Beads A suspension-based method was employed to obtain maximum surface coverage of beads using spectrofluorometry. The concentration of 5-FAM$(DA)_3$ was varied while maintaining the amount of beads constant (Fig. 24.5). The analysis gave an approximate value of binding sites on beads of about 50 million. The equilibrium binding of 5-FAM$(DA)_3$ to teic was analyzed using spectrofluorometry and flow cytometry. The mean value of each histogram in flow cytometry (Fig. 24.6a) was normalized to the value for blank beads to obtain each data point. The equilibrium binding data from fluorometry and flow cytometry were analyzed as sigmoidal response curves as shown in Fig. 24.6b and c. The sigmoidal analysis can be used to obtain binding kinetics for 5-FAM$(DA)_3$ and teic-beads. The dissociation constant (K_d) derived from flow cytometry data is about 3.04×10^{-6} M (Fig. 24.6b) and 6.5×10^{-6} M from spectrofluorometry (Fig. 24.6c). The close resemble of values indicate the correlation between two approaches. These data are consistent with capillary electrophoresis data obtained from solution-based homogeneous experiments (32). The micromolar range dissociation constant suggests the binding of peptides to the

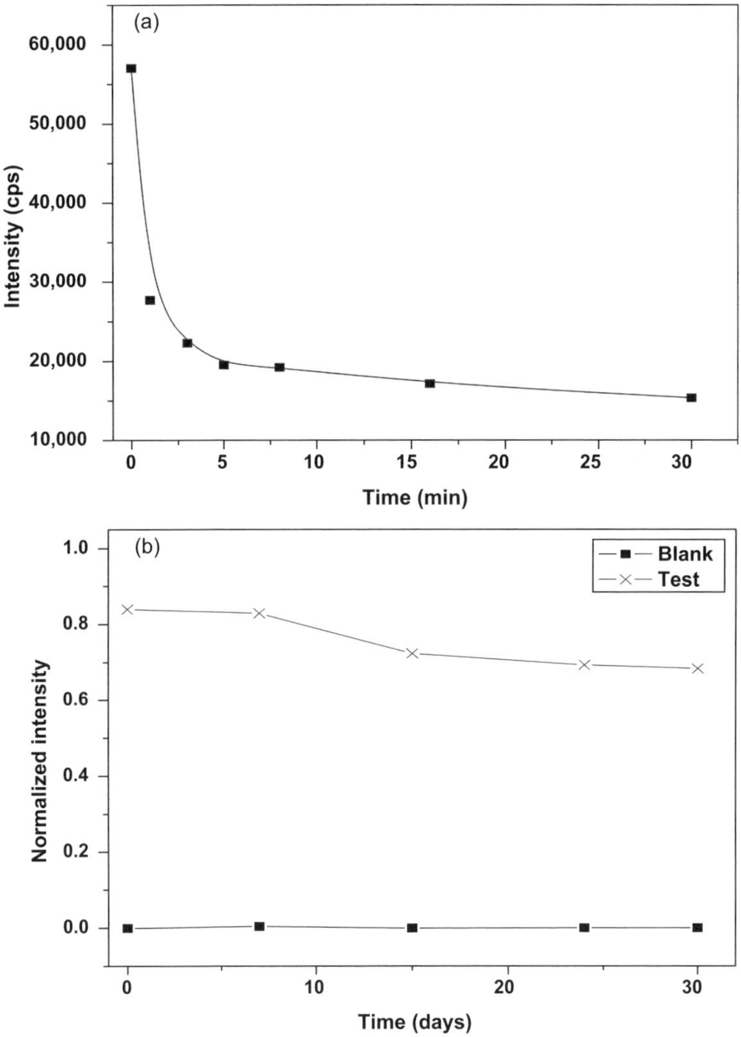

FIGURE 24.4 Spectrofluorometric studies of reactivity and stability of magnetic beads derivatized with teicoplanin. (**a**) Reactivity of teic-beads were analyzed by monitoring the change in fluorescence intensity of supernatant 5-FAM(DA)$_3$. (**b**) Stability of derivatized bead were analyzed by monitoring the reactivity of beads for 1 month.

teic-bead is not strong and a weak binding ligand can be replaced by a strong binding ligand.

24.3.2.3 Competitive Assays with NNL(DA)$_2$ Certain varieties of D-alanine (and glycine) containing peptides can have different affinities toward teicoplanin due to the differences in electrostatic interaction. One of the reasons for these differences in affinity is due to the change in number of hydrogen bonds formed between the

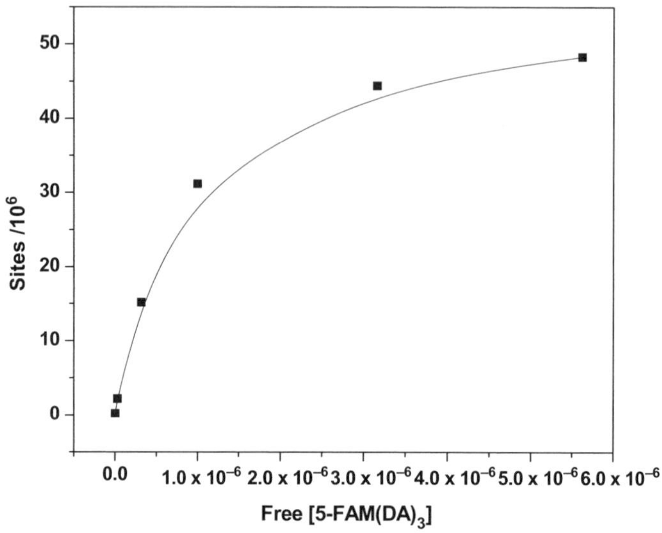

FIGURE 24.5 Graph of number of receptor sites versus free 5-FAM$(DA)_3$. Number of receptor sites on beads was calculated using Eq. 24.2 to obtain the maximum surface coverage of binding sites per bead. Data suggest $\sim 50 \times 10^6$ binding sites per bead.

peptide and the antibiotic (32). As a proof-of-concept study to demonstrate the feasibility of using teic-beads for competitive binding of peptides and to show peptides' different affinities toward beads, we chose unlabeled $(DA)_2$ and NNL$(DA)_2$ along with 5-FAM$(DA)_3$. The normalized intensity of the supernatant 5-FAM$(DA)_3$ in Fig. 24.7 was calculated using Eq 24.3. According to this expression, the graph shows the degree of binding of 5-FAM$(DA)_3$ in the presence of two competing peptides. It is apparent that the fluorescent peptide binds more favorably at low concentrations than the two competing peptides. Even at higher concentrations of $(DA)_2$ the binding of 5-FAM$(DA)_3$ is still greater indicating the strong interaction of teic-beads with the fluorescent peptide. This binding is mainly due to the presence of an extra hydrogen bond between 5-FAM$(DA)_3$ and teicoplanin (26). In the case of NNL$(DA)_2$ the binding of fluorescent peptide is weaker. Both peptides have the same number of hydrogen bonds with teicoplanin on the beads but the bulkiness of the fluorescent tag can cause weaker binding. Further studies are needed to confirm this hypothesis.

24.3.3 Fluorescent Imaging

Figure 24.8a shows a fluorescent micrograph of teic-beads reacted with 10 μM 5-FAM$(DA)_3$. Fig. 24.8b is the fluorescent micrograph of blank beads after reacting with the same concentration of fluorescent ligand. For better image acquisition purposes a higher concentration of ligand was used. The micrographs show the absence of any significant nonspecific binding of ligands to the nonderivatized beads. The micrographs were taken using a 40× objective lens.

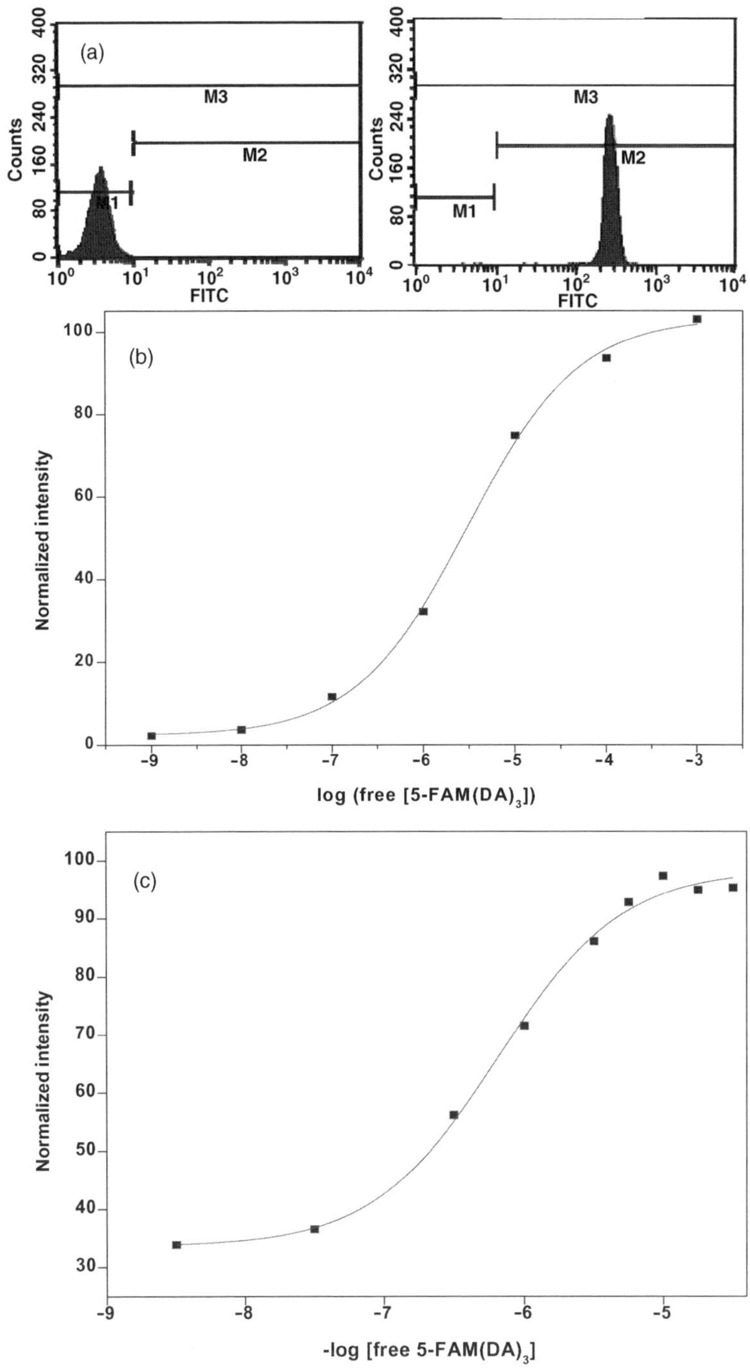

FIGURE 24.6 Equilibrium binding of 5-FAM(DA)$_3$ with teic-beads. (**a**) Mean value of each histogram for different concentration was used for flow cytometric analysis. (**b**) Sigmoidal analysis of flow cytometric data. A K_d of ~3.04×10^{-6} M was derived from the equilibrium binding curve. (**c**) Sigmoidal analysis of spectrofluorometric data. A K_d of ~6.5×10^{-6} M was derived from the equilibrium binding curve.

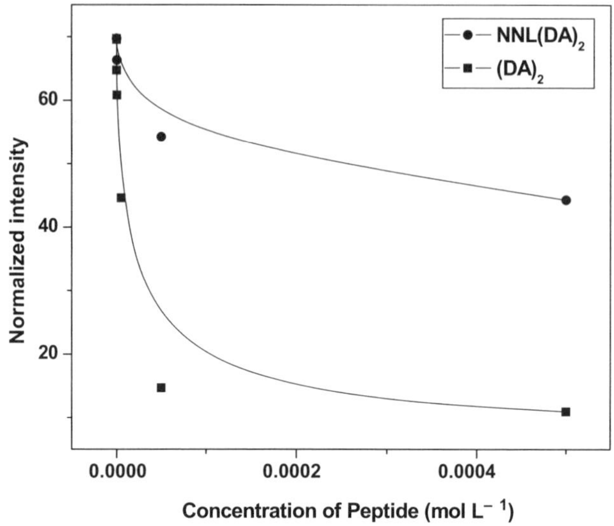

FIGURE 24.7 Competitive binding of 5-FAM$(DA)_3$ with structurally similar ligands, NNL $(DA)_2$ and $(DA)_2$.

24.3.4 Bacteria Binding to the Teic-Beads

The binding of bacteria to teic-beads was monitored via spectrofluorometry. When a known amount of teic-beads is incubated with a sample of *S. aureus* prior to the incubation with 5-FAM$(DA)_3$, if the bacteria bind to the beads, the number of receptors on the beads able to bind to 5-FAM$(DA)_3$ will be decreased (Fig. 24.9a). The bacteria binding can be observed indirectly by monitoring the degree of labeled ligand binding. In Fig. 24.9b, normalized intensity (Eq. 24.3) of control is taken as 100% binding of 5-FAM$(DA)_3$ to teic-beads and compared with normalized intensities for samples containing teic-beads that were preincubated with different concentrations of bacteria prior to reacting with 5-FAM$(DA)_3$. Each analysis was done in triplicate. A portion of each solution was analyzed by the colony counting method to determine the concentration of bacteria. The limit of detection of *S. aureus* in said method was approximately 200. We did not observe any significant difference in signal for the bacteria samples of ~1600 and ~200 cfu. We tested bacteria samples

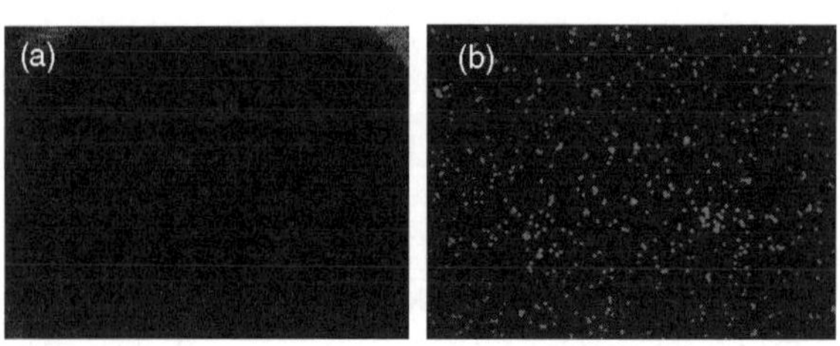

FIGURE 24.8 Fluorescence micrographs of magnetic beads after reacting with 5-FAM $(DA)_3$. (**a**) Non-derivatized and (**b**) derivatized with teicoplanin. (See color insert)

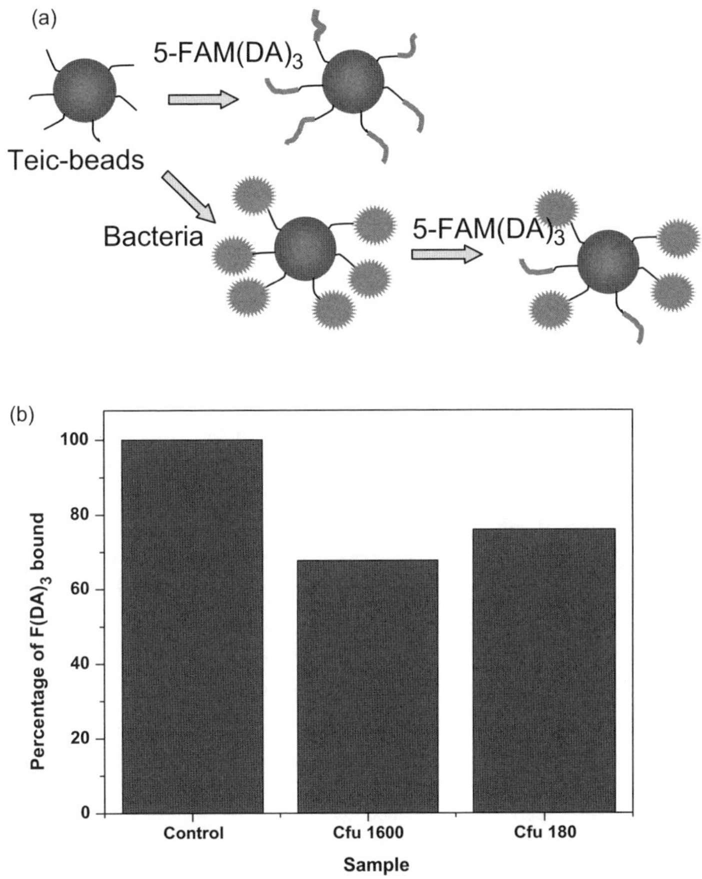

FIGURE 24.9 Gram-positive bacteria binding to teicoplanin bearing magnetic beads. (**a**) Schematic diagram showing the basis of bound bacteria determination. Teic-beads will be incubated with 5-FAM$(DA)_3$ in control. Teic-beads will be incubated with bacteria first and then with 5-FAM$(DA)_3$ for test samples. (**b**) Degree of bacteria binding is determined by the percentage of fluorescence decrease with comparison to control.

with less than ~180 cfu, but could not distinguish the presence of bacteria. This observation may be mainly due to the strong attraction toward 5-FAM$(DA)_3$ that can replace the bound bacteria to the beads. The concept we demonstrate here is simple and the response is fast in comparison to conventional methods. We believe this approach can be further utilized in developing new types of biosensors for pathogen detection.

24.3.5 Microfabrication of Microfluidic Chips

Magnetic beads were captured by aligning two small rare earth magnets externally to the microchannel. In unpublished work, we found that the thickness of the PDMS layer between a magnet and magnetic object can affect the strength of the magnetic field from the magnet. We could minimize the distance between the microchannel surface and the two magnets by creating a thin layer of PDMS on the photoresist mold (Fig. 24.3b) followed by alignment of the magnets and finally creating a thick layer of PDMS to make a more rigid device.

24.3.6 Microfluidic Assays on Teicoplanin-Ligand Binding

A simple microfluidic experiment was performed to demonstrate the binding of 5-FAM$(DA)_3$ to teic bearing beads in a microchip. Beads were confined to a region of approximately 800 μm in length. Due to the dead volume between the injection point and the packed segment, the initial zone of 5-FAM$(DA)_3$ took approximately 2 min to reach the beads. The initial increase in fluorescent intensity (shown as circles and squares in Fig. 24.10) is mainly due to the presence of excess ligand in the system. The beads were incubated for 10 min without flow of fluid and then washed for an additional 10 min to remove excess ligand from the system. The sharp decrease in intensity is indicative of the removal of excess ligand. The more stable fluorescent intensity after 50 min (circles) indicates the presence of 5-FAM$(DA)_3$ bound to teic beads. The gradual decrease in intensity from 55 min to the end of the curve is mainly due to the dissociation of bound 5-FAM$(DA)_3$. A control experiment was carried out by using nonderivatized magnetic beads (squares in Fig. 24.10). The amount of beads used was similar to that used in the test experiment. A significant retention

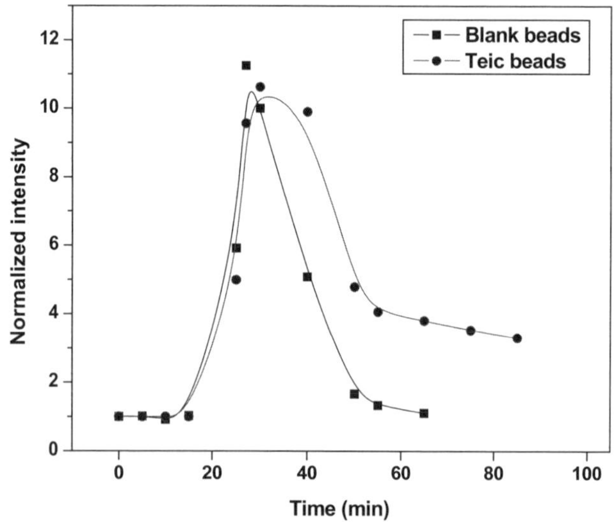

FIGURE 24.10 Analysis of 5-FAM$(DA)_3$ binding to magnetic beads packed in microfluidic channels. The slow decrease in fluorescence intensity in trace with circles indicative of binding of fluorescent ligand to teic beads where as rapid decrease in intensity in trace with squares indicative of the nonbinding of fluorescent ligand to nonderivatized beads.

of fluorescent intensity at the bead segment was not observed indicative of the nonbinding of fluorescent ligands to the nonderivatized beads. This simple microfluidic experiment demonstrates that bead-based methods can be developed in a microfluidic format to analyze antibiotic–ligand interactions. The major limitation is the weak binding of ligands to teic-beads, causing many initially bound complexes to dissociate during washing steps, thus decreasing the sensitivity and limit of detection.

24.4 CONCLUSIONS

A successful derivatization of magnetic particles with the antibiotic teicoplanin and its use in the detection of D-Ala-D-Ala terminus ligands has been described. The dissociation constants for 5-FAM$(DA)_3$ obtained from spectrofluorometry and flow cytometry are similar and comparable with known dissociation constants for similar peptides obtained from other techniques. We have also shown that a particle-based system can be employed to detect low levels of bacteria avoiding time consuming steps used in conventional techniques. The weak binding of ligand to teicoplanin-bearing beads is not suitable for developing microfluidic assays with direct fluorescence detection mainly due to the dissociation of bound complexes during the rinsing of unbound reagents. Further work will focus on examining tighter receptor–ligand binding models and or using other fluorescence based techniques such as fluorescence resonance energy transfer (FRET).

ACKNOWLEDGMENTS

The authors gratefully acknowledge Lilian J. Real and Dr. Howard Xu at Department of Biology at California State University, Los Angeles and Rochelle A. Diamond at Department of Biology at California Institute of Technology for their technical supports and useful discussions, and for the National Science Foundation (CHE-0515363, DMR-0351848 and DMR-0520565), and the National Institutes of Health (1R15AI65468-01) for financial supports for this research.

REFERENCES

1. Purcell EM. Am J Phys 1977;45:3–11.
2. Seong GH, Crooks RM. J Am Chem Soc 2002;124:13360–13361.
3. Sato K, Tokeshi M, Kimura H, Kitamori T. Anal Chem 2001;73:1213–1218.
4. Piyasena ME, Buranda T, Wu Y, Huang J, Sklar LA, Lopez GP. Anal Chem 2004;76: 6266–6273.
5. Chung W-J, Kim M-S, Chos, Park S-S, Kim J-H, Kim Y-K, Kim B-G, Lee Y-S. 2005;26:694–702.

6. Oleschuk RD, Shultz-Lockyear LL, Ning Y, Harrison DJ. Anal Chem 2000;72:585–590.
7. Bucak S, Jones DA, Laibinis PE, Hatton TA. Biotechnol Prog 2003;19:477–484.
8. Technical Information. Available at http://www.invitrogen/dynal.
9. Jordan A, Scholz R, Maier-Hauff K, van Landeghem F, Waldoefner N, Teichgraeber U, Pinkernelle J, Bruhn H, Neumann F, Thiesen B, von Deimling A, Felix R. J Neurooncol 2006;78:7–14.
10. Richardson J, Hawkins P, Luxton R. Biosens Bioelectron 2001;16:989–993.
11. Hayes MA, Polson NA, Phayre AN, Garcia AA. Anal Chem 2001;73:5896–5902.
12. Choi J-W, Oh KW, Thomas JH, Heineman WR, Halsall HB, Nevin JH, Helmicki AJ, Henderson HT, Ahn CH. Lab Chip 2002;2:27–30.
13. Georgieva J, Milling A, Orfanos CE, Geilen CC. Melanoma Res 2002;12:309–317.
14. Lea T, Vartdal F, Davies C, Ugelstad J. Scand J Immunol 1985;22:207–216.
15. Treleaven JG, Gibson FM, Ugelstad J, Rembaum A, Philip T, Caine GD, Kemshead JT. Lancet 1984;8368:70–73.
16. Fan ZH, Mangru S, Granzow R, Heaney P, Ho W, Dong Q, Kumar R. Anal Chem 1999;71:4851–4859.
17. Safarik I, and Safarikova M. BioMagn Res Technol 2004;2:7.
18. Jenkins C, Pearce MC, Smith AW, Kright HL, Shaw DJ, Cheasty T, Foster G, Gunn GJ, Dougan GJ, Smith HR, Frankel G. Lett Appl Microbiol 2003;37:207–212.
19. Gu H, Ho P-L, Tsang KWT, Wang L, Xu B. J Am Chem Soc 2003;125:15702–15703.
20. Kahne D, Leimkuhler C, Lu W, Walsh C. Chem Rev 2005;105:425–448.
21. Loll PJ, Axelsen PH. Annu Rev Biomol Struct 2000;29:265–289.
22. Dudley H, Williams BB. Angew Chem Int Ed 1999;38:1172–1193.
23. Bardone MR, Paternoster M, Coronelli C. J Antibiot 1978;31:276–283.
24. Parenti F, Beretta G, Berti M, Arioli V. J Antibiot 1978;31:276–283.
25. Barry AL, Thornsberry C, Jones RN. J Clin Microbiol 1986;23:100–103.
26. Soriano A, Popesau D, Garcia S, Bori G, Martinez JA, Balasso V, Marco F, Almela M, Mensa J. Eur J Clin Microbiol Infect Dis 2006;25:35–38.
27. Verbist L, Tjandramaga B, Hendrickx B, Van Hecken A, Van Melle P, Verbesselt R, Verhaegen J, De, Schapper PJ. Antimicrob Agents Chemother 1984;26:881–886.
28. Lin Y-S, Tsai P-J, Weng M-F, Chen Y-C. Anal Chem 2005;77:1753–1760.
29. Novick RP. Molecular Biology of the Staphylococci. New York NY: VCH Publishers; 1990.
30. Duffy DC, McDonald JC, Schueller OJA, Whitesides GM. Anal Chem 1998;70: 4974–4984.
31. Zavaleta J, Chinchilla DB, Kaddis CF, Martinez K, Brown A, Gomez A, Pao A, Ramirez A, Nilapwar S, Ladbury JE, Gomez FA. J Cap Elect Microchip Tech 2006;9:101–117.
32. Loll PJ, Kaplan J, Selinsky BS, Axelsen PH. J Med Chem 1999;42:4714–4719.

25

INTERFACING MICROCHANNEL ELECTROPHORESIS WITH ELECTROSPRAY IONIZATION MASS SPECTROMETRY

KATHLEEN KELLY, BRENT RESCHKE, AND AARON T. TIMPERMAN

25.1 INTRODUCTION

Capillary electrophoresis (CE) excels at the separation of charged species in solution, owing largely to its unparalleled separation efficiencies. The high separation efficiencies mean that CE is capable of producing very narrow peaks and separating a large number of components that are present in many biological samples. Mass spectrometry (MS) is used extensively for structural identification and characterization, and has very low detection limits, making it excellent for interfacing with microscale separations and the analysis of biological samples. MS separates ions in the gas phase and measures their mass-to-charge ratio, providing an additional dimension of separation. To obtain greater structural information, ions can be separated, isolated, fragmented, and the m/z of the fragment ion measured in the MS. Electrospray ionization (ESI) is the preferred method for producing gas-phase ions of biomolecules from the solution phase and on-line coupling of liquid-phase separations, such as CE, with MS. The combination of CE and MS provides a powerful analytical tool that can achieve highly efficient separations, structural identification, and characterization for compounds at trace levels. These characteristics make microchannel electrophoresis–MS well suited for the analysis of complex biological samples, including those encountered in metabolomics, genomics, and proteomics. An additional benefit of microfluidic devices for biological samples is that they have

Biological Applications of Microfluidics, Edited by Frank A. Gomez

the potential to be mass produced inexpensively, leading to disposable devices and eliminating the potential for sample carryover.

Microchannel electrophoresis and CE both use an electric field to separate charged compounds in a weak electrolyte solution and have the same mechanisms of separation. Because the separations are essentially the same, many of the same challenges are present for coupling microchannel electrophoresis and CE with ESI–MS. Physically, however, the systems are different: CE is performed in a tubular capillary with a circular cross-section, and microchannel electrophoresis is performed in channels formed on a planar substrate with various cross-sections that often approximate trapezoids or squares. Microchip or microfluidic platforms offer the unique advantage of having intersections that make it possible to steer sample plugs through intersections. For microchannel electrophoresis, this steering or directing of samples is normally achieved electrokinetically by controlling the voltages applied to electrodes in the buffer reservoirs at the ends of the channels. With intersections and band steering, much more can be achieved with microchannel electrophoresis than just separation; processes such as sample preparation, concentration, digestion, and purification can be performed in an automated fashion with minimal losses and contamination before or after the separation (1,2). Intersections and a wide array of microfabrication techniques for microfluidic devices increase possible ESI–MS interface configurations, which are being used to improve the coupling of microchannel electrophoresis and MS with ESI.

Another important characteristic of ESI is that it is a concentration sensitive technique, meaning a mass of sample diluted in a greater volume will yield a lower signal in the MS when compared to a more concentrated sample (3). This is a particularly salient feature when considering the use of microfluidic devices because microfluidic devices allow the sample to be constrained to a smaller volume, allowing the concentration to be maintained at a higher level. Thus, for a constant sample mass, microfluidic devices can be used to separate and process samples while minimizing dilution, which will ultimately improve the detection limits for ESI–MS. Smaller channel dimensions combined with lower volumetric flow rates provide smaller sample band volumes and higher concentrations. Therefore, it is important to use the lowest flow rates that are stable under routine conditions. With cross-sections smaller than a human hair, sample volumes of a few nanoliters are achievable for routine analyses.

Reversed phase-high performance liquid chromatography (RP-HPLC) has been the method of choice for online separations prior to MS analysis, mainly due to the simplicity and more rugged nature of the ESI–MS interface. CE–ESI–MS is complimentary to RP-HPLC and is well suited for the separation of compounds that have similar hydrophobicities but substantially different charges, such as phosphorylated and glycosylated peptides and proteins, prior to MS. Additionally, it is anticipated that microchannel electrophoresis will find increased usage in multidimensional separation platforms. In multidimensional combinations of RP-HPLC and CE, the CE separation is after the RP-HPLC separation because CE is much faster and RP-HPLC is a gradient method whose speed is easily controlled. Thus, in multidimensional combinations of RP-HPLC and

CE, the CE separation will be the second dimension and must be interfaced with ESI-MS.

Although CE–ESI–MS has many favorable characteristics, these characteristics have not yet translated into widespread use in industry and research applications. There are very few commercial CE–ESI–MS interfaces available, and the few that are available are not widely used. From this view of the field, it follows that many improvements are still needed for rugged CE–ESI–MS interfaces. However, to effectively develop better interfaces, it is a prerequisite to first discuss the limitations of CE–ESI–MS interfaces and the challenges involved with making a stable CE–ESI–MS interface. Although stability is critical, it is also important that the other desirable characteristics of CE, such as the extremely high separation efficiencies, are maintained by the interface. Interfaces that greatly diminish desirable characteristics of the CE separation or the detection limits in the MS are less than ideal solutions.

A brief discussion of the most important processes and features that lead to the instability of the electrospray will be discussed, followed by a general overview of ESI, and finally a more comprehensive discussion of the factors that lead to the instability of electrospray. The first problem is the instability of the electroosmotic flow (EOF). The EOF generated in the separation channel typically provides the bulk flow that is used to produce the electrospray, and the instabilities in the EOF cause the electrospray to be unstable as well. To achieve stable EOF, surface adsorption on the walls of the microchannels must be minimized. Surface adsorption is minimized by coating the microchannel walls, which is particularly important with sticky biological samples, such as proteins, that are prone to surface adsorption. The second challenge is associated with coupling or fabricating an ESI tip on the microfluidic chip. The appropriate ESI tip orifice inner diameter (i.d.) is proportional to the flow rate, decreasing as the flow rate decreases. As the ESI tip orifice decreases, the frictional resistance to bulk flow increases. A mismatch in flow rates or small dead volumes can increase band broadening and eliminate the high separation efficiencies achievable with CE. The third problem for CE–ESI–MS is that an intermediate electrode must be inserted in the circuit to separate the electric field for the separation and the potential for the ESI. The analyte must not flow to this intermediate electrode and perturbation of the separation must be minimal. Additionally, increased flexibility is needed in CE–ESI–MS interfaces to change the buffer/solvent composition after the separation and before the ESI. It is anticipated that microchannel electrophoresis–ESI–MS will become a well used method for the analysis of complex samples as clever methods are found to overcome the challenges encountered with the interface design.

25.2 ELECTROSPRAY IONIZATION

ESI is the method of choice for creating gas-phase ions of biomolecules in the solution phase. A soft ionization process that minimizes ion fragmentation is required for the analysis of large biomolecules. The fragmentation of large molecules in the ionization source can produce a myriad of product ion peaks that can make the spectra nearly impossible to interpret and is consequently avoided. A basic understanding of the chemical processes occurring is important when designing an effective CE–MS interface. The following discussion will be limited to positive ion mode since it is most commonly used for proteins and peptides and has lower limits of detection (LODs) than the negative ion mode.

An overview of the ESI process in the cone-jet mode is shown in Fig. 25.1. In the positive ion mode, a positive potential is applied to the emitter tip and the MS inlet is held at or near ground. The electric field repels the solution from the emitter tip and draws it toward the MS, forming a Taylor cone (4). When Coulombic repulsion exceeds the surface tension of the solution, a filament emerges from the tip of the Taylor cone that breaks into droplets (5). These droplets rapidly shrink as the solvent evaporates, increasing the charge density of the droplets until a Coulombic explosion, known as droplet fission, ejects more small highly charged droplets (4,6,7). The drying process and repeated droplet fission eventually lead to the creation of multiply charged gas-phase ions (8,9). In a typical ESI source, most of the droplets are not fully desolvated, and only a small fraction of the sample produces bare ions in the gas phase. Consequently, sources that produce smaller droplets are most efficient at producing dry gas-phase ions (10). Nanospray sources with lower flow rates produce droplets that are significantly smaller than that of other ESI sources (11,12). As the droplets move toward the MS, their Coulombic repulsions cause the droplet trajectories to diverge producing a broad spray plume. Unfortunately, the smaller droplets diverge the most from the center of the plume, while the larger droplets remain closer to the center. Orthogonal sources sample the small droplets at the periphery of the plume while avoiding the large solvent droplets.

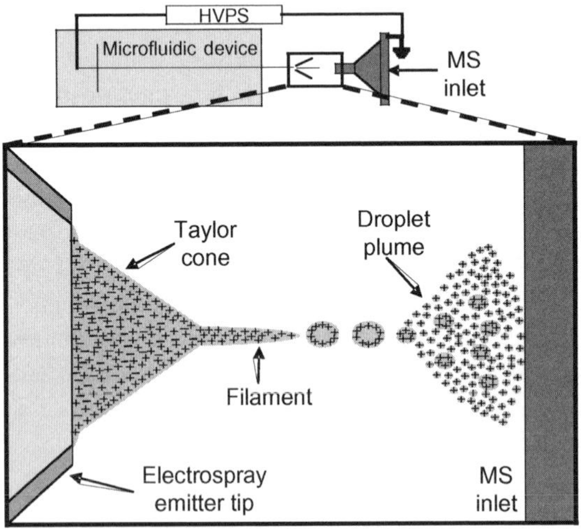

FIGURE 25.1 A schematic is shown of the microfluidic chip with the basic electric circuit and the high voltage power supply (HVPS) applying the potential. In the expanded view, more detail is shown of the electrospray process. The appropriate voltage to maintain stable electrospray is determined by a number of parameters including the electrospray emitter orifice i.d. and the air gap distance. In the cone jet mode, larger droplets are ejected from the filament and dry down through an iterative fission process that eventually yields bare ions.

The creation of the multiply charged gas-phase ions characteristic of the electrospray process has been explained by two competing mechanisms. The first mechanism states that ions are created by consecutive droplet fission processes and solvent evaporation. Droplets become smaller with each fission process, ultimately resulting in the proton being transferred to a completely desolvated analyte (8). The second mechanism is quite similar to the first, with respect to the fission process. However, in this mechanism the protonated analyte is ejected from the droplet once the thermodynamic and kinetic conditions become favorable, rather than the solvent evaporating leaving charged ions (9).

If the ESI voltage is initially set to zero and slowly increased, the electrospray will only be observed once a minimum potential for onset is surpassed. As the ESI voltage is increased past this minimum, the ESI process progresses through different modes that have different ion yields and degrees of stability (13). The cone-jet mode, which is present at intermediate voltages, provides the greatest stability and the highest ion currents in the MS (10,14). For the cone-jet mode to be stable, a stable flow rate from the upstream sample delivery system or separation is required.

While insufficient flow rates are often problematic, low flow rates are typically preferred. Low flow rates minimize the elution volume, which improves the limits of detection because ESI is concentration sensitive. One method of producing efficient electrospray while minimizing the flow rate required is to use a spray tip with a small i.d. as will be discussed later in more detail. However, incorporating a narrow i.d. spray tip increases the resistance to bulk flow, which can result in flow irregularities caused by back pressure, making sample injection and manipulation difficult. An additional drawback to using narrow i.d. tips is that they are susceptible to clogging because small particles can easily become lodged in the tip, which effectively serves to stop the electrospray. Rather than altering the required flow rate, the flow rate can be effected using surface modification techniques such as coatings.

It is also important to note that the ESI system is also an electrolytic cell with a spray voltage that is typically greater than 1000 V. At this voltage, nearly all electrochemically active species can be oxidized. In the positive ion mode, the emitter tip is the anode and electroactive compounds are oxidized at this electrode. Because water is the main solution component, most of the Faradaic current oxidizes water leaving oxygen gas and protons in solution. To prevent bubbling, acidification, and analyte oxidization, many ESI interfaces isolate the electrode from the solvent/buffer solution using a liquid junction. When the ESI electrode is in direct contact with the solution in a CE channel, the bubbling makes the ESI very unstable and frequently causes current breakdown, which stops the separation.

25.3 COATINGS

In free zone electrophoresis separations, the analytes are separated on their inherent electrophoretic mobilities and not on interactions with the channel walls. Microchannels and capillaries have high surface area to volume ratios, thus increasing the probability of nonspecific surface interactions with the channel walls. Nonspecific interactions of the analyte molecules with the microchannel surface can cause significant band broadening that destroys the excellent separation efficiencies obtainable with microchannel electrophoresis. Therefore, to obtain the best separations these nonspecific interactions must be minimized. Nonspecific interactions are minimized through the use of coatings that modify the surface and make it less prone to adsorption. Numerous coatings have been developed, and the best coating for a particular application is largely dependent on the molecular characteristics of the analytes to be separated.

In electrokinetically driven systems, coatings also greatly affect the bulk flow by altering the EOF. If the microchannel or capillary wall has a negative surface charge when it is filled with solution, a double layer forms at the wall/solution interface to balance the excess surface charge of the solid. The double layer is composed of the stern layer and the diffuse layer; the stern layer consists of immobile cations while the diffuse layer is mobile. When a tangential electric field (along the length of the microchannel) is applied, the mobile diffuse layer migrates and drags the bulk solution along through viscous coupling. In glass microchannels or capillaries, the negative surface charge imparted by the surface silanol groups is compensated by the formation of a double layer with a net positive charge in solution. In glass, the surface charge is typically large enough to create EOF that is greater than the electrophoretic mobility of the analytes, sweeping both cations and anions toward the cathodic outlet. When a capillary with a positive surface charge is used, the EOF will be reversed so the polarity must be switched to result in sample reaching the detector. In a capillary with a neutral surface charge, EOF is eliminated and positive or negative analytes migrate in opposite directions. Therefore, changing the surface chemistry greatly affects the analyte migration.

When coupled with ESI–MS, maintaining a consistent flow rate is required to obtain a stable electrospray. In addition to a stable flow, stable electrospray requires the appropriate voltage and tip orifice i.d. If any of these parameters is inappropriate, then electrospray will breakdown or be sporadic yielding poor or no signal in the MS. Therefore, a key requirement when coatings are used is stability; if the coating starts to degrade, the flow rate can be adversely affected.

Highly charged surface coatings can be used to produce high EOF to generate adequate and stable bulk flow through the ESI emitter and to separate cations and anions simultaneously. However, a high EOF system is not suitable for proteins because the surface charge increases the unwanted protein adsorption. It is particularly challenging to prevent the nonspecific adsorption of proteins because of the high diversity of their structures that results from the heterogeneity of the amino acids from which they are made. The characteristic side chain of an amino acid can be basic or acidic or neutral, and it can be hydrophobic or hydrophilic. The solubility and tendency of proteins to adsorb to surfaces varies widely from protein to protein. A single protein usually contains many acidic and basic amino acids producing regions of the structure that will bind to either positive or negative sites on the microchannel wall.

One approach to minimize surface adsorption of proteins and to create significant EOF is to use a positive microchannel coating and a low pH buffer. The general idea is that decreasing the pH increases the positive charge on the proteins, which repels them from the positive microchannel surface. However, lowering the pH decreases protein solubility, particularly for proteins that are greater than 20–30 kDa, which prevents their analysis and clogs the microchannels (15).

The problem of surface adsorption pertaining to proteomics associated with high EOF systems can be avoided by minimizing surface charge, but the lack of surface charge also minimizes EOF. A low EOF CE-MS interface is more challenging to design because electrophoresis does not supply sufficient flow to maintain stable ESI. As the EOF approaches zero, the bulk flow also approaches zero and no solution is available for droplet formation. This difficulty can be alleviated by using a makeup or sheath flow, but this addition increases the complexity of the system.

Surface coatings can also be classified in one of two categories: dynamic coatings or permanent coatings. Dynamic coatings adsorb to the surface through interactions such as hydrogen bonding or electrostatic interactions while permanent coatings interact through covalent bonds. Dynamic coatings frequently require addition of the coating molecules to the running buffer to maintain the coating and ensure that there is a significant amount of exchange with the molecules in solution. Others bind strongly enough that after coating no additional coating molecules need to be added to the running buffers. The elution of the coating molecules into the mass spectrometer creates an excessive background, preventing their use with MS.

Permanent covalent coatings are more desirable for applications involving mass spectrometry. Whitesides' group demonstrated that neutral, polyethylene glycol (PEG)-based covalent coatings attached to glass surfaces were capable of reducing surface adsorption of proteins (16). Razunguzwa et al. characterized the compatibility of similar PEG-based coatings with MS and found very low background interference (17). These coatings are also stable at intermediate pH values, which can be used for tryptic digestions, allowing for sequential digestion followed by separation.

25.4 SPRAY EMITTERS

A crucial aspect of the microfluidic ESI–MS interface is the spray emitter from which the highly charged droplets are generated. The spray emitter is formed as a pointed tip. To maximize the performance, the spray emitter or tip should physically constrain the solution from spreading to minimize the Taylor cone volume. Minimizing the Taylor cone volume prevents band broadening at the tip and produces smaller droplets with higher charge density. The characteristics of the microchannel surface also affect the characteristics of the Taylor cone by affecting the contact angle of the solution. Additionally, the spray emitter should have a short length, minimal resistance to flow, not be easily clogged, and have no dead volume. Thus, there are numerous characteristics that must be considered, and the microscale dimensions have made fabrication challenging.

There have been many different approaches used for emitter fabrication to overcome these challenges. The initial approach was rather straightforward as electrospray was formed directly at the end of a flat microchip (18,19). While the design of these devices was simple, the Taylor cone that formed at the channel orifice became large as it spread out across the flat surface of the device. The accumulation and spreading cause sample dilution and band broadening in the case of an on-chip separation. In addition to the band broadening and sample dilution, the larger Taylor cone produced larger droplets leading to less efficient electrospray, as discussed previously.

A subsequent approach was to combine the commonly pulled capillary spray emitter tips with the microfluidic chip. The pulled capillary tips provide a narrow i.d. orifice that confines the size of the Taylor cone, allowing the use of a lower spray voltage and lower flow rates. The pulled glass or fused-silica tips can be pulled to provide a small orifice and excellent spatial constraint to keep the Taylor cone volume minimal. Numerous interfaces have incorporated fused-silica spray tips with a narrow i.d. (2,20–37). However, integrating a pulled capillary tip with the microfluidic chip presents the challenge of minimizing the dead volume that easily forms at the pulled capillary/microfluidic chip interface. While low dead volume connections are needed, they can be difficult to obtain in practice mainly because of the nanoliter sample volumes being manipulated. Any dead volume greater than picoliters in volume destroys the high separation efficiencies achievable with free solution electrophoretic separations. The dead volume problem can be minimized by using higher flow rates, but this reduces the concentration and increases the droplet size. The first successful method for coupling glass microfluidic chips with pulled capillary emitters employed direct drilling of the channel end, providing separation efficiencies of greater than 100,000 (37).

As a method to circumvent connections, several references have incorporated the spray emitter into the chip design, thereby eliminating the need to connect a spray emitter to the microfluidic device (38–51). Many of the references utilized a polymeric microfluidic device with the microfluidic channel terminating at the end of a patterned emitter tip (46–51). Le Gac et al. created a unique design called a "nib-tip" (39–43). In this design, the emitter is formed from a polymeric planer device but is open on both the top and the bottom and functions similar to a fountain pen. Liquid in the reservoir flows between the tips via capillary action and is then electrosprayed from the tip when under the influence of an applied potential. A nib-tip with an adjustable gap was fabricated by Ek et al. and is stable at flow rates down to 75 nL/min, making it well suited for microchannel electrophoresis (52). Another method used to create an on-chip electrospray emitter is a deep reactive ion etching procedure (44,45). The on-chip electrospray emitters fabricated using reactive ion etching by Sjodal et al. have a shape similar to fused-silica tips with a final i.d. of 10 μm and height of 70 μm.

Several authors have used tips coated with various types of conductive materials like silver (53) and carbon (54) for CE–MS and microchannel electrophoresis–MS interfaces. However, a commonly reported problem associated with using metal-coated tips is that the coating degrades over time leading to unstable electrospray.

To remedy this, conductive polymers have been used to coat the spray tips (55,56). These coatings provide stable electrospray for a longer time resulting in a more robust tip.

25.5 CE AND ESI ELECTRODE CONNECTIONS

25.5.1 Two-Electrode Systems

The simplest method of applying the separation and spray voltages can be achieved using a two-electrode system. In this configuration, the high voltage is applied to one electrode upstream of the sample inlet in the microchannel CE, and the grounded MS inlet serves as the second electrode. Although this approach is straightforward and simple, there are many problems associated with its operation. With the two-electrode design, the voltage applied at the inlet generates both the ESI potential and the bulk flow through electroosmotic flow. Therefore, the flow rate and ESI potential can not be adjusted independently, which does not allow enough flexibility to maintain stable spray. Additionally, the majority of the voltage is dropped across the air gap and not the separation channel, as shown in Fig. 25.2. In this scenario, when

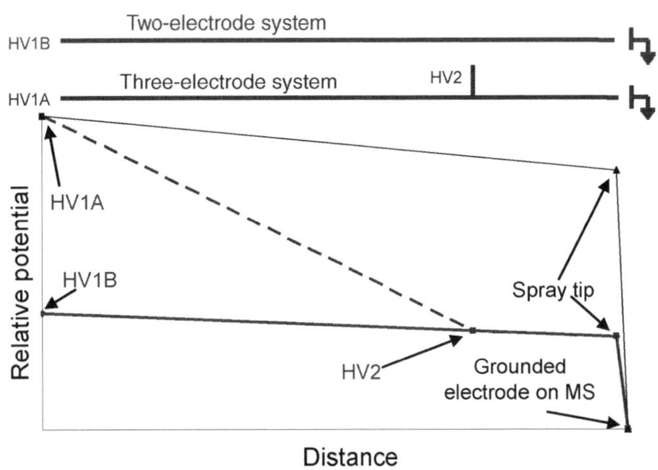

FIGURE 25.2 A diagram showing the simple electrical circuit (top) and the voltage drop as a function of distance along the length of the channel (bottom) for the two- and three-electrode systems. In the three-electrode system, the electrode through which HV2 is applied separates the microfluidic channel into two regions with a greater electric field for the separation and a lower electric field in the downstream region before the electrospray emitter. The intermediate electrode defines the intermediate voltage and acts as a current sink with this polarity. The potentials correspond to a low EOF system where cation migration to the outlet is dominated by their electrophoretic mobility. In the two-electrode system, the voltage drop across the channel and air gap is determined by their relative resistances, and the voltage drop across the channel is low with the appropriate spray voltage. Increasing the inlet potential (HV1B) to increase the voltage across the separation channel raises the electrospray voltage above the appropriate level (as shown by the black line).

the appropriate voltage is used to produce stable electrospray, the electric field across the separation capillary is much less than the electric field typically used for a CE separation, thus resulting in a poor separation. Resolution in CE and microchannel electrophoresis is directly proportional to the applied voltage. With the lower voltage drop, the separation occurs over a longer period of time due to decreased velocities allowing more time for diffusion. Raising the voltage of the first electrode at the sample inlet side increases the electric field for the microchannel electrophoresis separation, but it raises the electrospray potential above the optimal or working range.

25.5.2 Three-Electrode Systems

To increase the reliability and to improve performance of the system, a three-electrode system is usually employed. Using this approach, an intermediate electrode is inserted that divides the system into two regions: the first defines the separation and the second defines the ESI voltage. Thus, a high electric field can be maintained in the separation channel with independent adjustment of the ESI voltage. In the first region, a high voltage is applied to the first electrode and the second electrode is configured to ensure that large voltage drop occurs between the first and the second electrodes (Fig. 25.2). With the configuration shown in Fig. 25.3, the intermediate electrode lowers the voltage before the electrospray and consequently acts as a current sink (57,58). With the intermediate electrode acting as a current sink, ions, including the analytes, migrate toward this electrode. The challenge is to have most of the analyte ions migrate through the ESI emitter and to prevent most of the analyte

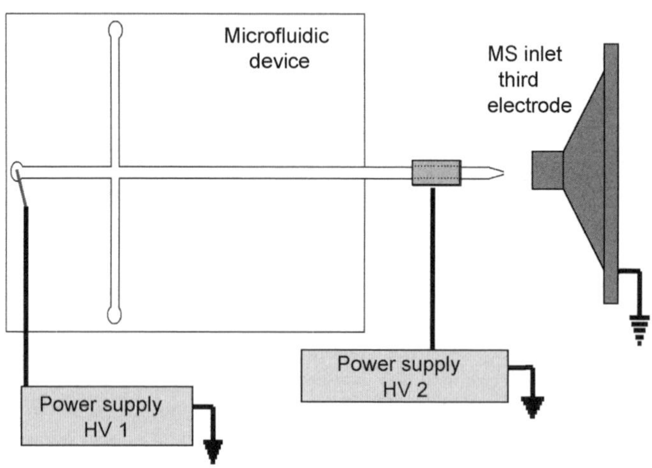

FIGURE 25.3 A schematic showing the application of an intermediate electrode through direct contact. The electrode is in direct contact with the solution/buffer in the separation channel and the electrochemical processes that occur at this electrode cause bubble formation, pH shifts, and potentially sample oxidation reduction. In this figure, a metal union is used as the electrode, although a wire electrode could be used as well.

ions from migrating to the intermediate electrode. In an attempt to overcome these challenges, there have been a number of designs and approaches used for fabrication of three-electrode systems with many combinations of similar elements, making for a large number of potential classification schemes.

25.5.3 Direct Contact

The most straightforward means of including the conductive electrode is to place it in direct contact with the separation buffer/solvent solution within or at the end of the microfluidic separation channel using a conductive union or a conductive spray tip, as diagrammed in Fig. 25.3 or 25.4. Although this method is relatively easy to fabricate, it suffers from having all of the electrochemical processes caused by the high potential occurring in or at the end of the separation channel. In the positive ion mode, oxidation of water, the metal electrode, and oxidizable analytes occurs. The bubbles caused by electrolysis will decrease the stability of the spray as they elute and can end the separation if they fill the microchannel and break the electrical circuit. Using the coated tips, the buffer contacts the electrode at the end of the electrospray emitter, preventing the formation of bubbles inside the microchannel or the emitter tip. The electrolysis of water also produces protons that decrease the pH of the buffer, causing instabilities in the CE separation. Because the dimensions of the microchannel are small compared to a reservoir, the system is more susceptible to buffer depletion when the intermediate electrode is in direct contact with the buffer solution of the separation channel. Hydroquinone has a lower oxidation potential

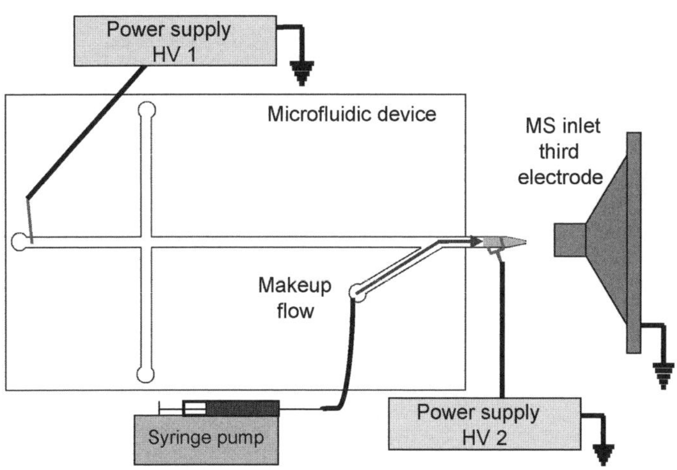

FIGURE 25.4 A makeup solution is shown in use with a conductively coated tip for direct electrode contact. The makeup flow differs from a sheath flow because it is not used for electrode connection. The makeup flow must be prevented from applying a backpressure to the separation channel and stopping the separation. The makeup flow can be combined with other approaches, and it does not have to be used in combination with a conductive electrospray trip.

than water and has been used as a buffer additive to suppress bubble formation and has been shown to be MS compatible (59).

An integrated capillary/PDMS microfluidic device with a molded electrospray tip was developed by Dahlin et al. (60). The sheathless design relied on electroosmotic flow to maintain stable spray and a tip coated with graphite/PDMS was used as the intermediate electrode. The device was reported to operate at flow rates in the range between 100 and 1000 nL/min. Five peptides were separated in less than 1 min with separation efficiencies between 8000 and 25,000 depending on the peptide.

25.5.4 Liquid Junctions

Because of the many nonideal processes that can occur when the electrode is held in direct contact with the buffer solution, a liquid junction is often used to isolate the intermediate electrode from being in direct contact with the buffer solution in the microfluidic separation channel. The liquid junction uses a microchannel or a capillary (as shown in Fig. 25.5) to remove the electrode/solution interface from the microfluidic separation channel. A larger reservoir at the electrode/solution interface provides a large buffer volume to minimize any buffer changes and physically separates the bubbles from the microfluidic separation channel. There are a variety of methods for making the liquid junction that can be used with or without a sheath flow or makeup flow system.

In a sheathless system, such as the one shown in Fig. 25.3, the bulk flow that supplies the solution for droplet formation in the ESI is produced by EOF. By using

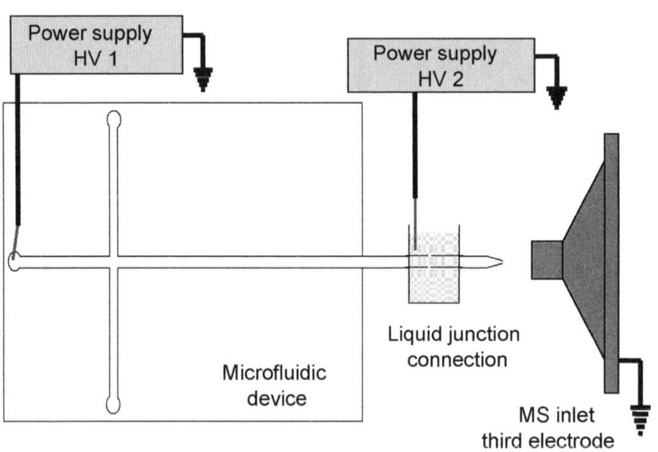

FIGURE 25.5 Another approach for connecting the intermediate electrode is to use a liquid junction. The electrode is contained in a relatively large buffer reservoir that removes the bubbles from the channel and has adequate buffering capacity to suppress pH changes. Current flows through the junction, and if not impeded, analyte ions migrate to the intermediate electrode. A membrane or frit-like material can be used to impede the bulk flow through the junction. The EOF generated upstream of the liquid junction can be used to carry the analytes through this intersection and toward the electrospray emitter.

EOF to provide bulk flow, a relatively simple interface design is achieved. To prevent loss of analyte at the liquid junction or microchannel intersection, a frit, membrane, or similar material is used. The frit-like material preferentially allows the passage of smaller ions. It is important to note that these designs rely on high EOF to move the analyte molecules through the liquid junction. If the EOF is low, then the system resembles analyte concentrators similar to those reported by Yeung and Ramsey (61,62). This concentration process will clearly interfere with the electrophoretic separation and degrade the separation efficiency. A benefit of this approach is that the sample band is usually not diluted, which benefits concentration-sensitive ESI.

A unique glass membrane liquid junction was developed by Lazar et al. (29). In this work, a thin glass membrane between a side channel and the main channel served as the electrical contact for the second electrode. The thin glass membrane, only 4–6 μm wide, allowed ions to migrate but prevented high EOF bulk flow from entering the side channel. Significant band broadening occurred and the influence of mismatches in flows on band broadening is discussed.

A liquid junction interface for use with low EOF systems was developed by Zhang et al. that incorporated a nebulizing gas to apply a vacuum at the ESI emitter. The vacuum at the ESI emitter increased the flow rate aiding in the desolvation process (26,28). This liquid junction shares many similarities with the sheath-flow interfaces because solution is added at the liquid junction to prevent the ions from migrating to the intermediate electrode and to stabilize the flow. A notable difference with sheath-flow systems is that the device does not apply any back pressure to the electrophoretic separation. This interface provided a universal coupling method for any mode of CE and could also be used with liquid chromatography (LC) or capillary electrochromatography (CEC). In the earlier publication, the liquid junction was formed off-chip between a transfer capillary and the pulled capillary spray tip (26,28). In the latter publication, the nebulizing gas and the makeup flow were integrated onto the chip (26,28). With the off-chip capillary liquid junction 313,000 plates/m was obtained, while 300,000 plates/m was obtained for the on-chip liquid junction of eight different angiotensin peptides. With the on-chip liquid junction, a separation of eight different angiotensin peptides was carried out resulting in 31,000 theoretical plates with attomole detection limits.

Tachibana et al. and Akashi et al. developed microchip electrophoresis systems that utilized high EOF and a liquid junction for application of the intermediate potential (31,32). One of the more novel aspects of the work is that it utilized a microfluidic device holder constructed of polychlorotrifluoroethylene with integrated fluid and reservoir connections, and quartz insert that contained the separation channel. Using the liquid junction interface and MS detection, the reported separation efficiency of 2600 was quite lower than the 10,000 reported for fluorescence detection under comparable conditions. It is interesting to note that the peak width more than doubled when the orifice i.d. was increased from 20 to 50 μm. Presumably, the larger Taylor cone associated with the large orifice causes significant band broadening. Using the same basic design, Akashi et al. performed a separation of basic proteins at pH 5.7 in positively coated channels (32). The separation

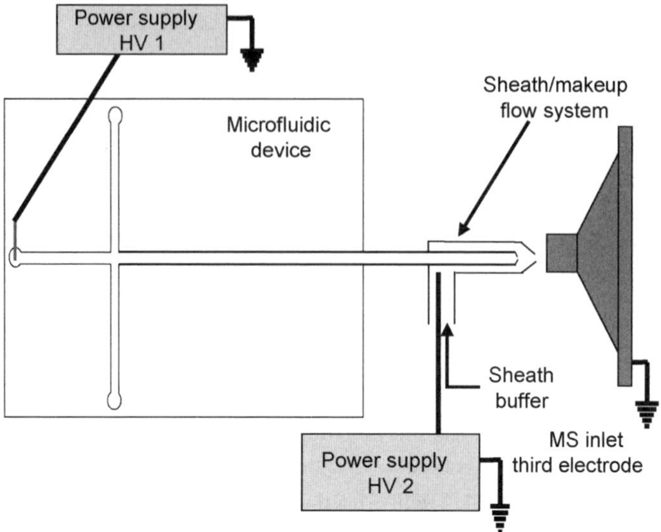

FIGURE 25.6 A sheath-flow design can be used to incorporate the intermediate electrode. The sheath buffer flows toward the inlet and is usually driven by applying a positive pressure at the sheath buffer inlet. The sheath buffer also carries the current, and its velocity toward the inlet prevents ions from migrating toward the electrode in the sheath solution.

efficiencies were not reported but appeared to be low in the electropherogram presented.

Sheath-flow systems have also been used as a type of liquid junction. This liquid junction could be termed a dynamic liquid junction because it relies on two moving flow streams as shown in Fig. 25.6. The sheath flow is pressure driven and is in contact with the electrode somewhere upstream of the interface between the sheath liquid and the microchannel electrophoresis separation buffer. With the potentials set for a low EOF configuration (as shown in Fig. 25.2), the hydrodynamic flow velocity must be greater in magnitude than the electrophoretic velocity to prevent the migration of the analytes to the intermediate electrode. Finding the appropriate flow rate can be a careful balancing act. If the sheath flow rate is too high, excess back pressure is created inside the sheath tip that can force the analytes that should be migrating through the microfluidic separation channel back out through the inlet, thus destroying the separation. If the sheath flow rate is too low, then the analytes will migrate toward the intermediate electrode. The additional solution provided by the sheath flow can be beneficial if the stability of the bulk flow through the ESI emitter is improved, and it can be used to optimize the electrospray solution composition after the separation.

A makeup flow system is similar to the sheath, but it is not in contact with the intermediate electrode and does not carry the current supplied by the intermediate electrode. Because the electrode is not an integral part of the design, a makeup flow can be used with either a three-electrode or two-electrode design. A diagram of a three-electrode system with a makeup flow and the intermediate potential applied through a coated tip is shown in Fig. 25.4. Generally, the makeup flow starts in a side channel, which then merges with the main channel. This side channel forms an acute angle with the main channel and is directed toward the outlet, as shown in Fig. 25.4. Therefore, the makeup flow mixes with the separation buffer prior to exiting the emitter tip.

Systems utilizing both a sheath and a makeup flow have the advantages of improved flow stability through the ESI emitter and the ability to change the solution composition before the electrospray. The stability of the flow is improved because the pressure-driven flow, with inherently better stability than the EOF, provides a significant portion of the flow. Thus, the portion of the flow generated by the EOF is reduced and the overall stability is improved. The addition of the sheath or the makeup flow allows the solution composition to be made suitable for ESI. For example, protein separations can be performed at moderate or higher pHs, before the pH is dropped to about 3, which enhances the signal in the MS. Additionally, methanol or other organic modifiers can be added to decrease the surface tension and decrease the electrospray droplet size.

The undesirable affects associated with the sheath and the makeup flow are the dilution of the sample and the extra back pressure induced at the outlet of the separation channel. The sheath liquid causes some dilution of the sample, but the sheath flow rate need not be greater than typical EOF flow rates. Therefore, the reduction in the signal should be minimal when compared to systems that have higher flow rates. Another undesirable effect of using a sheath or a makeup flow system occurs if the flow rate is too high. In this situation, the additional flow encounters high resistance from the emitter tip causing the sheath flow to travel back through the inner capillary resulting in decreased separation efficiencies. To remedy this problem, Razunguzwa et al. introduced hydrodynamic restrictors (HDRs) into chip designs (63). These HDRs prevented the makeup solvent from flowing backward while having little effect on the electrophoretic separation efficiency. The HDRs consisted of a series of high aspect ratio channels that were created using laser micromachining.

The Harrison and Thibault groups developed a number of different interfaces that incorporated makeup flows, metal-coated electrospray emitters, and sheath flow systems (34–37,64). The sheath interface with a nebulizing gas provided some of the best separation efficiencies reported, ranging from 26,000 to 58,000 with detection limits as low as 45 nM (37).

25.6 INTEGRATED APPLICATIONS

Numerous approaches have been used to develop microchannel electrophoresis ESI–MS interfaces. Many of the approaches perform moderately and rely on high EOF. However, the power of microfluidics lies in the ability to integrate multiple techniques or processes on the same device. A few reports that show the promise of integrating additional processes with microchip ESI–MS by incorporating a sample pretreatment method or a two-dimensional separation before the MS analysis will be highlighted in the following section.

Wang et al. performed an on-chip protein digestion followed by an electrokinetic injection and an electrophoretic separation of the digested peptides (33). In this work, a flat bed was created on the chip and was then packed with trypsin-coated beads. Next, the sample was infused onto the bed, which was then collected in a reservoir at the end of the bed. Afterward, the digested sample was electrophoretically separated and sprayed into the MS. For this work, two different proteins were digested: cytochrome C and bovine serum albumin (BSA). The MS analysis produced an amino acid sequence coverage of 92% for cytochrome C. When BSA was digested and separated, the sequence coverage was 71%, which was comparable to previous reports.

In another report, isoelectric focusing (IEF) was performed on a chip followed by ESI–MS (38). Here, the sample was mixed with Pharmalyte, a commercially available ampholyte, and pumped onto the chip. Next, the anolyte and the catholyte were added to opposing reservoirs and a potential difference of −8 kV was applied between the two reservoirs. Once the focusing step was completed, the analytes were hydrodynamically mobilized and sprayed into the MS while maintaining the potential difference to ensure that the proteins would stay focused. The separation efficiencies acquired in this work were lower than those achieved with capillary isoelectric focusing (CIEF). These lower efficiencies may have been a consequence of interactions between the protein and the surface of the microfluidic device. It was hypothesized that the interactions were either due to the nature of the polymer surface or because of a roughened channel surface as a result of the laser micromachining process. Because of these reasons, it was concluded that with proper care the interactions could be minimized and the separation efficiencies ultimately increased.

Finally, Lazar et al. created a microfluidic device capable of performing CEC followed by ESI–MS (20). The stationary phase was a UV-cured monolith that was first pumped throughout the device and then exposed to UV radiation to polymerize the monolith. After the device had been fabricated, the sample was loaded, injected, separated, and sprayed into the MS. A tryptic digest was analyzed in this work, which resulted in numerous peaks, and many of which were not fully resolved in the total ion chromatogram. Since not all of the peaks were sufficiently resolved, it was suggested that either a longer separation channel or a more hydrophobic stationary phase would be required to satisfactorily resolve the sample constituents. Since all the channels were filled with the monolithic stationary phase, it was concluded that a longer separation channel would be the best choice to increase the separation efficiency.

25.7 CONCLUSIONS

Coupling microchannel electrophoresis with ESI–MS provides fast separations with the online structural identification and characterization. Although the interface may seem to be fairly straightforward, maintaining a high electric field in the separation channel and transporting the separated bands through a small tip orifice have proven to be challenging. Many different interface designs have been developed, and an important consideration when choosing an ESI interface is whether a high or a low EOF system is required. Many high EOF systems have been developed that are analogous to an EOF pump. If positive ion mode MS is used for detection and the microchannel walls are positively coated, the cationic analytes will not migrate toward the intermediate electrode. Reliable microchannel electrophoresis-ESI–MS interfaces have been developed, while development of new approaches continues.

REFERENCES

1. Yu C, Davey MH, Svec F, Frechet JM. Anal Chem 2001;73:5088–5096.
2. Fortier M-H, Bonneil E, Goodley P, Thibault P. Anal Chem 2005;77:1631–1640.
3. Tang L, Kebarle P. Anal Chem 1993;65:3654–3668.
4. Taylor G. Proc Royal Soc London Series A Math Phys Sci 1964;280:383–397.
5. Mutoh M, Kaieda S, Kamimura K. J App Phys 1979;50:3174–3179.
6. Kebarle P, Tang L. Anal Chem 1993;65:972A–986A.
7. Gomez A, Tang K. Phys Fluids 1994;6:404–414.
8. Dole M, Mack LL, Hines RL, Mobley RC, Ferguson LD, Alice MB. J Chem Phys 1968;49:2240–2249.
9. Iribarne JV, Thomson BA. J Chem Phys 1976;64:2287–2294.
10. Nemes P, Marginean I, Vertes A. Anal Chem 2007;79:3105–3116.
11. Wilm MS, Mann M. Int J Mass Spec Ion Proc 1994;136:167–180.
12. Wilm M, Mann M. Anal Chem 1996;68:1–8.
13. Cloupeau M, Prunet-Foch B. J Aerosol Sci 1994;25:1021–1036.
14. Cloupeau M. J Aerosol Sci 1994;25:1143–1157.
15. Schiffer E, Mischak H, Novak J. Proteomics 2006;6:5615–5627.
16. Ostuni E, Chapman RG, Holmlin RE, Takayama S, Whitesides GM. Langmuir 2001;17: 5605–5620.
17. Razunguzwa TT, Warrier M, Timperman AT. Anal Chem 2006;78:4326–4333.
18. Xue Q, Foret F, Dunayevskiy YM, Zavracky PM, McGruer NE, Karger BL. Anal Chem 1997;69:426–430.
19. Ramsey RS, Ramsey JM. Anal Chem 1997;69:1174–1178.
20. Lazar IM, Li L, Yang Y, Karger BL. Electrophoresis 2003;24:3655–3662.
21. Chan JH, Timperman AT, Aebersold R. Anal Chem 1999;71:4437–4444.
22. Figeys D, Gygi SP, McKinnon G, Aebersold R, An integrated microfluidics-tandem mass spectrometry system for automated protein analysis. Anal Chem 1998;70:3728–3734.
23. Figeys D, Aebersold R. Anal Chem 1998;70:3721–3727.
24. Figeys D, Ning Y, Aebersold R. Anal Chem 1997;69:3153–3160.
25. Pinto DM, Ning Y, Figeys D. Electrophoresis 2000;21:181–190.
26. Zhang B, Liu H, Karger BL, Foret F. Anal Chem 1999;71:3258–3264.
27. Lazar IM, Karger BL. Anal Chem 2002;74:6259–6268.
28. Zhang B, Foret F, Karger BL. Anal Chem 2000;72:1015–1022.

29. Lazar IM, Ramsey RS, Jacobson SC, Foote RS, Ramsey JM. J Chromatogr A 2000;892: 195–201.
30. Lazar IM, Ramsey RS, Sundberg S, Ramsey JM. Anal Chem 1999;71:3627–3631.
31. Tachibana Y, Otsuka K, Terabe S, Arai A, Suzuki K, Nakamura S. J Chromatogr A 2003;1011:181–192.
32. Akashi S, Suzuki K, Arai A, Yamada N, Suzuki E-I, Hirayama K, Nakamura S, Nishimura Y. Rapid Commun Mass Spectrosc 2006;20:1932–1938.
33. Wang C, Oleshuk RD, Ouchen F, Li J, Thibault P, Harrison DJ. Rapid Commun Mass Spectrom 2000;14:1377–1383.
34. Li J, Wang C, Kelly JF, Harrison DJ, Thibault P. Electrophoresis 2000;21:198–210.
35. Li J, LeRiche T, Tremblay T-L, Wang C, Bonneil E, Harrison DJ, Thibault P. Mol Cell Proteomics 2002;1:157–168.
36. Li J, Kelly JF, Chernushevich I, Harrison DJ, Thibault P. Anal Chem 2000;72:599–609.
37. Li J, Thibault P, Bings NH, Skinner CD, Wang C, Colyer C, Harrison DJ. Anal Chem 1999;71:3036–3045.
38. Wen J, Lin Y, Xiang F, Matson DW, Udseth HR, Smith RD. Electrophoresis 2000;21:191–197.
39. Brinkman M, Blossey R, Arscott S, Druon C, Tabourier P, Gac SL, Rolando C. Appl Phys Lett 2004;85:2140–2142.
40. Le Gac S, Cren-Olive C, Rolando C, Arscott S. Am Soc Mass Spectrom 2004;15: 409–412.
41. Le Gac S, Arscott S, Rolando C. Electrophoresis 2003;24:3640–3647.
42. Le Gac S, Arscott S, Cren-Olive C, Rolando C. J Mass Spectrom 2003;38:1259–1264.
43. Le Gac S, Arscott S, Rolando C. In: 7th International Conference on Miniaturized Chemical and Biochemical Analysis Systems; Squaw Valley, California, USA; 2003.
44. Griss P, Melin J, Sjoedahl J, Roeraade J, Stemme G. J Micromech Microeng 2002; 12:682–687.
45. Sjodahl J, Melin J, Griss P, Emmer A, Stemme G, Roeraade J. Rapid Comm Mass Spectro 2003;17:337–341.
46. Kim JS, Knapp DR. J Chromatogr A 2001;924:137–145.
47. Kim JS, Knapp DR. J Amer Soc Mass Spectro 2001;12:463–469.
48. Kim JS, Knapp DR. Electrophoresis 2001;22:3993–3999.
49. Licklider L, Wang XQ, Desai A, Tai YC, Lee TD. Anal Chem 2000;72:367–375.
50. Yuan C-H, Shiea J. Anal Chem 2001;73:1080–1083.
51. Gobry V, Oostrum Jv, Martinelli M, Rohner TC, Reymond F, Rossier JS. Girault HH. Proteomics 2002;2:405–412.
52. Ek P, Sjoedahl J, Roeraade J. Rapid Commun Mass Spectrosc 2006;20:3176–3182.
53. Chen YR, Her GR. Rapid Commun Mass Spectrosc 2003;17:437–441.
54. Chang YZ, Her GR. Anal Chem 2000;72:626–630.
55. Chao B-F, Chen C-J, Li F-A, Her GR. Electrophoresis 2006;27:2083–2090.
56. Smith DR, Moy MA, Dolan AR, Wood TD. Analyst 2006;131:547–555.
57. Jackson GS, Enke CG. Anal Chem 1999;71:3777–3784.
58. Hau J, Roberts M. Anal Chem 1999;71:3977–3984.

59. Moini M, Cao P, Bard AJ. Anal Chem 1999;71:1658–1661.
60. Dahlin AP, Wetterhall M, Liljegren G, Bergstroem SK, Andren P, Nyholm L, Markides KE, Bergquist J. Analyst (Cambridge U.K.) 2005;130:193–199.
61. Wei W, Yeung ES. Anal Chem 2002;74:3899–3905.
62. Foote RS, Khandurina J, Jacobson SC, Ramsey JM. Anal Chem 2005;77:57–63.
63. Razunguzwa TT, Lenke J, Timperman AT. Lab Chip 2005;5:851–855.
64. Deng Y, Henion JD. Anal Chem 2001;73:639–646.

INDEX